Protein–Protein Recognition

About the Author

Camilla Stivers is Levin Professor of Urban Studies and Public Service at the Levin College of Urban Affairs, Cleveland State University. She is Associate Editor *of Public Administration Review*, author of *Bureau Men, Settlement Women: Constructing Public Administration in the Progressive Era*, and a coauthor of *Government Is Us: Public Administration in an Anti-government Era*. She received her Ph.D. in public administration and policy from Virginia Polytechnic Institute and State University. She was a practicing administrator in nonprofit and public agencies for nearly two decades.

Frontiers in Molecular Biology

SERIES EDITORS

B. D. Hames

Department of Biochemistry
and Molecular Biology
University of Leeds, Leeds LS2 9JT, UK

D. M. Glover

Department of Genetics,
University of Cambridge, UK

TITLES IN THE SERIES

1. Human Retroviruses
Bryan R. Cullen

2. Steroid Hormone Action
Malcolm G. Parker

3. Mechanisms of Protein Folding
Roger H. Pain

4. Molecular Glycobiology
Minoru Fukuda and Ole Hindsgaul

5. Protein Kinases
Jim Woodgett

6. RNA–Protein Interactions
Kyoshi Nagai and Iain W. Mattaj

7. DNA–Protein: Structural Interactions
David M. J. Lilley

8. Mobile Genetic Elements
David J. Sherratt

9. Chromatin Structure and Gene Expression
Sarah C. R. Elgin

10. Cell Cycle Control
Chris Hutchinson and D. M. Glover

11. Molecular Immunology (Second Edition)
B. David Hames and David M. Glover

12. Eukaryotic Gene Transcription
Stephen Goodbourn

13. Molecular Biology of Parasitic Protozoa
Deborah F. Smith and Marilyn Parsons

14. Molecular Genetics of Photosynthesis
Bertil Andersson, A. Hugh Salter, and James Barber

15. Eukaryotic DNA Replication
J. Julian Blow

16. Protein Targeting
Stella M. Hurtley

17. Eukaryotic mRNA Processing
Adrian Krainer

18. Genomic Imprinting
Wolf Reik and Azim Surani

19. Oncogenes and Tumour Suppressors
Gordon Peters and Karen Vousden

20. Dynamics of Cell Division
Sharyn A. Endow and David M. Glover

21. Prokaryotic Gene Expression
Simon Baumberg

22. DNA Recombination and Repair
Paul J. Smith and Christopher J. Jones

23. Neurotransmitter Release
Hugo J. Bellen

24. GTPases
Alan Hall

25. The Cytokine Network
Fran Balkwill

26. DNA Virus Replication
Alan J. Cann

27. Biology of Phosphoinositides
Shamshad Cockcroft

28. Cell Polarity
David G. Drubin

29. Protein Kinase Functions
Jim Woodgett

30. Molecular and Cellular Glycobiology
Minoru Fukuda and Ole Hindsgaul

31. Protein–Protein Recognition
Colin Kleanthous

Protein–Protein Recognition

EDITED BY

Colin Kleanthous

School of Biological Sciences
University of East Anglia

OXFORD
UNIVERSITY PRESS
Great Clarendon Street, Oxford OX2 6DP
Oxford University Press is a department of the University of Oxford
It furthers the University's objective of excellence in research, scholarship,
and education by publishing worldwide in
Oxford New York
Athens Auckland Bangkok Bogotá Buenos Aires Calcutta
Cape Town Chennai Dar es Salaam Delhi Florence Hong Kong Istanbul
Karachi Kuala Lumpur Madrid Melbourne Mexico City Mumbai
Nairobi Paris São Paulo Singapore Taipei Tokyo Toronto Warsaw
with associated companies in
Berlin Ibadan

Published in the United States
by Oxford University Press Inc., New York

First published 2000

British Library Cataloguing in Publication Data
Data available

Library of Congress Cataloging in Publication Data
1 3 5 7 9 10 8 6 4 2

ISBN 0 19 963761 X (Hbk)
ISBN 0 19 963760 1 (Pbk)

Typeset by Footnote Graphics, Warminster, Wilts

Printed in Great Britain on acid free paper by
The Bath Press Ltd, Avon

Preface

The purpose of this book is simple, to bring together specialists in the field of protein-protein recognition to discuss recent advances in particular areas of biology with a view to understanding the wider issues of how and why proteins form complexes with each other. With that in mind, one could ask the following questions of the book itself:

1. why 'protein-protein recognition'?
2. why now?
3. why the particular areas that have been included and what ties them together ?

The first question is easy to answer. Protein-protein recognition occurs in all forms of life and underpins a wide array of biological function, from the mammalian immune system to bacterial cell division. Given then the fundamental nature of protein-protein interactions in biology a preface for a book on the subject might reasonably be expected to begin with an apology, explaining why yet another weighty volume has appeared to take-up space on a bookshelf already burgeoning with other examples. Except that this is not the case. While there have been many reviews and book chapters there have been no volumes, to my knowledge, dedicated to protein-protein recognition as a subject in its own right. Hence, my belief that this book may well find a vacant spot on a biologists albeit cramped bookshelf.

The answer to the second question concerns the state-of-play of biological science itself. The completion of the human genome, along with those of many other organisms, has revolutionised the way biologists think about their subject not least because of the sheer numbers of genes and proteins that have become accessible for study. Coupled with developments in array technology and high-throughput two-hybrid screens[1], analysis of organismal gene expression patterns and protein-protein associations on a cellular scale have become a reality. The next few years will likely be dominated by large-scale research efforts aimed at constructing protein 'circuit diagrams' that describe complex biological processes such as limb development. These advances place protein-protein recognition centre stage and so it is timely to take stock of our current level of understanding of the field, highlighting what we know and, in particular, what we do not.

Answers to the final questions are manifold. Four criteria were used to identify subject areas for inclusion in the book (the submission of a completed manuscript on time being a 5th). (1) All the chapters deal with protein-protein recognition from a structural perspective, where structures exist for bound complexes and/or for one or both of the isolated partners; to understand molecular recognition requires knowledge of molecular structure. Many of these systems have in addition also been the subject of kinetic and thermodynamic studies which provide some understanding of the forces involved in stabilising the complexes and how specificity is achieved. (2) For the most part only water-soluble, heteromeric protein complexes are discussed; in other words, only complexes from aqueous environments in which the constituent proteins can exist in isolation of their partner(s). (3) The systems represent a wide spectrum of biology, including eukaryotic proteins from the extracellular matrix, the immune system and intracellular signalling pathways and prokaryotic proteins involved in respiration and cell death. (4) A property of protein-protein complexes that is poorly understood is their ability to form complexes of very diverse thermodynamic stability, with affinities as low as millimolar or as high as femtomolar. The systems that have been included in the book span this entire thermodynamic range to reflect this diversity, the hope being that by juxtaposing systems of such varied stability the underlying reasons for their differences might begin to emerge. The reader will have to judge how successful this has been.

Having addressed the reasons for the book, I should say a little about its organisation. It should be viewed as being in four sections, the first (Chapters 1 and 2) dealing with issues generic to all protein-protein interactions (kinetics, thermodynamics and structural classification) with the subsequent three sections dealing with specific examples of weak (Chapters 3 and 4; K_d, mM-μM), intermediate (Chapters 5-7; K_d, μM-nM) and high (Chapters 8 and 9; K_d, nM-fM) affinity complexes. Since many biological associations occur at concentrations of micromolar to nanomolar this group dominate the book. (That this is a truism should come as no surprise when one remembers that the concentration of a single molecule in a bacterial cell is of the order of nanomolar). However, valuable information about protein-protein interactions can also be obtained by looking at the extreme ends of the stability scale and so the reader is urged not ignore these.

Finally, one apology that I would like to make is to those working on other protein-protein interaction systems whose work has not been included in the book. While I would have wished to have included many more chapters the resulting opus would surely have put paid to that creaking bookshelf.

Norwich CK
August 2000

Reference

[1]Uetz *et al.* (2000) A comprehensive analysis of protein-protein interactions in *Saccharomyces cerevisae*. *Nature* **403**, 623-627.

Contents

2. Analysis and classification of protein–protein interactions from a structural perspective 33

SUSAN JONES AND JANET M. THORNTON

Contributors

JAKE BEGUN
Department of Biochemistry, University of Cambridge, 80 Tennis Court Road, Cambridge CB2 1GA, UK.

TOM BLUNDELL
Department of Biochemistry, University of Cambridge, 80 Tennis Court Road, Cambridge CB2 1GA, UK.

BRADFORD C. BRADEN
Department of Natural Sciences, Bowie State University, Bowie, MD 20715, USA.

TIM R. DAFFORN
Division of Structural Medicine, Department of Haematology, Cambridge Institute for Medical Research, University of Cambridge, Wellcome–MRC Building, Hills Road, Cambridge CB2 2XY, UK.

MARTIN J. HUMPHRIES
Wellcome Trust Centre for Cell-Matrix Research, School of Biological Sciences, University of Manchester, 2.205 Stopford Building, Oxford Road, Manchester, M13 9PT, UK.

MARKO HYVÖNEN
Department of Biochemistry, University of Cambridge, 80 Tennis Court Road, Cambridge CB2 1GA, UK.

SUSAN JONES
Biomolecular Structure and Modelling Unit, Biochemistry and Molecular Biology Department, University College, Gower St, London WC1E 6BT, UK.

JOËL JANIN
Laboratory of Structural Enzymology and Biochemistry, UPR9063 CNRS, 91198-Gif-sur-Yvette, France.

COLIN KLEANTHOUS
School of Biological Sciences, University of East Anglia, Norwich NR4 7TJ, UK.

MICHAEL LASKOWSKI, JR
Department of Chemistry, Purdue University, W. Lafayette, Indiana 47907-1393, USA.

ARTHUR M. LESK
Division of Structural Medicine, Department of Haematology, Cambridge Institute for Medical Research, University of Cambridge, Wellcome–MRC Building, Hills Road, Cambridge CB2 2XY, UK.

ROBERT C. LIDDINGTON
The Burnham Institute, 10901 N. Torrey Pines Road, La Jolla, CA 92037, USA.

STEPHEN M. LU
Department of Chemistry, Purdue University, W. Lafayette, Indiana 47907-1393, USA.

F. SCOTT MATHEWS
Department of Biochemistry, Washington University Medical School, St Louis, Missouri, USA.

A. GRANT MAUK
Department of Biochemistry and Molecular Biology, University of British Columbia, Vancouver, British Columbia V6T 1Z3, Canada.

GEOFFREY R. MOORE
School of Chemical Sciences, University of East Anglia, Norwich NR4 7TJ, UK.

ROBERTO J. POLJAK
University of Maryland Biotechnology Institute, Center for Advanced Research in Biotechnology, 9600 Gudelsky Drive, Rockville, MD 20850, USA.

ANSGAR J. POMMER
School of Biological Sciences, University of East Anglia, Norwich NR4 7TJ, UK.

M. A. QASIM
Department of Chemistry, Purdue University, W. Lafayette, Indiana 47907-1393, USA.

JANET M. THORNTON
Biomolecular Structure and Modelling Unit, Biochemistry and Molecular Biology Department, University College, Gower St, London WC1E 6BT, UK.

Abbreviations

Abu	α-aminobutyric acid
Ahp	α-aminoheptanoic acid
Ahx	α-aminohexanoic acid
Ape	α-aminopentanoic acid
APPI	Alzheimer's amyloid precursor protein (Kunitz) inhibitor
Arf	ADP ribosylation factor
ASA	accessible surface area
AST	anionic salmon trypsin
β-APP	β-amyloid precursor protein
BBI	Bowman–Birk inhibitor
BLIP	β-lactamase protein inhibitor
BPTI	bovine pancreatic trypsin inhibitor
BT	bovine trypsin
BWQEL	bob-white quail lysozyme
C	immunoglobulin constant domain
CAD	caspase-activated DNase
CAK	Cdk-activating kinase
CaM	calmodulin
cam	camphor
CaMK	Ca^{2+}/CaM-dependent protein kinase
CARL	subtilisin Carlsberg (from *Bacillus lichenformis*)
CATH	Structural classification scheme based on: class, architecture, topology, homology
CCBD	central cell-binding domain
C*c*P	cytochrome *c* peroxidase
CD	Circular dichroism
Cdk	cyclin-dependent kinase
CdkI	Cdk inhibitor
CDR	complementarity-determining regions
CH	horse-heart cytochrome *c*
CHYM (BCHYM)	bovine chymotrypsin Aα
CLIP	class II invariant peptide
ColE9	colicin E9
CY	yeast cytochrome *c*
Dim-CH	N^{ε}, N^{ε}-dimethylated-CH
Dim-CY	N^{ε}, N^{ε}-dimethylated-CY
EDC	1-ethyl-3-(dimethylaminopropyl)carbodiimide
EDTA	ethylenediaminetetraacetic acid

EGF	epidermal growth factor
ELISA	enzyme-linked immunoabsorbent assay
ER	endoplasmic reticulum
Fab	antigen-binding fragment
Fc	crystallizable fragment of immunoglobulin
FSH	follicle-stimulating hormone
Fv	variable portion g fab
GAP	GTPase activating protein
GDI	guanine nucleotide-dissociation inhibitors
GEF	guanine nucleotide-exchange factors
Giα	inhibitory Gα-subunit
GluSGP	glutamate-specific *Streptomyces griseus* proteinase
GPCR	G protein-coupled receptors
Gsα	stimulatory Gα-subunit
HEL	hen egg-white lysozyme
Hep II/IIICS	heparin-II/type III connecting fragment
hGH	human growth hormone
HIV	human immunodeficiency virus
HLE (HNE)	human leucocyte (neutrophil) elastase
hRI	human ribonuclease inhibitor
Hse	homoserine
I	ionic strength
ICAD	inhibitor of caspase-activated DNase
ICAM	intercellular cell-adhesion molecule
ICE	interleukin converting enzyme
Ig	immunoglobulin
IL	interleukin
Im	immunity protein
$Ins(1,4,5)P_3$	inositol triphosphate
Ii	invariant chain
ITC	isothermal titration calorimetry
K_a	equilibrium association constant
K_d	equilibrium dissociation constant
k_{et}	rate constant for unimolecular electron transfer within a bimolecular complex
K_{hyd}	the equilibrium constant for reactive-site peptide bond hydrolysis
K_i	inhibition constant
LDTI	leech-derived trypsin inhibitor
LFA	leucocyte function-associated antigen
LH	luteinizing hormone
LRR	leucine-rich repeat
mAb	monoclonal antibody
MAdCAM	mucosal addressin cell-adhesion molecule
MADH	methylamine dehydrogenase
MHC	Major Histocompatibility Complex

MAPK	mitogen-activated protein kinase
Mcg	a human myeloma immunoglobulin
MD	molecular dynamics
MIDAS	metal ion-dependent adhesion site
NIF	neutrophil inhibitory factor
NMR	nuclear magnetic resonance
NOE	nuclear Overhauser enhancement
NPGB	*p*-nitrophenylguanidinobenzoate
OMTKY3	ovomucoid third domain from turkey
PDB	(Brookhaven) Protein Data Bank
PH	pleckstrin homology
PLC	phospholipase C
PPE	porcine pancreatic elastase
PPII	polyproline type-II
pRI	porcine ribonuclease inhibitor
PSTI	pancreatic secretory trypsin inhibitor
PTB	phosphotyrosine binding (domain)
PTI	pancreatic trypsin inhibitor
PTK	protein tyrosine kinase
pY	phosphorylated tyrosine
RBD	Ras-binding domain
RGS	regulators of G-protein signalling
RI	ribonuclease inhibitor
rms	root mean square
Sc	shape complementarity
SGPA	*Streptomyces griseus* proteinase A
SGPB	*Streptomyces griseus* proteinase B
smMLCK	smooth muscle myosin light-chain kinase
Sos	Son of sevenless
SSI	*Streptomyces* subtilisin inhibitor
STI	soybean trypsin inhibitors
TCR	T-cell receptor
TAP	tick anticoagulation protein
Tb	terbium
TEL	turkey egg-white lysozyme
TFPI	tissue-factor pathway inhibitor
T_m	melting temperature
TSH	thyroid-stimulating hormone
TTQ	tryptophan tryptophylquinone
V	immunoglobulin variable domain
VCAM	vascular cell-adhesion molecule
vWf	von Willebrand factor

1 Kinetics and thermodynamics of protein–protein interactions

JOËL JANIN

1. Introduction

The physical chemistry of non-covalent protein–protein interactions aims to answer some fundamental questions: what is the origin of the affinity which enables biological macromolecules to form stable, complex, self-assemblies; what governs the specificity which makes these assemblies unique; which physical features give them functions that their individual components do not have? Given the ubiquitous character of protein–protein interactions, these questions are relevant to nearly all fields in biology, and they have attracted much interest over the years. An outline of the answer was given long ago by Linus Pauling and Max Delbrück (1):

> It is our opinion that the processes of synthesis and folding of highly complex molecules in living cells involve, in addition to covalent bonds, only the intermolecular interactions of van der Waals attraction and repulsion, electrostatic interactions, hydrogen-bond formation, etc., which are now well understood. These interactions are such as to give stability to a system of two molecules with *complementary* structures in juxtaposition... In order to achieve maximum stability, the two molecules must have complementary surfaces, like die and coin, and also a complementary distribution of active groups.

At the time when these lines were written, the physics of non-covalent forces was established, but there was little biochemical data and no structural information at all that could support the statement that known forces suffice to explain self-assembly in biology. Yet, it has stood unchanged since then, except for the addition of the hydrophobic effect (2–3) to the list of contributing processes. Now, 60 years later, we ought to be able not only to verify it but to add hard numbers into the answer. Proteins are now available in large quantities via recombinant DNA techniques and efficient overexpression systems. We can modify them at will by site-directed mutagenesis. We have detailed atomic models for at least 100 specific protein–protein complexes that have been studied by X-ray crystallography. Affinity, conveniently estimated by the dissociation constant of a protein–protein complex, can be approached by sensitive techniques, either at thermodynamic equilibrium or by analysing the kinetics of association. Recent progress in calorimetry has given access to changes in the thermodynamic-state functions that accompany association: the Gibbs energy (free enthalpy), enthalpy, and entropy. Can we now say precisely *how* 'van der Waals attraction and repulsion, electrostatic interactions, hydrogen-bond formation, etc.…' stabilize protein–protein interactions and determine specificity? We are still far from it despite a wealth of data, yet a confrontation of the structural approach with the thermodynamic and kinetic data ought to yield elements of the answer.

Such a confrontation is attempted here. It is centred on studies of the self-assembly of a pair of soluble proteins—such as an enzyme and a protein inhibitor, or an antibody and its cognate antigen—to form non-covalent heterodimeric complexes. It leaves aside the assembly of subunits in oligomeric proteins, even though this assembly is also based on non-covalent protein–protein interactions, because subunit folding and assembly are usually tightly coupled. In protein–protein complexes, the two processes are dissociated and may be studied separately. After a presentation of the formalism and of the methodology, a survey of the structural basis of protein—protein interactions will lead to an analysis of its energetics. Functional inferences made from the structure should be tested by experiment, with site-directed mutagenesis being a major tool. Many of the conclusions drawn from the study of complexes between soluble proteins may be extended to systems where structural data are scarce, namely membrane proteins, and to larger assemblies.

2. The thermodynamic and kinetic formalism

2.1 Affinity at thermodynamic equilibrium

Two parameters characterize protein–protein interactions at equilibrium: the stoichiometry and the affinity of the partners for each other. One-to-one association is the most common case. The formation of a complex AB from two components A and B may be written as:

$$A + B \underset{k_d}{\overset{k_a}{\longleftrightarrow}} AB. \qquad \text{[Scheme 1]}$$

This simple scheme is sufficient in most cases of one-to-one association. It can be extended to reactions with more than two components by breaking them into successive bimolecular steps. Here, k_d is the first-order rate constant for the unimolecular dissociation reaction, k_a the second-order rate constant for the bimolecular association reaction; their ratio is the equilibrium constant noted K_a for association or

K_d for dissociation. K_d is related to the concentrations of A, B, and AB at thermodynamic equilibrium by the law of mass action. In the way the law of mass action is written in Table 1, K_d has the dimension of a concentration and is expressed in mol litre^{-1} (noted M); k_d may be expressed in s^{-1} and k_a in M^{-1} s^{-1}. Commonly accepted in biology, this formulation is not shared by physical chemists, who replace concentrations with chemical affinities (4). Affinities differ from concentrations in two ways: they are dimensionless quantities, and they are corrected for non-ideality. The correction for non-ideality is small when concentrations are kept low, which is often the case in biochemical experiments, although there are situations where it cannot be ignored, both *in vitro* and *in vivo*. Ideality will be assumed below. The affinity of an ideal solute is the ratio of its concentration to the standard-state concentration ($c°$). For reactions in solution, the convention $c° = 1$ M is so usual that it is rarely made explicit, but other references are used in gas phase. In all cases $c°$ is required in the formula that relates K_d to ΔG_d, the free enthalpy change upon dissociation (Table 1), since the quantity under the logarithm must be dimensionless. The value of ΔG_d depends on the choice of $c°$, and so does the entropy of dissociation ΔS_d. Among the thermodynamic-state functions listed in Table 1, two are independent of the standard-state concentration: ΔH_d, the enthalpy of dissociation, and ΔC_d, the heat-capacity change at constant pressure, which is its temperature derivative.

The free enthalpy (or Gibbs energy) of dissociation is directly related to K_d. It is often called a binding energy although it is not an energy proper, as it determines the stability of the complex AB and characterizes the affinity of the two components for

Table 1 Thermodynamic state functions and equilibrium constants†

Law of mass action	$\frac{[A]\,[B]}{[AB]} = \frac{1}{K_a} = K_d = \frac{k_d}{k_a}$
Standard state free enthalpy change	$\Delta G_d = -RT \ln \frac{K_d}{c°} = \Delta H_d - T\Delta S_d$
Enthalpy change (van't Hoff)	$\Delta H_d = -R \frac{d(\ln K_d)}{d(1/T)}$
Entropy change	$\Delta S_d = -\frac{d(\Delta G_d)}{dT}$
Heat capacity change	$\Delta C_d = \frac{d(\Delta H_d)}{dT} = T \frac{d(\Delta S_d)}{dT}$
Gas constant	$R \approx 2$ cal mol^{-1} K^{-1}
Standard state‡:	pressure $p° = 1$ bar, concentration $c° = 1$ M

† Change in state functions upon dissociation on one mol of complex. Values referring to association differ only by the sign from those given here.
‡ The values of ΔG_d and ΔS_d depend on the choice of c°.

each other. Assuming that B is in excess over A, the law of mass action may be rewritten to yield the ratio of the equilibrium concentrations of bound and free A:

$$\frac{bound}{free} = \frac{[AB]}{[A]} = \frac{[B]}{K_d}.$$

A high affinity implies that most of A is in the bound state and that the bound:free ratio is large. Then, K_d must be small relative to [B], which indicates that affinity depends on the ligand concentration as well as on K_d. Thus, affinity is a relative concept, not an intrinsic property of molecule A. A hormone receptor with a $K_d \approx 1$ μM may be considered as low affinity and unlikely to be functional, because hormone concentrations are well below the micromolar range and the receptor will bind very little of it at equilibrium. Yet, a $K_d \approx 1$ μM is also the equilibrium constant for dissociation of the human haemoglobin tetramer into αβ-dimers in the presence of oxygen. The dimers cannot bind oxygen cooperatively and are functionally different from the tetramer. However, their presence can be neglected in red blood cells in spite of the high value of K_d, since haemoglobin is well above millimolar concentrations and dimer–dimer affinity is quite sufficient under normal physiological conditions.

2.2 Determination of the dissociation constant

The determination of the dissociation constant and of the reaction stoichiometry is an important step in the study of protein–protein interactions. It generally relies on measuring the equilibrium concentrations of the complex and its components. This requires having pure proteins and a means of detecting their association at concentrations around the K_d (5–6). The range of K_d values observed in biologically relevant processes that rely on protein–protein interactions is extremely wide, and extends over at least 12 orders of magnitude from 10^{-4} to 10^{-16} M (see Chapter 9). A number of physicochemical methods are available when the K_d is in the micromolar range or above, including fluorescence quenching, equilibrium ultracentrifugation, and microcalorimetry. Data are typically obtained by titrating a fixed concentration of component B with increasing concentrations of component A. They can be analysed by drawing the binding isotherm which relates the concentration y of bound A to that of the free species x. Calling y_0 the total concentration of B, or more correctly, the total concentration of sites that bind A, the graph of y versus x (*bound* versus *free*) should be a rectangular hyperbola. Its asymptote at $y = y_0$ represents saturation, and half-saturation is achieved for $x = K_d$. Alternatively, a plot of y/x versus y yields the linear Scatchard diagram, which has a slope equal to $-1/K_d$ and intersects the horizontal axis at $y = y_0$ from which the stoichiometry may be derived.

When K_d is in the nanomolar range, more sensitive detection methods are needed. Binding isotherms may be obtained after radioisotope labelling of one of the components, or with antibodies. In the conventional ELISA (enzyme-linked immunoabsorbent assay) method for measuring antigen–antibody interactions, the antigen is adsorbed on the plastic walls of a titration plate, the specific antibody is added, and the amount that is bound is determined with the help of a second antibody coupled to an enzyme. The method is unlikely to yield correct values of K_d as the conformation of the antigen may be perturbed by the adsorption process; and, in any case, the experiment is performed at a two-dimensional interface, whereas K_d is defined in homogeneous solution. However, the method can be adapted to give the proper result by performing the antigen–antibody association reaction in solution, and using the ELISA procedure to estimate the concentration of the free instead of the bound antibody (7).

2.3 Determination of the rate constants

Below the nanomolar range, artefacts such as adsorption on to glassware make the determination of protein concentrations unreliable. Thus, the methods mentioned above fail for very tight complexes. Kinetic measurements are then a useful alternative to equilibrium methods. They can be performed at concentrations well above the K_d, which is derived as a ratio of rate constants. Association and dissociation rate constants can be measured in the same or in separate experiments (8–10). In a typical mixing experiment, B is added in excess to A and pseudo first-order association kinetics are recorded by following a spectroscopic signal. The kinetics should follow a single exponential curve with a relaxation rate $1/\tau$ that increases linearly with the concentration x of excess ligand (Table 2, Scheme 1). The slope of the plot of $1/\tau$ versus x yields the k_a. As k_a values are typically in the range 10^5–10^8 $M^{-1} s^{-1}$, one expects τ to be less than a second when reagent concentrations are in the micromolar range or above. The experiment must be performed in a fast-mixing device, a stopped-flow apparatus for instance. Solutions of A and B are stored in two syringes and delivered into a mixing chamber by depressing the plungers. The evolution of the mixture is followed optically (UV absorbance or fluorescence) during the following milliseconds or seconds. Commercial devices achieve mixing in 2–5 milliseconds, making it possible to follow reactions with rates up to 500 s^{-1}.

The rate of relaxation $1/\tau$ is at least equal to the k_d, which may be derived from the kinetic data by extrapolating the plot of $1/\tau$ versus x to $x = 0$. However, the extrapolation is unreliable when reagent concentrations are much above the K_d. Then, the k_d should be measured in a separate experiment. Generally, it is impractical to follow the dissociation kinetics by diluting the complex below the K_d due to the low protein concentration that would be required. The experiment can, however, be performed by chasing a radioactive ligand with an excess of cold ligand. Dissociation is rate-limiting under these conditions, and the rate of release of the radioactivity yields the k_d. Values of k_d observed in protein–protein complexes can be as low as 10^{-7} s^{-1}, which implies that it takes months for very tight complexes to dissociate, or as high as 10^3 s^{-1} for weak complexes, which have a life time of milliseconds. Slow and fast dissociating complexes behave very differently under many experimental conditions, upon chromatography for instance, and it is extremely useful to have some idea of the value of k_d when preparing a complex.

Table 2 Kinetic models and rate constants

Reaction scheme†	Equilibrium/rate constants‡
1. Bimolecular association	
$\mathrm{A} + \mathrm{B} \underset{k_d}{\overset{k_a}{\rightleftharpoons}} \mathrm{AB}$	$K_d = \frac{k_d}{k_a}$ $1/\tau \approx k_a x + k_d$
2. Association–isomerisation ($k_2 \gg k_2$)	
$\mathrm{A} + \mathrm{B} \underset{k_{-1}}{\overset{k_1}{\rightleftharpoons}} \mathrm{AB'} \underset{k_{-2}}{\overset{k_2}{\rightleftharpoons}} \mathrm{AB}$	$K_d = \frac{k_{-1}}{k_1} \frac{k_{-2}}{k_2} = K_1 K_2$ $1/\tau \approx \frac{k_2 x}{x + k_{-1}/k_1}$ $\tau \approx \frac{1}{k_2} (1 + \frac{K_1}{x})$

† k_1 and k_a are second-order microscopic rate constants; all others are first-order.
‡ The rate of relaxation $1/\tau$ is expressed in terms of the microscopic rate constants and of a concentration noted x. In mixing experiments, x is the concentration of the species in excess. In near-equilibrium relaxation experiments (temperature jump), x is the sum of the concentrations of free A and free B.

2.4 Coupled association–isomerization

The relationship between the equilibrium and rate constants depends on the reaction mechanism, which we have assumed to be a single step up to now. More elaborate mechanisms include unimolecular steps that precede or follow the bimolecular association step (8–10). In principle, these steps may be detected in kinetic experiments, either directly or by exploring the concentration-dependence of the rate of reaction and analysing its departure from second-order kinetics. The most frequently encountered alternative to single-step binding is association followed with isomerization, possibly a conformational change induced by the interaction of the two components.

In Table 2, Scheme 2 comprises an intermediate AB′ isomerizing to the final complex AB, and four microscopic rate constants; k_2 is large relative to k_{-2}, driving the reaction towards the right. If in addition k_2 is large relative to k_{-1}, isomerization is fast relative to the dissociation of the intermediate complex, so AB′ never accumulates and Scheme 2 reduces to Scheme 1. If isomerization is slow or comparable in rate, a mixing experiment may detect a single relaxation step with a rate $1/\tau$ that increases linearly with the concentration x of the excess reagent, just as in Scheme 1. However, the observed bimolecular rate constant is reduced by a factor k_2/k_{-1} relative to its true value k_1 and the linear relationship between $1/\tau$ and x breaks down at high concentrations. Then, isomerization becomes rate-limiting and $1/\tau$ approaches an asymptotic value equal to k_2. Provided the kinetic measurements are performed over a sufficient concentration range, a plot of τ versus $1/x$ should yield both k_2 and the k_{-1}/k_1 ratio, which is the dissociation constant of the intermediate complex AB′ (Table 2). Individual values of k_1 and k_{-1} are not accessible unless the (presumably fast) bimolecular step is independently detected. The k_{-2} rate constant may be measured in a chase experiment just like k_d in Scheme 1, for the rate-limiting step for dissociation is now the slow isomerization of the stable complex back to the fast-dissociating intermediate. Finally, K_d may be independently determined under equilibrium conditions, and its value compared with the product of the k_{-1}/k_1 and k_{-2}/k_2 ratios. A satisfactory agreement between kinetic and equilibrium data is essential for establishing which scheme correctly represents the reaction mechanism. Scheme 2 has often been called upon just to explain low experimental values of bimolecular rate constants, but these may have other origins as will be discussed in Section 5.2.

In all the above experiments, association is induced by mixing two components. This is not possible if the components are identical, i.e. for a dimerization reaction. Dimerization kinetics can be followed by coupling with another reaction, subunit folding for instance. An alternative is to perturb a pre-existing thermodynamic equilibrium by a jump of the pressure or temperature. The rate of relaxation towards a new equilibrium is related to the microscopic rate constants by the same equations as in mixing experiments where the perturbation is induced by changing concentrations, except that equilibrium concentrations replace initial concentrations (9–10). Pressure or temperature jumps may also be used when A and B are different species, but the experiment is practical only for weakly associating species—for it must be performed under conditions where the complex coexists with its free components, and therefore with both components at concentrations in the order of the K_d.

2.5 Surface plasmon resonance

Surface plasmon resonance is a very sensitive optical method that measures local changes in the refractive index at the surface of a gold chip (11). Most molecules including proteins have a higher refractive index than water, and any increase in their local concentration affects the refractive index in proportion to the added mass. The biosensor device uses surface plasmon resonance to detect that change. It is well adapted to kinetic studies of protein–protein association. In a typical experiment, purified partner A is first chemically immobilized on the surface of the chip. Then, partner B is added to the buffer, and its binding to the immobilized component is detected by recording the optical signal due to the increased refractive index. After binding is completed, a switch is made back to pure buffer and the dissociation of B yields a signal of opposite sign.

The measurement gives immediate access to the kinetics of association and of dissociation (12). Moreover, protein B does not need to be pure as long as no other component of the solution binds to A. However, any estimate of the stoichiometry, the equilibrium and the rate constants, must allow for possible artefacts of the immobilization procedure. When the protein is immobilized by chemically bridging its amino groups with reactive groups on the sensor chip, the product is highly heterogeneous. Some molecules will have lost their affinity completely, which affects the stoichiometry, or partly, which leads to binding isotherms that deviate from a rectangular hyperbola. To limit heterogeneity, one may replace the chemical modification with non-covalent immobilization procedures using high-affinity antibodies or affinity tags (13). A useful tag is biotin, which tightly binds to immobilized streptavidin. Biotinylation or antibody association may still perturb the affinity of the immobilized partner and the kinetics of binding.

Remarks made above concerning the difference between a reaction in homogeneous solution and after surface adsorption, apply to the biosensor. In addition, the continuous flow of reagent or buffer implies that the system cannot reach equilibrium, only a steady state. This can be far from the true equilibrium unless the flow rate is very low compared to the reaction rate. Under standard conditions, ligand binding may well be limited by the transport of the ligand to the sensor surface (14). Then, its rate saturates as it does in Scheme 2 (see Table 2) above, but the process no longer represents the association reaction. Mass transport may perturb the measurement when k_a is above a value in the order of $10^6\ M^{-1} s^{-1}$. In all cases, equilibrium and kinetic parameters derived after immobilization cannot be attributed to the reaction in homogeneous solution without careful consideration. The agreement is often good when biosensor data are compared with kinetic and equilibrium data obtained by conventional techniques, yet discrepancies of an order of magnitude or more are observed in specific cases (15).

3. The structural basis for recognition

3.1 Examples of protein–protein complexes

Many non-covalent protein–protein complexes have been analysed by X-ray crystallography, yielding atomic coordinate data sets that are available through the Brookhaven Protein Data Bank (PDB) (16). These data provide a structural basis for understanding protein–protein interactions (17–22). Except for oligomeric proteins, the best represented categories are complexes between antibodies and cognate protein antigens and between proteases and their protein inhibitors. There are also examples of enzymes other than proteases that form complexes with protein substrates or inhibitors, and of complexes involved in signal transduction within cells. Representative examples of these categories are shown in Plates 1–4: hen egg lysozyme bound to the variable domains of the HyHEL5 monoclonal antibody (23); the pancreatic trypsin inhibitor (PTI) bound to trypsin (24); the barstar inhibitor bound to the bacterial ribonuclease barnase (25–26); transducin, a heterotrimeric guanine nucleotide binding protein from the bovine retina (27). The complexes are drawn as protein backbones in two orthogonal directions. The side view shows the two components together; in addition to the backbone, the front view displays the region of the surface of one component that is in contact with the other. In these systems, the geometric complementarity is excellent, the chemical complementarity is expressed in the presence of 11–13 hydrogen bonds or salt bridges. In trypsin–PTI, the convex end of PTI fits into the concave surface of the protease active site; a binding loop on the PTI surface forms β-sheet-like, main chain-to-main chain hydrogen bonds with trypsin; it provides a lysine side chain to fill the primary specificity pocket. In lysozyme–HyHEL5 and barnase–barstar, the interacting surfaces are fairly flat in their general appearance, but they are bumpy on the atomic scale with a precise knob-into-hole fit; more of the polar interactions involve side chains, especially the charged side chains of arginine (in lysozyme–HyHEL5) or acidic residues (in barnase–barstar).

These complexes are specific and stable, with dissociation constants in the order of 10^{-10} M for lysozyme–HyHEL5 (23; see Chapter 5), 10^{-14} M for trypsin–PTI (28; see Chapter 8) and barnase–barstar (29; see Chapter 9). The lysozyme–HyHEL5 complex illustrates antibody–antigen recognition as selected by the immune system of vertebrates, the trypsin–PTI and barnase–barstar, enzyme–inhibitor interactions that undergo strong Darwinian selection. PTI prevents trypsin from being activated in the pancreas. Barstar prevents barnase from degrading cellular RNA before it is excreted into the surrounding medium, and any mutation that raises the dissociation constant above a certain threshold is lethal (30, 31).

Transducin belongs to a different system. It is a heterotrimeric, guanine nucleotide binding protein (G-protein) that mediates the signal initiated at the primary light receptor in vision, rhodopsin. Upon interaction with the photoactivated receptor at the membrane, transducin dissociates into soluble G_{α}- and $G_{\beta\gamma}$-subunits, and the G_{α}-subunit exchanges a bound GDP molecule for GTP; then, GTP is hydrolysed and the trimer reassembles. To perform its function, the G_{α}-subunit must cycle between an activated GTP-bound monomeric form and an inactive GDP-bound trimeric form (32). The affinity of G_{α} for $G_{\beta\gamma}$ depends on the nature of the bound nucleotide and on the attachment of the protein to the membrane via acyl groups present on chain termini. It cannot be very high, for the heterotrimer must dissociate much faster and much more easily than for trypsin–PTI or barnase–barstar.

3.2 Rigid-body association versus conformation changes

The X-ray structures of the lysozyme–antibody, trypsin–PTI, and barnase–barstar complexes illustrated in Plates 1–4 may be compared to those of their free components which are independently known. They are found to differ only by small changes, typically 1 Å in amplitude, in the polypeptide main-chain fold, and by the rotation of a few amino-acid side chains. The mobility of atoms at the interface, described in PDB files by the crystallographic temperature factor, is often reduced in the complex relative to the free protein, but their position changes little (17). Specific recognition takes place between complementary protein surfaces that are largely preformed; it requires only small adjustments. To a good approximation, the complexes form by rigid-body association.

Rigid-body association is frequent, but not general, in protein–protein recognition. The PDB shows examples of significant or major conformation changes upon association. The complex formed between trypsinogen and PTI is a case in point. It is essentially identical to the complex with trypsin shown in Plate 2 where the two components undergo little conformation change. However, free trypsinogen is unlike free trypsin. In addition to an N-terminal peptide that is cleaved upon activation, the catalytically inactive precursor of trypsin contains disordered loops of polypeptide chain. These are fully ordered in the mature enzyme, and also in trypsinogen–PTI. PTI binding induces a disorder-to-order transition in trypsinogen that converts it to the conformation normally found after activation (33). The entropic (and possibly enthalpic) cost of the transition is reflected in the dissociation constant, which is several orders of magnitude higher for trypsinogen–PTI than for trypsin–PTI. Nevertheless, the trypsinogen–PTI complex is of physiological significance, for no trypsin is normally present in the pancreas and the primary function of PTI is to prevent the accidental activation of the precursor.

Conformation changes upon protein–protein association are illustrated by transducin. The heterotrimer with bound GDP shown in Plate 4 is the inactive form of the protein. X-ray structures are known for the trimer, the free G_α subunit in the activated (GTP-bound) or inactive (GDP-bound) form, and also for the free $G_{\beta\gamma}$-subunit (34–36). A comparison shows that important conformation changes occur in G_α both upon GTP hydrolysis and upon binding $G_{\beta\gamma}$. In either case, they concern segments of the polypeptide chain called Switch I and Switch II. These segments are close to the phosphate groups of the bound nucleotide and they form most of the interface with $G_{\beta\gamma}$ in the trimer. Switch I and Switch II undergo main-chain movements of up to 10 Å upon GTP hydrolysis and upon $G_{\beta\gamma}$ binding, different in the two cases. In the latter, even the secondary structure is modified; rearrangements of both the main chain and side chains completely remodel the protein surface. Additional interactions with $G_{\beta\gamma}$ involve an N-terminal segment that forms an α-helix in the heterotrimer whilst it is fully disordered in the free subunit. Thus, transducin undergoes both a conformation change and a disorder-to-order transition in the regions of G_α that interact with $G_{\beta\gamma}$. Other parts of G_α remain the same, and lesser changes affect $G_{\beta\gamma}$. Instead of being preformed as in trypsin–PTI or barnase–barstar, the surfaces that form the G_α–$G_{\beta\gamma}$ interface acquire their complementary shapes only when the two components associate.

3.3 Measurement of interface areas

The extent of the interaction between two proteins can be quantified when atomic coordinates are available. A convenient measure is the interface area, which is obtained by comparing the solvent-accessible surface of the complex to that of its components (37):

$$B = A_\mathrm{A} + A_\mathrm{B} - A_\mathrm{AB}.$$

where A_AB is the solvent accessible surface area of the complex, A_A and A_B, those of the dissociated components. Accessible surface areas are estimated by rolling a spherical probe representing a solvent molecule over the protein surface (38–39). They may be measured either at the van der Waals surface of protein atoms or one probe-radius away from that surface, yielding different sets of values. To avoid confusion, here we use only the second convention (see note 1). *B* represents the area of the surface of the two component proteins that is buried in their contact. This surface is illustrated in Plates 1–4 for one of the two partners of each complex. It can also represent the area of the protein that is buried when the proteins associate, but only if there is no conformation change. *B* is proportional to the number of atom pairs in contact and provides a convenient measure of the extent of the contact between the two partners (40). The contribution of each partner to the interface area is very close to one-half of *B* when the surfaces are flat. In trypsin–PTI, the interface is made by a convex surface on the inhibitor fitting into a concave surface on the protease. Due to curvature effects, the second area is 10% less than the first.

Figure 1 is a histogram of interface areas observed for a sample of 75 protein–-protein complexes of known three-dimensional structure. No interface area is less than 1100 Å^2, suggesting that proteins cannot form a stable complex unless at least 1100 Å^2 of the protein surface is removed from contact with water. Each of the two partners must provide at least 550 Å^2 of complementary surface. The histogram displays a well-defined peak around B = 1600 Å^2 which comprises 70% of the complexes, and a set of larger interfaces with *B* ranging from 2000 to 5000 Å^2. Most

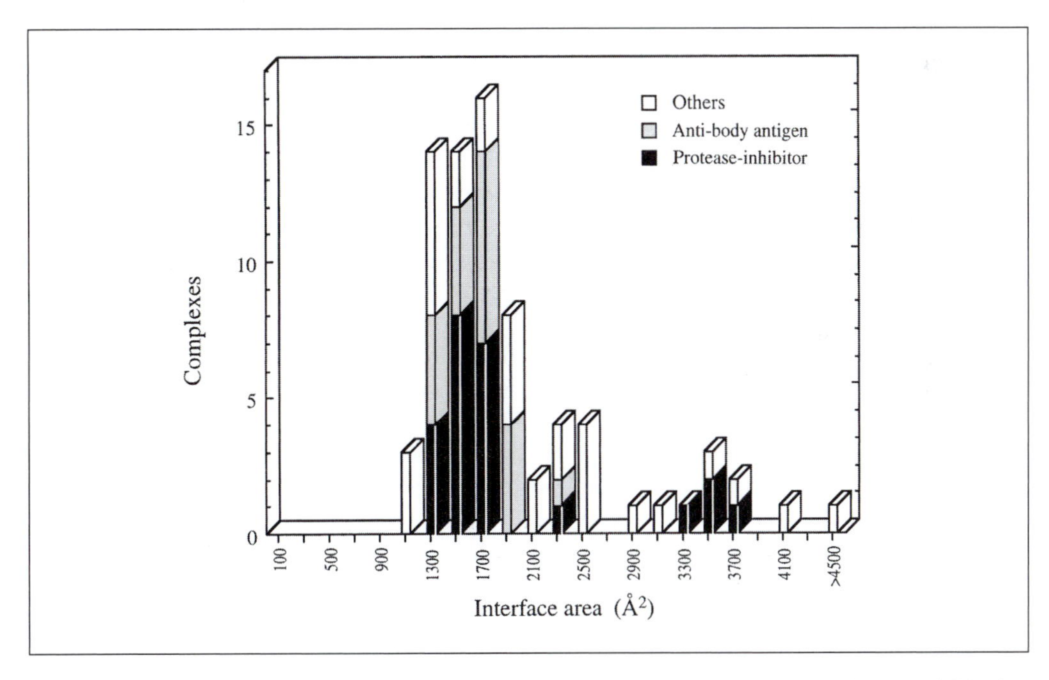

Fig. 1 Histogram of interface areas. Data are for 20 antigen–antibody, 24 protease–inhibitor, and 31 other protein–protein complexes of known three-dimensional structure. Solvent accessible surface areas are calculated from the atomic coordinates of the complexes as deposited in the PDB, and the interface area is derived by difference. The calculation uses a program by M. Gerstein (Yale University) that implements the Lee–Richards (38) rolling sphere algorithm and a probe radius of 1.4 Å. (Adapted from Lo Conte *et al.* (118) where a detailed list of the complexes may be found.)

antigen–antibody and protease-inhibitor complexes are in the main peak. The peak includes lysozyme–HyHEL5, trypsin–PTI, and also barnase–barstar, but not transducin which has a larger interface. Each partner in one of these standard-size interfaces loses about 800 Å^2 of accessible surface, contributed by some 20 residues. The average interface residue loses 40 Å^2, but individual values range from a few Å^2 to over 150 Å^2, which is the area lost when a fully accessible arginine or tryptophan side chain becomes buried. The ratio of about 40 Å^2 per residue is maintained in the larger interfaces, some of which involve over 50 residues on each partner.

Out of 75 complexes, three have interface areas just below 1200 Å^2 at the lower end of the histogram. Of these, two involve fragments of larger proteins where some of the contact region may be missing. Another is the cytochrome *c*–cytochrome *c* peroxidase complex (41), a redox enzyme–substrate complex which must be only transiently stable for the reaction to undergo turnover. Several other redox complexes have small interfaces (18). The cytochrome–peroxidase complex may illustrate a type of protein–protein association where function requires the rate constant k_d and the equilibrium constant K_d to be high, and which is poorly represented in the PDB at present.

Beyond the remark that the cytochrome–peroxidase complex has both a small interface and a high K_d, there is no simple correlation between B and the affinity. Standard-size interfaces at $B \approx 1600$ Å^2 occur in complexes typically with K_d values ranging from 10^{-7} to 10^{-14} M. Nevertheless, there seems to be a relationship between interface area, affinity, and the occurrence of conformation changes. The lysozyme–HyHEL5, trypsin–PTI, and barnase–barstar complexes, all of which associate as quasi-rigid bodies, have interface areas near 1500 Å^2. The trypsinogen–PTI complex also, and the much higher affinity of PTI for trypsin, can safely be attributed to the lack of a conformation change in trypsin–PTI. Conversely, very large interfaces often concur with major conformation changes. A first example is hirudin (from the medicinal leech) binding to the blood protease thrombin (42). The complex has an interface area of 3300 Å^2, more than twice that of trypsin–PTI, distributed in two separate patches. One is at the active site of the protease, and one at a different site where the C-terminal tail of hirudin binds; the tail is disordered in free hirudin. Like the trypsinogen–PTI association mentioned above, hirudin–thrombin association induces a disorder-to-order transition in hirudin. Transducin provides a second example. The G_α–$G_{\beta\gamma}$ interface is distributed in two patches as can be seen in Plate 4. The patch that involves Switch I–Switch II of G_a covers 1500 Å^2, an area equivalent to that of the trypsin–PTI and barnase–barstar interfaces. The other, smaller (1000 Å^2) patch involves the N-terminal helix. When this is cleaved, the affinity of G_α for $G_{\beta\gamma}$ drops considerably (43), and K_d becomes at least six orders of magnitude larger than for trypsin–PTI or barnase–barstar. Most the difference should reflect the cost of the large conformation changes described above.

3.4 Non-polar and polar interactions

Interfaces bury non-polar groups, mostly from amino-acid side chains, and polar groups originating from both the main chain and the side chains. As discussed below, both types of surfaces contribute to stabilizing the complexes: non-polar groups through the hydrophobic effect; polar groups through hydrogen bonds. In lysozyme–HyHEL5 and barnase–barstar complexes, non-polar and polar groups contribute approximately equally to the interface area. In trypsin–PTI and transducin, the non-polar : polar area ratio is about 56 : 44, which is the average value for interfaces found in a larger sample of protein–protein complexes. The average surface of soluble proteins has essentially the same non-polar : polar ratio (44). Thus, the average protein–protein interface is not less polar or more hydrophobic than the surface remaining in contact with the solvent. It differs from interfaces between the subunits of oligomeric proteins, which generally are both larger and more hydrophobic (19, 44, 45). This is apparent in the amino-acid composition of the interfaces: the most abundant residue is leucine at subunit interfaces, arginine at protein–protein interfaces. In most protein–protein complexes, there is no indication of non-polar patches serving as binding sites, and polar groups always contribute at least 30% of the interface area.

With very few exceptions, polar groups that become buried at the interface form hydrogen bonds linking the components of the complex. The number of hydrogen bonds per unit buried surface area is similar in all complexes, about one per 170 Å^2. A standard-size interface with $B \approx 1600$ Å^2 buries about 900 Å^2 of the non-polar surface, 700 Å^2 of polar surface, and contains 10 ± 5 hydrogen bonds (17–22). Larger interfaces have more hydrogen bonds: 18 link G_{α} to $G_{\beta\gamma}$ in transducin for instance. In addition to direct bonds between protein groups, one finds water molecules bridging polar groups on the two partners. Water is usually excluded from the contact region and crystallographic water molecules tend to line its periphery, but there are examples of 'wet' interfaces penetrated by water molecules (46, 47). In lysozyme–HyHEL5, trypsin–PTI, and barnase–barstar complexes, a majority of the hydrogen bonds at the interface are charged, that is, either the donor or the acceptor group is an anion or a cation. Fewer are salt bridges, hydrogen bonds between an anionic and a cationic side chain. These features are present in many other protein–protein complexes (48).

4. Energetics of protein–protein interactions

4.1 Calorimetric measurement of the binding enthalpy and heat capacity

The free enthalpy of dissociation ΔG_d of a complex can be calculated directly from its dissociation constant (Table 1). The enthalpy component ΔH_d may, in principle, be derived by measuring the K_d at several temperatures and applying van't Hoff's law. However, the accuracy is poor and a van't Hoff plot (plot of $\ln K_d$ versus $1/T$) need not be linear, because ΔH_d is itself temperature-dependent. In recent years, the development of isothermal titration calorimetry (ITC) has provided direct access to the enthalpy change (49–51). Sensitive microcalorimeters measure the heat evolved upon mixing the two components: it is equal to $-\Delta H_d$ per mol of complex formed. Measurements at several temperatures then yield ΔC_d. The calorimetric enthalpy should agree with the value derived from a van't Hoff plot of the temperature dependence of K_d if the reaction mechanism is that of Scheme 1, and it should be more accurate. The two values may differ when the reaction mechanism is more complex. In addition, the heat released can be used to monitor association and to titrate one component against the other. A calorimetric experiment can give both the ΔH_d and the ΔG_d, but only if the K_d is fairly large (10^{-7} M), since protein concentration must be sufficient for the heat released to be detected. If the affinity is high, K_d must be independently measured.

With the K_d ranging from 10^{-4} to 10^{-14} M, ΔG_d ranges from 6 to 19 kcal mol^{-1}. The free enthalpy change breaks down into an enthalpic contribution, ΔH_d, and an entropic contribution, $-T\Delta S_d$, the sign and magnitude of which depend on the system. Association is commonly said to be 'enthalpy driven' when ΔH_d is positive (favouring association) and ΔS_d is also positive (unfavourable); it is 'entropy driven' when ΔS_d is negative and large and when ΔH_d is small or negative. However, these words poorly describe the underlying physics, for both ΔH_d and ΔS_d vary strongly with temperature. The temperature derivative of ΔH_d is ΔC_d, which is almost always positive and large. The consequences are illustrated in Fig. 2 for the lysozyme–HyHel5 antigen–antibody interaction observed in ITC measurements (see also Chapter 5). At 25 °C, association is enthalpy driven: ΔH_d is large and positive, which is sufficient to explain the stability of the complex (52). When the temperature is lowered, ΔH_d and ΔS_d become smaller due to a large heat-capacity change, while ΔG_d remains fairly constant because enthalpy and entropy compensate each other. ΔS_d actually changes

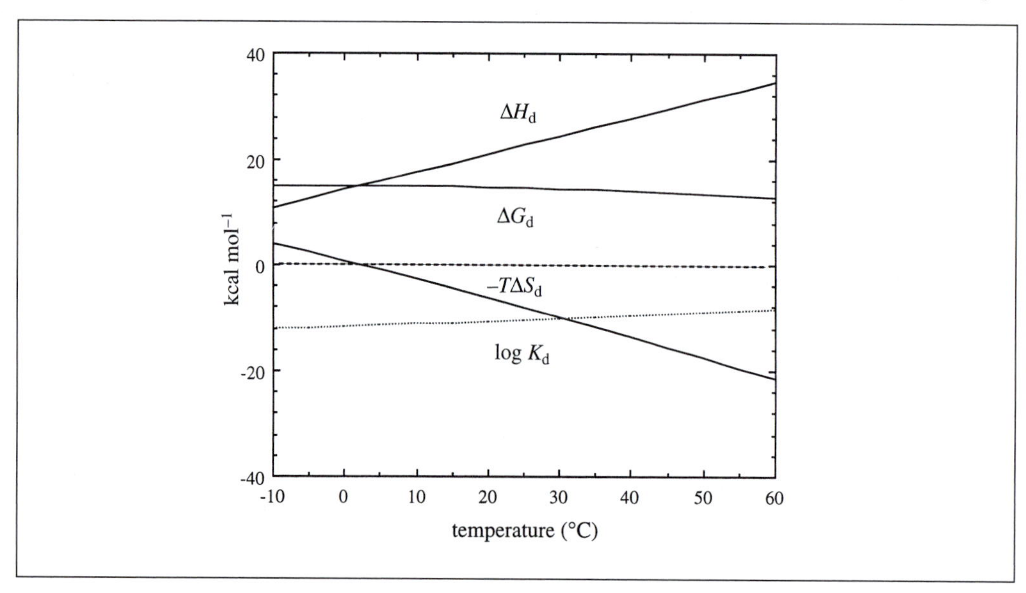

Fig. 2 Thermodynamics of association in an antibody–antigen complex. The dependence on temperature of the state functions for dissociation of the lysozyme–HyHEL5 complex, is derived by applying the relationships given in Table 1 to the value at 25 °C of the dissociation constant K_d which yields ΔG_d. ITC measurements performed in the range from 10 to 37 °C range (52) yield ΔH_d and ΔC_d. The observed value $\Delta C_d = 0.34$ kcal mol^{-1} K^{-1} is assumed to be constant over the larger temperature range of the figure.

sign near 0 °C and below, lysozyme–HyHEL5 association could be said to be entropy driven (but only at the standard-state concentration $c^\circ = 1$ M), although the forces involved are certainly the same at 0 and 25 °C.

Similar studies have been performed on other antigen–antibody complexes, protease–inhibitor complexes (46, 53), and barnase–barstar (54). Measured values of ΔC_d in these systems are remarkably close to that of lysozyme–HyHEL5, implying that the temperature-dependence is the same. Whichever sign ΔH_d and ΔS_d have at 25 °C, a large positive ΔC_d is a sure indication that the hydrophobic effect plays an important role. The non-polar and polar chemical groups that are buried at the interface become hydrated when a complex dissociates. Non-polar group hydration has a large positive heat capacity and is the dominant term in the heat capacity change. Polar group hydration may involve more enthalpy than for non-polar groups, but it is much less temperature-dependent and therefore contributes little to ΔC_d (55–59).

4.2 The contribution of hydration to affinity

Experimental values of ΔC_d in systems such as lysozyme–HyHEL5 or barnase–barstar, where the three-dimensional structure is known, can be fitted reasonably well (to within ± 25%) by applying a proportionality coefficient to the non-polar component of their interface area. Proportionality coefficients that relate solvent-accessible surface areas to the hydration heat capacity, enthalpy, and free enthalpy, are also available. Several sets exist in the literature, all derived from physico-chemical data on small molecules serving as models for chemical groups found in proteins (60–63). They are poorly consistent with each other (64) and their application to proteins assumes additivity of the contribution of individual atoms, which is debatable (65). Nevertheless, they are useful tools in relating structures to thermodynamics. The sets are of two different types: they refer to the transfer of the groups to water from either the gas phase or from a non-polar organic solvent (often octanol). In the first case, all water–solute interactions including van der Waals contribute to the coefficients; in the second case, only those that differ between the polar (water) and non-polar (octanol) solvent contribute. The popular Eisenberg–McLachlan (60) set is of the second type.

Given a set of coefficients, the free enthalpy of hydration of the protein surface is calculated as:

$$\Delta G^{hyd} = \Sigma\, \gamma_i A_i;$$

where A_i is the area contributed by chemical groups of type i and γ_i is the relevant coefficient. By difference, dehydrating a protein–protein interface provides:

$$\Delta G^{hyd} = \Sigma\, \gamma_i\, B_i;$$

which is a weighted sum of B_i, the groups' contribution to the interface area. Processes such as folding or association reduce the protein-accessible surface area causing large changes in ΔG^{hyd}. The value of γ for the transfer of alkanes from octanol to water is 25 cal mol^{-1} Å^{-2} (66; possibly an underestimate: 67, 68). The $\approx$ 800 Å^2 of non-polar protein surface buried at a standard-size protein–protein interface imply $\Delta G^{hyd} \approx 20$ kcal mol^{-1}. Making the assumption that the environment inside a protein or a protein–protein interface resembles octanol, ΔG^{hyd} measures the contribution of the hydrophobic effect to ΔG_d. The assumption is certainly incorrect, for the protein interior is not a liquid, it is ordered and compact with a density 60% larger than for octanol. Organic solids are probably a better model than the liquids. Nevertheless, there is no doubt that the hydrophobic effect stabilizes self-assembly just as it does protein folding (2, 3) and that the dehydration of non-polar groups at the interface is essential for stable association, a point to which we shall return.

It has been suggested that the formula giving ΔG^{hyd} can yield ΔG_d and not just the contribution of hydration. By modifying the Eisenberg–McLachlan coefficients, Horton and Lewis (69) obtained a linear correlation between ΔG^{hyd} and experimental ΔG_d values in a sample of 15 protein–protein and protein–peptide complexes, including trypsin–PTI and lysozyme–HyHEL5. The experimental values could be fitted to within 2 kcal mol^{-1} for all 15 complexes. However, the correlation must break down for complexes such as thrombin–hirudin or transducin that have much larger interfaces and lower affinities than trypsin–PTI. Moreover, point mutations at an interface have been shown to change the ΔG_d by several kcal mol^{-1} while keeping ΔG^{hyd} unchanged (70). The correlation may nevertheless hold outside the original sample for those protein–protein complexes that associate as rigid bodies and have interfaces that have been optimized by natural selection. Under these conditions, van der Waals and polar interactions may be expected to scale-up with the interface area just as hydration effects do.

4.3 Modelling affinity from structure

The basis for using hydration coefficients is the thermodynamic cycle shown in Fig. 3. In a thought experiment, association is first performed in a gas phase or an organic solvent (octanol for instance), then the complex and its components are transferred to water. The free enthalpy of dissociation in water may then be estimated as:

$$\Delta G^{wat} = \Delta G^{gas} + \Delta G^{hyd}$$

or:

$$\Delta G^{wat} = \Delta G^{oct} + \Delta G^{tra}.$$

where ΔG^{gas} and ΔG^{oct} are the values of ΔG_d in a gas phase or in octanol and the free enthalpy of the octanol-to water transfer is denoted ΔG^{tra} to distinguish it from the gas-to-water transfer. Both quantities can be derived from the interface area with appropriate coefficients. Similar expressions yield the enthalpy, entropy, and heat capacity of dissociation.

Most empirical models of affinity (71–74) use the organic-solvent version of this cycle. It is rarely made explicit, and ΔG^{oct} is never estimated. Indeed, its calculation would be just as difficult as in water, because it requires enthalpy and entropy values for protein–solvent and solvent–solvent interactions as well as for protein–protein

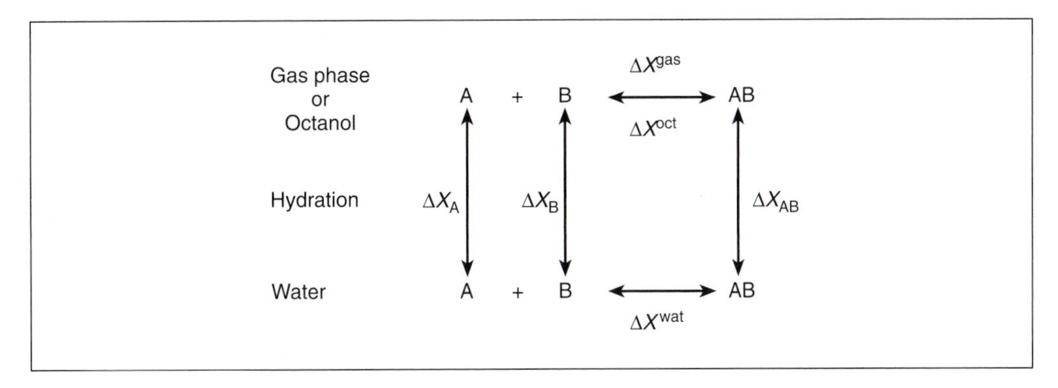

Fig. 3 A thermodynamic cycle for protein–protein association. The change ΔX^{wat} of thermodynamic state function X upon dissociation of AB in water is equal to the sum of ΔX^{gas}, the change upon dissociation in gas phase, and of ΔX^{hyd} for hydrating the interface when the complex dissociates. This is given by:

$$\Delta X^{hyd} = \Delta X_A + \Delta X_B - \Delta X_{AB}$$

X can be the free enthalpy, enthalpy, entropy, or heat capacity and dissociation may be performed in an organic solvent (octanol) instead of gas phase.

interactions. In gas phase, only the latter need be evaluated, since protein–solvent and solvent–solvent interactions are dealt with during the hydration step. In addition, the organic solvent-to-water cycle yields only free enthalpies due to a lack of consistent sets of proportionality coefficients for other state functions. With the gas phase-to-water cycle, enthalpy/entropy changes and their temperature dependence may also be modelled.

This has been attempted for the lysozyme–HyHEL5 complex (75–76) and the results are shown in Fig. 4. The calculated enthalpy and free enthalpy changes reproduce experimental data at 25 °C reasonably well, but they are small differences between large numbers with large uncertainties. The calculation does no more than indicate the relative importance of individual terms contributing to ΔG_d and ΔH_d. In gas phase, the two proteins make strong van der Waals and electrostatic interactions. Their energy, estimated by molecular mechanics, yields a large positive ΔH^{gas}. In water, ΔH^{gas} is largely offset by the enthalpic cost of dehydrating groups at the interface, especially polar groups that interact strongly with water. The value of ΔH^{hyd} derived from the interface area is comparable to ΔH^{gas}, but of opposite sign.

The entropic term $-T\Delta S$ behaves in an opposite manner to ΔH. In the gas phase, enthalpy favours and entropy opposes association: the number of external (rotational and translational) degrees of freedom is reduced as two molecules become one; side-chain and main-chain atoms become immobilized at the interface (77). Empirical estimates of these effects may be used to predict the value of $-T\Delta S^{gas}$ (78–81). In contrast, the hydration entropy favours association which releases water molecules at the interface into bulk solution. For non-polar groups, this more than compensates for the loss of van der Waals interactions with water, and the free enthalpy of non-polar group hydration favours binding. Its contribution, which measures the hydrophobic effect, is about the same as that derived above by applying the octanol-to-water γ coefficient. The entropy of polar-group hydration also favours association, but as it is far from offsetting the enthalpy of polar interactions with water, the free enthalpy remains negative and large: dehydration of polar groups at the interface strongly opposes association.

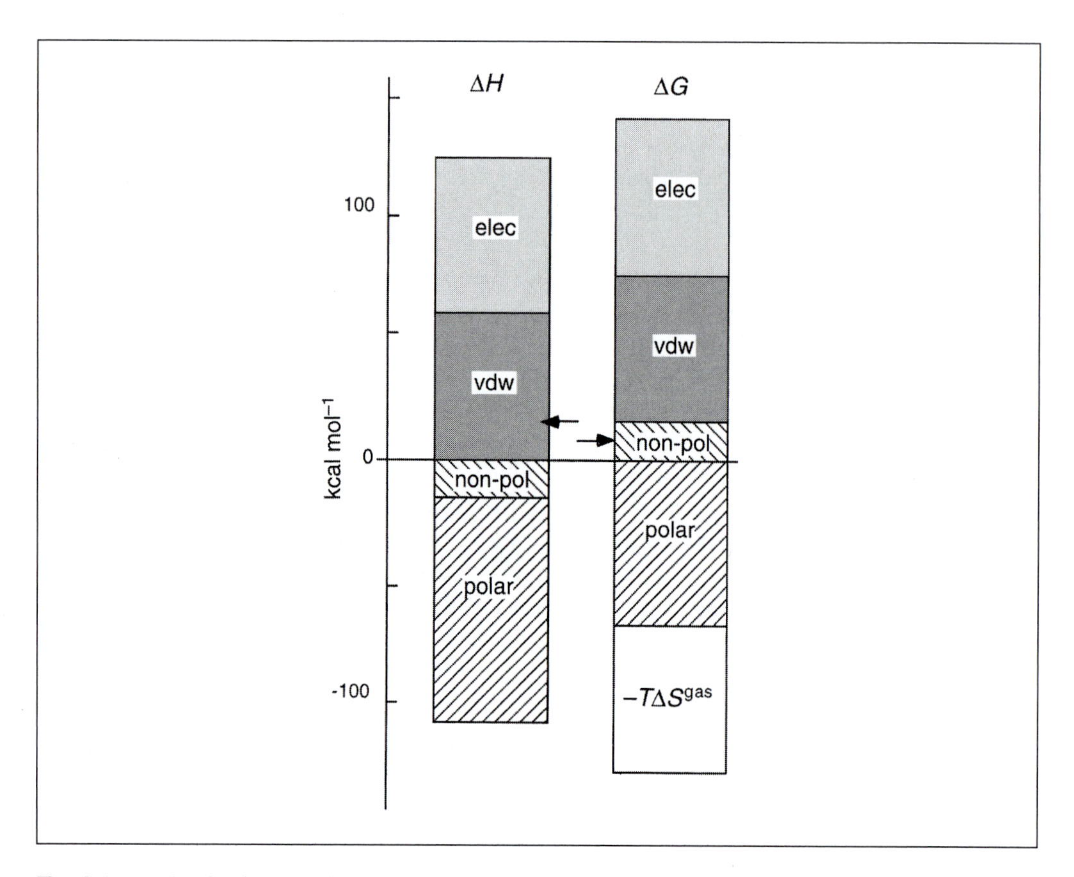

Fig. 4 Accounting for ΔH_d and ΔG_d. The enthalpy and free enthalpy of dissociation of the lysozyme–HyHEL5 antibody complex are first estimated in gas phase, then the value in water is obtained by adding hydration components as in Fig. 3. Positive terms in ΔH and ΔG stabilize the complex, negative terms favour dissociation. The derivation of each term is described in ref. 75. In gas phase, the energy of van der Waals (vdw) and electrostatic (elec) interactions makes a large positive contribution to ΔH; the entropy of degrees of freedom lost upon association makes a negative contribution to ΔG noted $-T\Delta S^{gas}$. These degrees of freedom are both external (rotational/translational) and internal (vibrations and side-chain rotations). Though the enthalpy of dehydration of non-polar groups at the interface (non-pol) is negative, the free enthalpy is positive and the hydrophobic effect contributes to the stability of the complex. For polar groups, both the enthalpy and the free enthalpy of dehydration are negative and large, and they fully compensate the stabilizing electrostatic interactions observed in the gas phase. Arrows indicate the balance of the calculated values: 19 kcal mol^{-1} for ΔH and 8 kcal mol^{-1} for ΔG. These may be compared to the experimental ΔH_d = 22.6 kcal mol^{-1} and ΔG_d = 14.5 kcal mol^{-1} at 25 °C (52), with the reserve that they are small differences between large numbers with large error bars.

5. From affinity to kinetics and specificity

5.1 Site-directed mutagenesis studies of affinity

Site-directed mutagenesis is often used to evaluate the contribution to affinity made by individual amino-acid side chains and residue pairwise interactions at interfaces (82). Because in many protein–protein complexes, the interface comprises only about 20 amino-acid residues of each partner, they can be systematically tested by substitution on the basis of an X-ray structure. This has been done on barnase–barstar (29, 83), several antibody–antigen complexes (84–87), the human growth hormone–hormone receptor complex (88–90), and the complex between Factor VIIA and soluble tissue factor (91).

The effect of a mutation on the stability of the complex is evaluated as the ratio of the dissociation constants of the wild type and the mutant, or equivalently, as the variation of the dissociation free enthalpy, denoted $\Delta\Delta G_d$:

$$\begin{aligned}\Delta\Delta G_d &= \Delta G_d^{wt} - \Delta G_d^{mut}\\ &= RT \ln (K_d^{mut}/K_d^{wt}).\end{aligned}$$

Large ratios define important residues delineating what has been called a functional epitope (89). The systematic substitution to alanine, termed alanine scanning (92), effectively deletes a side chain and all its interactions, usually with little structural perturbations. A factor of two in K_d is equivalent to a $\Delta\Delta G_d = 0.4$ kcal mol^{-1} and can be measured with confidence. Occasionally, a mutation marginally increases the affinity and $\Delta\Delta G_d$ is negative. Factors larger than 10 ($\Delta\Delta G_d = 1.4$ kcal mol^{-1}) are observed upon the substitution of some of the interface residues defined in the three-dimensional structure, but not outside. Less expectedly, alanine scanning shows that a large majority of these interface residues can be mutated to alanine with little effect on K_d: the functional epitope is a subset of the structural epitope (92). In the systems mentioned above, the average number of residues that yield a factor of 10 is only six, one-quarter of those that contribute to the interface area; zero to four residues yield a factor of 10^3 ($\Delta\Delta G_d = 4$ kcal mol^{-1}).

The value of $\Delta\Delta G_d$ derived from the mutant dissociation constant points to the importance of the residue, but it should not be confused with the energy of interactions made by the deleted side chain. When an alanine is substituted for a polar residue at an interface, the dramatic drop in affinity that may be recorded reflects both the loss of its hydrogen bonds and the cost of burying other residues in an unfavourable environment. Suppose that an arginine side chain in partner A forms a salt bridge with an aspartate in partner B. When either the arginine or the aspartate is deleted, a charged group becomes dehydrated in the complex causing a large increase in K_d. To recover the free enthalpy of the arginine–aspartate interaction, the dissociation constant of a complex between the two mutant proteins should be measured and a 'double-mutant cycle' performed (93–94; see also Chapters 5 and 9):

$$\Delta G^{inter} = \Delta\Delta G_d\,(A^{mut}B) + \Delta\Delta G_d\,(AB^{mut}) - \Delta\Delta G_d\,(A^{mut}B^{mut}).$$

It is related to the four dissociation constants by:

$$\Delta G^{\text{inter}} = RT \ln \frac{(K_d{}^{\text{AmutB}})\,(K_d{}^{\text{ABmut}})}{(K_d{}^{\text{AB}})\,(K_d{}^{\text{AmutBmut}})}.$$

If there is no interaction between the two mutated residues, the effects of the two substitutions should be additive and $\Delta G^{\text{inter}} \approx 0$. If they form a salt bridge, its energy is lost in all three mutant complexes and this contributes directly to ΔG^{inter}, whereas the cost of dehydrating the two side chains cancels out (Table 3). Mutagenesis data collected in several systems (83, 85, 86) reveal that a single pairwise interaction may account for as much as 6 kcal mol^{-1}. Residue pairs that form salt bridges and charged hydrogen bonds yield the largest values; pairs making neutral hydrogen bonds or non-polar interactions are in the 0–3 kcal mol^{-1} range. This is much less than the energy of a hydrogen bond, and implies that interaction between the two residues in the complex is only marginally stronger than the interactions with water they make in the free proteins.

It should be kept in mind that polar interactions also include a van der Waals' component and that site-directed mutagenesis cannot assess the contribution of the main chain. In complexes of known three-dimensional structure, the peptide group is part of at least half the hydrogen bonds at protein–protein interfaces (17). Side chain-to-main chain bonds are especially common, and main chain-to-main chain bonds do occur.

5.2 Kinetic effects of long-range interactions

From a kinetic point of view, a mutation that reduces affinity and increases K_d may do so either by increasing the rate of dissociation (k_d) or by reducing the rate of association (k_a). In the systems described above, association/dissociation rate measurements have been performed on many mutants in order to determine affinity changes (83–85, 87, 91, 95). The data show that in most mutants, k_d increases in parallel to K_d, as mutations at the interface weaken interactions and make dissociation easier. Unless they destabilize the protein itself, most point substitutions have little effect on k_a. Those which do are therefore interesting, for they carry information on the association process. Most concern charged residues and their effects prove that electrostatic interactions play an important part in the kinetics (see Chapters 3 and 9).

Table 3 Double-mutant cycle

Complex	**Residue substitution**		**Contribution to $\Delta\Delta G$ (a)**
1. Wild type AB	Arg	Asp	$-\Delta G^{\text{int}} + \Delta G^{\text{hyd}}$ (Arg) $+ \Delta G^{\text{hyd}}$ (Asp)
2. A^{mut} B	Ala	Asp	ΔG^{hyd} (Asp)
3. A B^{mut}	Arg	Ala	ΔG^{hyd} (Arg)
4. A^{mut} B^{mut}	Ala	Ala	(0)

(a) The free enthalpy changes are relative to the A^{mut} B^{mut} complex.

Charged molecules feel each other's electric charge and dipole moment long before they make contact. Electrostatic attraction at distances of 10–50 Å can accelerate association by several orders of magnitude, even though the force is much weaker than at short range (96, 97). Long-range interactions may be less critical for affinity and specificity than the short-range hydrogen bonds or salt bridges which we see in X-ray structures, but they do contribute to the kinetics. In addition, they may explain why the substitution of charged amino-acid residues outside the interfaces sometimes affect affinity.

Kinetic effects may be understood with the help of a simple model of rigid-body association (98). Two protein molecules collide by translational diffusion. If they happen to do so in the right position and orientation, they form a precomplex, the transition state that quickly evolves into a stable complex; otherwise, the encounter pair immediately dissociates (Fig. 5). The model takes into account the precise geometry of the interfaces illustrated in Plates 1–4, and the fact that the rotational diffusion of large molecules is too slow for proteins to adjust their orientation during the nanosecond lifetime of the encounter pair. Unlike Scheme 2 in Table 2 above, which has two distinct steps each with its own transition state and activation barrier, there is a single bimolecular step and a single transition state beyond which only very fast events take place: water molecules are removed from the interface and local adjustments optimize short-range interactions.

The rate of collision (k_{coll}) of two freely diffusing spheres of equal size is given by the Einstein–Smoluchowski equation of collision theory (4, 99). It depends on temperature and the viscosity of the solution, but not on the size of the spheres, for the larger target size of a bigger sphere compensates for its slower diffusion. With proteins, collision theory must be adapted on two points. First, orientation matters, which reduces the rate of productive collisions by the probability p_r that the encounter pair is a precomplex. Second, as proteins carry electric charges and dipoles, attractive electrostatic interactions make collisions more frequent, repulsive ones, less frequent; charge–dipole and dipole–dipole interactions orient molecules and modulate the probability p_r. Diffusion may be facilitated and guided by electrostatic forces (96). This is expressed by writing:

$$k_a = \kappa\, q_t\, q_r\, p_r\, k_{coll}.$$

At 300 K and the viscosity of pure water, k_{coll} is $6.6.10^9\ M^{-1}\,s^{-1}$ from collision theory, $\kappa \approx 1$ is a transmission coefficient; p_r is the probability defined above; q_t and q_r express the effect of long-range interactions, on the collision rate and the orientation respectively. The value of p_r is determined by the geometry of the interaction and depends on how accurately the two molecules must be positioned and oriented in the precomplex. The probability p_r is small: values as low as 10^{-7} have been proposed, implying very low rates of association. Docking simulations support a value in the order of 10^{-5}, predicting the bimolecular rate constant to be $p_r k_{coll} \approx 10^5\ M^{-1}\,s^{-1}$ in the absence of long-range electrostatic effects.

Observed rates of protein–protein association fall mostly in a 10^5–$10^7\ M^{-1}\,s^{-1}$ range. For instance, trypsin–PTI and a lysozyme–antibody complex have a $k_a = 10^6\ M^{-1}\,s^{-1}$

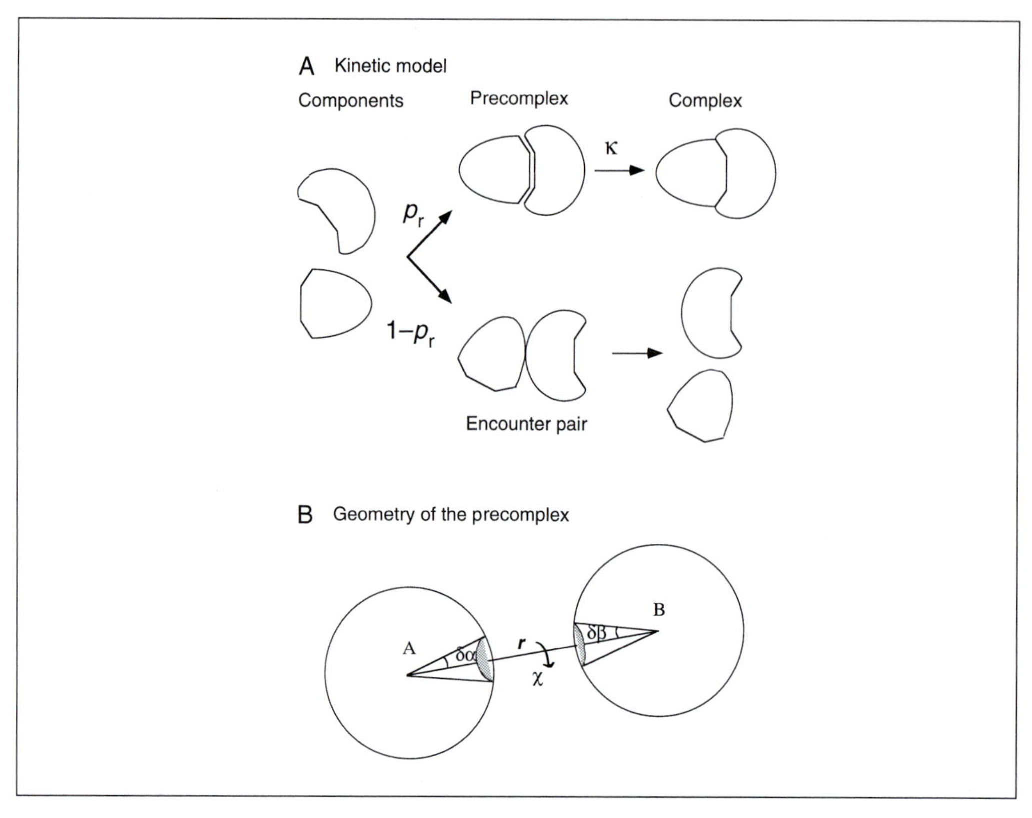

Fig. 5 A kinetic model of protein–protein association. (A) The two components of a protein–protein complex undergo random collision yielding a short-lived encounter pair. With a small probability p_r, they are positioned to form a proper interface. Then, they constitute a precomplex, which is the transition state for association and has a high probability $\varkappa$ to turn into a stable complex. Otherwise, the encounter pair quickly dissociates. (B) The probability p_r depends on how accurately the two molecules must be positioned and oriented in the precomplex. In this figure, the centre of the two complementary regions of the protein surface is positioned to within an angle $\delta\alpha$ of its correct position on molecule A and $\delta\beta$ on molecule B. The probability of each positioning is proportional to the shaded solid angle. In addition, the orientation about the line of centres is defined to within. For small angles, the combined probability evaluates to:

$$p_r \approx 1/16\pi\, \delta\alpha^2\, \delta\beta^2\, \delta\chi.$$

Docking simulations and experimental data on barnase–barstar suggest that $p_r \approx 10^{-5}$. This is equivalent to saying that, in the transition state, all orientation angles must be within ≈ 14 degrees of their final value (98).

(28, 100, 101). Barnase–barstar association is much faster: $k_a = 3.10^8\ M^{-1}\,s^{-1}$ at medium ionic strength (95). Yet, the barnase–barstar, trypsin–PTI, or lysozyme–Fab interfaces are very similar in size and complementarity, and the geometric requirements that determine p_r should be equally stringent. Fast association is largely due to the electrostatic attraction between the positively charged barnase, and barstar which carries both a large negative net charge and a large dipole moment. The latter has a steering effect on barstar as it interacts with the net charge of barnase. Accordingly, k_a is strongly ionic strength-dependent, and amino-acid substitutions which increase the net charge of barnase accelerate association. At low ionic strength, some of the high-charge mutants bind barstar faster than allowed by collision theory. In contrast, mutants with a zero net charge bind barstar at only 10^5–10^6 M^{-1} s^{-1} whatever the ionic strength. At high ionic strength, all rates extrapolate to a common value of 10^5 M^{-1} s^{-1} independent of the net charge of the two proteins. Under these conditions, long-range electrostatic interactions are fully screened, $q_t = q_r = 1$, and the value of k_a is equal to $p_r\, k_{coll}$ yielding $p_r \approx 10^{-5}$.

The conclusion is that the basal rate for protein–protein association is $\approx 10^5$ M^{-1} s^{-1} and acceleration factors as high as 10^3 can occur at medium ionic strength. This conclusion has been confirmed by computer simulations using Brownian dynamics (102). A simulation of cytochrome *c* binding to cytochrome *c* peroxidase (103) suggests that dipolar steering contributes more than charge–charge attraction, e.g. $q_r >> q_t$. The role of electric dipoles is confirmed by a model study of the dynamics of the O_2^- ion binding to superoxide dismutase, which yields an acceleration factor of only one order of magnitude (104). This small ligand has a full charge but no dipole, and orientational effects are negligible. With barnase–barstar, Brownian dynamics can be made to reproduce the rates observed with various charge mutants and their ionic strength dependence, by supposing that a stable complex forms each time two pairs of interacting amino-acid residues at the interface come within about 6 Å of each other (105). This condition could pass as a definition of the precomplex shown in Fig. 5.

5.3 Modelling complementarity and specificity by docking and statistical mechanics

Complementarity is an essential feature for both stability and specificity. Interfaces are complementary from both geometric and chemical points of view: the shapes of the two protein surfaces fit, and polar groups pair to form hydrogen bonds. Shape complementarity can be quantified (20, 106–108). It is generally excellent in protease–inhibitor complexes, possibly less so in antigen–antibody complexes (107). Nevertheless, packing defects do occur, especially in the larger interfaces (109). Chemical complementarity shows up in the distribution of polar and non-polar surface groups. Whereas there is little correlation in the distribution of charged groups on the two sides of an interface, electrostatic potentials calculated at the protein surface are complementary in protease–inhibitor and antigen–antibody complexes (110). However, that particular complementarity may not pre-exist association, for many of the charged groups are carried by the long flexible side chains of Glu, Lys, and Arg residues. Between the free proteins and the complex, the position of the charge can move by 10 Å just through a side-chain rotation.

Two extreme models of shape complementarity are the lock-and-key fit of two rigid preformed shapes, and the hand-into-glove fit where one partner adapts to the shape of the other. Whereas neither model is a satisfactory description of reality, the first is a better approximation for trypsin–PTI, lysozyme–HyHEL5, or barnase–barstar, the second for transducin. In lysozyme–HyHEL5 and barnase–barstar, shape complementarity is sufficient for a computer search to select the right mode of association by systematically docking lysozyme on to the combining site of the antibody or barstar on to barnase, and evaluating the extent of their interaction with a simple score function related to the interface area or the hydration energy. Docking algorithms have been reviewed (111–113). The search succeeds when docking is performed on proteins from the dissociated complex, and also on the free protein structures, which is a more difficult problem because side-chain conformations are not optimized for association. Thus, the mode of association of a β-lactamase and its protein inhibitor BLIP was correctly predicted by docking the two components prior to elucidating the crystal structure of the complex (114). Successful docking procedures used in this prediction differed by the search algorithm and the type of score they used, but all relied on shape complementarity. In contrast, a search based only on electrostatic complementarity failed to reproduce the native mode of association.

When interface residues are mutated, complementarity is perturbed. Deleting a large side chain creates a hole that water may or may not fill depending on its size; removing (or adding) a polar group leaves a hydrogen-bond donor or acceptor unsatisfied. This is experimentally achieved by site-directed mutagenesis or by using natural variants, for instance bird lysozymes that are closely related to hen lysozyme and bind to the same monoclonal antibody with different affinities. The results are sometimes interpreted in terms of specificity. Indeed, a point mutation can alter specificity: in a double-mutant cycle, the $A^{mut}B^{mut}$ complex is often more stable than the single-mutant species $A^{mut}B$, in which case A^{mut} could be said to be specific for B^{mut} instead of B.

However, point mutants are not a sufficient test of the specificity required for function either *in vivo* or, for antibodies, in immunochemical experiments. Specificity is both the capacity to discriminate between closely related molecules and to identify a given molecular species among many others. The study of point mutants deals with the first property. The second is relevant both *in vivo* and *in vitro* to the situation where an epitope competes for binding to an antibody with features on the surface of many other proteins, not with a mutant version of itself. In an immunoassay, how many chemical species present in the sample have sufficient complementarity with the combining site of the antibody to compete with the epitope? The answer to this question obviously depends on the composition of the solution, and it determines the response to the test. Random contacts in such a mixture create many non-specific protein–protein interactions having a large spectrum of energies. These may be described with the help of an appropriate statistical model (115) or simulated in the computer by systematic docking as above. The results indicate that most randomly generated complexes have small interfaces and very few reach 1100 Å^2, the lower limit in specific complexes. Thus, the presence of regions covering a sufficient area on the surface of two proteins (of the order of 800 Å^2 per partner in our examples) and having complementary shapes and chemical composition, appears to be both necessary and sufficient for specific recognition by rigid-body association. How specificity is achieved in systems like transducin which undergo large conformation changes and disorder-to-order transitions remains to be understood.

6. Conclusions

Protein–protein interactions may be compared with the interaction of proteins with other biological macromolecules, and also with protein folding. The forces involved are the same: those cited by Pauling and Delbrück (1) and the hydrophobic effect (2, 3). All these processes involve the dehydration of protein groups, the formation of novel van der Waals and electrostatic interactions, and the loss of degrees of freedom as atoms or molecules are immobilized. Almost all the methods described above are also used to study protein–nucleic acid interactions and protein folding, including fast kinetics, microcalorimetry, and site-directed mutagenesis.

Nevertheless, structural and thermodynamic data point to specific features in each process. Protein–DNA recognition, for instance, repeatedly uses some structural motifs, e.g. the insertion of an α-helix into the major groove of the DNA double helix or of extended polypeptide chain into the minor groove (116). On the DNA side, a single type of chemical group, the phosphate moiety, provides for the majority of the interface area and polar interactions. In protein–protein recognition, the diversity of the surfaces involved is much greater and there is no recurrent structural binding motif. There is no equivalent to the specific recognition of DNA bases by amino-acid side chains, the recognition of a guanine base by an arginine for instance, often observed in protein–DNA complexes. Arginine residues are major contributors to protein–protein interfaces as they are to protein–DNA interfaces, but they indifferently donate hydrogen bonds to all hydrogen-bond acceptor groups including the carbonyl of peptide groups. The average protein surface is less polar and carries fewer electric charges than that of DNA or RNA. From the point of view of hydrophobicity, protein–protein interfaces are in between protein–DNA interfaces and the interfaces between subunits of oligomeric proteins. Consequently, the number of polar interactions is less at protein–protein than at protein–DNA interfaces and the role of the hydrophobic effect more essential, even though it should not be neglected in protein–DNA interaction (56).

At present, physical and computer models of protein–protein association have been developed only within the rigid-body approximation. Beyond, the problem becomes as complex as protein folding, no theoretical estimate of the enthalpy/entropy changes or of the rate constants can be made, and present-day docking algorithms fail. Rigid-body association is a valid approximation in most protein–protein complexes that form standard-size interfaces with $B \approx 1600$ Å^2. It almost never applies to protein–DNA complexes and to oligomeric proteins. In the former, the protein, the DNA, or both, frequently undergo conformation changes and disorder-to-order transitions are common (116). In the latter, folding generally accompanies association. In counterpart, most subunit interfaces in oligomeric proteins and most protein–-DNA interfaces are larger than standard-size protein–protein interfaces (20, 75) and more similar in size to that of transducin. In transducin, conformation changes and disorder-to-order transitions do accompany protein–protein association, and their occurrence in other systems appears to correlate with the formation of interfaces larger than in antigen–antibody and protease–inhibitor complexes (118). The existence of protein–protein complexes with large interfaces and conformation changes has been established by structural studies, but their mode of formation has not yet been analysed by kinetic or calorimetric methods. Thus, we ignore the rate at which binding takes place and how much energy and entropy it involves. Because many processes involved in cellular regulation are expected to rely on interactions of this type, the experimental study of non-rigid body, protein–protein associations is of major importance and its modelling will be a challenge for physical chemists in coming years.

References

1. Pauling, L. and Delbrück, M. (1940). The nature of the intermolecular forces operative in biological processes. *Science* **92**, 77–79.
2. Kauzmann, W. (1959). Some factors in the interpretation of protein denaturation. *Adv. Protein Chem.* **14**, 1–63.
3. Tanford, C. (1979). Interfacial free energy and the hydrophobic effect. *Proc. Natl. Acad. Sci. USA* **76**, 4175–6.
4. Atkins, P. W. (1990). *Physical chemistry* (4th edn). Oxford University Press.
5. Creighton, T. E. (1993). *Proteins: structures and molecular properties* (2nd edn). W. H. Freeman, New York.
6. Phizicky, E. M. and Fields, S. (1995). Protein–protein interactions: methods for detection and analysis. *Microbiol. Rev.* **59**, 94–123.
7. Friguet, B., Chaffotte, A. F., Djavadi-Ohaniance, L., and Goldberg, M. E. (1985). Measurements of the true affinity constant in solution of antigen–antibody complexes by Enzyme-Linked Immunoabsorbent Assay. *J. Immunol. Meth.* **77**, 305–19.
8. Fersht, A. R. (1985). *Enzyme structure and mechanism* (2nd edn). W. H. Freeman, New York.
9. Eigen, M. and Hammes, G. G. (1963). Elementary steps in enzyme reactions. *Adv. Enzymol.* **25**, 1–39.
10. Fierke, C. A. and Hammes, G. G. (1995). Transient kinetics approaches to enzyme mechanisms. In *Methods in enzymology*, Vol. 249 (ed. D. L. Purich), pp. 3–37. Academic Press, New York.
11. Garland, P. B. (1996). Optical evanescent wave methods for the study of biomolecular interactions. *Q. Rev. Biophys.* **29**, 91–117.
12. Schuck, P. (1997). Use of surface plasmon resonance to probe the equilibrium and dynamic aspects of interactions between biological macromolecules. *Annu. Rev. Biophys. Biomol. Struct.* **26**, 541–66.
13. Karlsson, R. and Fält, A. (1997). Experimental design for kinetic analysis of protein–-protein interactions with surface plasmon resonance biosensors. *J. Immunol. Meth.* **200**, 121–33.
14. Schuck, P. and Minton, A. P. (1996). Analysis of mass-transport limited binding kinetics in evanescent wave biosensors. *Anal. Biochem.* **240**, 262–72.
15. Nieba, L., Krebber, A., and Plückthun, A. (1996). Competition BIAcore for measuring true affinities: large differences from values determined from binding kinetics. *Anal. Biochem.* **234**, 155–65.
16. Bernstein, F. C., Koetzle, T. F., Williams, J. B., Meyer, E. F. Jr, Brice, M. D., Rodgers, J. R., Kennard, O., Shimanouchi, T., and Tasumi, M. (1977). The protein data bank. A computer-based archival file for macromolecular structures. *J. Mol. Biol.* **112**, 535–42.

17. Janin, J. and Chothia, C. (1990). The structure of protein–protein recognition sites. *J. Biol. Chem.* **265**, 16027–30.
18. Janin, J. (1995). Principles of protein–protein recognition from structure to thermodynamics. *Biochimie* **77**, 497–505.
19. Jones, S. and Thornton, J. M. (1996). Principles of protein–protein interactions. *Proc. Natl. Acad. Sci. USA* **93**, 13–20.
20. Jones, S. and Thornton, J. M. (1995). Protein–protein interactions: a review of protein dimer structures. *Progr. Biophys. Mol. Biol.* **63**, 131–65.
21. Davies, D. R. and Cohen, G. H. (1996). Interactions of protein antigens with antibodies. *Proc. Natl. Acad. Sci. USA* **93**, 7–12.
22. Chothia, C. (1997). Protein–protein and protein–carbohydrate recognition. In *Molecular aspects of host–pathogen interactions* (ed. M. A. McCrae, J. R. Saunders, C. J. Smyth, and N. D. Stow), pp. 1–22. Cambridge University Press.
23. Sheriff, S., Silverton, E. W., Padlan, E. A., Cohen, G. H., Smith-Gill, S. J., Finzel, B. C., and Davies, D. R. (1987). Three-dimensional structure of an antibody–antigen complex. *Proc. Natl. Acad. Sci. USA* **84**, 8075–9.
24. Huber, R., Kukla, D., Bode, W., Schwager, P. Bartlels, K., Deisenhofer, J., and Steigemann, W. (1974). Structure of the complex formed by bovine trypsin and bovine pancreatic trypsin inhibitor. *J. Mol. Biol.* **89**, 73.
25. Guillet, V., Lapthorn, A., Hartley, R. W., and Mauguen, Y. (1993). Recognition between a bacterial ribonuclease, barnase, and its natural inhibitor, barstar. *Structure* **1**, 165–77.
26. Buckle, M., Schreiber, G., and Fersht, A. R. (1994). Crystal structural analysis of a barnase–barstar complex at 2.0 Å resolution. *Biochemistry* **33**, 8878–89.
27. Lambright, D. G., Sondek, J., Bohm, A., Skiba, N. P., Hamm, H. E., and Sigler, P. B. (1996). The 2.0 Å crystal structure of a heterotrimeric G-protein. *Nature* **379**, 311.
28. Vincent, J. P. and Lazdunski, M. (1972). Trypsin–pancreatic trypsin inhibitor association. Dynamics of the interaction and the role of disulfide bridges. *Biochemistry* **11**, 2967.
29. Schreiber, G. and Fersht, A.R. (1993). Interaction of barnase with its polypeptide inhibitor barstar studied by protein engineering. *Biochemistry* **32**, 5145–50.
30. Hartley, R. W. (1989). Barnase and barstar: two small proteins to fold and fit together. *Trends Biochem. Sci.* **14**, 450–4.
31. Jucovic, M. and Hartley, R. W. (1996). Protein–protein interactions: a genetic selection for compensating mutations at the barnase–barstar interface. *Proc. Natl. Acad. Sci. USA* **93**, 2343–7.
32. Conklin, B. R. and Bourne, H. R. (1993). Structural elements of $G\alpha$ subunits that interact with $G\beta\gamma$, receptors and effectors. *Cell* **73**, 631–41.
33. Marquart, M., Walter, J., Deisenhofer, J., Bode, W., and Huber, R. (1983). The geometry of the reactive site and of the peptide groups in trypsin, trypsinogen and its complexes with inhibitors. *Acta Crystallogr.* **B39**, 480–92.
34. Noel, J. P., Hamm, H. E., and Sigler, P. B. (1993). The 2.2 Å crystal structure of transducin complexes with GTPγS. *Nature* **366**, 654–62.
35. Lambright, D. G., Noel, J. P., Hamm, H. E., and Sigler, P. B. (1994). Structural determinants for activation of the α-subunit of a heterotrimeric G protein. *Nature* **369**, 621.
36. Sondek, J., Bohm, A., Lambright, D. G., Hamm, H. E., and Sigler, P. B. (1996). Crystal structure of a G_A protein $\beta\gamma$ dimer at 2.1 Å resolution. *Nature* **379**, 369.
37. Chothia, C. and Janin, J. (1975). Principles of protein–protein recognition. *Nature* **256**, 705–8.

38. Lee, B. K. and Richards, F. M. (1971). Solvent accessibility of groups in proteins. *J. Mol. Biol.* **55**, 379–400.
39. Shrake, A. and Rupley, A. (1973). Environment and exposure to solvent of protein atoms. Lysozyme and insulin. *J. Mol. Biol.* **79**, 351–71.
40. Wodak, S., de Crombrugghe, M., and Janin, J. (1987). Computer studies of interactions between macromolecules. *Progr. Biophys. Mol. Biol.* **49**, 29–63.
41. Pelletier, H. and Kraut, J. (1992). Crystal structure of a complex between electron transfer partners, cytochrome c peroxidase and cytochrome c. *Science* **258**, 1748–55.
42. Rydel, T. J., Tulinsky, A., Bode, W., and Huber, R. (1991). The refined structure of the hirudin–thrombin complex. *J. Mol. Biol.* **221**, 583–601.
43. Bigay, J., Faurobert, E., Franco, M., and Chabre, M. (1994). Roles of lipid modifications of transducin subunits in their GDP-dependent association and membrane binding. *Biochemistry* **33**, 14081–90.
44. Janin, J., Miller, S., and Chothia, C. (1988). Surface, subunit interfaces and interior of oligomeric proteins. *J. Mol. Biol.* **204**, 155–64.
45. Tsai, C. J. and Nussinov, R. (1997). Hydrophobic folding units at protein–protein interfaces: implications to protein folding and to protein–protein association. *Protein Sci.* **6**, 1426–37.
46. Baht, T. N., Bentley, G. A., Boulot, G., Greene, M. I., Tello, D., all'Acqua, W., Souchon, S., Schwarz, F. P., Mariuzza, R. A., and Pljak, R. J. (1994). Bound water molecules and conformational stabilization help mediate an antigen–antibody association. *Proc. Natl. Acad. Sci. USA* **91**, 1089–93.
47. Covell, D. G. and Wallquist, A. (1997). Analysis of protein–protein interactions: effect of amino acid mutations on their energetics. The importance of water molecules in the binding epitope. *J. Mol. Biol.* **269**, 281–97.
48. Xu, D., Tsai, C. J., and Nussinov, R. (1997). Hydrogen bonds and salt bridges across protein–protein interfaces. *Protein Eng.* **10**, 999–1012.
49. Wiseman, T., Williston, S., Brandts, J. F., and Lin, L. N. (1989). Rapid measurement of binding constants and heats of binding using a new titration calorimetry. *Anal. Biochem.* **179**, 131–7.
50. Fisher, H. F. and Singh, N. (1995). Calorimetric methods for interpreting protein–ligand interactions. In *Methods in enzymology*, Vol. 259 (eds M. L. Johnson & G. K. Ackers), pp. 194–221. Academic Press, New York.
51. Ladbury, J. E. and Chowdhry, B. Z. (1996). Sensing the heat: the application of isothermal titration calorimetry to thermodynamic studies of biomolecular interactions. *Chem. Biol.* **3**, 791–801.
52. Hibbits, K. A., Gill, D. S.,and Willson, R. C. (1994). Isothermal titration calorimetric study of the association of hen egg lysozyme and the anti-lysozyme antibody Hy–HEL5. *Biochemistry* **33**, 3584–90.
53. Baker, B. M. and Murphy, K. P. (1997). Dissecting the energetics of protein–protein interactions: the binding of ovomucoid third domain to elastase. *J. Mol. Biol.* **268**, 557–69.
54. Frisch, C., Schreiber, G., Johnson, C. M., and Fersht, A. R. (1997). Thermodynamics of the interaction of barnase and barstar: changes in free energy *versus* changes in enthalpy on mutation. *J. Mol. Biol.* **267**, 696–706.
55. Spolar, R. S., Ha, J.-H., and Record, M. T. Jr (1989). Hydrophobic effect in protein folding and other noncovalent processes involving proteins. *Proc. Natl. Acad. Sci. USA* **86**, 8382–5.
56. Spolar, R. S. and Record, M. T. Jr (1994). Coupling of local folding to site-specific binding of proteins and DNA. *Science* **263**, 777–83.

57. Privalov, P. L. and Gill, S. J. (1988). Stability of protein structure and hydrophobic interaction. *Adv. Protein Chem.* **39**, 191–234.
58. Privalov, P. L. and Makhatadze, G. I. (1992). Contribution of hydration and non-covalent interactions to the heat capacity effect on protein unfolding. *J. Mol. Biol.* **224**, 715–23.
59. Oobatake, M. and Ooi, T. (1993). Hydration and heat stability effects on protein unfolding. *Progr. Biophys. Mol. Biol.* **59**, 237–84.
60. Eisenberg, D. and McLachlan, A. D. (1986). Solvation energy in protein folding and binding. *Nature* **319**, 199–203.
61. Ooi, T., Oobatake, M., Némethy, G., and Scheraga, H. A. (1987). Accessible surface areas as a measure of the thermodynamic parameters of hydration of peptides. *Proc. Natl. Acad. Sci. USA* **84**, 3086–93.
62. Spolar, R. S., Livingstone, J. R., Record, M. T. Jr (1992). Use of liquid hydrocarbon and amide transfer data to estimate contributions to thermodynamic functions of protein folding from the removal of nonpolar and polar surface from water. *Biochemistry* **31**, 3947–55.
63. Makhatadze, G. I. and Privalov, P. L. (1994). Hydration effects in protein unfolding. *Biophys. Chem.* **51**, 291–309.
64. Juffer, A. H., Eisenhaber, F., Hubbard, S. J., Walther, D., and Argos, P. (1995). Comparison of atomic solvation parametric sets: applicability and limitations in protein folding and binding. *Protein Eng.* **4**, 2499–509.
65. Ben-Naim, A. (1994). Solvation: from small to macro molecules. *Curr. Opin. Struct. Biol.* **4**, 264–8.
66. Chothia, C. (1974). Hydrophobic bonding and accessible surface area in proteins. *Nature* **248**, 338–9.
67. Sharp, K. A., Nicholls, A., Fine, R. F., and Honig, B. (1991). Reconciling the magnitude of the microscopic and macroscopic hydrophobic effects. *Science* **252**, 106–9.
68. Vajda, S., Weng, Z., and DeLisi, C. (1995). Extracting hydrophobicity parameters from solute partition and protein mutation/unfolding experiments. *Protein Eng.* **8**, 1081–92.
69. Horton, N. and Lewis, M. (1992). Calculation of the free energy of association for protein complexes. *Protein Sci.* **1**, 169–81.
70. Chacko, S., Silverton, E., Kam-Morgan, L., Smith-Gill, S., Cohen G., and Davies, D. (1995). Structure of an antibody–lysozyme complex. Unexpected effect of a conservative mutation. *J. Mol. Biol.* **245**, 261–74.
71. Novotny, J., Bruccoleri, R. E., and Saul, F. A. (1989). On the attribution of binding energy in the antigen–antibody complexes McPC603, D1.3 and HyHEL-5. *Biochemistry* **28**, 4735–48.
72. Krystek, S., Stouch, T., and Novotny, J. (1993). Affinity and specificity of serine endopeptidase–protein inhibitor interactions. Empirical free energy calculations based on X-ray crystallographic structures. *J. Mol. Biol.* **234**, 661–79.
73. Tulip, W. R., Harley, V. R., Webster, R. G., and Novotny, J. (1994). N9 neuraminidase complexes with antibodies NC41 and CN10: empirical free energy calculations capture specificity trends observed with mutant binding data. *Biochemistry* **33**, 7986–97.
74. Vajda, S., Weng, Z., Rosenfeld, R., and DeLisi, C. (1994). Effect of conformational flexibility and solvation on receptor-ligand binding free energies. *Biochemistry* **33**, 13977–88.
75. Janin, J. (1995). Elusive affinities. *Proteins: Struct., Funct., Genet.* **21**, 30–9.
76. Janin J. (1997). Angstroms and calories. *Structure* **5**, 473–9.
77. Gilson, M. K., Given, J. A., Bush, B. L., and McCammon, J. A. (1997). The statistical-thermodynamics basis for computation of binding affinities: a critical review. *Biophys. J.* **72**, 1047–69.

78. Finkelstein, A. V. and Janin, J. (1989). The price of lost freedom: entropy of bimolecular complex formation. *Protein Eng.* **3**, 1–3.
79. Erickson, H. P. (1989). Cooperativity in protein–protein association. The structure and stability of the actin filament. *J. Mol. Biol.* **206**, 465–74.
80. Lee, K. H., Xie, D., Freire, E., and Amzel, L. M. (1994). Estimation of changes in side chain configurational entropy in binding and folding: general methods and application to helix formation. *Proteins: Struct., Funct., Genet.* **20**, 68–84.
81. Sternberg, M. J. E. and Chickos, J. S. (1994). Protein side-chain conformational entropy derived from fusion data—comparison with other empirical scales. *Protein Eng.* **7**, 149–55.
82. Pace, C. N. (1995). Evaluating contribution of hydrogen bonding and hydrophobic bonding to protein folding. In *Methods in enzymology*, Vol. 259 (eds M. L. Johnson and G. K. Ackers), pp. 538–54. Academic Press, New York.
83. Schreiber, G. and Fersht, A. R. (1995). Energetics of protein–protein interactions: analysis of the barnase–barstar interface by single mutations and double mutant cycles. *J. Mol. Biol.* **248**, 478–86.
84. Dall'Acqua, W., Goldman, E. R., Eisenstein, E., and Mariuzza, R. A. (1996). A mutational analysis of the binding of two different proteins to the same antibody. *Biochemistry* **35**, 9667–76.
85. Dall'Acqua, W., Goldman, E. R., Lin, W., Teng, C., Tsuchiya, D., Li, H., Ysern, X., Braden, B. C., Li, Y., Smith-Gill, S. J., and Mariuzza, R. A. (1998). A mutational analysis of binding interactions in an antigen–antibody protein–protein complex. *Biochemistry* **37**, 7981–91.
86. Goldman, E. R., Dall'Acqua, W., Braden, B. C., and Mariuzza, R. A. (1997). Analysis of binding interactions in an idiotope-antiidiotope protein–protein complex by double mutant cycles. *Biochemistry* **36**, 49–56.
87. Dougan, D. A., Malby, R. L., Gruen, L. C., Kortt, A. A., and Hudson, P. J. (1998). Effects of substitutions in the binding surface of an antibody on antigen affinity. *Protein Eng.* **11**, 65–74.
88. Wells, J.A. (1996). Binding in the growth hormone receptor complex. *Proc. Natl. Acad. Sci. USA* **93**, 1–6.
89. Clackson, T. and Wells, J. A. (1995). A hot spot of binding energy in a hormone-receptor interface. *Science* **267**, 383–6.
90. Clackson, T., Ultsch, M. H., Wells, J. A., and de Vos, A. M. (1998). Structural and functional analysis of the 1:1 growth hormone:receptor complex reveals the molecular basis for receptor affinity. *J. Mol. Biol.* **277**, 1111–28.
91. Kelley, R. F., Costas, K. E., O'Connell, M. P., and Lazarus, R. A. (1995). Analysis of the factor VIIa binding site on human tissue factor: effects of tissue factor mutations on the kinetics and thermodynamics of binding. *Biochemistry* **34**, 10383–92.
92. Cunningham, B. C. and Wells, J. A. (1989). High resolution epitope mapping of hGH-receptor interactions by alanine-scanning mutagenesis. *Science* **244**, 1081–5.
93. Fersht, A., Shi, J. P., Knill-Jones, J., Lowe, D. M., Wilkinson, A. J., Blow, D. M., Brick, P., Carter, P., Waye, M. M. Y., and Winter, G. (1985). Hydrogen bonding and biological specificity analysed by protein engineering. *Nature* **314**, 235–8.
94. Fersht, A. R. (1988). Relationships between the apparent binding energies measured in site-directed mutagenesis experiments and energetics of binding and catalysis. *Biochemistry* **27**, 1577–80.
95. Schreiber, G. and Fersht, A. R. (1996). Rapid, electrostatically assisted association of proteins. *Nature Struct. Biol.* **3**, 427–31.

96. von Hippel, P. H. and Berg, O. G. (1989). Facilitated target location in biological systems. *J. Biol. Chem.* **264**, 675–8.
97. Zhou, H. X. (1993). Brownian dynamics study of the influence of electrostatic interaction and diffusion on protein–protein association kinetics. *Biophys. J.* **64**, 1711–16.
98. Janin, J. (1997). The kinetics of protein–protein recognition. *Proteins: Struct., Funct.,Genet.* **28**, 153–61.
99. Berg, G. B. and von Hippel, P. H. (1985). Diffusion-controlled macromolecular interactions. *Annu. Rev. Biophys. Biophys. Chem.* **14**, 131–60.
100. Castro, M. J. and Anderson, S. (1996). Alanine point-mutations in the reactive region of bovine pancreatic trypsin inhibitor: effects on the kinetics and thermodynamics of binding to β-trypsin and α-chymotrypsin. *Biochemistry* **35**, 11435–46.
101. Hawkins, R. E., Russell, S. J., Baier, M., and Winter, G. (1993). The contribution of contact and non-contact residues of antibody in the affinity of binding to antigen. The interaction of mutant D1.3 antibodies with lysozyme. *J. Mol. Biol.* **234**, 958–64.
102. Northrup, S. H. and Erickson, H. P. (1992). Kinetics of protein–protein association explained by Brownian dynamics computer simulations. *Proc. Natl. Acad. Sci. USA* **89**, 3338–42.
103. Northrup, S. H., Reynolds, J. C. L., Miller, C. M., Forrest, K. J., and Boles, J. O. (1986). Diffusion-controlled association rate of cytochrome c and cytochrome c peroxydase in a simple electrostatic model. *J. Am. Chem. Soc.* **108**, 8162–70.
104. Getzoff, E. D., Cabelli, D. E., Fisher, C. L., Parge, H. E., Viezzoli, M. S., Banci, L., and Hallewell, R. A. (1992). Faster superoxide dismutase mutants designed by enhancing electrostatic guidance. *Nature* **358**, 347–351.
105. Gabdoulline, R. R. and Wade, R. C. (1997). Simulation of the diffusional association of barnase and barstar. *Biophys. J.* **72**, 1917–29.
106. Connolly, M. L. (1986). Shape complementarity at the hemoglobin a1b1 subunit interface. *Biopolymers* **25**, 1229–47.
107. Lawrence, M. C. and Colman, P. M. (1993). Shape complementarity at protein/protein interfaces. *J. Mol. Biol.* **234**, 946–50.
108. Norel, R., Lin, S. L., Wolfson, H. J., and Nussinov, R. (1994). Shape complementarity at protein–protein interfaces. *Biopolymers* **34**, 933–40.
109. Ysern, X., Li, H., and Mariuzza, R. A. (1998). Imperfect interfaces. *Nature Struct. Biol.* **5**, 412–14.
110. McCoy, A. J., Epa, V. C., and Colman, P. M. (1997). Electrostatic complementarity at protein/protein interfaces. *J. Mol. Biol.* **268**, 570–84.
111. Janin, J. and Cherfils, J. (1993). Protein docking algorithms: simulating molecular recognition. *Curr. Opin. Struct. Biol.* **3**, 265–9.
112. Janin, J. (1996). Protein–protein recognition. *Progr. Biophys. Mol. Biol.* **64**, 145–65.
113. Sternberg, M. J. E., Gabb, H. A., and Jackson, R. M. (1998). Predictive docking of protein–protein and protein–DNA complexes. *Curr. Opin. Struct. Biol.* **8**, 250–6.
114. Strynadka, N. C. J., Eisenstein, M., Katchalski-Katzir, E., Shoichet, B., Kuntz, I., Abazyan, R., Totrov, M., Janin, J., Cherfils, J., Zimmermann, F., Olson, A., Duncan, B., Rao, M., Jackson, R., Sternberg, M., and James, M. N. G. (1996). Molecular docking programs successfully determine the binding of a β-lactamase inhibitory protein to tem-1 β-lactamase. *Nature Struct. Biol.* **3**, 233–9.
115. Janin, J. (1996). Quantifying biological specificity: the statistical mechanics of molecular recognition. *Proteins: Struct., Funct., Genet.* **25**, 438–45.
116. Lilley, D. M. J. (1995). *DNA–protein structural interactions.* Oxford University Press, Oxford.

117. Nicholls, A., Sharp, K., and Honig, B. (1992). Protein folding and association: insights from the interfacial and thermodynamic properties of hydrocarbons. *Proteins: Struct., Funct., Genet.* **11**, 281–96.
118. Lo Conte, L., Chothia, C., and Janin, J. (1999). The atomic structure of protein–protein recognition sites. *J. Mol. Biol.* **285**, 2177–98.

Notes

1. Different types of surfaces are described by Jones and Thornton in Chapter 2 where interface areas are cited per component of the complexes, or approximately $B/2$.

2 Analysis and classification of protein–protein interactions from a structural perspective

SUSAN JONES and JANET M. THORNTON

1. Introduction

Interactions between macromolecules are fundamental to many biological processes. Such interactions are central to the functioning of the vertebrate immune system, enzyme regulation, neurotransmitter release in the nervous system, and cell adhesion. The characterization of the interfaces formed when proteins interact is important if these and other systems are to be fully understood, and if the potential to control these interactions for medical and industrial applications is to be explored.

The Brookhaven Protein Data Bank (PDB) (1, 2) now contains the three-dimensional coordinates of more than 12 000 structures; an ever-increasing number of which are protein–protein complexes, including: oligomeric proteins; enzyme–inhibitor complexes; and antibody–protein complexes. Their deposition allows the analysis of large numbers of these structures, and such computational studies have provided important overviews of the physical and chemical properties of these associations (see, for example, refs 3–6). Protein–protein complexes with different interaction histories and functions require different modes of interaction. This chapter begins with the classification of these associations, considering functional classes of complexes, permanent and non-obligate complexes, and biological and crystallographic associations. The physical and chemical characteristics of biological associations are then discussed, presenting an overview of the properties favouring associations, namely hydrophobic and electrostatic interactions, and shape complementarity. These are then considered in more detail with reference to computational studies conducted on large data sets of protein complexes. With a clearer picture of the nature of these interactions, the potential to use this to predict the location of binding sites on protein surfaces is discussed. A number of methods based on searching for hydrophobic patches on protein surfaces are presented, together with a method of searching for patches with multiple properties matching specific criteria. The potential for predictions are important in the light of recent applications of protein-interaction research, and examples are discussed that are relevant to disease processes.

2. Classification of protein–protein complexes

2.1 Functional classes of complexes

Protein associations are often discussed in terms of their function. Common functional classes include enzyme–inhibitor complexes, antibody–protein complexes, and protein–receptor complexes. The structures of a number of enzyme–inhibitor complexes have been solved, including a large group of trypsin-like serine proteinases and their inhibitors (7; see Chapter 8). The subtilisins, which are extracellular alkaline proteinases from *Bacillus* spp., are a further group of protein–inhibitor complexes for which structures have been solved (e.g. refs 8–10) (Fig. 1a). The proteolytic functions of these enzymes are important as they have industrial applications, e.g. additives in laundry detergents to facilitate the removal of proteinaceous stains. Antibody–protein associations play a fundamental role in the vertebrate immune system. The structures of a number of antibody–protein complexes (Fig. 1b) have been solved (e.g. refs 11–13), and have been the target of many studies (14). One major area of controversy is how recognition is achieved in this specific system, and what part is played by the conformational changes that are observed on antigen binding (e.g. refs 15–17). The structures of a number of protein–receptor complexes have also been solved: one example is human growth hormone complexed with the extracellular domain of its receptor (hGHbp) (18). This complex comprises one molecule of hormone and two receptor molecules (Fig. 1c). The receptor is activated on binding the hormone, an interaction fundamental to the growth and metabolism of muscle, bone, and cartilage cells, all requirements for normal growth and development.

2.2 Permanent and non-obligate complexes

Within the functional classes of protein–protein complexes discussed above two further classes can be distinguished. We have termed these permanent and non-obligate complexes (6). Permanent complexes comprise protomers that only occur in the complexed state, an example is the homodimer interleukin 8 (Fig. 2a). This cytokine is a chemotactic factor released by monocytes in response to inflammatory stimuli (19). Its biological function of neutrophil activation is integrally related to its dimeric structure, with the proposed receptor site being composed of helices from each protomer (20). Human immunodeficiency virus (HIV) protease is a further example of a permanent complex (Fig. 2b). This protein exhibits an active site bisected by the symmetry axis, with catalytic residues contributed from each protomer (21). Most homodimers are only observed in the multimeric state, and it is often impossible to separate them without denaturing the individual monomer structures. Many

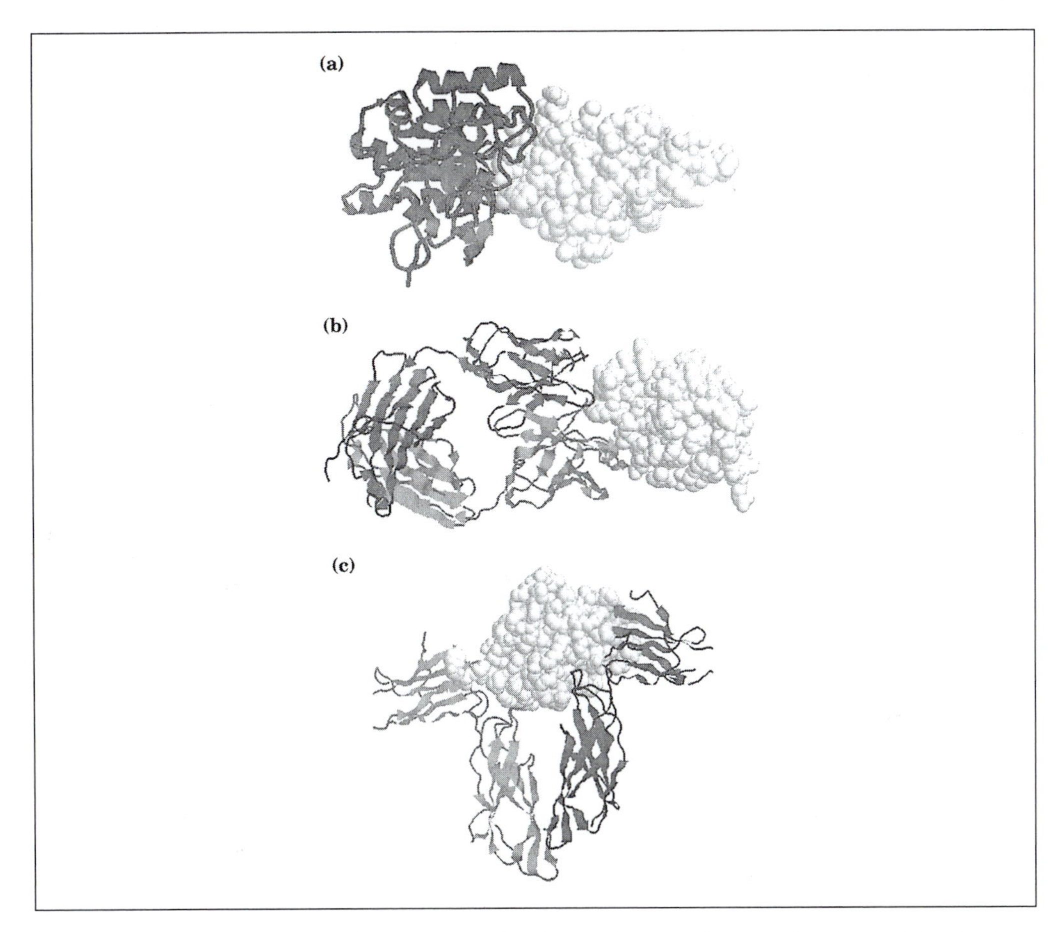

Fig. 1 Example structures from functional classes of protein–protein complexes. (a) Subtilisin with streptomyces subtilisin inhibitor* (PDB code 2six). (b) Immunoglobulin FAB fragment complexed with lysozyme* (PDB code 2hfl). (c) Human growth hormone* complexed with the extracellular domain of two receptor molecules hGHbp. The component depicted as a Corey–Pauling–Koltan (CPK) model is indicated with a '*' symbol, other components are shown as ribbon diagrams.

homodimers also have twofold symmetry, which places additional constraints on their intersubunit relationship. Many heterocomplexes (i.e. those comprising different protomeric units) can also form permanent associations. An example of such a permanent interaction is seen in human chorionic gonadotropin. This hormone, one that is involved in the early stages of pregnancy in humans, is a heterodimer with two protomers (α and β) each comprising ribbons of β-sheets in a similar topology, with three disulfides forming a cystine knot (22) (Fig. 2c). A further example of a heterodimeric permanent complex is seen in human immunodeficiency virus type-1 reverse transcriptase. This enzyme comprises one subunit of 66 kDa, which has five structural domains, and a 51 kDa subunit, which has four structural domains (CATH classification (23)) (Fig. 2d). The two subunits interact to form a single polymerase cleft which binds, among other molecules, transfer RNA (24).

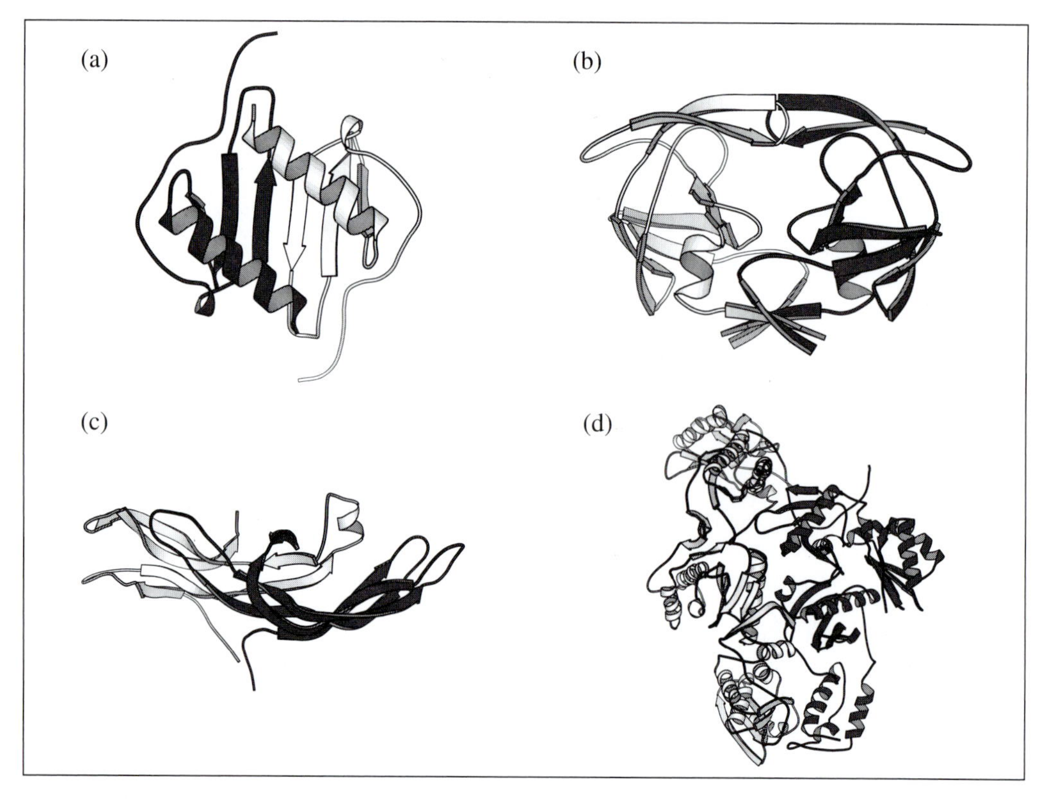

Fig. 2 MOLSCRIPT diagrams of permanent protein–protein complexes. (a) Interleukin 8 (PDB code 1il8). (b) HIV protease (PDB code 5hbp). (c) Human chorionic gonadotropin (PDB code 1hrn). (d) HIV reverse transcriptase (PDB code 3hvt).

Non-obligate complexes are built from units that can exist both as part of the complex and separately in the cell. Enzyme–inhibitor complexes are a good example of such associations. Despite their non-obligate status, once such complexes are formed many enzymes and inhibitors are strongly associated, with association constants ranging from 10^7 M^{-1} to 10^{13} M^{-1} (25). Antibody–protein complexes are further examples of non-obligate associations (see Chapter 5). As well as these two functional classes of non-obligate associations, there are hundreds more having diverse roles within the cell. The structure of a profilin–β-actin heterodimer has been solved (26) (Fig. 3a) and there is evidence that this is a cellular storage form of monomeric filamentous actin (27). This association is important in regulating actin-filament assembly. A non-obligate association is also observed in the complex formed between cytochrome *c* peroxidase (CCP) and cytochrome *c* in yeast (Fig. 3b). CCP is a biological redox partner of cytochrome *c*, and the complex they form provides the framework for an electron transfer pathway important in the respiratory process (28; see also Chapter 3). The fact that the components of non-obligate complexes need to exist alone in the cell at some stage in their life history imposes constraints on the

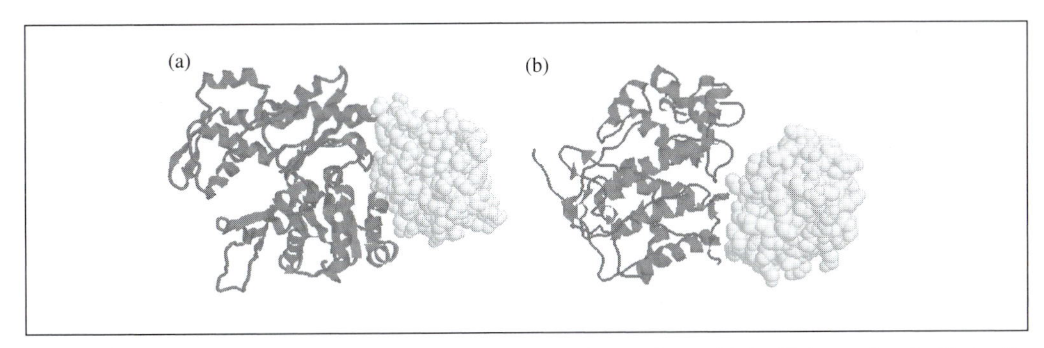

Fig. 3 Ribbon and CPK diagrams of non-obligate protein–protein complexes. (a) Profilin* complexed with β-actin (PDB code 2btf). (b) Cytochrome *c* peroxidase (CCP) complexed with cytochrome *c** (PDB code 2pcb). The component depicted as a CPK model is indicated with a '*' symbol.

nature of the surfaces they use to interact. As will be discussed later, the interfaces of proteins forming non-obligate associations are very different from those that form permanent associations.

2.3 Crystallographic vs. biological protein–protein complexes

All the classes of protein–protein complexes discussed so far can be termed biological complexes, in that they are associations known to exist in solution and hence, by inference, in the cell. A completely different set of protein–protein complexes are those represented by associations observed in crystal packing, termed crystallographic complexes. The problem of distinguishing between crystallographic complexes and true biological complexes is a difficult one. Protein–protein interactions in crystal packing differ significantly from both permanent and non-obligate biological complexes. Crystal packing contacts have no biological role and hence are not subject to evolutionary pressures (29). Recent comparative studies of crystal complexes and biological complexes have highlighted key structural differences between the two. Dasgupta and co-workers (30) studied 58 oligomeric complexes and 223 protein crystal structures, and found that the crystal contacts featured a large number of small polar areas of contact compared to the smaller number of large hydrophobic areas of contact in the oligomers. Contacts in oligomeric associations are more extensive than in the crystal contacts, with only 1% of crystal contacts exhibiting interfaces of $\geqslant 1600\text{Å}^2$ (29, 31). In a similar study, Carugo and Argos (32) analysed 962 protein–protein crystal contacts in 78 monomeric proteins and compared them to physiological protein–protein contacts in oligomerization. They found that, relative to biological contacts, the crystal contacts were small, with 45% having $< 100\ \text{Å}^2$ and only 8% having contacts larger than $500\ \text{Å}^2$. The sequence segmentation of the crystal contact interface was found to be less than that observed in biological contacts. From such analysis it is suggested that certain crystal contacts could be functionally relevant in an as yet unidentified oligomeric protein structure. Janin (29) cites the example of the C domain of Smad4 tumour suppressor, that was observed to be a trimer in the crystal and which subsequently proved to be a trimer in solution.

Whilst there appears to be sharp contrasts in the features of crystal and biological contacts, their differentiation remains a difficult process. The only certain way of isolating sets of specific classes of contact is to review the literature. However, with the ever-increasing number of complexes in the PDB, automatic methods are needed to ascertain whether the contents of the asymmetrical unit (deposited in the PDB) represent biological or crystallographic associations. The automatic evaluation of classes of contacts has been conducted as part of a Protein Quaternary Structure File Server (PQS) (http://pqs.ebi.ac.uk). PQS is an Internet resource that makes available coordinates for the likely quaternary states of structures contained in the PDB as determined by X-ray structure. In the PDB the coordinates deposited for each structure usually represent those contained within the crystallographic asymmetrical unit. The server provides a means of identifying where multiple copies of the biological complex exist, and where symmetry operations are required to generate the complete coordinate set. The method of discriminating between crystal and biological complexes involved the use of an empirical weighted score, with contributions from the accessible surface area (ASA) of the interface, a delta-solvation energy of folding (33), and the number of intermolecular salt bridges and disulfide bonds. Some of the server's results for specific classes of protein were checked against on-line annotation (e.g. ref. 34) to ascertain if the server's allocation of 'biological' or 'crystal' complex was correct. Using this limited checking procedure it was observed that, whilst it was correct in a large percentage of cases, the empirical weighted score could not differentiate between biological and crystal complexes unambiguously. One example of a misassigned complex was T4 lysozyme; this was assigned as a biological dimer, when in fact the protein is known to be a monomer and hence the contacts observed between protomers in the crystal were crystallographic contacts. As there are over 12 000 structures in the PDB a complete check of the literature for each structure is a practical impossibility, and therefore the overall accuracy of the PQS is difficult to ascertain. Hence, this useful tool requires further validation of its methodology.

3. Structural properties of protein–protein interfaces

Protein–protein association involves the specific complementary recognition of two macromolecules to form a stable assembly (35). The recognition process involves factors favouring and opposing the stable association. Hydrophobic and electrostatic interactions favour the association. The loss of translational and rotational freedom of amino acids on binding opposes the association. The affinity for two molecules is determined by the change in energy and entropy of a system that contains the two proteins and solvent, and the complex and solvent (36). Recent work has focused on the kinetics of protein associations (36–39; see also Chapter 1), proposing new kinetic models of association. However, the lack of experimental binding-association data has meant that the relative contributions of factors contributing to the binding energy of association remains unclear.

The number and type of interactions in protein–protein complexes are considered in

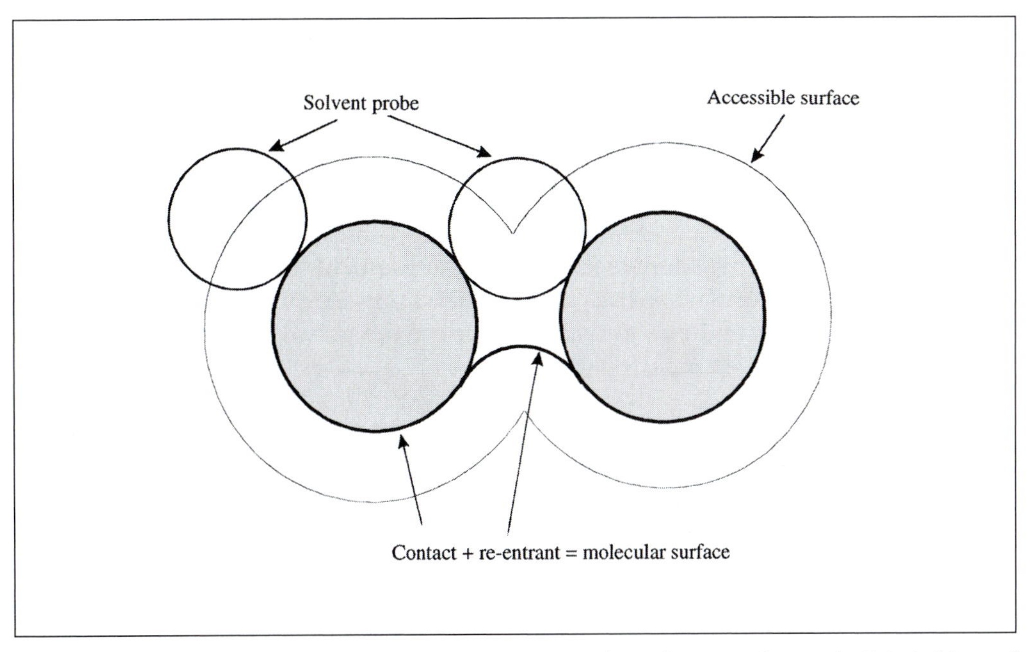

Fig. 4 The definition of accessible surface area (ASA) and molecular surface area of a protein. (Adapted from ref 53).

relation to the area of the interface (e.g. refs 40, 41). This is measured in terms of either accessible surface area or molecular surface area. The native structure of proteins exists only in the presence of water, and the ASA describes the extent to which protein atoms can form contacts with water. Lee and Richards (42) were the first to propose the concept of ASA, defining it as the area of a sphere of radius R, on each point of which the centre of a solvent molecule can be placed in contact with an atom without penetrating any other atoms of the molecule. The radius, is given by the sum of the van der Waals radius of the atom and the chosen radius of the solvent molecule (Fig. 4). Subsequent to the initial definition of surface accessibility, Richards (43) proposed an alternative: that of the molecular surface. The molecular surface consists of the part of the van der Waals surface of the atoms that are accessible to a probe sphere (termed contact surfaces), connected by a network of surfaces (termed re-entrant surfaces) which smooth over the crevices and pits between the atoms. This surface is the boundary of the volume from which a probe sphere is excluded if it is not to experience van der Waals overlap with the atoms (Fig. 4). However, the methods of ASA molecular surface calculation imply that the system is static; they do not account for any movement or flexibility an atom or group may possess within the molecule.

3.1 Hydrophobicity

The term 'hydrophobic interaction' is used to describe the gain in free energy upon the association of non-polar protein residues in an aqueous environment (44).

Hydrophobic interactions are considered to be the driving force in the stabilization of protein associations (40, 45). The quantitative evaluation of exactly how much hydrophobic interactions contribute to the stabilization of protein–protein associations is controversial (45–50). The controversy is based on different definitions and interpretations of the hydrophobic effect in proteins (49, 50). It has been estimated, from an empirical correlation, that for non-polar surfaces there is an energy gain of approximately 25 calories per $Å^2$. However energy values as high at 72 calories per $Å^2$ have been reported (51). More recently, the significance of correlating the hydrophobic effect with the molecular surface area, rather than accessible surface area, has become apparent (52); and energy values of approximately 47 calories per $Å^2$ of molecular surface have been reported (53).

3.2 Electrostatic interactions

Electrostatic interactions, including hydrogen bonds and van der Waals interactions, also make important contributions in protein associations (40, 54). Hydrogen bonds between protein molecules are more favourable than those made with water (55), and hence intermolecular hydrogen bonds contribute to the binding energy of association. It has been proposed that, whilst hydrophobic forces drive protein–protein interactions, hydrogen bonds and salt bridges confer specificity (55, 56). Van der Waals interactions occur between all neighbouring atoms, but these interactions at the interface are no more energetically favourable than those made with the solvent. However, they are more numerous, as the tightly packed interfaces are more dense than the solvent (35) and hence they contribute to the binding energy of association. A detailed study of electrostatic interactions in 319 protein–protein complexes found that the geometry of hydrogen bonds across protein–protein interfaces are generally less optimal and have a wider distribution than those observed in the interior of proteins (57). This work suggested that due to constraints of the interface, the hydrogen bonds in them were weaker than those in protein interiors. A recent study found that the number of hydrophilic bridges across the binding interface showed a strong positive correlation with the free energy of the interaction (58). This study also showed that salt bridges across the binding interface can significantly enhance stability in some complexes.

3.3 Shape complementarity

The complementarity of protein interfaces is derived from both electrostatic interactions and shape. Shape complementarity has been characterized by the size of the buried surface and the packing density of interface atoms. The close packing of protein interiors has been analysed (59), and this method has been used to measure the packing density of protein–protein interfaces, revealing that such sites are tightly packed (40). The close-packed nature of interfaces in antibody–antigen complexes has also been observed using volume calculations (60). More recently, a shape correlation statistic (Sc) has been defined to measure packing and shape complementation in protein–protein interfaces (61). This statistic has a number of advantages, in that it is relatively insensitive to the precise atomic radii used in defining the molecular surface and to atomic coordinate errors typical of refined X-ray structures. The statistic was used to measure the complementarity of interfaces in a series of enzyme–-inhibitor complexes, oligomeric proteins, and antibody–antigen complexes. This analysis revealed that antibody–protein interfaces were less well packed than other types of complex.

4. Analysis of protein–protein complexes

Chemical and physical aspects of protein–protein recognition have been analysed by a number of research groups using large data sets of protein structures from the PDB. Studies have principally detailed hydrophobicity, amino-acid composition, hydrogen bonding, secondary structure, and the shape of the protein–protein interfaces on both permanent multimeric complexes (3–5, 62) and non-obligate complexes (6, 35, 40, 41, 63). Using the method of defining interface atoms and residues, based on the ASA they lose when changing from the isolated to the complexed state, general analyses of protein–protein interfaces have been conducted by Janin *et al.* (4) on a data set of 23 oligomeric proteins (including dimers, tetramers, hexamers, and octamers) and by Argos (3) who studied a data set of 24 oligomeric proteins (including dimers, trimers, and tetramers). We have conducted similar studies on a more specific data set of 32 homodimeric proteins (5), and on five different categories of complex (homodimeric proteins, heterodimeric proteins, homotrimeric proteins, enzyme–inhibitor complexes, and antibody–protein complexes) (6). Within these five categories, permanent and non-obligate complexes were also identified and analysed (63). All these studies have added to our current knowledge of protein binding sites, and can be summarized in the five points below.

4.1 Size and shape

On average, dimers contribute 12% of their ASA to the contact interface, trimers 17.4%, and tetramers 20.9% (3). However, variations are large and total interface areas range from 670 Å^2 (9%) in dimeric superoxide dismutase to 10 570 Å^2 (40%) in tetrameric catalase (4). In a data set of 28 homodimers, we observed interface ASAs of between 368 Å^2 for 434 repressor and 4746 Å^2 in citrate synthase, with the percentage ASA per subunit ranging from 6% in inorganic pyrophosphatase to 29% in Trp repressor. In non-obligate complexes the interface ASA is limited by the size of the smallest component of the complex. For a data set of 10 enzyme–inhibitor complexes the mean interface ASA was 785 Å^2 and for a data set of six antibody–protein complexes a similar figure of 777Å^2 was calculated (6). In homodimers the ASA is approximately linearly related to the molecular weight of the protomer (5) (Fig. 5), so the larger the protomer the larger the interface site required to stabilize its interaction with a second protein molecule. Argos (3) suggests a 5–6% ASA of the monomer as a minimum requirement for stabilization of an interface.

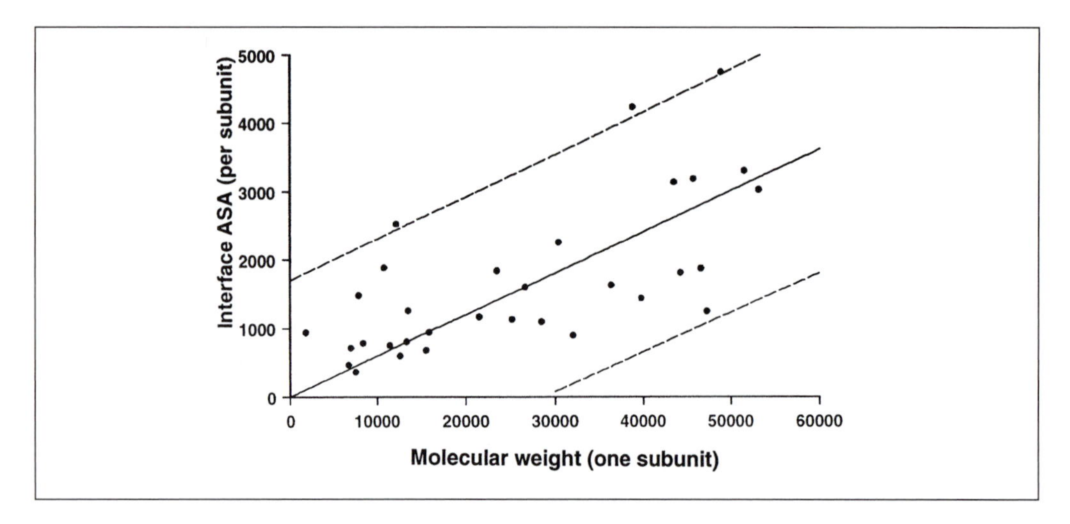

Fig. 5 Relationship between the molecular weight of the protomer and the interface ASA (per subunit). The solid line is the fitted line of the equation $y = 0.06x$ and the dashed lines are the 95% confidence limits of the individual *y* values. The correlation coefficient (*r*) is 0.69.

When viewed as an overall or global cross-section, interfaces are generally flat. Argos (3) found that 83% of interfaces in a data set of 24 oligomers were flat, and we found a similar figure of 84% in a data set of 28 homodimers (5). Exceptions were those proteins in which the two subunits were twisted together across the interface (e.g. isocitrate dehydrogenase) (Fig. 6a) or proteins having subunits with 'arms' that clasp the two halves of the structure together (e.g. aspartate aminotransferase) (Fig. 6b). Interfaces were found to be similarly flat in non-obligate associations such as enzyme–inhibitor and antibody–protein complexes (6). The circular nature of interfaces has also been analysed (5, 6) finding that, with few exceptions, the interfaces are approximately circular areas on the protein surface in both permanent and non-obligate complexes.

4.2 Amino-acid composition

The hydrophobic nature of protein–protein interfaces has been extensively documented for data sets of oligomeric proteins (3–6, 62). The interfaces have been shown to be more hydrophobic than the exterior but less hydrophobic than the interior. In our study of homodimers 32% of interface atoms were polar (5); a similar figure of 35% of the atoms were found to be oxygens or nitrogens in a data set of oligomeric proteins including dimer, and higher multimers (3). At the residue level, 47% of interface residues were hydrophobic, 31% polar, and 22% charged (5). By calculating propensities, which give an indication of the relative importance of different amino acids in the interface compared to the protein surface as a whole (6), it was found that specific amino acids had high probabilities to be present in protein–protein interfaces (compared with their frequency on the exposed surface of the protein) (Fig. 7). In

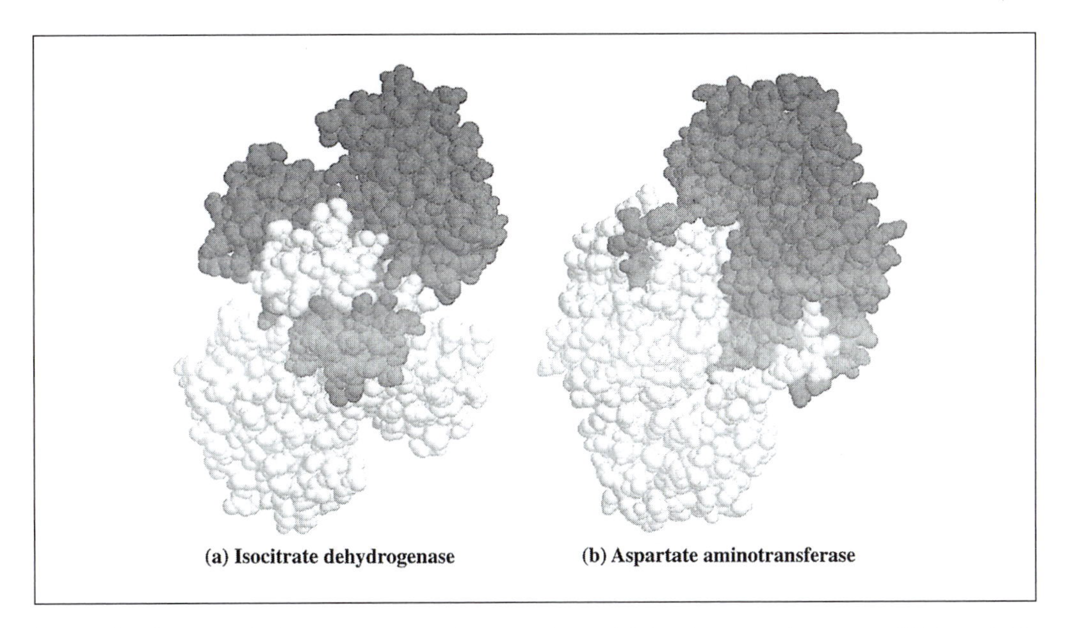

Fig. 6 CPK diagrams of structures with non-planar interfaces. (a) Isocitrate dehydrogenase (PDB code 3icd). (b) Aspartate aminotransferase mutant (PDB code 3aat). In each diagram one subunit is coloured dark grey and one light grey.

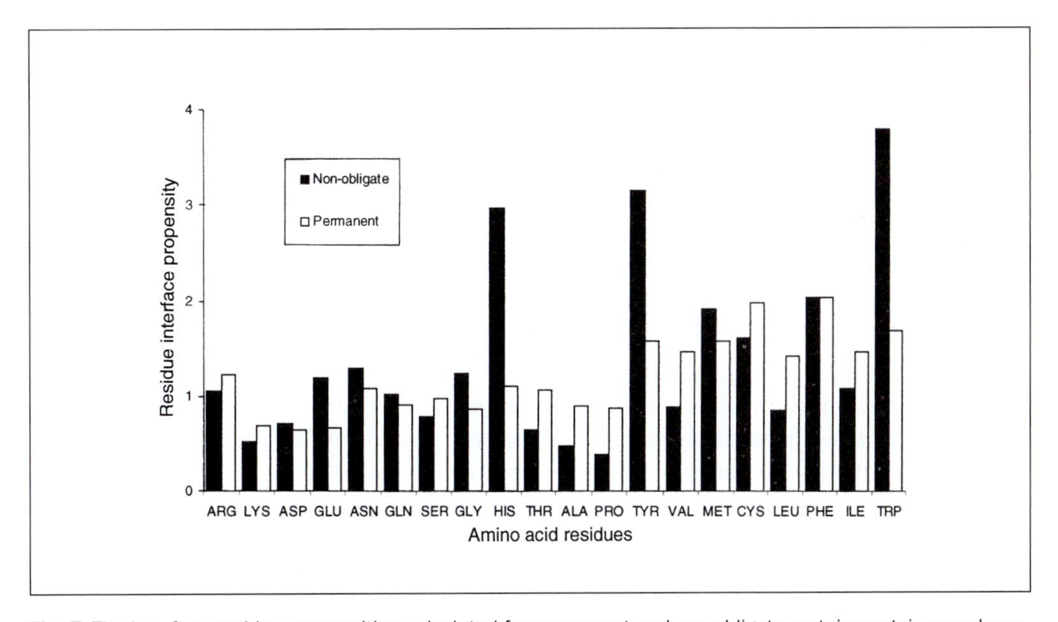

Fig. 7 The interface residue propensities calculated for permanent and non-obligate protein–protein complexes. Propensities are calculated for each amino acid (AAj) based on the fraction of ASA that AAj contributed to the interface compared with the fraction of ASA that AAj contributed to the whole surface. A propensity of > 1 denotes that the amino acid occurs more frequently in the interface than on the protein surface.

general, the permanent complexes have interfaces that contain hydrophobic residues, whilst the interfaces in non-obligate complexes favour the more polar residues. The single aromatic residues-histidine, tyrosine, and phenylalanine—are common in the interfaces of both permanent and non-obligate complexes. It was suggested that these residue types make a particularly good 'glue' for sticking protein subunits together (4).

4.3 Electrostatic interactions

The number of intermolecular hydrogen bonds is approximately proportional to the ASA buried in the surface (3, 5). In permanent homodimers there are, on average, 0.88 hydrogen bonds per 100 $Å^2$ of ASA buried (for interfaces covering > 1500 $Å^2$ per subunit); but the number of hydrogen bonds varies from zero in some complexes (e.g. uteroglobin) to 46 in variant surface glycoprotein (5). Side-chain hydrogen bonds represent approximately 76–78% of the interactions (3, 5). However intermolecular hydrogen bonds are more prevalent in non-obligate complexes, with values of 1.4 bonds per 100 $Å^2$ of ASA buried for enzyme–inhibitor complexes and 1.1 bonds per 100 $Å^2$ of ASA buried for antibody–protein complexes (6). Salt bridges have also been observed between subunits of oligomeric proteins, but only 56% of homodimeric proteins were found to possess such interactions—many having none, and at the most five. Intermolecular disulfides are rarely seen in dimeric proteins (5), as they only occur in oxidizing environments. However, when intermolecular disulfides do occur they often play an important role in structural stabilization. Protein engineering experiments on two structures, platelet-derived growth factor B (64) and thymidylate synthase (65), have both shown that the introduction of intermolecular disulfides increases the stability of the complex.

4.4 Packing

Hubbard and Argos (66) analysed the internal packing between protein subunits in multi-subunit proteins, between domains and within domains. They found that, on average, 10.7% of the total surface area was occupied by intersubunit cavities. This compared with 9.6% in interdomain interfaces and 2.8% within domains. The cavity volume increased in absolute value with protein size, and the intersubunit and interdomain cavities were, on average, larger in size and were more often solvated than the intradomain cavities. In studies of homodimeric proteins (5) and hetero-complexes (6) the packing of subunit interfaces was evaluated by calculating a gap-volume index. This index was calculated using a procedure developed by Laskowski (67) that estimated the volume enclosed between any two molecules using a series of spheres, delimiting the boundary by defining a maximum allowed distance between both interfaces. The total volume of the spheres was calculated and normalized by the ASA of the contact region to give a gap-volume index for each interface. By calculating mean gap-volume indices for different types of protein–protein complex it was found that interfaces in homodimers, enzyme–inhibitor complexes, and permanent heterocomplexes were the most complementary, whilst the antibody–protein complexes and the non-obligate heterocomplexes were the least complementary (6). This confirmed a similar analysis made using a shape complementarity statistic (61).

4.5 Secondary structure

Interfaces occur between helix, sheet, and coil motifs, with both like and non-like interactions across the interface in permanent and non-obligate complexes. Miller (68) examined nine dimeric and nine tetrameric proteins and found that, commonly, an interface had a central area of either an extended β-sheet, helix–helix packing or sheet–sheet packing decorated at the edges by loop interactions. The loop interactions contributed, on average, 40% of the interface contacts. The loops were found to interact with other loops and with the ends of secondary structures and were stabilized by large numbers of hydrogen bonds. Motifs are often shared across interfaces, and stability within interfaces is enhanced by converting loops within motifs into linkers across interfaces (69). In our analysis of a data set of 28 homodimeric proteins, 53% of the interface residues were classed as α-helix, 22% as β-sheet, and 12% as αβ, with the rest being coils (5). The β-sheet motifs at interfaces could be divided into stacked (Fig. 8a), extended (Fig. 8b), and complex categories; the extended structures being comparable to those observed by Miller (68).

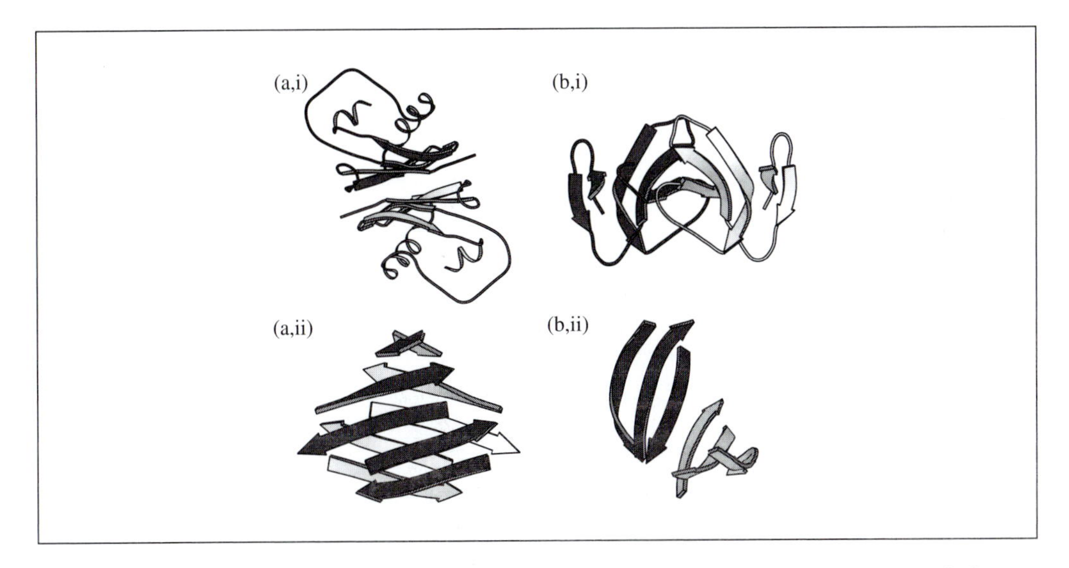

Fig. 8 MOLSCRIPT diagrams of extended and stacked β-sheets in protein–protein interfaces. In each diagram one subunit is shaded dark grey and the other light grey. For each structure (i) shows the complete dimer structure and (ii) shows just the interface β-strands. (a) Stacked β-interface in Streptomyces subtilisin inhibitor (PDC code 2ssi). (b) Extended β-interface in cardiotoxin (PDB code 1cdt).

4.6 Comparing protein-binding sites

Protein–protein complexes with different functions and interaction histories have binding sites with different characteristics. A comparison of permanent and non-obligate protein–protein associations has been made (6, 63) and is summarized in Table 1. These data clearly show that the interfaces in permanent protein–protein associations tend to be larger, less planar, more highly segmented (in terms of sequence), and closer packed, but that they have fewer hydrogen bonds than interfaces in non-obligate associations. The non-obligate complexes tend to be more hydrophilic in comparison, as each component has to exist independently in the cell. To

Table 1 Results of structural analysis for two classes of protein–protein complexes

Interface characteristic		Permanent dimer complexes	Non-obligate dimer complexes
Number of examples		36	23
ΔASA (Å^2)			
	Mean	1722	804
	σ	1985	147
Planarity (Å)			
	Mean	3.56	2.57
	σ	1.65	0.52
Circularity			
	Mean	0.69	0.65
	σ	0.18	0.14
Sequence segmentation			
	Mean	5.30	4.83
	σ	2.80	2.24
Hydrogen bonds (per 100 Å^2 ΔASA)			
	Mean	0.74	1.13
	σ	0.48	0.45
Gap index (Å)			
	Mean	2.12	2.66
	σ	0.94	0.87

i. ΔASA: The change in accessible surface area for one subunit on complexation. In the heterodimers, enzyme–inhibitor and 'other' heterocomplexes, the mean ASA is that buried by each subunit on complexation. For the antibody–protein complexes the mean ASA is that on the protein (antigen) surface buried on complexation. In the trimers the mean ASA shown is the mean ASA buried by each pair of interacting subunits.

ii. Planarity: Root-mean-squared (rms) deviation of all the interface atoms from the least-squares plane through all these atoms.

iii. Circularity: Ratio of the lengths of the principal axes of the least-squares plane through the atoms in the interface.

iv. Segmentation: It was defined that interface residues separated by more than five residues were allocated to different segments.

v. Intermolecular hydrogen bonds: The number of intermolecular hydrogen bonds per 100 Å^2 Δ ASA were calculated using HBPLUS (70), in which hydrogen bonds were defined according to standard geometric criteria.

vi. Gap-volume index: The gap volume between the two components of the complexes was calculated using SURFNET (67).

possess exposed hydrophobic patches on the protein surfaces would be energetically unfavourable. Permanent homodimer complexes are among the most well-packed associations, whilst the antibody–protein complexes are amongst the least well-packed associations (6, 61). Such differences can be explained if the complexes are considered with respect to their binding constants and evolutionary history. Antibodies bind antigens, initially recognizing surfaces never previously encountered. Most antibody–antigen interactions have an initial association constant of around 10^7 M^{-1}, but in subsequent immune responses this may increase as somatic mutations improve recognition and the strength of binding. In contrast, multimeric proteins have been subject to selective evolutionary pressures. Many dimeric interactions, that can have associations constants greater than 10^{16} M^{-1}, are so strong that the monomers have to be denatured to separate the complex.

Differences have also been observed in the size and hydropathy of biological protein–protein associations and crystallographic associations (30, 31, 62) (see Section 2.3). Comparisons have also been made between intersubunit associations and interdomain interactions (3). Hence, although protein–protein interfaces share common characteristics, the type of protein–protein association has an important influence on the specific nature of the interaction.

4.7 Protein interface analysis as an Internet resource

A server on the WWW (http://www.biochem.ucl.ac.uk/bsm/PP/server) containing the major elements of the computer software used in our analysis of protein–protein complexes (5, 6) is now available. The protein interaction server enables the user to subunit the 3-dimensional coordinates of a protein complex to obtain a set of physical and chemical parameters that describe the nature of the protein–protein interface. The server provides information on the size of the protein interface in terms of the accessible surface area, shape, intermolecular bonding, polarity, bridging water molecules, and packing as described in the previous section. A listing of the residues involved in the protein–protein interface (those that have an ASA that decreases by > $1Å^2$ on complex formation) is also given, indicating the relative importance of each residue. The parameters for individual structures can be compared to the distributions obtained from data sets of known protein–protein complexes (6). Such comparisons allow an estimation of the validity of interfaces in new protein complexes.

5. Prediction of protein–protein interaction sites

The reliable prediction of protein–protein interaction sites is an important goal in the field of molecular recognition. It is of direct relevance to the design of drugs for blocking or modifying protein–protein interactions. Predictions can be divided into two main areas. The first area is the docking of two proteins of known structure. The second, and the one discussed here, is the identification of interaction sites on the surface of an isolated protein known to be involved in protein–protein interactions, but where the structure of the complex is unknown. A number of methods fall into this second area, including the algorithm HSITE (71, 72) that maps hydrogen-bonding regions on protein surfaces. This algorithm was designed as an interactive tool to build a knowledge-base of protein surfaces for the subsequent automatic construction of novel ligands to fit specific binding sites. However, the hydrophobic nature of protein interaction sites has meant that the majority of the methods focus on the location of hydrophobic surface clusters on proteins (62, 73–75). Our method of searching for potential interaction sites also uses hydrophobicity, but combines this with a number of other structural features to isolate putative interface sites (76, 77).

5.1 Prediction of protein–protein interaction sites using hydrophobic clusters

Korn and Burnett (62) were amongst the first to realize the potential of using interface hydropathies to predict binding-site locations. They calculated hydropathies for subsets of atoms in exterior contact surfaces and exterior non-contact surfaces. The hydropathy values were based on a normalized, consensus hydropathy scale (78). They surveyed 40 multimeric proteins and 2 protein–protein complexes to find that the hydropathy of the contact surfaces was higher than the exterior non-contact surfaces. They then went on to predict the location of the protein-binding sites on dimeric inorganic pyrophosphatase, by manually searching for concentrations of hydrophobic residues. In this way they also differentiated between the biological and the crystallographic dimer in wheatgerm agglutinin.

Young and co-workers (74) took this work much further and developed an algorithm to search for hydrophobic clusters on the surface of proteins. The method sums hydrophobic parameters of residues within a certain distance from a solvent-accessible grid point to obtain the hydrophobicity of the cluster formed by these residues. They then ranked clusters of surface residues on the basis of the hydrophobicity of their constituent amino acids. They found, using nine classes of enzymes, seven antibody fragments, and a number of other complexes, that the location of the co-crystallized ligand is most commonly found to correspond to one of the strongest hydrophobic clusters on the surface of the target molecule. In 25 out of 29 cases they found that the position of the most hydrophobic cluster corresponded to the position of the surface buried by the bound ligand.

Most recently, Lijnzaad and co-workers (79) have developed a new technique for delineating explicit contiguous hydrophobic regions of protein surfaces. Using a dot representation of the solvent-accessible surface area, the method identifies regions of a surface of arbitrary size and shape consisting solely of carbon and sulfur atoms. Only the largest patches, with a size exceeding random expectation, are considered meaningful. Hence, unlike previous methods, the size of the hydrophobic regions identified are not prelimited by the method employed to find them. Using this technique on eight multimeric proteins they found that, in general, only the largest patches coincided with regions of known biological relevance. This methodology has also been used to predict the location of 59 multimeric protein interface sites (75). They found that the largest, or second largest, hydrophobic patch on the accessible surface was involved in the protein–protein interfaces of 90% of the structures studied.

5.2 Prediction of protein–protein interaction sites using multiple characteristics of surface patches

From an initial analysis of a number of data sets of protein–protein complexes, including permanent and transient associations, it was found that it was possible to distinguish interaction sites from other similar sites on the protein surface by analysing a series of chemical and structural parameters (76). The parameters used were solvation potential, residue interface propensity, hydrophobicity, planarity, protrusion, and ASA. The solvation potentials quantified how much a patch preferred to be exposed to solvent, based on a knowledge-based potential derived as part of a threading algorithm (80). The residue interface propensities (as discussed in Section 4.2) quantify the preference for certain amino acids to be in interface sites. Hydrophobicity was estimated using the experimentally derived scale of Fauchère and Pliska (81). Planarity was estimated by calculating the root mean square of a best-fit plane through the residues in the patch. Protrusion was quantified by calculating a protrusion index; this gives an absolute value for the extent to which a residue protrudes from the surface of a protein. In this method a patch was simply defined as a central surface-accessible residue and n nearest neighbours, as defined by $C\alpha$ positions. In the analysis, n (the size of the patch in terms of the number of residues) was defined as the number of residues in the observed interface site. A number of simple vector constraints were imposed on the patches to prevent patches sampling through the centre of the protein or forming rings around the surface.

By defining a whole series of surface patches on a protein, and calculating six parameters for each patch, it became evident that the known interface sites of individual structures generally occurred at the extremes of the parameter distributions (Fig. 9). By combining distributions in a data set of 28 homodimers, the known interface site was found to be amongst the most planar, protruding, and accessible patches; and amongst the patches with highest interface propensity. The distributions varied for different types of complexes. For example, the binding sites of a series of antibodies were found to be amongst the most hydrophilic and had low residue interface propensities, whilst still being the most planar, accessible, and protruding.

With this knowledge of how interface sites differed from other similar sites on the protein surface, a predictive algorithm (PATCH) was developed that used multiple parameters of patches to identify those most likely to form putative interface sites. The method involved calculating a relative combined score that gave the probability of a surface patch forming protein–protein interactions. The score combined up to six parameters, with different parameters used for different types of protein–protein complexes. Using the PATCH algorithm it was possible to predict the location of interface sites in known complexes for 66% of the structures. In many cases, when the known interface site was not predicted this could be rationalized in terms of the

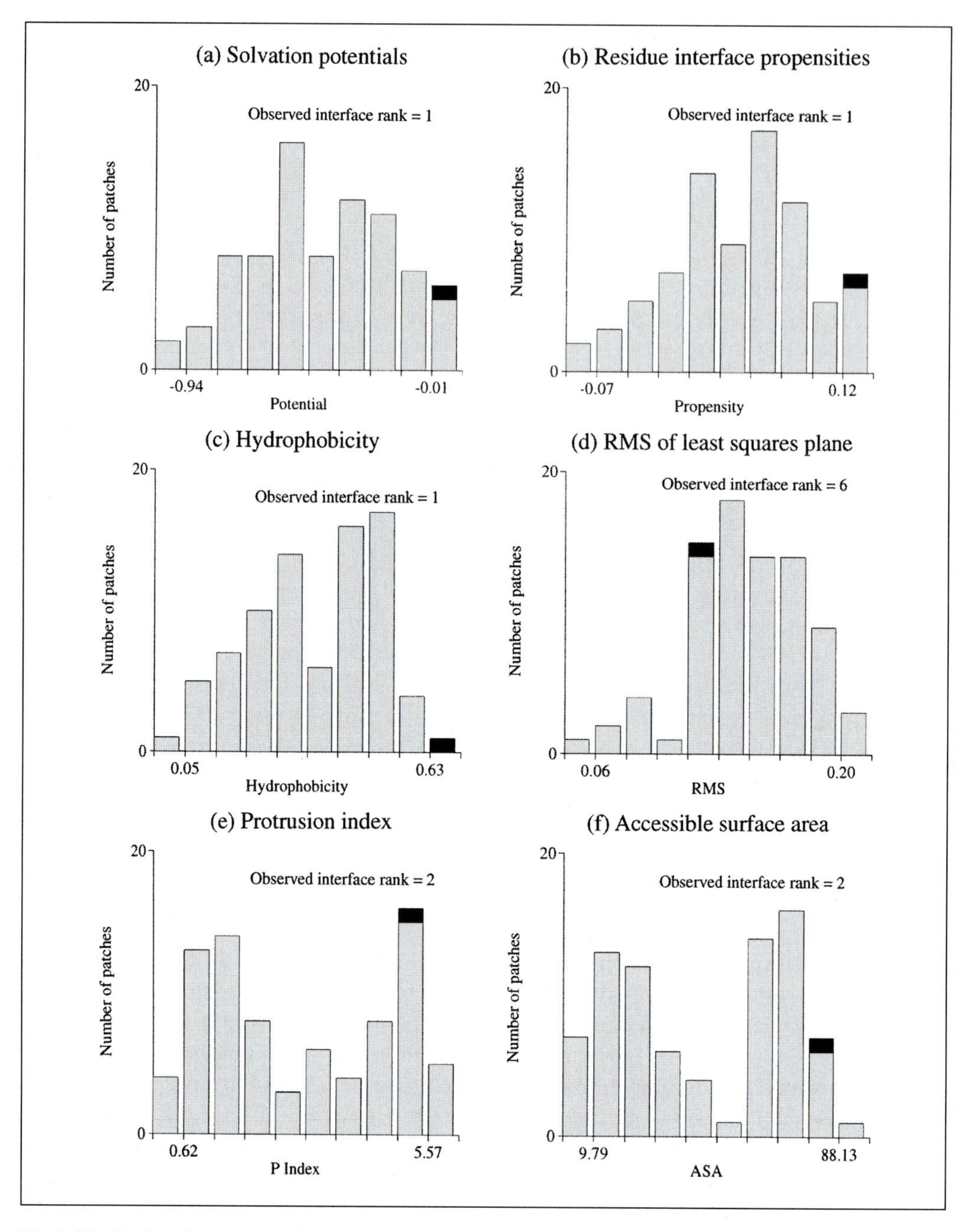

Fig. 9 Distribution of parameters for all patches in one protomer of HIV protease (PDB code 5hvp). Distributions are shown for: (a) solvation potential; (b) residue interface propensity; (c) hydrophobicity; (d) root-mean-square (rms) deviation of atoms from the least-squares plane through the interface propensities; (e) protrusion index; (f) accessible surface area. On each graph the surface patches are presented by grey bars and the observed interface patch is shown as a black bar. Relative rankings (on a scale of 1–10) were calculated from each distribution and are indicated on each graph.

presence of additional interaction sites on the protein surface. A PATCH profile is shown in Fig. 10 for mannose-binding protein. This structure has 88 surface patches, and the one that had the highest combined probability score (based on searching for the most planar, hydrophobic, accessible, and protruding patch; and the patch with the highest solvation potential and highest residue interface propensity) mapped to the known interface site.

6. Applications for protein–protein interaction prediction and analysis

Detailed knowledge of protein associations and the ability to predict the location of potential binding sites on protein surfaces, could prove to be important for drug design or for the optimization of drug therapies already in use. Detailed knowledge of specific protein–protein interfaces has provided information for redesigning insulin therapy for the treatment of diabetes, and in the search for HIV protease inhibitors that can be used as drugs. The potential for using prediction software is also evident in the case of designing potential antagonists and agonists to human chorionic gonadotropin. Each of these cases is discussed below to give an insight into the application of research into protein–protein interfaces.

6.1 Redesigning insulin therapy

Insulin is a peptide hormone involved in the transport of glucose, and hence it has an essential metabolic role. An inability to produce sufficient levels of insulins can result in the disease diabetes mellitus. Insulin has a quaternary structure in which the monomers associate to form dimers and then hexamers. When insulin is injected as a therapeutic agent the rate of absorption is slow, as the molecules undergo self-assembly into zinc-containing hexameric structures. This limited rate of absorption from the injection site means that it does not lead to normal concentrations in the blood, and this is thought to cause of the detrimental health effects which result from long-term diabetes. The protein associations in the hexamer have been analysed, with the aim of producing monomeric insulin which would be absorbed at a faster rate and be more effective at controlling glucose levels in diabetic patients (82).

With the detailed knowledge of the structure of both heterodimeric and hexameric insulin, a monomeric insulin that retains biological activity has now been manufactured through protein engineering (83). It has been found that the strongest subunit interactions are between dimerizing monomers, and that if these interactions could be disrupted, then the association to hexamers could be stopped. The dimer-forming surfaces displayed the characteristic properties of all protein–protein interactions, being planar and largely non-polar with a number of intermolecular hydrogen bonds that conferred specificity. Mutations were made to residues located in the dimer interface of one subunit, but on the periphery of the receptor-binding region of the hormone (82, 83). Substitutions were made to introduce charge repulsion in the

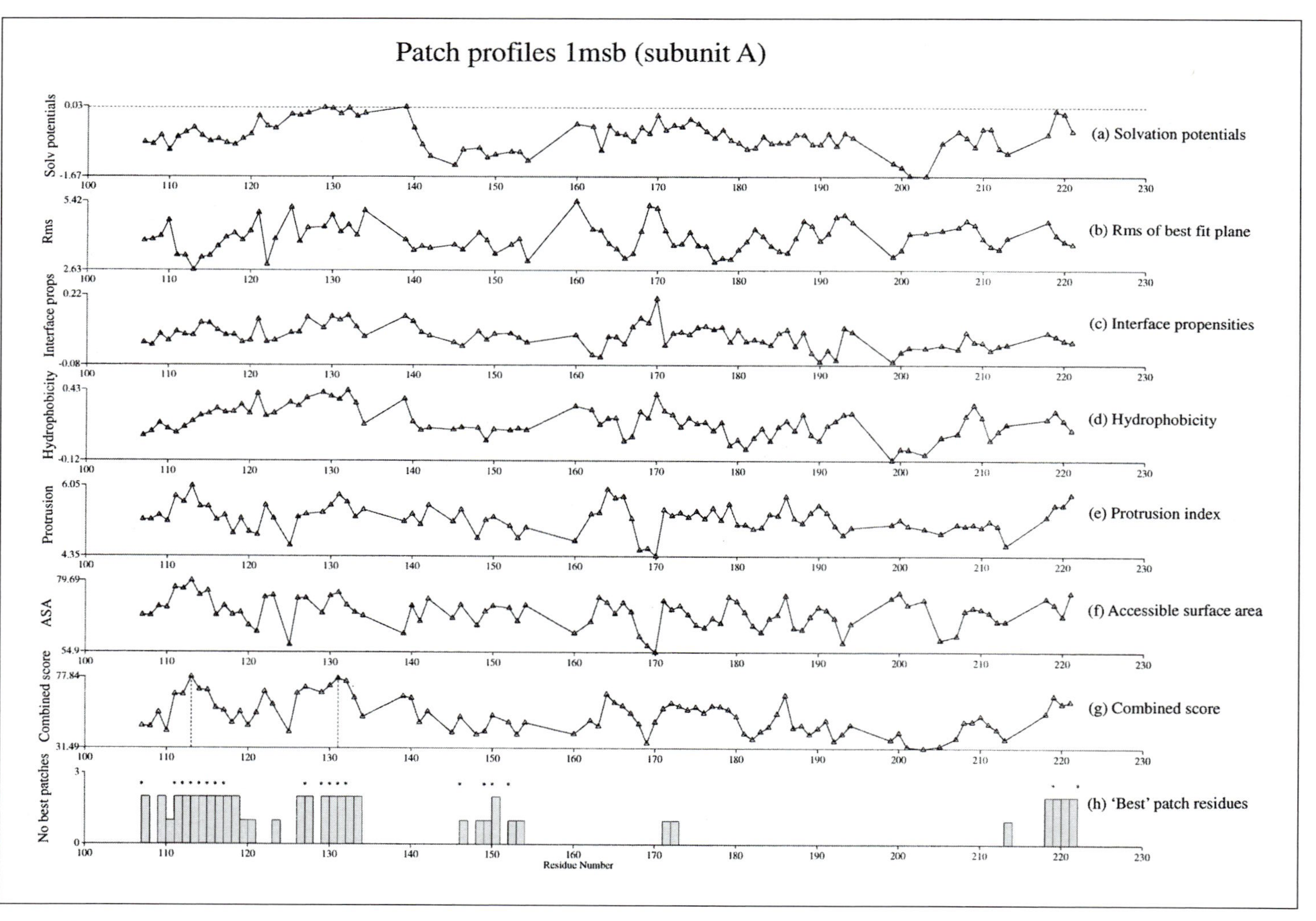

Fig. 10 PATCH profile for subunit A of mannose binding protein (1msb). The patch size used was 28 residues and the best patches (i.e. those most likely to form interaction sites based on the highest combined score) were selected using a 15% cut-off value (i.e. those patches in the top 15% of the combined score distribution). The frequency of surface residues in the best patches is indicated in the histogram (h) where '*' symbols mark the residues in the observed interface.

interface, with negative charges inserted where they would oppose negative charges on the complementary interface. This strategy proved successful in creating a series of monomeric insulins, many of which had full or nearly full activity and comparable or increased rates of absorption to those observed in human insulin. These insulins gave improved control of glucose levels in diabetic patients, but their potential side-effects are still to be evaluated.

6.2 HIV-protease inhibitors

HIV protease has a fundamental role in the proteolytic processing of viral polyproteins, and viral progeny without functional proteases are not infectious (84). It is also known that chemical inhibitors of HIV protease stop replication of the virus *in vitro* (85). When the structure of HIV-1 protease was first solved in 1989 (21, 86) it was suggested that a possible blueprint for a drug designed to combat the HIV virus would be a molecule that could disrupt the dimerization of this enzyme (86, 87).

Detailed structural studies of the active site at the dimer interface of HIV protease has led to the development of a whole series of potential inhibitors, many acting as substrate or transition-state analogues, rather than disrupting the dimer interface. However, it is only recently that analogues have been found that also have a long half-life in the body and enhanced oral availability (88, 89). Some have even reached the stage of advanced clinical trials (88, 89), whilst three—saquinavir, indinavir, and ritonavir—have been approved for clinical use (90).

6.3 Prediction of structural epitopes on human chorionic gonadotropin

Human chorionic gonadotropin (hCG) is a glycoprotein hormone responsible for the maintenance of the early stages of pregnancy in humans (91). Its functional significance in the human reproductive cycle means that agonists are of potential importance for infertility treatments, and antagonists are candidates for contraceptive vaccines (92). The protein is a member of a family of heterodimeric glycoprotein hormones, which includes luteinizing hormone (LH), follicle-stimulating hormone (FSH), and thyroid-stimulating hormone (TSH). These hormones function by binding to specific cell-surface receptors; hCG and LH bind to the same cell-surface receptor, but their secretion occurs under different conditions (92). Both the α- and β-subunits are required in each glycoprotein for receptor binding, but the β-subunit is the one that determines the specific activity of the hormone (22).

Work is currently being conducted into the production of a highly specific immunological contraceptive vaccine (93). This involves the initial identification of epitopes on the surface of the hCG β-subunit using a combination of site-directed and random mutagenesis (93; T. Lund, personal communication). The aim is to find and remove epitopes on the surface of hCG that crossreact with those on LH. This experimental work prompted the application of the PATCH software to the problem. The hCG

Table 2 List of substitutions made in the β-subunit of hCG and their position [a,b]

Mutant	Amino-acid substitution	Colour code (see Plate 5b)
1	^{20}Lys – Asn	Turquoise
	^{21}Glu – Arg	
	^{22}Gly – Glu	
2	^{24}Pro – His	Green
	^{25}Val – Tyr	
3	^{68}Arg – Glu	Purple
	^{74}Arg – Ser	
	^{75}Gly – His	
	^{79}Val – His	

[a] The colours in the third column refer to their location as indicated in Plate 5b.
[b] T. Lund, unpublished results

system provided an opportunity to compare a theoretical prediction of structural epitopes with experimentally determined functional epitopes derived from mutagenesis studies.

The prediction algorithm was used to select those patches that were the most hydrophilic, planar, accessible, and protruding on the surface of the β-subunit of hCG (βhCG). These were the criteria that matched interfaces in known antibody–protein complexes (63). A number of overlapping patches were selected that included two main segments, i.e. residues 19–27 and 70–78 (Plate 5a); these were the putative epitope sites. Panels of monoclonal antibodies (mAbs) have been used to identify epitope sites on the surface of the βhCG (93). Site-specific mutations in βhCG have been found to affect the binding of some monoclonal antibodies (T. Lund, unpublished results), and these include the three substitutions shown in Table 2 and Plate 5b. It is clearly evident that the putative structural epitopes from the prediction overlap with the functional epitopes identified from the mutagenesis studies.

7. Conclusions

Protein–protein associations are fundamental to so many biological processes that much research, both experimental and computational, has been conducted in an effort to understand their complexities. This has revealed that protein–protein associations share common characteristics. A protein-binding site is, in general, hydrophobic, planar, and complementary, and has the potential to form electrostatic interactions that confer specificity. However, the importance of each characteristic varies and is dependent upon the system in question.

The protein associations analysed and predicted by the methods discussed here are relatively simple systems. There are a vast number of important, highly complex protein associations, such as the ribosome and molecular chaperones that are both essential for the production of correctly folded proteins. The components of the ribosome must provide highly specific binding sites for the growing peptide chain (94). Molecular chaperones, such as GroEL (95), not only exhibit complex protein–protein interactions within their oligomeric conformations, but must also provide binding sites for partially folded proteins. Computational tools designed for interface analysis will aid the understanding of the recognition process as more new and complex protein associations emerge. The implications of a better understanding of these interactions for the design of new drugs and environmental products are immediately apparent.

References

1. Bernstein, F. C., Koetzle, T. F., Williams, G. J. B., Meyer, E. F., Brice, M. D., Rodgers, J. R., Kennard, O., Shimanouchi, T., and Tasumi, M. (1977). The Protein Data Bank: a computer-based archival file for macromolecular structures. *J. Mol. Biol.*, **112**, 535.
2. Abola, E. E., Bernstein, F. C., Bryant, S. H., Koetzle, T. F., and Weng, J. (1987). Protein Data Bank. In *Crystallographic databases—information content, software systems, scientific applications*, (ed. F. H. Allen, G. Bergerhoff, and R. Sievers), p. 107.
3. Argos, P. (1988). An investigation of protein subunit and domain interfaces. *Protein Eng.*, **2**, 101.
4. Janin, J., Miller, S., and Chotia, C. (1988). Surface, subunit interfaces and interior of oligomeric proteins. *J. Mol. Biol.*, **204**, 155.
5. Jones, S. and Thornton, J. M. (1996). The principles of protein–protein interactions. *Proc. Natl. Acad. Sci. USA*, **193**, 13.
6. Jones, S. and Thornton, J. M. (1995). Protein–protein interactions: a review of protein dimer structures. *Progr. Biophys. Mol. Biol.*, **63**, 31.
7. Bode, W. and Huber, R. (1992). Natural protein proteinase inhibitors and their interaction with proteinases. *Eur. J. Biochem.*, **204**, 433.
8. Bode, W., Papamokos, E., and Musil, D. (1987). The high resolution X-ray crystal structure of the complex formed between subtilisin Carlsberg and Eglin C, an elastase inhibitor from the leech *Hirudo medicinalis. Eur. J. Biochem.*, **166**, 673.
9. Dauter, Z. and Betzel, C. (1991). Complex between the subtilisin from a mesophilic bacterium and the leech inhibitor Eglin-C. *Acta Crystallog.*, **B47**, 707.
10. Takeuchi, Y., Satow, Y., Nakamura, K. T., and Mitsui, Y. (1991). Refined structure of the complex of subtilisin BPN′ and Streptomyces subtilisin inhibitor at 1.9 Å resolution. *J. Mol. Biol.*, **221**, 309.
11. Fischmann, T. O., Bently, G. A., Bhat, T. N., Boulot, G., Mariuzza, R. A., Phillips, S. E. V., Tello, D., and Poljak, R. J. (1991). Crystallographic refinement of the 3-dimensional structure of the FAB D13 lysozyme complex at 2.5 Angstroms. *J. Biol. Chem.*, **266**, 12915.
12. Padlan, E. A., Solverton, E. W., Sheriff, S., Cohen, G. H., Smith-Gill, A. J., and Davies, D. R. (1989). Structure of an antibody-antigen complex. Crystal structure of the HY/HEL10 FAB lysozyme complex. *Proc. Natl. Acad. Sci. USA*, **86**, 5938.
13. Sheriff, S., Silverton, E. W., Padlan, E. A., Cohen, G. H., Smith-Gill, S. J., Finzel, B. C., and Davies, D. R. (1987). Three-dimensional structure of an antibody-antigen complex. *Proc. Natl. Acad. Sci. USA*, **84**, 8075.
14. Wilson, I. A. and Stanfield, R. L. (1994). Antibody-antigen interactions: new structures and new conformational changes. *Curr. Opin. Struct. Biol.*, **4**, 857.

15. Herron, J. N., He, X. M., Ballard, D. W., Blier, P. R., Pace, P. E., Bothwell, A. L. M., Voss, E. W., and Edmundson, A. B. (1991). An autoantibody to single-standard DNA: comparison of the three-dimensional structures of the unliganded Fab and a deoxynucleotide-Fab complex. *Proteins*, **11**, 159.
16. Rini, J. M., Schulze-Gahmen, U., and Wilson, I. A. (1992). Structural evidence for induced fit as a mechanism for antibody-antigen recognition. *Science*, **255**, 959.
17. Friedman, A. R., Roberts, V. A., and Tainer, J. A. (1994). Predicting molecular interactions and inducible complementarity: fragment docking of Fab complexes. *Proteins*, **20**, 15.
18. de Vos, A. M., Ultsch, M., and Kossiakoff, A. A. (1992). Human growth hormone and extracellular domain of its receptor; crystal structure of the complex. *Science*, **255**, 306.
19. Baldwin, E. T., Weber, I. T., Charles, R., Xuan, J., Appella, E., Yamada, M., Matsuishima, K., Edwards, B. F. P., Clore, G. M., Gronenborn, A. M., and Wlodawer, A. (1991). Crystal structure of interleukin 8: synbiosis of NMR and crystallography. *Proc. Natl. Acad. Sci. USA*, **88**, 502.
20. Clore, M. G., Appella, E., Yamada, M., Matsushima, K., and Gronenborn, A. M. (1990). Three dimensional structure of interleukin 8 in solution. *Biochemistry*, **29**, 1689.
21. Navia, M. A., Fitzerald, P. M. D., Mckeever, B. M., Leu, C. T., Heimback, J. C., Herber, W. K., Sigal, I. S., Darke, P. L., and Springer, J. P. (1989). Three-dimensional structure of aspartyl protease from human immunodeficiency virus HIV-1. *Nature*, **337**, 615.
22. Lapthorn, A. J., Harris, D. C., Littlejohn, A., Lustbader, J. W., Canfield, R. E., Machin, K. J., Morgan, F. J., and Isaacs, N. W. (1994). Crystal structure of human chorionic gonadotropin. *Nature*, **369**, 455.
23. Orengo, C. A., Michie, A. D., Jones, S., Jones, D. T., Swindells, M. B., and Thornton, J. M. (1997). CATH—a hierarchic classification of protein domain structures. *Structure*, **5**, 1093.
24. Wang, J., Smerdon, S. J., Jager, J., Kohlstaedt, L. A., Rice, P. A., Friedman, J. M., and Steitz, T. A. (1994). Structural basis of asymmetry in the human immunodeficiency virus type 1 reverse transcriptase heterodimer. *Proc. Natl. Acad. Sci. USA*, **91**, 7242.
25. Laskowski, M. and Kato, I. (1980). Protein inhibitors of proteinases. *Annu. Rev. Biochem.*, **49**, 593.
26. Schutt, C. E., Myslik, J. C., Rozycki, M. D., Goonesekere, N. C. W., and Lindberg, U. (1993). The structure of crystalline profilin-—actin. *Nature*, **365**, 810.
27. Southwick, F. S. and Young, C. L. (1990). The actin released from profilin actin complexes is insufficient to account for the increase in F-actin in chemoattractant-stimulated polymorphonuclear leukocytes. *J. Cell. Biol.*, **110**, 1965.
28. Pelletier, H. and Kraut, J. (1992). Crystal structure of a complex between electron transfer partners, cytochrome c peroxidase and cytochrome c. *Science*, **258**, 1748.
29. Janin, J. (1997). Specific versus non-specific contacts in protein crystals. *Nature Struct. Biol.*, **4**, 973.
30. Dasgupta, S., Iyer, G. H., Bryant, S. H., Lawrence, C. E., and Bell, J. A. (1997). Extent and nature of contacts between protein molecules in crystal lattices and between subunits of protein oligomers. *Proteins*, **28**, 494.
31. Janin, J. and Rodier, F. (1995). Protein–protein interactions at crystal contacts. *Proteins*, **23**, 580.
32. Carugo, O. and Argos, P. (1997). Protein–protein crystal-packing contacts. *Protein Sci.*, **6**, 2261.
33. Eisenberg, D. and McLachlan, A. D. (1986). Solvation energy in protein folding and binding. *Nature*, **319**, 199.

34. Walsh, L. L. (1994). Navigating the Brookhaven Protein Data-Bank. *CABIOS*, **10**, 551.
35. Duquerroy, S., Cherfils, J., and Janin, J. (1991). Protein–protein interaction: an analysis by computer simulation. In *Protein conformation* (Ciba Foundation Symposium) p. 237, John Wiley, Chichester.
36. Janin, J. (1995). Elusive affinities. *Proteins*, **21**, 30.
37. Janin, J. (1995). Protein–protein recognition. *Progr. Biophys. Mol. Biol.*, **64**, 145.
38. Janin, J. (1996). Quantifying biological specificity: the statistical mechanics of molecular recognition. *Proteins*, **25**, 438.
39. Janin, J. (1997). The kinetics of protein–protein recognition. *Proteins*, **28**, 153.
40. Chothia, C. and Janin, J. (1975). Principles of protein–protein recognition. *Nature*, **256**, 705.
41. Janin, J. and Chothia, C. (1990). The structure of protein–protein recognition sites. *J. Biol. Chem.*, **265**, 16027.
42. Lee, B. and Richards, F. M. (1971). The interpretation of protein structures: estimation of static accessibility. *J. Mol. Biol.*, **55**, 379.
43. Richards, F. M. (1977). Areas, volumes, packing, and protein structure. *Annu. Rev. Biophys, Bioeng.*, **6**, 151.
44. Kauzmann, W. (1959). Some factors in the interpretation of protein denaturation. *Adv. Protein Chem.*, **14**, 1.
45. Dill, K. A. (1990). Dominant forces in protein folding. *Biochemistry*, **29**, 7133.
46. Privalov, P. L. and Gill, S. J. (1988). Stability of protein structure and hydrophobic interactions. *Adv. Protein Chem.*, **39**, 191.
47. Lesser, G. J. and Rose, G. D. (1990). Hydrophobicity of amino acid subgroups in proteins. *Proteins*, **8**, 6.
48. Murphy, K. P., Privalov, P. L., and Gill, S. (1990). Common features of protein unfolding and dissolution of hydrophobic compounds. *Science*, **247**, 559.
49. Dill, K. A. (1990). The meaning of hydrophobicity. *Science*, **29**, 1990.
50. Muller, N. (1993). Hydrophobicity of stability for a family of model proteins. *Biopolymers*, **33**, 1185.
51. Sharp, K. A., Nicholls, A., Friedman, R., and Honig, B. (1991). Extracting hydrophobic free energies from experimental data: relationship to protein folding and theoretical models. *Biochemistry*, **30**, 9686.
52. Nicholls, A., Sharp, K. A., and Honig, B. (1991). Protein folding and association: insights from the interfacial and thermodynamic properties of hydrocarbons. *Proteins*, **11**, 281.
53. Jackson, R. M. and Sternberg, M. J. E. (1994). Application of scaled particle theory to model the hydrophobic effect: implications for molecular association and protein stability. *Protein Eng.*, **7**, 371.
54. Fersht, A. R., Shi, J., Knill-Jones, J., Lowe, D. M., Wilkinson, A. J., Blow, D. M., Brick, P., Carter, P., Waye, M. M. Y., and Winder, G. (1985). Hydrogen bonding and biological specificity analysed by protein engineering. *Nature*, **314**m 235.
55. Fersht, A. R. (1984). Basis of biological specificity. *TIBS*, **9**, 145.
56. Fersht, A. R. (1987). The hydrogen bond in molecular recognition. *TIBS*, **12**, 301.
57. Xu, D., Lin, S. L., and Nussinov, R. (1996). Role of hydrophilic bridges in protein associations—a difference between protein–protein interfaces and the interior of proteins. *Biophys. J.*, **70**, 193.
58. Xu, D., Lin, S. L., and Nussinov, R. (1997). Protein binding versus protein folding: the role of hydrophilic bridges in protein associations. *J. Mol. Biol.*, **284**, 68.
59. Richards, F. M. (1974). The interpretation of protein structures: total volume, group volume distributions and packing density. *J. Mol. Biol.*, **82**, 1.

60. Walls, P. H. and Sternberg, M. J. E. (1992). New algorithm to model protein–protein recognition based on surface complementarity: applications to antibody-antigen docking. *J. Mol. Biol.*, **228**, 277.
61. Lawrence, M. C. and Colman, P. M. (1993). Shape complementarity at protein/protein interfaces. *J. Mol. Biol.*, **234**, 946.
62. Korn, A. P. and Burnett, R. M. (1991). Distribution and complementarity of hydropathy in multisubunit proteins. *Proteins*, **9**, 37.
63. Jones, S. (1996). Protein–protein interactions, University College, London, PhD Thesis.
64. Prestrelski, S. J., Arakawa, T., Duker, K., Kenny, W. C., and Narhi, L. O. (1994). The conformational stability of a non-covalent dimer of a platelet-derived growth factor-B mutant lacking the two cysteines involved in interchain disulphides. *Int. J. Pept. Protein Res*, **44**, 357.
65. Gokhale, R. S., Agarwalla, S., Francis, V. S., Santi, D. V., and Balaram, P. (1994). Thermal stabilization of thymidylate synthase by engineering two disulphide bridges across the dimer interface. *J. Mol. Biol.*, **235**, 89.
66. Hubbard, S. J. and Argos, P. (1994). Cavities and packing at protein interfaces. *Protein Sci.*, **3**, 2194.
67. Laskowski, R. A. (1991). SURFNET, University College, London. (http: www.biochem.ucl.ac.uk/~roman/surfnet/surfnet.html)
68. Miller, S. (1989). The structure of interfaces between subunits of dimeric and tetrameric proteins. *Protein Eng.*, **3**, 77.
69. Slingsby, C., Bateman, O. A., and Simpson, A. (1992). Motifs involved in protein–protein interactions. *Mol. Biol. Rep.*, **17**, 185.
70. McDonald, I. K. and Thornton, J. M. (1994). Satisfying hydrogen bonding potential in proteins. *J. Mol. Biol.*, **238**, 777.
71. Danziger, D. J. and Dean, P. M. (1989). Automated site-directed drug design: a general algorithm for knowledge acquisition about hydrogen-bonding regions at protein surfaces. *Proc. Roy. Soc. London B*, **236**, 101.
72. Danziger, D. J. and Dean, P. M. (1989). Automated site-directed drug design: the prediction and observation of ligand point positions at hydrogen-bonding regions on protein surfaces. *proc. R. Soc. London B*, **236**, 115.
73. Vakser, I. A. and Aflalo, C. (1994). Hydrophobic docking: a proposed enhancement to molecular recognition techniques. *Proteins*, **20**, 320.
74. Young, L., Jernigan, R. L., and Covell, D. G. (1994). A role for surface hydrophobicity in protein–protein recognition. *Protein Sci.*, **3**, 717.
75. Lijnzaad, P. and Argos, P. (1997). Hydrophobic patches on protein subunit interfaces: characteristics and prediction. *Proteins*, **28**, 333.
76. Jones, S. and Thornton, J. M. (1997). Analysis of protein–protein interaction sites using surface patches. *J. Mol. Biol.*, **272**, 121.
77. Jones, S. and Thornton, J. M. (1997). Prediction of protein–protein interaction sites using patch analysis. *J. Mol. Biol.*, **272**, 133.
78. Eisenberg, D. (1984). Three-dimensional structure of membrane and surface proteins. *Annu. Rev. Biochem.*, **53**, 595.
79. Lijnzaad, P., Berendsen, H. J. C., and Argos, P. (1996). A method for detecting hydrophobic patches on protein surfaces. *Proteins*, **26**, 192.
80. Jones, D. T., Taylor, W. R., and Thornton, J. M. (1992). A new approach to protein fold recognition. *Nature*, **358**, 86.
81. Fauchere, J. L. and Pliska, V. (1983). Hydrophobic parameters π of amino acid side chains from the partitioning of *N*-acetyl-amino-acid amides. *Eur. J. Med. Chem.*, **18**, 369.

82. Dodson, E. J., Dodson, G. G., Hubbard, R. E., Moody, P. C. E., Turkenburg, J., Whittingham, J. Xiao, B., Brange, J., Kaarsholm, N., and Thorgersen, H. (1993). Insulin assembly: its modification by protein engineering and ligand binding. *Philos. Trans. R. Soc. Lond.: Series A*, **345**, 153.
83. Brange, J., Ribel, U., Hansen, J. F., Dodson, G., Hansen, M. T., Havelund, S., Melberg, S. G., Norris, F., Norris, K., Snel, L., Sorensen, A. R., and Voight, H. O. (1988). Monomeric insulins obtained by protein engineering and their medical implications. *Nature*, **333**, 679.
84. Kohl, N. E., Emini, E. A., Schleif, W. A., Davis, L. J., Heimbach, J. C., Dixon, R. A., Scolnick, E. M., and Sigal, I. S. (1988). Active human immunodeficiency virus protease is required for viral infectivity. *Proc. Natl. Acad. Sci. USA* **92**, 2484.
85. Huff, J. R. (1991). HIV protease—a novel chemotherapeutic target for AIDS. *J. Med. Chem.*, **34**, 2305.
86. Wlodawer, A., Miller, M., Jaskolski, M., Sathyanarayana, B. K., Baldwin, E., Weber, I. T., Selk, L. M., Clawson, L., Schneider, J., and Kent, S. B. H. (1989). Conserved folding in retroviral proteases: crystal structure of a synthetic HIV-1 Protease. *Science*, **245**, 616.
87. Blundell, T. L. and Pearl, P. (1989). A second front against AIDS. *Nature*, **337**, 596.
88. Kempf, D. J., Marsh, K. C., Denissen, J. F., McDonald, E., Vasavanonda, S., Flentge, C. A., Green, B. E., Fino, L., Park, C. H., Kong, X., Wideburg, N. E., Saldivar, A., Ruiz, L., Kati, W. M., Sham, H. L., Robins, T., Stewart, K. D., Hsu, A., Plattner, J. J., Leonard, J. M., and Norbeck, D. W. (1995). ABT-538 is a potent inhibitor of human immunodeficiency virus protease and has high oral bioavailability in humans. *Proc. Natl. Acad. Sci. USA*, **92**, 2484.
89. Reich, S. H., Melnick, M., Davies, J. F., Appelt, K., Lewis, K. K., Fuhry, M. A., Pino, M., Trippe, A. J., Nguyen, D., Dawson, H., Wu, B. W., Musick, L., Kosa, M., Kahil, D., Webber, S., Gehlhaar, D. K., Andrada, D., and Shetty, B. (1995). Protein structure-based design of potent orally bioavailable, nonpeptide inhibitors of human immunodeficiency virus protease. *Proc. Natl. Acad. Sci. USA*, **92**, 3298.
90. Vondrasek, J. and Wlodawer, A. (1997). Database of HIV proteinase structures. *TIBS*, **22**, 183.
91. Pierce, J. G. and Parsons, T. F. (1981). Glycoprotein hormones: structure and function. *Annu. Rev. Biochem.*, **50**, 465.
92. Wu, H., Lustbader, J. W., Liu, Y., Canfiled, R. E., and Hendrickson, W. A. (1994). Structure of human chorionic gonandotropin at 2.6Å resolution from MAD analysis of the selenomethionyl protein. *Structure*, **2**, 545.
93. Jackson, A. M., Klonisch, T., Lapthorn, A. J., Berger, P., Isaacs, N. W., Delves, P. J., Lund, T., and Roitt, I. M. (1996). Identification and selective destruction of shared epitopes in human chorianic gonadotropin beta subunit. *J. Reprod. Immunology*, **31**, 21–36.
94. Zimmermann, R. A. (1995). Ins and outs of the ribosome. *Nature*, **376**, 391.
95. Hartl, F., Hlodan, R., and Langer, T. (1994). Molecular chaperones in protein folding: the art of avoiding sticky situations. *TIBS*, **19**, 20.

3 Protein–protein complexes formed by electron transfer proteins

F. SCOTT MATHEWS, A. GRANT MAUK, and GEOFFREY R. MOORE

1. Introduction

Transfer of an electron from one protein to another is one of the most common biochemical reactions (1). Not only are such reactions a primary feature of respiratory chains, such as the aerobic pathways terminating with the reduction of O_2 to H_2O by cytochrome *c* oxidase, but they are also critical to many other metabolic processes. Two examples of such processes are the conversion of non-functional methaemoglobin to deoxyhaemoglobin by cytochrome b_5 and cytochrome b_5 reductase (2), and the complex organic biotransformations catalysed by cytochrome P450 (3). The widespread occurrence of protein–protein electron transfer reactions and their apparent simplicity has provided an attractive subject for extensive theoretical and experimental studies. A major objective of many such studies concerns the structural definition of the electron transfer process and identification of those mechanistic factors that are rate-limiting.

In this chapter, we review several intermolecular electron transfer systems to elucidate the role that protein–protein recognition plays in their function. The systems we have chosen are not only well characterized from a structural perspective, but they also illustrate various general aspects of the role of protein–protein recognition in biological electron transfer reactions, and the application of different techniques in defining protein–protein complexes. Furthermore, each system is part of a physiological process and not just a set of proteins that react together *in vitro*. Class I cytochromes *c* occupy a particularly important position in this field, so before turning to general aspects of electron transfer and protein–protein interactions, we provide a brief description of this class of proteins.

Class I cytochromes *c* (Fig. 1) occur in all eukaryotes and in many prokaryotes. They possess a covalently bound haem group with the haem iron axially coordinated to a histidyl residue near the N-terminus and to a methionyl residue near the C-terminus (4–6). Covalent attachment of the haem macrocycle to the polypeptide

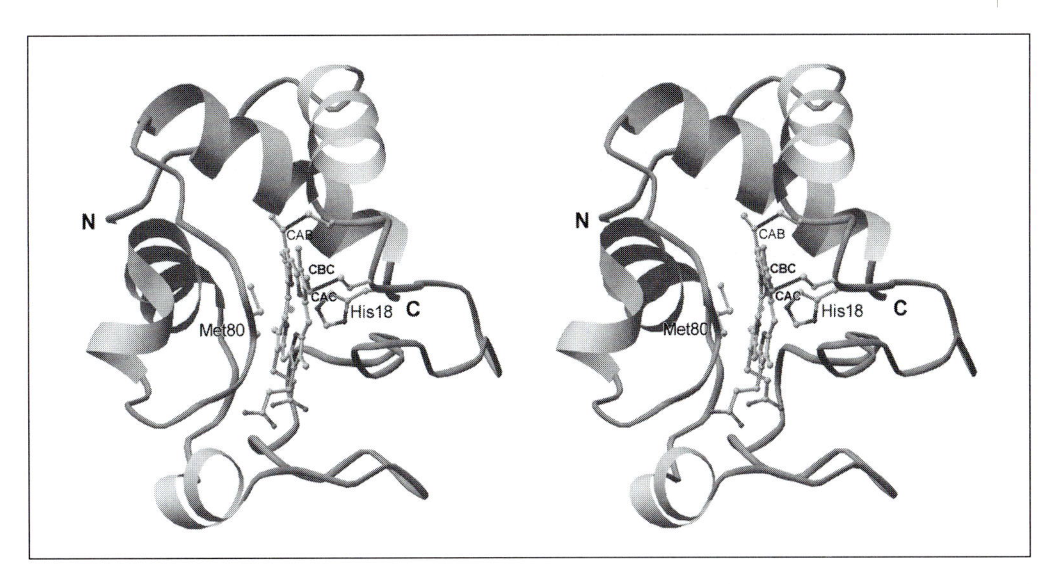

Fig. 1 Stereo diagram of cytochrome *c* from *Saccharomyces cerevisiae*. The α-helices characteristic of eukaryotic cytochromes are indicated by coils. The covalent attachment sites on the haem group, CAB and CAC as well as the solvent-exposed methylene carbon atom, CBC, are indicated.

chain is achieved by two thioether bonds that involve the vinyl methylene atoms, CAB and CAC, and two cysteine side chains. In fact, the presence of a haem group bound covalently to the cysteine side chains of a Cys–X–X–Cys–His motif is the *sine qua non* of a *c*-type cytochrome. Most of the haem group is in a hydrophobic environment, often with one or both propionic acid groups shielded from the solvent but with the CBC methyl group exposed to the solvent (7, 8 and references therein). The protein secondary structure is simple, and usually consists of three or four α-helices and no β-strands or sheets. The first helix is located before the histidine ligand, and the last helix occurs after the methionine ligand. With the exception of *Tetrahymena* cytochrome *c*, mitochondrial cytochromes *c* are very basic and have several positively charged side chains clustered around the exposed edge of the haem. Bacterial class-I cytochromes *c* are often neutral or acidic, but in some cases they also maintain a patch of positive side chains surrounding the edge of the haem that is partially exposed to solvent (6). Several structural families of cytochromes *c* have been identified including the cytochromes *f* discussed below.

2. Factors influencing bimolecular rates of electron transfer

2.1 Background

The minimal reaction scheme for a bimolecular electron transfer reaction between an oxidised electron acceptor (A_o) and reduced donor (B_r) to produce the reduced acceptor (A_r) and oxidised donor (B_o) involves three main stages (eqns 1–5) (9): formation of a bimolecular precursor complex (eqn 1), followed by electron transfer (eqn 3) and dissociation of the bimolecular product complex (eqn 5).

Association:	$A_o + B_r$	$\rightarrow$	$(A_o\text{---}B_r)$;	[1]
Equalization of energy levels:	$(A_o\text{---}B_r)$	$\rightarrow$	$(A_o\text{---}B_r)^*$;	[2]
Electron transfer:	$(A_o\text{---}B_r)^*$	$\rightarrow$	$(A_r\text{---}B_o)^*$;	[3]
Relaxation:	$(A_r\text{---}B_o)^*$	$\rightarrow$	$(A_r\text{---}B_o)$;	[4]
Dissociation:	$(A_r\text{---}B_o)$	$\rightarrow$	$A_r + B_o$	[5]

For the actual transfer of an electron (reaction 3) to occur isoenergetically the bimolecular collision complex must be activated, and this process gives rise to the activation and relaxation steps 2 and 4. The nature of the activated complex is an important determinant of the rate of electron transfer, but a number of factors may be rate-determining. These include the following:

1. *The reactant rearrangement energy*: this may be electronic or conformational in origin and not restricted to events at the electron transfer centres.
2. *The thermodynamic driving force for the reaction*: this is usually the difference between the reduction potentials of the electron acceptor and donor couples.
3. *Electron transmission within the activated complex*: this depends upon the overlap between the donor and acceptor electronic wave-functions and is determined by the electronic properties of the donor and acceptor centres, the medium between them, and their separation.

The influence of these factors on the rate of electron transfer is generally considered within the framework of the theory pioneered by Marcus and others (9–13 and references therein). However, in many systems the overriding factors determining the rate of electron transfer between A_o and B_r are the interaction work terms governing the protein association step, reaction 1 (6, 9, 14, 15).

A key feature of many electron transfer systems is that electrons are passed along a chain of proteins before being delivered to a reaction centre where the chemical transformation of a substrate occurs. Thus, the electron donors and acceptors for some electron transfer proteins are two different proteins. However, sites of electron entry to, and exit from, the redox centres of most proteins that have been adequately characterized appear to be close together, so that the donor and acceptor protein interaction surfaces overlap. This characteristic introduces a feature not present for many of the other systems described in this book: the attractive forces for productive complex formation must be sufficient to allow interprotein electron transfer to occur but not great enough to impede prompt dissociation of the product complex.

2.2 The protein association step

The importance of the protein association step is reflected in the overall rate equation:

$$\text{rate} = k\,[A_o][B_r]; \qquad [6]$$

where:

$$k = K_a \times k_{et} \qquad [7]$$

and:

$$K_a = k_{on}/k_{off} \qquad [8]$$

In these relationships, k_{et} is the rate constant for unimolecular electron transfer within the bimolecular complex (A_o---B_r), K_a is the association constant for the (A_o---B_r) complex, k_{on} is the rate constant for formation of the (A_o---B_r) complex, and k_{off} is the rate constant for dissociation of the (A_o---B_r) complex.

Self-exchange reactions, those in which the oxidant and reductant are different oxidation states of the same protein, illustrate the significance of the protein–protein association step. With increasing ionic strength, self-exchange rate constants generally increase as the electrostatic repulsion between A_o and B_r is diminished (6, 16–18) (Fig. 2). Perhaps even more striking is the pH-dependence of the rate of electron self-exchange, which for horse cytochrome *c* reaches a maximum close to the isoelectric point of 10.4 (Fig. 3). What is surprising about this observation is that ferricytochrome *c* undergoes an oxidation state-linked conformational change above pH ~8 (6, 19), which produces a significant rearrangement energy that will affect k_{et}. However, the effect of pH on K_a overwhelms the effect of the conformation change on k_{et}. Cross-exchange reactions also illustrate the importance of the protein–protein association step, as observed in the ionic strength-dependence of the bimolecular rate constant for the reduction of ferricytochrome *c* by ferrocytochrome b_5 (20, 21) (Fig. 4). In this case, increasing ionic strength reduces the favourable electrostatic interactions that help to bring the positively charged cytochrome *c* and negatively charged cytochrome b_5 together, and so the rate of reaction decreases.

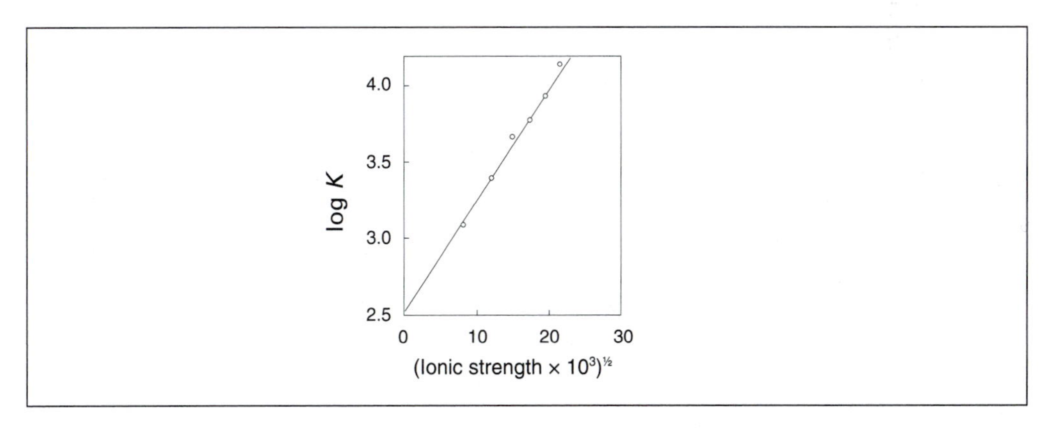

Fig. 2 Ionic strength-dependence of the electron transfer self-exchange rate of cytochrome *c* at pH 7 (see Gupta *et al.*, ref. 16).

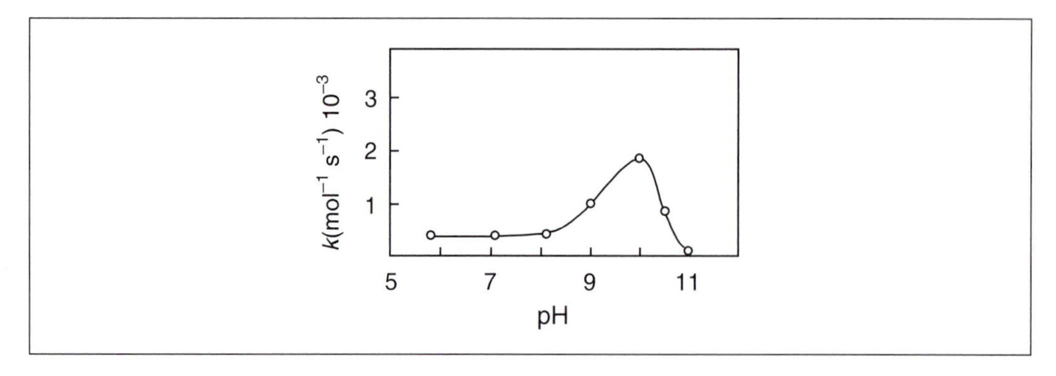

Fig. 3 pH-dependence of the electron transfer self-exchange rate of cytochrome *c* in 50 mM Hepes buffer (see Gupta *et al.*, ref. 16).

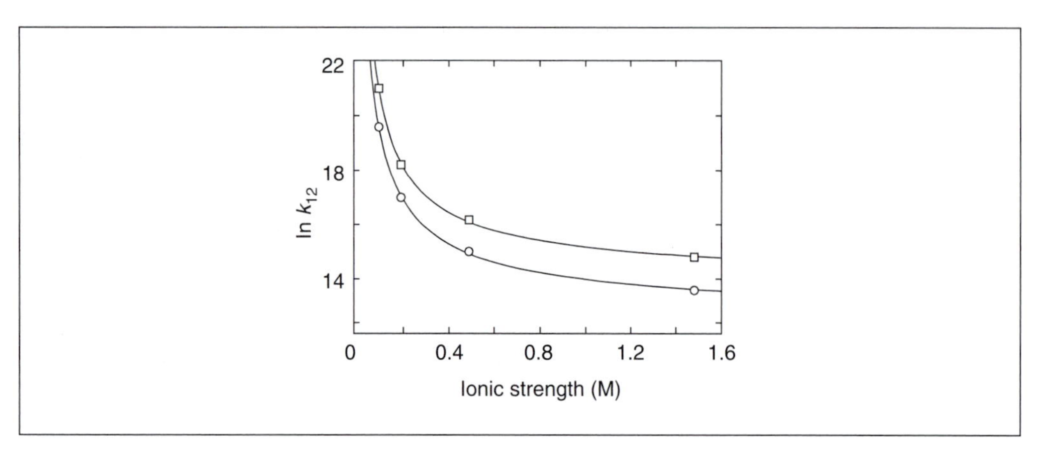

Fig. 4 Dependence of the second-order rate constant for reduction of native (□) and dimethylated ferricytochrome c (○) by tryptic ferrocytochrome b_5 on ionic strength (25 °C, pH 7, sodium phosphate buffer) (see Moore *et al.*, ref. 60).

2.3 Protein–protein association constants

As noted above, the association constant for the (A_o---B_r) complex is a key determinant of the overall second-order rate for a bimolecular electron transfer reaction (eqn 7). In general, binding affinities for complexes of electron transfer proteins are relatively small, and since association rate constants tend to be high it is the high value of the dissociation rate constants, k_{off}, that produce the low K_a values. The factors determining K_a for protein–protein complexes have been discussed by several authors (22–24). A key observation is that complementary binding surfaces with interprotein hydrogen bonds, van der Waals contacts, and salt bridges are needed to obtain high binding affinities. The size of the interface area should therefore be an important parameter, and Janin shows for a sample of 75 structurally characterized protein–protein complexes that each protein contributes at least 550 Å^2 to the interface region (24). However, Janin remarks that there is no simple correlation between the size of the interface surface and the value of K_a, other than noting that the crystallographically defined cytochrome *c*–C*c*P complex (Section 5) has both a small interface and a low K_a. We can add to this that the plastocyanin–cytochrome *f*, plastocyanin–cytochrome *c*, MADH/amicyanin and amicyanin/cytochrome c_{551}, and putidaredoxin–P450$_{cam}$ complexes described in the sections below have interface areas ranging from 425 to 750 Å^2 per protein and all have relatively small values of K_a ranging from 10^3 to 10^6 M^{-1} depending on ionic strength, confirming that the size of the interface surface is not correlated with K_a.

2.4 Oxidation-state dependence of protein–protein association constants

The dependence of K_a on the oxidation state of one of the proteins is a potentially important physiological factor. The redox potential of an electron transfer protein will be perturbed if there is differential binding of the protein in its two oxidation states to another protein (6, 25, 26). This follows from consideration of the equilibria in the square scheme:

$$\begin{array}{ccc} \mathrm{Pr-Ox} & \overset{E_m}{\rightleftharpoons} & \mathrm{Pr-Red} \\ K_a \updownarrow & & \updownarrow K_a^* \\ \mathrm{Ox} & \underset{E_m^*}{\rightleftharpoons} & \mathrm{Red} \end{array}$$

where: Ox and Red represent a redox couple in equilibrium; Pr—Ox and Pr—Red the same redox couple bound to a partner protein; K_a is the association constant for protein Pr binding to Ox and K_a^* for Pr binding to Red; and E_m and E_m^* are the midpoint redox potentials for the Ox/Red couple on its own and bound to protein Pr, respectively. When $K_a \neq K_a^*$, $E_m \neq E_m^*$.

There is an extensive body of literature on this effect as a possible control element for biological electron transfer, but in most characterized systems the difference in redox potential between a bound and unbound redox couple is rather small: for example, the shift in reduction potential of mitochondrial cytochrome *c* bound to its oxidase, reductase, or peroxidase is only 0 to –40 mV (6, 26). However, for some systems the change in redox potential is much larger. For example, the potential of the Fe_2S_2 centre of adrenodoxin is shifted by about –100 mV on binding to its flavoprotein reductase, consistent with reduction of the adrenodoxin causing K_a to be reduced by a factor of about 50, thereby assisting dissociation of the complex formed between adrenodoxin and adrenodoxin reductase (27). This is important because the reduced adrenodoxin has to bind to the steroidogenic cytochrome P450 in order to pass its electron on, but the affinity of adrenodoxin for its reductase is higher than for cytochrome P450. By having excess adrenodoxin present, and with reduced adrenodoxin binding more weakly to the reductase than oxidized adrenodoxin, there is sufficient free reduced adrenodoxin available to sustain cytochrome P450 activity (27).

The midpoint potential of the copper centre of the blue copper protein amicyanin is reduced by 73 mV on binding to methylamine dehydrogenase (MADH) (Section 7), consistent with the oxidized form of amicyanin binding to MADH more tightly than the reduced form (28). The molecular origin of this binding-affinity difference has been shown to be related to a pH-dependent, oxidation state-linked conformation change involving His95 of amicyanin, which is one of the copper ligands (29). The reduction potential of free amicyanin is pH-dependent between pH 6.5 and 8.5, resulting from the protonation of His95 and its dissociation from Cu(I) coordination. However, the potential of MADH-bound amicyanin is pH-independent from pH 6.5 to 8.5, and His95 is prevented from moving away from the Cu(I) coordination sphere by steric constraints at the MADH–amicyanin interface. As far as we are aware, this is the first mechanistic rationale for a protein complexation-induced potential change at the structural level, but such potential variation need not always involve protein conformational changes. Electrostatic effects are capable of producing sizeable perturbations of redox potentials (6, and references therein).

3. Electrostatic and dynamic aspects of redox protein interactions

3.1 Background

During the period 1965–1985 reactions between electron transfer proteins and small, non-physiological electron transfer reagents such as $[Fe(CN)_6]^{3-/4-}$, $[Ru(NH_3)_6]^{2+/3+}$, $[Co(1,10\text{-phenanthroline})_3]^{2+/3+}$ and $[Fe(EDTA)\,(H_2O)_n]^{-/2-}$ were extensively studied (9, 14, 15, 30). An appeal of this approach was that it provided a means of simplifying investigations of electron transfer reactions involving proteins, particularly where the aim was to apply the Marcus cross-reaction equation to the protein reactions. For reactions in which precursor complex formation (reactions 1 and 2) is not kinetically detectable, interpretation of the kinetics was relatively straightforward. In several cases, however, precursor complex formation was kinetically detectable (e.g. ref. 31), but even when it was not there was often considerable difficulty in dealing with the electrostatic terms responsible for formation of the transiently stable precursor complexes (14). In addition, many small molecule reagents bound at more than one site on the protein, so that measured kinetic parameters could have been averages from a number of different complexes (15).

The reaction of eukaryotic cytochrome *c* with $[Fe(CN)_6]^{3-/4-}$ is a prime example of these problems (15, 31, and references therein). With a pI ~ 10, eukaryotic cytochrome *c* has a high positive charge that leads to detectable binding for small anions, including $[Fe(CN)_6]^{3-}$, to at least three binding sites at low ionic strength. All the sites could be active in electron transfer, and, since they are relatively close together, negative co-operative binding interactions could prevent the simultaneous occupation of some or all of the anion-binding sites on the small cytochrome. The possibility that the bound $[Fe(CN)_6]^{3-}$ ions migrated between these sites was also considered on the basis of the nuclear magnetic resonance (NMR) data used to characterize the structures of the complexes formed between cytochrome and the redox-inactive $[Cr(CN)_6]^{3-}$ (18). This consideration led to the suggestion that for electron transfer to occur between a small molecule donor-acceptor and the cytochrome *c* haem, it was only necessary to get the donor-acceptor close to the haem and not to a precise part of the surface. Subsequently, studies of complexes formed by pairs of electron transfer proteins considered mobility within the reactive complex to be important (5, 32–34), and this possibility was discussed in terms of the diffusion of one protein across the surface of another.

From the mid-1980s onwards, the focus of the study of protein electron transfer moved away from weakly associated complexes of proteins and small molecules to the study of multicentre proteins that contain both electron donor and acceptor sites. Such systems do not entail the complexities of the reactant association steps and thus they allow the electron transfer step to be examined directly. According to the analysis by Dutton and colleagues (12), the distance separating the electron donor and acceptor centres appears to be the main determinant of the speed of electron transfer. This conclusion implies that for maximal rates of electron transfer to be obtained in interprotein reactions the donor and acceptor sites should be brought as close together as possible. However, sizeable rates are obtained for systems where the donor and acceptor groups are not in van der Waals contact.

3.2 Binding surfaces of redox proteins

Protein–protein interactions occur at the surfaces of proteins, so to understand protein–protein complex formation it is obviously important to understand the characteristics of these surfaces. However, the dynamic properties of protein surfaces make a precise definition of their structure more complex than a definition of the internal protein structure. Indeed, as many surface residues may experience a number of nearly isoenergetic conformations, representation of the protein surface as a unique structure is an oversimplification. Moreover, structures derived from X-ray diffraction analysis may be distorted at the surface by crystal packing effects, and structures derived from NMR spectroscopy may be poorly defined at the surface by the lack of sufficient experimental structural constraints.

Again, eukaryotic cytochrome *c* is instructive. Most attention concerning the surface of this protein has been focused on the lysyl residues because these are the residues that are responsible for interaction of the cytochrome with the physiological oxido-reductases (5, 15, 36, 37, and references therein). The lysines are distributed in a highly asymmetrical manner with the majority clustered on one side of the protein near the exposed edge of the haem group. This abundance of lysyl residues relative to carboxylate groups on this side of cytochrome *c* endows the protein with a large dipole moment, the positive end of which is near the probable electron entry site (36, 38). Bearing in mind that the lysine side chain from α-C to ε-N is about 10 Å in length when fully extended and that the chain can have considerable flexibility, structures derived from X-ray diffraction analysis do not always define the positions of the amine groups. Only 7 of the 16 lysines of tuna cytochrome *c* are crystallographically well defined (7, 15); of the remainder, some have blurred images in the electron density maps and some have different positions in what should be identical X-ray structures. Of the seven well-defined lysines, four are involved in interactions with other residues that presumably reduce their conformational flexibility. Apart from anything else, the flexibility of the lysines will lead to a time-dependent variation in the electrostatic field surrounding the cytochrome. However, Northrup (34) has investigated the role of lysine side-chain flexibility in reactions of cytochrome *c* through Brownian dynamics simulations of the diffusional association of this cytochrome with reaction partners, and found that side-chain flexibility alone makes little difference compared to a rigid-body calculation where the positions of the lysine charges are fixed. Other modelling studies, using molecular dynamics calculations to simulate side-chain flexibility, indicate that the altered dielectric environment of titratable groups caused by side-chain flexibility may result in significant perturbation of the electrostatic character of protein surfaces with which small molecules and macromolecules interact (39, 40). Using simulations of Brownian dynamics to direct the initial approach of the interacting proteins in conjunction with molecular dynamics calculations and energy minimization of the resulting encounter complexes may capture the dynamics of complex assembly and its electrostatic consequences in a computationally economical fashion.

3.3 Electrostatic steering and preorientation of interacting proteins

The preorientation of reaction partners, as a significant factor in bimolecular electron transfer reactions, was proposed by Matthew *et al.* (41) to account for the ionic strength-dependence of the non-physiological reaction between bacterial flavodoxin and mitochondrial cytochrome *c*. These authors suggested that electrostatic interactions between the proteins preorient the molecules before they achieve physical contact, thus increasing k_{on} and K_a. The importance of this effect can be gauged from the observation that the haem group of cytochrome *c* occupies just 0.6% of the molecular surface of the protein. If the exposed edge of the haem represents the electron injection site, as seems probable, and interprotein electron transfer reactions of cytochrome *c* proceed at a rate close to the diffusion controlled rate, electrostatic interactions must play a role in aiding the formation of the optimal reactive complex. Koppenol and Margoliash (36) had already demonstrated that the asymmetrical charge distribution of cytochrome *c* results in a large dipole moment for the protein (> 300 Debye) and that this dipole moment aids the formation of the reactive complex, so the preorientation model seemed plausible. However, following on from studies of protein–DNA interactions (42) and other interprotein complexes (24, 43), together with theoretical calculations of the dynamic aspects of protein–protein complexation (39, 43, 44), an alternative mechanism must be considered whereby electrostatic interactions may increase the rate of association of productive complexes; namely, the diffusion of one protein along the surface of another.

3.4 Diffusional dynamics of interacting proteins

As we have seen, migration of bound ions on proteins was considered to be important for the reactions of electron transfer proteins with small inorganic oxidants and reductants, and if diffusion across a surface plays a role in such non-physiological reactions, why not in protein–protein interactions? Northrup and co-workers (34, 45–47) developed a firm foundation for such considerations in calculations that presumed that proteins initially associate in unproductive orientations held together by relatively non-specific electrostatic interactions. The loosely associated proteins could undergo rotational diffusion with respect to each other until a productive orientation was formed or the assembly dissociated. In this approach it is important that the initial non-specific attractive forces are sufficiently strong to allow loose association of proteins without the production of stable, unproductive orientations. The preorientation steering effect described by Matthew *et al.* (41) would then be an additional factor to assist with the formation of the initial complex. Northrup (34, 47) has shown that treating protein association events with Brownian dynamics simulations allows the association rate constants to be computed under a variety of conditions. Protein–protein association kinetics are also addressed by Janin in Chapter 1 of this volume.

Historically, the reduction of ferricytochrome *c* by ferrocytochrome b_5 has attracted considerable attention (48), partly because the transient complex formed by these two proteins prior to electron transfer was the first for which even a hypothetical structural model was proposed (49). This model has been supported by subsequent experimental data (32, 48, 50). The reaction of these two proteins was the first to be studied by a combination of experimental and Brownian dynamics simulation. In this work, the ionic strength-dependence of the second-order rate constant for the reduction of ferricytochrome *c* by ferrocytochrome b_5 was adequately simulated with a simple model, in which each protein is represented as a sphere with a reactive surface patch and an asymmetrical surface-charge distribution represented by an appropriately aligned dipole moment (20). Moreover, this simple approach was sufficient to simulate the pH-dependence of the rate constant and the kinetic consequences of esterifying the haem propionate groups of cytochrome b_5 (50), which are believed to be involved in the interaction of the protein with cytochrome *c* (49). On the other hand, the experimentally observed temperature-dependence for the reaction was not simulated adequately by any spherical model. This result was interpreted to indicate that the electron transfer event is primarily responsible for the temperature-dependence rather than the protein association.

Subsequent Brownian dynamics simulations of the reduction of ferricytochrome *c* variants, bearing altered surface charges, by ferrocytochrome employed a procedure that replaced the spherical models for the proteins with models that account for:

(1) the topography of the protein surface as represented by the crystallographically determined structures of the interacting proteins;

(2) the screening influence of diffusible ions in the solvent; and

(3) the discontinuity of the dielectric constant across the surfaces of the proteins.

Analysis of a few thousand docking trajectories calculated in this manner predicted which surface residues of the two proteins are most likely to be involved in stabilizing the encounter complexes formed upon initial contact between the two proteins. Interestingly, the energetically most stable complex identified in these simulations was an orientation of the cytochrome *c*–cytochrome b_5 complex that had not been recognized previously. Moreover, the second most favoured complex identified by these simulations was, in fact, the model that Salemme proposed (49) solely on the basis of interactive inspection of the electrostatic potential surfaces of the two proteins. These simulations revealed that several orientations of the cytochrome b_5–cytochrome *c* complex appear to have reasonable stability. As a previous study of the pH-dependence of cytochrome b_5–cytochrome *c* complex formation suggested that the distribution between complexes of similar stability may vary with pH (50), it would be interesting to extend the Brownian dynamics simulations for this system as a function of pH. Regardless of this however, simulations of this type are useful, not least because they provide a predictably systematic means of identifying energetically feasible encounter complexes that might not be readily apparent simply on the basis of visual inspection.

Further experimental data suggesting that reactive complexes have dynamic structures come from a variety of sources. There are many cases in which electron transfer within protein complexes has been found to be slow at very low ionic strength (I) and to increase as I is raised, before falling off sharply as I is increased further, as shown above for the cytochrome *c*–b_5 reaction (Fig. 4). For example, electron transfer from cytochrome *c* to cytochrome *c* peroxidase (Section 5) is slower at I = 8 mM than at I = 275 mM (51), and reaction between plastocyanin and cytochrome *f* (Section 6) is slower at I = 5 mM than at I = 40 mM (52). Though these low ionic strengths are not physiologically relevant, these studies do indicate that the complex of greatest thermodynamic stability is not necessarily the best for electron transfer. A final example comes from work on reactions of cytochrome *c* bound to phosvitin, a highly negatively charged protein that can bind ~ 20 molecules of cytochrome *c* at pH 7 via electrostatic interactions. The rate of electron transfer of phosvitin-bound cytochrome *c* with reagents such as $[Fe(CN)_6]^{3-}$ were found to be strongly inhibited, while reaction with reagents such as $[Co(1,10\text{-phenanthroline})_3]^{2+}$ were facilitated (53). These observations point to the strong influence of electrostatic interactions involving the phosvitin template affecting the second-order cytochrome reactions. On the other hand, the self-exchange reaction of cytochrome *c* was considerably enhanced (~ 100-fold) by binding to the phosvitin (54). As the positive surface of cytochrome *c* that binds to phosvitin is also the interaction surface for electron self-exchange, the cytochrome must reorient on the phosvitin surface to allow reaction to occur. The decrease in the dimensionality of diffusion (55), from the three-dimensional diffusion of uncomplexed cytochrome in bulk solution to the two-dimensional diffusion of cytochrome *c* across the phosvitin surface, presumably accounts for the enhancement of rate induced by this binding.

4. Characterization of electron transfer complexes

4.1 Background

No experimental structures for (A_o---B_r) complexes for any pair of electron transfer proteins have been reported. Not surprisingly, unimolecular electron transfer processes are much too fast to permit the determination of such structures. In fact, the structures of only a small number of product complexes or non-productive complexes, such as those formed by two proteins in the same oxidation state, have been reported (56, and references therein). Because intermolecular electron transfer complexes are generally only weakly associated, the crystallization of such complexes is a difficult process and has suffered a low rate of success. The isolation of such complexes requires identification of those conditions where the complex is stabilized. If the association is stabilized by the interaction of oppositely charged side chains, then a low dielectric medium such as polyethylene glycol might favour crystal growth. If complex formation is driven by hydrophobic interactions of complementary surfaces, then high concentrations of an electrolyte might stabilize the complex and lead to crystallization. If a mixture of hydrophobic and charged interactions is present, then the best medium to favour crystallization might be some unpredictable combination of these precipitants.

4.2 Structural information from NMR spectroscopy

The use of NMR spectroscopy to study complex formation is, in principle, more straightforward than X-ray crystallography. This is because the components can be studied in the solution state and the need to identify optimal crystallization conditions is, therefore, eliminated. However, complexation-induced changes in NMR chemical shifts do not provide unambiguous information regarding interprotein interaction surfaces, as they are highly sensitive to the local magnetic environments of nuclei and may be altered when two proteins form a complex. This alteration may occur because the nuclei being detected are located at one of the interacting surface regions or because the environments of nuclei not located at the interprotein interface region are altered by a conformational or electronic change induced by complex formation. Firm structure determination requires the measurement of intermolecular nuclear Overhauser enhancements (NOEs). However, intermolecular NOEs are rarely observed for weakly associated complexes of the type commonly formed by electron transfer proteins, limiting the information gained from NMR to the less-useful, complexation-induced, chemical shift perturbations. We are not aware of a complex formed by electron transfer proteins that has been characterized using intermolecular NOEs.

The importance of lysines in electrostatically driven interprotein interactions of the type prevalent for electron transfer proteins has led to the development of various chemical modification procedures to assist their detection by NMR in interprotein complexes (e.g. 57–60). N^{ε}, N^{ε}-dimethylation appears to be one of the best modifications since this maintains the positive charge on the derivatized lysyl residues and provides each derivatized residue with two methyl groups, the resonances of which can be observed in 2D ^{1}H–^{13}C correlation spectra. Perturbations to the derivatized lysine methyl resonances resulting from complex formation can then be used to identify intermolecular binding surfaces (57–60). Functional assays such as those used in the study of N^{ε}, N^{ε}-dimethylated cytochrome *c* and cytochrome b_5 (60) indicate that N^{ε}, N^{ε}-dimethylation of lysines is a relatively benign modification.

4.3 Physical characterization of electron transfer complexes

Additional methods used to characterize complexes formed by electron transfer proteins, described below, have generally provided information concerning the stoichiometry and stability of complex formation. In some cases, information has been derived regarding the changes in protein solvation that accompany complex formation or estimates of the number of electrostatic interactions that stabilize complex formation. Related studies of the interaction between proteins, in which specific amino-acid residues have been modified by chemical modification or site-directed mutagenesis, provide information regarding the residues on the surfaces of the proteins that are involved in complex formation.

4.3.1 Electronic difference spectroscopy

Measurements of electronic difference spectra of complex formation by haem proteins involves monitoring a small change in the electronic spectrum of one or both of the interacting proteins induced by complex formation (50, 61–64). The change in absorbance is determined as the ratio of one protein to the other is increased. Unless complex formation involves significant structural changes, such as a change in spin-state of one or both proteins (61), the change in absorption observed is often just 2–3%. The absorbance change probably results from a change in the dielectric environment of the edge of the haem group that is accessible to solvent in the unbound protein. The region where the haem is exposed to solvent is usually in the area involved in binding other proteins, so the environment of the haem edge is changed when the complex forms. Presumably, the change in spectrum is small because the total surface of the haem that is exposed to solvent in the monomeric protein is relatively small. Since these experiments usually involve the measurement of a small absorbance change over a large background absorbance, the precision of the result is low, and it is generally impossible to study systems that form complexes with either a high or low affinity. Use of circular dichroism and magnetic circular dichroism spectroscopy to monitor the formation of complexes formed by electron transfer proteins has been used sparingly (65) and merits further evaluation.

4.3.2 Fluorescence quenching and energy transfer

One approach to overcoming the limited sensitivity and precision of electronic difference spectroscopy to monitor complex formation has been the use of fluorescence quenching titrations. In general, the intrinsic fluorescence of haem proteins is efficiently quenched by the haem group, so fluorescent derivatives are prepared by removal of the haem iron (66–68, and references therein), replacement of the haem iron with zinc (e.g. 69, 70), or the use of fluorescent reporter groups by modification of surface cysteinyl residues (71). In principle, each of these modifications has the potential to influence the nature of complex formation to varying degrees, but there is no reason at present to question the results of such studies with the systems described herein. More recently, Hoffman and co-workers have developed the basis for fluorescence titrations in multiple dimensions (72) and reported what they refer to as a 'reverse' Stern–Volmer plot.

4.3.3 Hyperbaric spectroscopy

The change in volume that accompanies the formation of a protein–protein complex may be determined by studying the effect of elevated hydrostatic pressure on the stability of the complex. Both electronic (73, 74) and fluorescence emission spectroscopy (70, 71) have been used to monitor the dissociation of electron transfer complexes at elevated pressure. For the reasons discussed above, these measurements have monitored fluorescence quenching. Studies of this type concerning the cytochrome *c*–cytochrome b_5 complex have indicated that the change in partial specific volume that accompanies the formation of this complex ranges from –50 to –100 ml mol^{-1} (68, 73). Formation of a complex with a change in volume of –50 ml mol^{-1}, for example, would be consistent with the formation of about four or five ion pairs at the protein–protein interface, a value that would be consistent with the hypothetical structural model for the complex proposed by Salemme (49). Subsequent work (75) extended these observations to consider the disruption of this complex by both pressure and glycerol, and led to the conclusion that about three water molecules are displaced at the interface of the two proteins during complex formation.

4.3.4 Potentiometric titrations

Stability of complex formation by electron transfer proteins is generally dependent on pH—as was first noted for the cytochrome *c*–cytochrome b_5 complex (63), for which optimal complex stability was observed at a pH that was intermediate between the isoelectric pH values of the two proteins. This dependence arises from the contribution of interactions between charged, titratable surface residues to complex stabilization and the changes in pK_a values of these and other adjacent titratable groups that occur upon complex formation. As developed in detail by Laskowski and Finkenstadt (76), linkage of any binding interaction to pH implies that complex formation involves the uptake or release of protons. This change in proton binding provides a means of characterizing the electrostatics of complex formation.

As demonstrated by analysis of the cytochrome *c*–cytochrome b_5 complex, two complementary types of potentiometric experiment may be employed in the characterization of sufficiently well-behaved systems (77). The first involves a modified pH-stat experiment in which one protein is titrated against the other, and the change in proton binding, H_e^+, as titrant is added to the titrand monitored. The resulting titration curve provides the stoichiometry, q, of proton liberation or uptake that is linked to complex formation and the association constant for complex formation. The dependence of q on pH comprises the change in the titration curves of the proteins in the complex relative to their titration curves in the absence of the other protein. The second experiment involves direct measurement of the complexation-induced titration curve. For this, the titration curve of each of the two proteins in the absence of the other and the titration curve of the complex are determined. Subtraction of the titration curves of the proteins from that of the complex then defines the change in titration curves of the two proteins that occurs upon complex formation. Integration of this difference titration curve provides a continuous plot of the dependence of K_d on pH (76). The integration constant required to place this plot on the ordinate can be obtained by determining the association constant at one or two pH values by titrating one protein against the other, as described above.

Data from a full potentiometric analysis of the type described here should be useful in detailed theoretical analyses that endeavour to simulate the changes in pK_a of titratable groups that accompany complex formation. Initial attempts at simulating the results for the cytochrome *c*–cytochrome b_5 complex indicate that 10–12 titratable groups change pK_a upon complex formation, but the identity of the groups that change pK_a remains unknown (77). The possibility that such a large number of groups change pK_a may provide further evidence for the presence of more than one structure of the complex and a pH-dependent distribution of these structures. Finally, it should be noted that the sensitivity and precision that can be achieved by potentiometric titrations permits the detection of low-affinity binding sites in the presence of high-affinity sites. This capability is exemplified by detection of the low-affinity binding site for cytochrome *c* on the surface of cytochrome *c* peroxidase (78)

4.3.5 Chemical methods

The first, and perhaps most extensively applied, method used to define the surface residues of interacting electron transfer proteins concerned in recognition and binding interactions involved the preparation of specifically modified proteins. In such studies, a protein is modified with a subsaturating amount of a modifying reagent that reacts with a specific type of amino-acid residue. In most examples of such work, lysyl residues are the target of modification, but, in some cases, acidic residues have been modified. Under appropriate reaction conditions, the resulting product is a mixture of proteins, each of which possesses a single, modified amino-acid residue. This mixture is then resolved by ion-exchange chromatography to produce a family of purified, singly modified proteins. The site of modification in each purified component is identified by peptide mapping techniques. As implemented by Margoliash, Millett, and their co-workers (reviewed in ref. 79), the consequences of these modifications on the ability of the modified protein to function in steady-state kinetics assays were used to evaluate the involvement of the modified residues in protein–protein interactions. In some cases, however, complex formation was studied directly.

The technique of differential chemical modification developed by Bosshard and colleagues avoids the need to prepare specifically modified forms of either protein constituent of a complex (35). In this type of experiment, the relative rates of chemical modification of bound and unbound forms of a protein are used to deduce the identities of the residues involved in complex formation. Residues at the interface between the two proteins are presumed to be protected from modification in the complex and so should react more slowly.

The other major chemical method used to study complexes of electron transfer proteins has been chemical cross-linking. Following the initial observation by Seiter *et al.* (80) that carbodiimides should be highly appropriate reagents for converting electrostatic interactions between carboxylate and amino groups on the surfaces of interacting proteins into amide bonds, the water-soluble carbodiimide EDC (1-ethyl-3-(dimethylaminopropyl)carbodiimide) has been used most frequently for this purpose. In most cases, however, the resulting reaction products have not been characterized adequately, but they probably consist of a number of products that differ in the location and number of cross-linking sites. Furthermore, the value of cross-linkers, particularly EDC, for the identification of residues involved in the stabilization of protein–protein complexes involves several assumptions. For example, complex formation is generally optimal at low ionic strength; addition of EDC as the hydrochloride, therefore, involves a rapid increase in ionic strength. Consequently, any protein–protein cross-links resulting from the addition of EDC occur as the interacting proteins are dissociating, or by random collisions of the two dissociated proteins. Although specific sites of EDC-induced cross-linking have rarely been identified, examples of success with this reagent have been reported (e.g. refs 81, 82). A review of the use of carbodiimides in protein–protein cross-linking studies has been published (83), and though other cross-linking reagents have been used less frequently they may merit further consideration (e.g. refs 84, 85).

4.3.6 Other methods

In addition to the methods described above, other techniques have been used with protein complexes that merit comment. Although analytical ultracentrifugation is a classical method of studying weakly interacting macromolecular assemblies, its use in the characterization of electron transfer proteins has been limited. As a result, the one study of which we are aware concerned the cytochrome *c*–cytochrome *c* peroxidase complex (86). Similarly, only restricted use has been made of infrared spectroscopy in the study of electron transfer protein complexes. Hollaway and Mantsch used Fourier transform infrared-difference spectroscopy of the cytochrome *c*–cytochrome b_5 complex to detect complexation-induced changes in intensity of a region of the spectrum they assigned to carboxylate groups (87). They attributed these changes to perturbation of the environments of three acidic residues on the surface of cytochrome b_5 that are involved in interaction with cytochrome *c*. Finally, a novel form of surface plasmon resonance has been used to study the binding of plastocyanin to the membrane-bound cytochrome b_6f complex (88). The advantage of this method derives from the use of a waveguide for signal detection to increase sensitivity and precision of the data acquired, and the fact that the immobilized species, cytochrome b_6f, is bound to a membrane structure. This eliminates the need to immobilize one protein through covalent attachment to a solid matrix, avoiding a fundamental difficulty in using this method to obtain thermodynamic information.

5. The cytochrome *c*–cytochrome *c* peroxidase reaction

5.1 Background

Cytochrome *c* peroxidase (C*c*P) is a haemoprotein (34 kDa) found in yeast mitochondria. It is composed of a single polypeptide and one haem prosthetic group, which catalyses the reduction of alkyl hydroperoxides, after which the enzyme is regenerated through the oxidation of ferrocytochrome *c* (89, 90). The initial product of the 2-electron oxidation of C*c*P by peroxide is an intermediate referred to as 'compound I'. Compound I possesses an oxyferryl iron species ($Fe^{IV}=O$) and a tryptophan pi-cation radical, both of which are subsequently reduced by the ferrocytochrome in two 1-electron-transfer steps. Reduction of compound I is most efficiently achieved *in vitro* by yeast cytochrome *c*, but horse-heart cytochrome *c* is also able to reduce this intermediate, albeit at a somewhat slower rate.

C*c*P is a highly helical, two-domain protein (91). Domain-1 contains the first 144 and the last 40 residues, while domain-2 contains the intervening 109 residues. The haem group is located in a cavity between the two domains, with a helix from the second part of domain-1 forming the back wall of the cavity. The haem moiety, including its two propionic acid groups, is almost totally buried. The haem iron is 5-coordinate, with the N_ε atom of His175 providing the fifth ligand. The plane of the indole ring of Trp191, the site of the free radical intermediate of compound I (92), is in van der Waals contact with both groups.

5.2 X-ray crystallographic structures of C*c*P–cytochrome *c* complexes

Structures of C*c*P in complex with both yeast cytochrome *c* (CY) (the C*c*P-CY complex) and horse-heart cytochrome *c* (CH) (the C*c*P–CH complex) have been determined by X-ray crystallography (Plate 6) (93). Although only the former involves proteins that interact physiologically, both complexes have been thoroughly studied in solution. The C*c*P-CY crystals were grown at 150 mM NaCl (with polyethylene glycol as the precipitant) while the C*c*P-CH crystals contained 5 mM NaCl (with 2-methyl-2,4-pentanediol as precipitant). Interestingly, earlier attempts to crystallize the complex formed by the two yeast proteins produced crystals in which only the electron density of the peroxidase could be identified (94), while the crystals that provided useful diffraction information required partial dehydration to diffract properly (93). The interactions between the electron transfer partners are globally similar in the two structures.

The interface between C*c*P and CY is largely hydrophobic and has a surface area of about 650 Å^2. C*c*P interacts with CY through three salt bridges (Asp34–Lys87, Asp34–Arg13, and Glu290–Lys73; 4.0–4.2 Å separation) and two hydrogen bonds (Glu290–Asn70 and Glu32 carbonyl–Lys87; 3.1–3.2 Å) (Plate 6a). The substituent of the CY haem group that is nearest C*c*P in the complex is the methyl group on pyrrole ring C; this methyl group is in van der Waals contact with Ala193 and Ala194 of C*c*P. The planes of two haem groups are inclined by about 60° to each other, and the iron-to-iron distance is about 26 Å. A predominantly covalent pathway connecting the haem groups of C*c*P and CY through peroxidase residues Trp191 (the free radical site), Gly192, Ala193, and Ala194 has been suggested as a possible course of electron transfer (93).

The structure of the C*c*P-CH complex differs from that of the C*c*P-CY complex through a slight translation and rotation of the CH molecule relative to the C*c*P molecule (Plate 6a). The interface of the C*c*P-CH complex is less hydrophobic than that of the C*c*P-CY complex and involves a smaller area (~550 Å^2). The C*c*P-CH complex is stabilized by three salt bridges (Asp34–Lys13, Glu35–Lys87, and Glu290–Lys72; 3.2–4.3 Å separation) and two hydrogen bonds (Asn38–Lys8 and Asn95–Gln12; 3.0–3.3 Å separation). The remaining interactions are hydrophobic, although the interface contains a number of additional acidic and basic residues. All the hydrogen-bonding interactions between C*c*P and CH are located at one edge of the interface. At the other edge, the methyl of pyrrole ring C is about 7 Å from Ala194 of C*c*P, and the iron-to-iron distance in the complex is about 30 Å.

5.3 Is the interface observed in the crystals catalytically relevant?

While the crystallographic studies provide static structures for the C*c*P–cytochrome *c* complexes within a crystalline lattice, the existence of alternative, more dynamic, structures has been a topic of continuing debate (46, 60, 94), with much evidence pointing to the formation of 2:1 cytochrome *c*-C*c*P complexes in a wide range of solution conditions (72, 78, 95, 96). Association constants for the binding of cytochrome *c* to the two binding sites differ by two to three orders of magnitude, depending on the oxidation state of the cytochrome and the solution conditions. For example, at pH 6 the ferrocytochrome binds to C*c*P with K_a values of 1.8 ($\pm$ 0.3) $\times$ 10^5 M^{-1} and 1.0 ($\pm$ 0.2) $\times$ 10^3 M^{-1} (I = 0.1 M, KNO_3, 25 °C) (78). Further indication of the complexity of the solution-state interaction comes from energy transfer measurements (72, 96) and ^{1}H NMR studies of the shielding of amide ^{1}H–^{2}H exchange of cytochrome *c* by complex formation with C*c*P (97, 98). These results suggest the existence of several closely related, dynamically interconverting binding conformers, consistent with Brownian dynamics simulations (46) and with electron transfer kinetic measurements (95).

In this context of multiple, interconverting binding conformers, NMR findings with N^{ε}, N^{ε}-dimethylated-CH and N^{ε}, N^{ε}-dimethylated-CY (Dim-CH and Dim-CY) binding to C*c*P are revealing, in that they support a model in which there is more than one binding orientation with some interfacial movement of the two proteins within the complex (60). Significantly, NMR studies provide evidence that the chemical environments of at least seven lysyl residues on the surface of Dim-CH cytochrome *c* are affected by complexation with C*c*P, while at least six lysyl residues of Dim-CY are perturbed by binding to C*c*P. The crystal structures of the cytochrome *c*–C*c*P complexes (93) indicate interactions of just four horse-cytochrome lysyl residues (8, 13, 72, 87) with C*c*P and the involvement of only two lysines (73, 87) of the yeast protein; Lys86 of the yeast cytochrome is involved in an additional intermolecular van der Waals interaction. Neither interaction scheme is fully consistent with the NMR data. Brownian dynamics simulations, indicating that several sites may be visited during each encounter between the proteins, identified six surface lysyl residues of horse cytochrome *c* as being those most frequently involved in ionic contacts with C*c*P: i.e. lysines 13, 25, 27, 72, 79, and 86 (46). All three of the NMR signals tentatively assigned to lysines 13, 72, and 86 of CH were affected by the binding interaction, and the NMR signals of trimethyllysine 72 and dimethyllysine 79 of CY indicated that these two groups were involved in intermolecular contacts. Overall, the CY–C*c*P NMR data are more in keeping with the theoretical predictions of Northrup *et al.* (46) for the CH–C*c*P complex than with the crystallographic structure of the CY–C*c*P complex.

5.4 Electron transfer within the C*c*P–cytochrome *c* complex

Important mechanistic questions, such as whether there is more than one electron transfer site on C*c*P for the cytochrome *c* and whether the electron transfer route from cytochrome *c* to $Fe^{IV}{=}O$ is the same as to the Trp191 radical, have been addressed by a combination of mutagenesis and biophysical experiments. Stopped-flow kinetic experiments have shown that at high ionic strength (100 mM or greater) reduction of the Trp191 radical of compound I occurs first in a rapid, bimolecular reaction with cytochrome *c* (99–101). This process is followed by a second bimolecular electron transfer reaction with cytochrome *c* at a rate 5- to 10-fold slower than the first (102). Mutation of C*c*P residue Asp34 to Asn, Glu290 to Gln, or Ala193 to Phe reduces both the fast and the slow electron transfer rates from CY or CH by 2–4-fold. Both the carboxylates form salt bridges to cytochrome *c*, and Ala193 is in contact with the haem of cytochrome *c* in the X-ray structure of the C*c*P–CY complex. Substitutions of C*c*P residues that are outside the interface or that are within the interface but not directly linked to cytochrome *c* have little or no effect on the electron transfer kinetics. As the response of both the faster and slower electron transfer rates to the mutational changes and to the variation in ionic strength are similar, it is likely that under these conditions the binding site on C*c*P for cytochrome *c* is the same for both 1-electron reduction steps and that it corresponds to the binding sites observed in the crystal complexes.

At low ionic strength (<100 mM), a second binding site for cytochrome *c* on the surface of the peroxidase can be detected that exhibits an ~1000-fold lower affinity than for the first binding site (72, 95). Under these conditions, stopped-flow kinetic studies (101) indicate that the iron(IV) species in compound I is first reduced to the Fe(III)/Trp* form. Studies of photoinduced electron transfer kinetics involving metal-substituted C*c*P or cytochrome *c* (72, 102) at low ionic strength using a redox-inactive cytochrome as a competitive inhibitor, indicate that the low-affinity site for cytochrome *c* accommodates up to 1000-fold higher electron transfer rates than the high-affinity site. Thus, it may be that a second, crystallographically undetected, independent, low-affinity site for cytochrome *c* binding exists at low ionic strength and that this site is much more efficient in transferring an electron to the iron site of C*c*P than to the radical site. Although this low-affinity binding site for the cytochrome has now been detected by a variety of techniques (see above), the high efficiency of this site in electron transfer reactivity has been observed only in flash photolysis experiments involving metal-substituted forms of the cytochrome or the peroxidase. As a result, some have suggested that this observation is not relevant to the reaction between the two native proteins. However, mapping of the anticipated efficiency of electronic coupling of the haem centre to the surface of the peroxidase with the algorithm developed by Beratan and Onuchic and colleagues (72, 103), predicts the presence of two distinct regions on the peroxidase surface that should be efficient in electron transfer (72). One of these regions corresponds well to the primary docking site for cytochrome *c* identified by the Brownian dynamics simulations of Northrup and co-workers, and the other corresponds to the secondary site predicted by these simulations (46).

6. The plastocyanin–cytochrome *f* reaction

6.1 Background

Plastocyanin is a soluble cupredoxin (104) located in the thylakoid lumen of chloroplasts. During the dark phase of photosynthesis, plastocyanin accepts an electron from the membrane-bound cytochrome b_6f complex of photosystem II and transfers it to pigment P700+ of photosystem I (105). The single copper atom, which cycles between the cupric and cuprous states during the cupredoxin electron transfer cycle, is coordinated by the side chains of two histidines, a cysteine, and a methionine in a distorted tetrahedral geometry. The secondary structure of plastocyanin consists of 8 β strands that form two twisted β-sheets to surround a hydrophobic interior. The copper ion is held in a pocket formed by three loops at one end of the structure with ligands contributed from both β-sheets. Although the copper ion is shielded from solvent, one of the histidine ligand residues is on the surface of the protein and is surrounded by hydrophobic residues. This hydrophobic surface patch has been implicated as a major binding site for electron transfer partners (106). Two anionic surface patches have also been identified. One of these negative patches is located near Tyr83, adjacent to copper ligand Cys84, and the other is on the opposite side of the molecule. These patches each contain four carboxylate groups: Asp42, Glu43, Asp44, and Asp51 and Glu59, Glu60, Asp61, and Glu68, respectively. Using site-directed mutagenesis, the negative patch near Tyr83 has been implicated as a binding site for other electron transfer proteins (107).

The b_6f complex consists of four major protein components: cytochrome *f* (31 kDa); the 2Fe-2S Rieske protein (20 kDa); cytochrome b_6 (23 kDa); and subunit-IV (18 kDa). Of these major components, only subunit-IV lacks a cofactor (108). The plastocyanin reaction partner on the b_6f complex is cytochrome *f*, a *c*-type cytochrome which is anchored to the membrane by a C-terminal, transmembrane helical segment. However, an active, water-soluble, haem-containing fragment of cytochrome *f* can be isolated from the complex that is spectroscopically and kinetically identical to the intact cytochrome (108). This haem-containing protein lacks the 33 residues from the C-terminus that anchor the protein to the thylakoid membrane, and has a three-dimensional structure distinctly different from that of mitochondrial cytochrome *c* (109).

The haem-containing portion of cytochrome *f* consists of a large domain (residues 1–186 and 232–247) and a small domain (residues 187–231). The large domain contains two β-sheets arranged in a β-sandwich. The small domain contains a single, 3-stranded, antiparallel β-sheet plus two extended strands that cover the sheet and connect the small domain to the large domain. The haem binding site is located in a crevice between the large and small domains. The haem group is partially covered by two short helices and an intervening loop, residues 1–25, that provide the covalent haem attachment sites, Cys21 and Cys24. In line with other *c*-type cytochromes, the 5th iron ligand is provided by His25. However, the 6th ligand is provided by the free amino group of the N-terminal residue of the protein, Tyr1. The haem environment is essentially hydrophobic. Only the thioether methyl group of pyrrole ring C, which is covalently attached to Cys24, and the two propionic acid groups are exposed to solvent. In contrast to mitochondrial cytochrome *c*, the protein surface surrounding the exposed thioether group is mostly hydrophobic and lacks any local cluster of positively or negatively charged side chains. There is an elongated patch of positively charged residues which extends from the small domain to a β-hairpin that protrudes from the β-sandwich of the large domain. This cationic patch is created by Lys187 and Arg209 of the small domain, and Lys66, Lys65, and Lys58 of the β-hairpin. Cross-linking experiments have implicated an interaction between Lys187 (28 Å from the haem iron) and Asp44 of plastocyanin in stabilizing the interaction between these two proteins in solution (81). However, two experimental studies concerning competence of the complex cross-linked in this fashion to function in electron transfer, however, are in conflict with each other (110, 111).

In solution, native plastocyanin and cytochrome *f* form a complex capable of very high electron transfer rates (112) (2×10^8 M^{-1} s^{-1} at 25 °C, 100 mM ionic strength, pH 6), although the driving force for the reaction is only about 20 mV. Furthermore, the stability of the complex formed by the two proteins is low under these conditions (K_a

~ 10^3 M^{-1}) (88, 112, 113), in part because the stability of complex formation is inversely related to the ionic strength of the solution (114). These observations suggest that the two proteins are oriented by electrostatic interactions to form a complex in which the donor and acceptor sites are optimally aligned for reaction.

6.2 Structure of the plastocyanin–cytochrome *f* complex

Two approaches have been taken to develop hypothetical structural models for the complex formed during electron transfer from cytochrome *f* to plastocyanin. The first involved the use of a Monte-Carlo sampling method combined with molecular dynamics (MD) simulations to identify energetically minimized structures for this complex (114). The MD calculations accounted for potential contributions of protein flexibility to stabilization of the resulting complex(es). The effects of solvent polarization by the proteins deduced from numerical solutions of the Poisson–Boltzmann equation, and coulombic and non-electrostatic contributions to stability were also considered. This work led to the identification of an optimal electron transfer pathway involving plastocyanin residues His87 and Pro86 and a haem propionate group of cytochrome *f*. Another pathway predicted by this method involves a cation–π complex formed by the interaction of plastocyanin residue Tyr83 with Lys65 from cytochrome *f*. The second modelling approach involved the use of Brownian dynamics simulations developed by Northrup and co-workers as described in Section 3 (115). This strategy identified a single dominant complex arising from the interaction of Lys58, Lys65, Lys187, and Arg209 on cytochrome *f* with Glu43, Asp44, Glu59, and Glu60 on plastocyanin. This structure, however, did not support the formation of the cation–π complex, in so far as the two relevant residues were not sufficiently close to permit interaction. In addition to the electrostatic interactions indicated, hydrophobic stabilization of the complex was also predicted. Importantly, these calculations predict a role for the plastocyanin copper ligand His87 in accepting electrons from the haem of cytochrome *f* and predicts a copper-to-iron distance of 16–17 Å.

The Monte-Carlo and Brownian dynamics approaches differ in three principal but related ways:

1. The Monte-Carlo method endeavours to identify the thermodynamically most stable complex, while the Brownian dynamics simulations identify so-called encounter complexes.
2. The use of MD simulations in the Monte-Carlo strategy attempts to capture the contribution of flexibility in protein structure to the stability of the resulting complex, while the Brownian dynamics approach assumes rigid-body structures.
3. Because the Monte-Carlo approach is designed to identify the most stable complex, it does not simulate diffusion of the proteins toward each other as does the Brownian dynamics method.

Thus, the Monte-Carlo approach cannot be used to predict reaction rates. Moreover, in view of the uncertain role of the thermodynamically most stable complex in efficient electron transfer reactions (see above), it might be argued that the Monte-Carlo approach is more relevant to the interpretation of measurements of complex stability at equilibrium than to the kinetics of electron transfer. In any case, probably the most interesting result of these studies is that the general features of the models proposed by these quite different strategies are in good agreement, in terms of the general regions of the protein surfaces that are predicted to stabilize complex formation.

Structural analysis of the plastocyanin–cytochrome *f* complex has been carried out by NMR spectroscopy using Cd(II)-substituted plastocyanin (to avoid complications from redox processes and paramagnetic effects of Cu(II)) and oxidized and reduced cytochrome *f* (116). The plastocyanin was enriched in ^{15}N to render the cytochrome *f* invisible to ^{1}H–^{15}N correlation NMR experiments. Differences in chemical shifts of the fully bound plastocyanin arising from the oxidation-state change of the haem were ascribed to through-space paramagnetic shifts, and used as input data for the structure calculations. This interpretation assumes that the relative positions of the plastocyanin ^{1}H and ^{15}N nuclei with respect to the haem are not dependent on the haem oxidation state. The structural model derived from these experiments was constrained so that the acidic patch close to Tyr83 would interact with the elongated positive patch on the surface of cytochrome *f*. This constraint was probably a major influence on the resulting structural model for the complex. Restrained structure refinement for the complex yielded 10 similar structures, with an overall average root-mean-square (rms) deviation between C^{α} atoms of 1 Å.

The final model for the complex (Plate 7) predicts that the Tyr83-containing acidic patch of plastocyanin makes contact with the small domain of cytochrome *f* and that the hydrophobic patch of plastocyanin is in contact with the interdomain region of cytochrome *f* that contains the haem group. The total surface area of plastocyanin in contact with cytochrome *f* is about 500 Å^{2}. Asp42, Glu43, and Asp44, which form one of the acidic patches of plastocyanin, interact with Arg209, Lys187, and Lys185 on the small domain of cytochrome *f*. Glu59 and Glu60 of the second plastocyanin acidic patch interact with Lys65 and Lys58 of the β-hairpin of the large domain of cytochrome *f*; and the copper ligand, His87, is in van der Waals contact with Tyr1 (the amino-terminus of which is one of the iron ligands), and with Phe4 next to Tyr1. The distance from the iron to the copper in this complex is about 11 Å, and the shortest path for electron transfer is predicted to be from His87 NE2 of plastocyanin to Tyr1 CD2 of cytochrome *f*, a distance of about 3 Å. The eight-residue hydrophobic patch on plastocyanin surrounding His87 described above is in close contact with cytochrome *f*.

The NMR model of the plastocyanin–cytochrome *f* complex indicates that a single structure is present. Whether this is a true reflection of the solution behaviour of the protein or a consequence of the structure-determining procedure employed remains to be established. Similar structure calculations, employing NMR through-space paramagnetic shifts for the complex of ^{15}N-labelled ferricytochrome b_5 and unlabelled ferricytochrome *c*, have led to a well-defined orientation of cytochrome b_5 relative to cytochrome *c*. However, in this case, chemical shift data of the type used to obtain the structure of the plastocyanin–cytochrome *f* complex could not be fitted to a single structure, so the final product represents a dynamic average between a family of structures (117). Similarly, both NMR (118) and modelling studies (119) of the reaction between plastocyanin and mitochondrial cytochrome *c* have indicated that the interacting proteins adopt more than one orientation in the complex.

Whatever the precise character of the dynamics of the plastocyanin–cytochrome *f* structure, the orientation found for the complexed proteins does appear to explain many of the mechanistic observations made regarding electron transfer reaction within this complex. However, it is in contrast with kinetic and cross-linking data which suggest that the formation of the productive complex is a two-step process. This discrepancy can be explained (116) if the first step consists of a purely electrostatic attraction of the two proteins that results in a loose mixture of similar conformations of roughly similar energy, and if the second step is a rearrangement of the molecules to juxtapose the haem and copper sites while maintaining a large fraction of the electrostatic interactions of the initial complex. That is to say, plastocyanin and cytochrome *f* follow the kind of model put forward by Northrup (Section 3.3).

7. The methylamine-dehydrogenase electron transfer system

7.1 Background

The oxidation of methylamine in methylotrophic bacteria is coupled to the reduction of cytochrome aa_3 oxidase within the plasma membrane by a soluble electron transport chain. The electron transfer process is initiated by the quinoenzyme methylamine dehydrogenase (MADH), which catalyses the oxidative deamination of methylamine and the incorporation of a water molecule to yield formaldehyde, ammonia, two protons, and two electrons. The structural biology and kinetics of this process have been well studied with proteins isolated from *Paracoccus denitrificans*. In this organism, the electron acceptor from MADH is amicyanin, a cupredoxin (120). The next acceptor along the chain is an inducible class-I cytochrome *c*, called cytochrome c_{551i} (121). The probable sequence of electron transfer between proteins (122) is:

$$\text{MADH} \rightarrow \text{Amicyanin} \rightarrow \text{Cyt } c_{551i} \rightarrow \text{Cyt } c_{550} \rightarrow \text{Cyt } aa_3;$$

where Cyt c_{550} is a constitutive mitochondrial-like periplasmic class-I cytochrome *c*.

MADH is a heterotetramer of two identical pairs of heavy (H)- and light (L)-subunits of 47-kDa and 15-kDa molecular mass, respectively. The redox cofactor for MADH, tryptophan tryptophylquinone (TTQ, Fig. 5), is located on the L-subunits. It is derived from a pair of tryptophan side chains which become cross-linked during biogenesis; one of them is then oxidized to an orthoquinone (123). The H-subunit of MADH is a β-propeller consisting of seven antiparallel 4-stranded β-sheets (124). The L-subunit consists of five β-strands, which form two antiparallel β-sheets, and is held together by six disulfide bridges and the cross-link of TTQ.

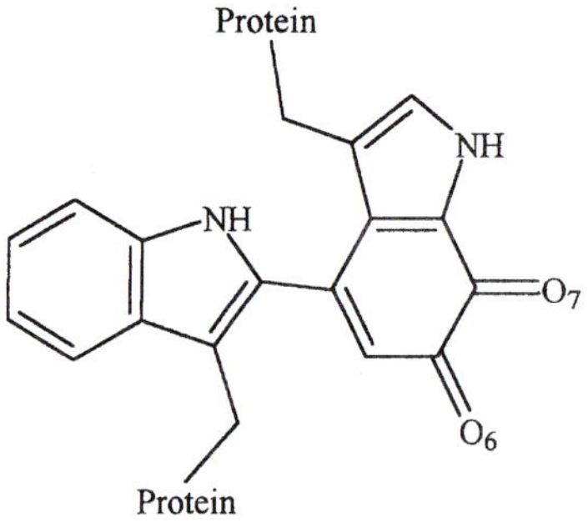

Fig. 5 Chemical structure of the tryptophan tryptophylquinol prosthetic group. The carbonyl oxygen at position 6 is involved in the catalysis of amine oxidation.

During steady-state turnover of MADH with substrate, the reaction follows a ping-pong mechanism with reductive and oxidative half-reactions (125). In the reductive half-reaction methylamine is oxidized to formaldehyde, whose release follows hydrolysis of the imine intermediate, with ammonia remaining bound to position 6 of the reduced TTQ (Fig. 5) in the form of an aminoquinol (126). In the oxidative half-reaction the TTQ is reoxidized in two sequential 1-electron transfer reactions with the formation of an aminosemiquinone intermediate (126). The two-electron redox potential for TTQ is +100 mV (127).

Amicyanin is a one-electron carrier with a redox potential of +294 mV (128) and a molecular mass of about 12.5 kDa (104). It is an obligatory intermediate in the transfer of electrons from methylamine dehydrogenase to cytochrome c_{551i} (127). The protein has a β-sandwich topology (129) with nine β-strands forming two mixed β-sheets. The copper atom is located in a pocket between the β-sheets and has four coordinating ligands supplied by two histidines, one cysteine, and one methionine. Its secondary and tertiary structures closely resemble those of plastocyanin, although it shares only about 20% sequence identity, and has an additional 20 residues at the N-terminus. Similarly to plastocyanin, one of the histidyl ligands to the copper is exposed to solvent and is surrounded by a ring of seven surface-exposed hydrophobic residues.

Cytochrome c_{551i} is highly acidic, containing 27 aspartate and glutamate residues and eight lysine and arginine residues. It has a molecular mass of 17.5 kDa and a redox potential of +190 mV (128). Its gene is a member of the methanol oxidation operon (130) and is also believed to be the primary acceptor of electrons from methanol dehydrogenase. Cytochrome c_{551i} contains five α-helices, the central three of which envelop the haem group and correspond to analogous helices in other prokaryotic *c*-type cytochromes (131). The haem is slightly exposed in the isolated cytochrome and is surrounded by a hydrophobic surface patch. Most of the charged residues are located on the protein surface, but are unevenly distributed over the protein surface, being located mainly on the side and back portions of the molecule, away from the haem group and its hydrophobic surroundings.

7.2 Solution studies of MADH complex formation

When MADH and amicyanin are mixed in equal proportions (with respect to copper and TTQ) the TTQ absorption spectrum is perturbed and the copper centre undergoes a decrease in redox potential, from ~295 mV to ~225 mV (28). This decrease in potential facilitates subsequent electron transfer to cytochrome c_{551i} (E_m = 190 mV). Kinetic (132) and chemical cross-linking (133) studies suggest a role for both electrostatic and hydrophobic interactions in stabilizing complex formation.

Under stopped-flow conditions, oxidation of the aminoquinol and aminosemiquinone forms of MADH by excess amicyanin occurs at about equal rates (50–100 s^{-1} at 30 °C, pH 7.5, 200 μM KCl $K_a \sim 1 \times 10^5$ M^{-1}) (134). A solvent deuterium kinetic-isotope effect for aminoquinol oxidation indicates that electron transfer is kinetically gated by a rate-limiting proton transfer step. The rate is strongly temperature-dependent, reaching ~600 s^{-1} at 40 °C. Analogous studies of MADH containing the *O*-quinol and *O*-semiquinone forms of TTQ show widely different rates of oxidation, with rate constants of ~10 s^{-1} and >1000 s^{-1}, respectively. A small solvent isotope effect for *O*-quinol oxidation indicates that the electron transfer step is rate-limiting. The temperature-dependence of the oxidation of the *O*- and the *N*-quinol forms of MADH also suggests that electron transfer to amicyanin is rate-limiting for the former but not the latter reaction.

Electron transfer from copper to haem has also been measured by stopped-flow kinetics in the MADH/amicyanin/cytochrome c_{551i} system (135). The observed rate of haem reduction is ~75 s^{-1} at 25 °C and is temperature-dependent. Thermodynamic analysis of the temperature-dependence of these rates indicates that electron transfer is rate-limiting. The observed K_a for cytochrome c_{551i} interacting with the MADH-amicyanin binary complex is ~5×10^4 M^{-1} (135).

7.3 Crystal structures of the MADH complexes

Crystals of a binary complex between MADH and amicyanin (136) and of a ternary complex between MADH, amicyanin, and cytochrome c_{551i} (131) were grown under nearly identical conditions using phosphate buffer as the precipitant (~2.4 M, pH 6.5). The two crystal forms are distinct from one another, the former being tetragonal and the latter orthorhombic. However, the amicyanin-MADH interactions are closely similar in the two crystal structures. Also, both complexes will form crystals, which are isomorphous with the native when copper-free apo-amicyanin is substituted for amicyanin.

The arrangement of the cytochrome, amicyanin, and one HL-dimer of MADH is shown in Fig. 6. The TTQ, copper, and haem groups are arranged in a linear fashion, placing the cytochrome and MADH on opposite sides of the amicyanin. The distance from the copper to the redox-active quinone oxygen of TTQ is about 17 Å, while the

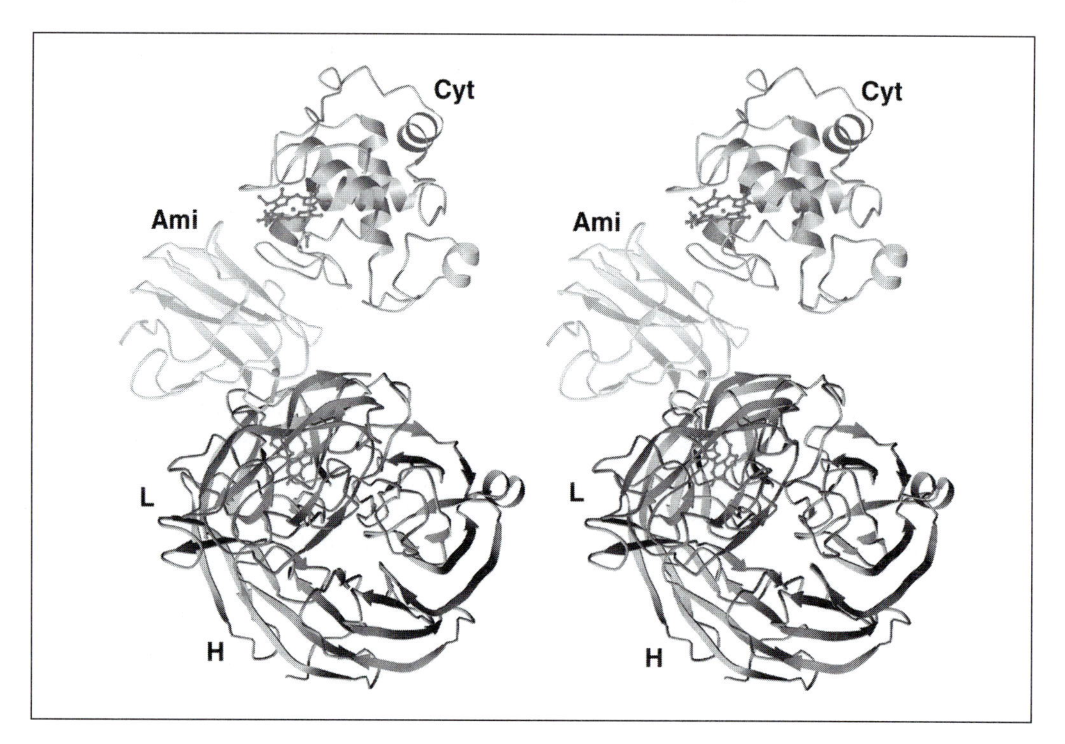

Fig. 6 Global stereo view of the MADH–amicyanin–cytochrome c_{551i} ternary complex. One HLAC heterotetrameric complex is shown. The H-, L-, and cytochrome-subunits are labelled.

distance from copper to the nearest atom of TTQ is 9 Å. The copper-to-iron distance is about 25 Å, and the distance from copper to the nearest atom of the haem is about 21 Å. The MADH/amicyanin interface has a surface area of about 750 $Å^2$ per subunit. About two-thirds of the MADH part of the interface is made up of residues from the L-subunit. On amicyanin, the interface is approximately centred on His95, the exposed copper ligand, and the seven surrounding hydrophobic residues forming the surface patch are largely occluded from solvent. The interaction between MADH and amicyanin is rather hydrophobic, despite the presence of eight charged residues and nine neutral hydrophilic residues in the interface; most of the charged side chains are directed into solution at the edge of the interface. There is one strong salt bridge (Arg99(A)[1]–Asp167(H), 2.7 Å, partially buried) and one weak salt bridge (Lys68(A)–Asp115(L), 4.4 Å, solvent-accessible) connecting MADH to amicyanin (Plate 8). In addition, three water molecules bridge MADH and amicyanin.

The amicyanin/cytochrome c_{551i} interface is smaller than the MADH/amicyanin interface, occupying about 425 $Å^2$ per monomer, and is more polar. Within the interface, 2 of the 10 amicyanin residues and 6 of the 12 cytochrome residues are hydrophobic, resulting in 65% of the interface residues being hydrophilic compared with about 40% in the MADH/amicyanin interface. The number of contacts between amicyanin and the cytochrome is also greater than the number between amicyanin and MADH (Plate 8), with one salt bridge (Lys29(A)[1]–Asp75(C), 2.9 Å), one side chain–side chain hydrogen bond (Asp24(A)–Thr79(C), 3.4 Å) and two main chain–main chain hydrogen bonds (Glu31(A)–Gly72(C), 3.1 Å and Glu34(A)–Lys68(C), 3.4 Å). Two solvent molecules also bridge the two proteins. In addition, there is a potential salt bridge between Glu34 of amicyanin and Lys68 of the cytochrome (5.9 Å) which can be modelled at a closer distance and might become important at lower ionic strength. The remaining charged or neutral groups at the interface point into solution.

Two prominent electron transfer paths (137) connect the copper atom to the indole ring of Trp57. One utilizes Trp108 of TTQ, while the other follows two main-chain residues of the L-subunit. The latter path is about 10-fold more efficient than the former, but depends on the presence of a bridging water molecule, which might not always be present, possibly reducing the relative efficiency of that path. In the case of the electron transfer from copper to iron, the two most prominent paths follow either Cys92 or Met98 (both of which are copper ligands) to a common point, and continue through three residues of amicyanin and two of the cytochrome. Both branches appear to be equally efficient.

7.4 Are the interfaces observed in the crystals physiologically relevant?

The importance of specific side-chain interactions for complex formation between MADH and amicyanin has been tested by mutagenesis experiments (138). This complex had earlier been shown to be stabilized by both electrostatic and hydrophobic forces (132, 133). In the crystalline complex, there is one short salt bridge connecting Arg99 of amicyanin with Asp167 of the H-subunit, and one long salt bridge connecting Lys68 of amicyanin with Asp115 of the L-subunit. To test the contribution of these paths to the electrostatic stability of the MADH–amicyanin complex, Arg99 of amicyanin was mutated to Asp and to Leu and Lys68 was mutated to Ala. Both Arg99 mutations resulted in an approx. 10- to 50-fold decrease in the affinity of amicyanin for MADH, but did not eliminate an inverse dependence on ionic strength. In contrast, the Lys68–Ala mutation had almost no effect on complex stability but did eliminate its dependency on ionic strength.

Amicyanin contains seven hydrophobic groups surrounding His95 which are buried in the interface, including Met71 and Phe97. When Met71 was changed to Arg, there was little effect on the binding affinity of amicyanin for MADH. This surprising result could mean that the long aliphatic side chain of arginine is able to replace the methionine side chain in the hydrophobic interface and still maintain the exposure of the guanidinium group to solvent on the periphery of the interface. Modelling of this mutation lends support to this possibility. Replacement of Phe97 by Glu, on the other hand, leads to an ~100-fold destabilization of the complex, indicating that the glutamic acid side chain disrupts the hydrophobic interface.

The catalytic competence of the crystalline binary and ternary complexes of MADH was analysed by single-crystal, polarized absorption microspectrophotometry (139).

This technique has been useful for probing the redox properties of proteins in the crystalline state (140). Spectra were recorded from crystals of both complexes prepared using either native amicyanin (holo) or copper-free amicyanin (apo). Since the crystals of these holo and apo complexes are isomorphous, they provide an internal control, allowing studies of the reactions of MADH in these crystalline complexes with and without the possibility of electron transfer through the copper atom.

Crystals of the apo-binary complex are reduced fully by methylamine, and the crystalline spectra in both oxidation states are similar to those observed in solution. When methylamine is added to crystals of the holo-binary complex, spectral changes indicate the formation of significant amounts of TTQ semiquinone (Fig. 7). This semiquinone signal can only arise from the transfer of one electron from TTQ to the copper of amicyanin after the TTQ has first been reduced fully to the aminoquinol form by substrate. This demonstrates that MADH in the crystalline holo-binary complex is competent to carry out both catalysis and electron transfer. The amount of semiquinone formed is dependent upon pH. At pH 5.7 a large fraction of the TTQ remains reduced and a significant absorbance by Cu(II) at about 600 nM can be observed (Fig. 7); at pH 9.0 the TTQ is mostly in the semiquinone form and the Cu(II) absorbance disappears. Furthermore, after the removal of excess methylamine, the ratio of the semiquinone to reduced forms of TTQ can be shifted reversibly by altering the pH. This indicates that the pH-dependencies of the redox potential for the semiquinone/reduced TTQ couple and for the Cu^{2+}/Cu^{+} couple are different. In fact, it has been shown that the two-electron redox potential of MADH is pH-dependent, with a slope of –30 mV per pH unit (141), whereas the redox potential of amicyanin within the binary complex in solution is independent of pH (28).

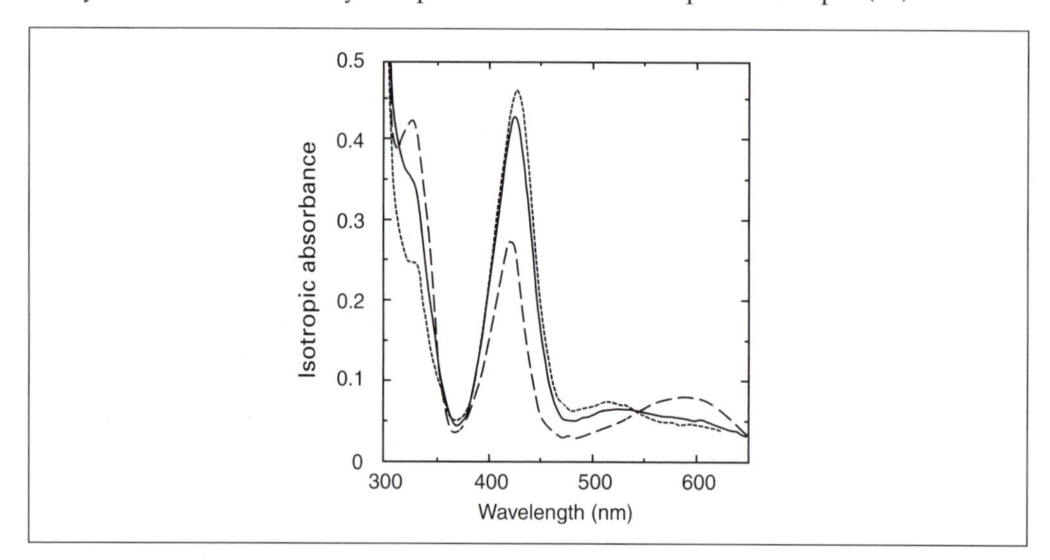

Fig. 7 Isotropic spectra of a single crystal of the MADH–amicyanin binary complex after reduction by 0.2 mM methylamine, observed at pH 5.7 (— — —), pH 7.5 (———), and pH 9.0 (··········). The three absorption maxima, at 330 nM, 420 nM and 590 nM correspond to the reduced TTQ, TTQ semiquinone, and Cu(II), respectively.

Within the holo-ternary complex, haem reduction occurs over about a 60-minute period when crystals are treated with methylamine at pH 7.5. At pH 5.7, very little haem reduction by substrate occurs, even after several days, but at pH 9 reduction is complete in about 30 minutes. This is consistent with the pH-dependency of the driving force for electron transfer observed in crystals of the binary complex described above. In the apo-ternary complex, haem reduction by methylamine can occur, but at about a 10-fold lower rate than in the holo-ternary complex. Since the overall distance from TTQ to haem is the same in the apo- and holo-ternary complexes, and pathways for electron transfer from TTQ to haem are undoubtedly available within the apo-ternary complex, the question arises as to how the presence of copper can enhance the electron transfer rate by an order of magnitude.

8. The putidaredoxin–cytochrome P450 reaction

8.1 Background

Cytochromes P450 (P450) catalyse the oxidation of hydrocarbons (3). Some are relatively non-specific towards the organic substrate, while others such as the camphor-hydroxylating P450 from *Pseudomonas putida* are specific. All P450s use O_2 as the oxidant and all require a supply of electrons, which may start on a reductant such as NADH but gets passed on to the P450 through the action of a short electron-transfer chain. A typical example is:

$$\text{NADH} \rightarrow \text{flavoprotein} \rightarrow \text{iron–sulfur protein} \rightarrow \text{cytochrome P450.}$$

In *P. putida* the iron–sulfur protein is putidaredoxin, a 106-residue Fe_2S_2-containing ferredoxin, and the P450 is a 45-kDa enzyme that catalyses the hydroxylation of camphor, and is thus known as $P450_{cam}$ (3, 142, 143). These form a binary complex with a K_a that is dependent on the oxidation states of the proteins, and on whether the P450 has substrate-bound. For reduced $P450_{cam}$ and oxidized putidaredoxin K_a is in the range from 1 to 10×10^5 M^{-1} (144, 145). The structures of both proteins have been determined: by NMR for putidaredoxin (142) and X-ray crystallography for $P450_{cam}$ (143). The complex between the two has been modelled using molecular dynamics docking procedures (146, 147) and studied using ^1H–^{15}N correlation NMR and ^{15}N-labelled putidaredoxin with unlabelled camphor-bound $P450_{cam}$ (148, 149).

The putidaredoxin structure comprises two domains, with the larger domain built around β-strands from the first ~50 and last ~20 amino acids, and the smaller domain consisting of the remainder (142). The Fe_2S_2 cluster is coordinated by two cysteine residues (numbers 45 and 86) and is contained within the larger domain, displaced to one side of the molecule but buried within the protein and not accessible to bulk solvent. The smaller domain contains two α-helices and three β-strands, with a well-conserved acidic region between residues Glu65 and Glu77. Despite expectations that this acidic patch would play a key role in interprotein interactions, single site-directed mutations suggest otherwise; thus, the Asp58Asn, Glu65Gln, Glu72Gln, and Glu77Gln mutants had almost the same K_M and V_{max} values as the wild-type putidaredoxin in $P450_{cam}$ assays (149).

Camphor-bound cytochrome $P450_{cam}$ consists of two tightly-associated domains: a helix-rich domain and a domain containing an extensive set of β-strands (143). The haem group is located at the edge of the helical domain wedged between two of the large helices of the α-domain, facing the β-domain and completely buried from solvent. The iron atom is 5-coordinate with the proximal fifth ligand position occupied by the S^{γ} atom of a cysteine side chain at the N-terminus of one of the helices. Both haem propionates are surrounded by hydrogen-bond donors, some of which are charged, and, importantly (see below), haem propionate-6 appears to interact with the side chain of Arg112. The camphor is located close to the haem (within 4 Å), on its distal side, in a pocket between the two domains, and interacts with both domains of the protein.

8.2 The putidaredoxin—$P450_{cam}$ complex

Models for the complex formed between putidaredoxin and camphor-bound $P450_{cam}$ have been obtained from MD simulations assuming that the haem iron of the P450 and an Fe of the Fe_2S_2 centre of putidaredoxin are brought as close together as possible (146, 147). An inter-Fe distance of about 12 Å was found in the final complex (Plate 9). The theoretical complexes had salt bridges between Arg79, Arg109, and Arg112 of P450 and the C-terminal carboxylate of Trp106 and side chains of Asp34 and Asp38, respectively, of putidaredoxin. Also identified in the complex were five intermolecular hydrogen bonds. NMR studies of putidaredoxin in the presence of camphor-bound P450 have not been as revealing as the corresponding studies of the plastocyanin complexes (Section 6), because the putidaredoxin resonances have been extensively broadened by the increased tumbling time resulting from complexation. Nevertheless, it is apparent that P450 does bind close to the Fe_2S_2 centre (147, 148). More convincing experimental evidence in support of the modelled complexes comes from mutagenesis studies. These show that the putidaredoxin mutations Asp38Asn (149, 150) and Asp34Asn (150), and the P450 mutations Arg112Met and Arg112Tyr (151), cause a sizeable reduction in activity. Importantly, K_M or K_d values increased, showing that the protein–protein interaction step was affected, as well as overall bimolecular electron transfer rates being reduced. Despite the three proposed intermolecular salt bridges and the electrostatic complementarity of the binding surfaces in the models for the complex, Aoki *et al.* (152) suggest from their calorimetric studies that the binding energy between putidaredoxin and P450 is dominated by van der Waals interactions and hydrogen bonds.

A pathways analysis of the putidaredoxin–$P450_{cam}$ complex suggests that Cys39 and Asp38 of putidaredoxin and Arg112 and a haem propionate of P450 form an appropriate electron transfer route (146). Thus some of the residues involved in intermolecular binding are also suggested to have a role in electron transfer. This should be detectable by mutagenesis experiments. Although the data described above for the Arg112 mutants of P450 (151) have been interpreted (147) in support of the pathways analysis, further work is needed as the Arg112 mutations are accompanied by a sizeable drop in the P450 haem redox potential (151)—perhaps caused by the Arg112/haem propionate-6 interaction being disrupted.

8.3 Putidaredoxin reductase versus cytochrome $P450_{cam}$

As with many electron transfer proteins, putidaredoxin has to interact with a reductase to gain the electrons it will pass on to its oxidase. The structure of the flavoprotein putidaredoxin reductase has not been determined, but there have been extensive studies of its reaction with putidaredoxin. The association-constant for putidaredoxin binding to $P450_{cam}$ is 10 times weaker than for binding to its reductase (152). Also, as with the putidaredoxin–$P450_{cam}$ complex, interprotein hydrogen bonds and van der Waals interactions are proposed to be more important than electrostatic interactions for forming the putidaredoxin–reductase complex. Holden *et al.* (150) report that the interaction sites on putidaredoxin for its two partner proteins are separate but overlapping, with Asp34 and Asp38, which are important for P450 interaction, not important for the reductase. The NMR studies of Aoki *et al.* (148) also indicate that the sites for the reductase and P450 are not identical. This raises the question of whether a ternary complex similar to that seen for the MADH system can be formed.

9. Conclusions

It is apparent that complexes formed by electron transfer proteins exhibit properties that distinguish them from complexes formed by subunits of oligomeric proteins, by antibodies and antigens, or by functional domains within multienzyme complexes, and that these distinguishing characteristics are related to the functional roles of the proteins. Perhaps their most unusual feature is that the form of the complex that exhibits the greatest thermodynamic stability is not necessarily the optimal complex for physiological function. As a result, complexes of electron transfer proteins in solution are labile, and the constituent proteins generally exhibit a dynamic equilibrium between bound and unbound states. Furthermore, in most cases a thermodynamically dominant, and therefore a unique, complex probably does not exist. Instead, 'the complex' consists of a family of similar structures of comparable stability. Complex formation is stabilized by a combination of electrostatic and hydrophobic interactions, but the electrostatic contribution is generally much greater than that exhibited by other types of protein–protein complexes. As a result, not only is complex formation highly dependent on ionic strength but is also dependent on pH. These features reflect the functional role of electron transfer proteins. In the simplest cases, such proteins provide a binding site for a protein that donates an electron and a binding site for a protein that accepts an electron. Particularly for smaller proteins, these sites frequently overlap with each other. Under turnover conditions, electron donor and acceptor proteins must continually dock and release quickly to permit efficient rates of reaction. The resulting reduction in specificity of docking interaction does not compromise the efficiency of electron transfer because there are low stereochemical requirements for such reactions compared to reactions that involve the making or breaking of chemical bonds.

The transient nature of complexes of electron transfer proteins makes their structural characterization highly challenging. While X-ray crystallography remains a powerful method for the investigation of such complexes, successful crystal preparation of complexes formed by electron transfer proteins has been rarely achieved. For those complexes that are stabilized by substantial contributions from electrostatic interactions, the relationship between the crystalline structure and the structure in solution remains unclear owing to uncertainties regarding:

(1) the ionic environment within the crystal and its relationship to the solution state;
(2) the role of crystal lattice forces in dictating the orientation of surface residues;
(3) the possibility that crystallization may select for just one complex conformer from an ensemble of related structures; and
(4) the role of solvent water (including its displacement from protein surfaces) in stabilizing protein–protein complexes and how this role may vary between the crystal and solution states.

As a result, the assessment of structures formed by these complexes in solution requires consideration of any crystallographic results in combination with information derived from methods such as NMR spectroscopy, potentiometric or microcalorimetric titrations, fluorescence quenching experiments, and Brownian dynamics simulations.

The importance of electrostatically guided diffusional processes in the association of electron transfer proteins is undeniable as far as the *in vitro* studies we have described above are concerned. The relevance of such processes is less clear *in vivo* where reaction components might be membrane-bound, the ionic strength may be relatively high (0.1–0.2 M), protein solutions are heterogeneous and viscosity may be relatively high. Berg and von Hippel (42) suggest that although three-dimensional diffusion may dominate in solutions of low viscosity, guided rotational diffusion may be more important under conditions of high viscosity. Of course, there is no doubt that the interprotein electron transfer reactions we describe are physiologically significant. Nevertheless, given the major influence that protein–protein association has on the rates of bimolecular electron transfer reactions and the complexities that exist in forming productive interprotein complexes, it is little surprise that multicentre electron transfer systems have evolved in which rapid electron transfer can occur between donor and acceptor centres within a single protein.

Acknowledgements

This work was supported by: the BBSRC/EPSRC, through their grant to the UEA Centre for Metalloprotein Spectroscopy & Biology, and the Wellcome Trust, through their support of Biomolecular NMR spectroscopy at UEA; MRC of Canada Operating

Grant MT-14021 (to A.G.M.), and NATO Travel Grant 0145/87 (to G.R.M. and A.G.M.); and by NSF Grant No. MCB972885 and NIH Grant GM20530 (to F.S.M.).

References

1. Bendall, D. S. (ed.) (1996). *Biological electron transfer*. Bios Scientific Publ., Oxford.
2. Hultquist, D. E., Sannes, L. J., and Juckett, D. A. (1984). Catalysis of methemoglobin reduction. *Curr. Top. Cell. Regul.*, **24**, 287–300.
3. Ortiz de Montellano, P. R. O. (ed.) (1996). *Cytochrome P-450: structure, mechanism, and biochemistry* (2nd edn). Plenum, New York.
4. Mathews, F. S. (1985). The structure, function and evolution of cytochromes. *Progr. Biophys. Mol.. Biol.*, **45**, 1–56.
5. Pettigrew, G. W. and Moore, G. R. (1987). *Cytochromes c: biological aspects*. Springer-Verlag, London.
6. Moore, G. R. and Pettigrew, G. W (1990). *Cytochromes c: evolutionary, structural and physicochemical aspects*. Springer-Verlag, London.
7. Takano, T. and Dickerson, R. E. (1981). Conformation change of cytochrome *c*. II. Ferricytochrome *c* refinement at 1.8 Å and comparison with the ferrocytochrome structure. *J. Mol. Biol.*, **153**, 95–115.
8. Brayer, G. D. and Murphy, M. E. P. (1996). Structural studies of eukaryotic cytochromes *c*. In *Cytochrome c: a multidisciplinary approach* (ed. R. A. Scott and A. G. Mauk), pp. 103–66. University Science Books, Sausilito, Ca.
9. Marcus, R. A. and Sutin, N. (1985). Electron transfers in chemistry and biology. *Biochim. Biophys. Acta*, **811**, 265–322.
10. DeVault, D. (1984). *Quantum mechanical tunnelling in biological systems* (2nd edn). Cambridge University Press, Cambridge.
11. Moser, C. C. and Dutton, P. L. (1996). Outline of theory of protein electron transfer. In *Protein electron transfer* (ed. D. S. Bendall), pp 1–21. Bios Scientific Publ., Oxford.
12. Moser, C. C., Page, C. C., Chen, X., and Dutton, P. L. (1997). Biological electron tunneling through native protein media. *J. Biol. Inorg. Chem.*, **2**, 393–8.
13. Mauk, A. G. Biological electron-transfer reactions. In *Essays in biochemistry*, Vol. 34 (ed. D. P. Ballou), pp. 101–24. Portland Press, London.
14. Wherland, S. and Gray, H. B. (1976). Electron transfer mechanisms employed by metalloproteins. In *Biological aspects of inorganic chemistry* (ed. A. W. Addison, W. Cullen, B. R. James, and D. Dolphin), pp. 289–368. Wiley-Interscience, New York.
15. Moore, G. R., Eley, C. G. S., and Williams, G. (1983). Electron transfer reactions of class I cytochromes *c*. *Adv. Inorg. Bioinorg. Mech.*, **3**, 1–96.
16. Gupta, R. K., Koenig, S. H., and Redfield, A. G. (1972). On the electron transfer between cytochrome *c* molecules as observed by nuclear magnetic resonance. *J. Magn. Res.* **7**, 67–73.
17. Dixon, D. W., Hong, X., and Woehler, S. E. (1989). Electrostatic and steric control of electron self-exchange in cytochromes c, c_{551}, and b_5. *Biophys. J.*, **56**, 3399–51.
18. Andrew, S. M., Thomasson, K. A., and Northrup, S. H. (1993). Simulation of electron-transfer self-exchange in cytochromes c and b_5. *J. Am. Chem. Soc.*, **115**, 5516–21.
19. Rosell, F. I., Ferrer, J. C., and Mauk, A. G. (1998). Protein-linked protein conformational switching: definition of the alkaline conformational transition of yeast *iso*-1-ferricytochrome *c*. *J. Am. Chem. Soc.*, **120**, 11234–45.

20. Eltis, L. D., Herbert, R. G., Barker, P. D., Mauk, A. G., and Northrup, S. H. (1991). Reduction of horse heart ferricytochrome *c* by bovine liver ferrocytochrome b_5. Experimental and theoretical analysis. *Biochemistry*, **30**, 3663–74.
21. Northrup, S. H., Thomasson, K. A., Miller, C. M., Barker, P. D., Eltis, L. D., Guillemette, J. G., Inglis, S. C., and Mauk, A. G. (1993). Effects of charged amino acid mutations on the bimolecular kinetics of reduction of yeast *iso*-1-ferricytochrome *c* by bovine ferrocytochrome b_5. *Biochemistry*, **32**, 6613–23.
22. Horton, N. and Lewis, M. (1992). Calculation of the free energy of association for protein complexes. *Protein Sci.*, **1**, 169–81.
23. Jones, S. and Thornton, J. M. (1995). Protein–protein interactions: a review of protein dimer structures. *Progr. Biophys. Mol. Biol.*, **63**, 31–65.
24. Janin, J. (1999). Kinetics and thermodynamics of protein–protein interactions. In Protein–protein recognition (ed. C. Kleanthous), pp. 1–32. IRL Press, Oxford.
25. Dutton, P. L. and Wilson, D. F. (1974). Redox potentiometry in mitochondrial and photosynthetic bioenergetics. *Biochim Biophys Acta* **346**, 165–212
26. Nicholls, P. (1974). Cytochrome c binding to enzymes and membranes. *Biochim. Biophys. Acta*, **346**, 271–310.
27. Lambeth, J. D., Seybert, D. W., and Kamin, H. (1979). Ionic effects on adrenal steroidogenic electron transport. *J. Biol. Chem.*, **254**, 7255–64.
28. Gray, K. A., Davidson, V. L., and Knaff, D. B. (1988). Complex formation between methylamine dehydrogenase and amicyanin from *Paracoccus denitrificans*. *J. Biol Chem.*, **263**, 13987–90.
29. Zhu, Z., Cunane, L. M., Chen, Z- W., Durley, R. C. E., Mathews, F. S., and Davidson, V. L. (1998). Molecular basis for interprotein complex-dependent effects on the redox properties of amicyanin. *Biochemistry*, **37**, 17128–36.
30. Bennett, L. E. (1973). Metalloprotein redox reactions. *Progr. Inorg. Chem.* **18**, 1–176.
31. Butler, J., Davies, D. M., and Sykes, A. G. (1981). Kinetic data for redox reactions of cytochrome *c* with $Fe(CN)_5X$ complexes and the question of association prior to electron transfer. *J. Inorg. Biochem.* **15**, 41–53.
32. Eley, C. G. S. and Moore, G. R. (1983). ^{1}H-n.m.r. investigation of the interaction between cytochrome *c* and cytochrome b_5. *Biochem. J.* **215**, 11–21.
33. Allison, S. A., Northrup, S. H., and McCammon, J. A. (1986). Simulation of biomolecular diffusion and complex formation. *Biophys. J.*, **49**, 167–75.
34. Northrup, S. H. (1996). Computer modeling of protein–protein interactions. In *Biological electron transfer*, (ed. D. S. Bendall), pp. 69–97. Bios Scientific Publ., Oxford.
35. Bosshard, H. R. (1996). Differential protection techniques in the analysis of cytochrome c interaction with electron transfer proteins. In *Cytochrome c: a multidisciplinary approach* (ed. R. A. Scott and A. G. Mauk), pp. 373–96. University Science Books, Sausilito, Ca.
36. Koppenol, W. H. and Margoliash, E. (1982). The asymmetric distribution of charges on the surface of horse cytochrome *c*. Functional implications. *J. Biol. Chem.*, **257**, 4426–37.
37. Salemme, F. R., Kraut, J., and Kamen, M. D. (1973). Structural bases for function in cytochromes *c*. *J. Biol. Chem.* **248**, 7701–16.
38. Koppenol, W. H., Rush, J. D., Mills, J. D., and Margoliash, E. (1991). The dipole moment of cytochrome *c*. *Mol. Biol. Evol.*, **8**, 545–58. (See erratum in *Mol. Biol. Evol.* (1991), **8**, 904.)
39. Wendoloski, J. J. and Matthew, J. B. (1989). Molecular dynamics effects on protein electrostatics. *Proteins*, **5**, 313–21.
40. Northrup, S. H., Wensel, T. G., Meares, C. F., Wendoloski, J. J., and Matthew, J. B. (1990).

Electrostatic field around cytochrome *c*: theory and energy transfer experiment. *Proc. Natl. Acad. Sci. USA.*, **87**, 9503–7.

41. Matthew, J. B., Weber, P. C., Salemme, F. R., and Richards, F. M. (1983). Electrostatic orientation during electron transfer between flavodoxin and cytochrome *c*. *Nature*, **301**, 169–71.
42. Berg, O. G. and von Hippel, P. H. (1985). Diffusion-controlled macromolecular interactions. *Annu. Rev. Biophys. Chem.*, **14**, 131–60.
43. Vijayakumar, M., Wong, K.-Y, Schreiber, G., Fersht, A. R., Szabo, A., and Zhou, H.-X (1998). Electrostatic enhancement of diffusion controlled protein–protein association: comparison of theory and experiment on barnase and barstar. *J. Mol. Biol.*, **278**, 1015–24.
44. Gabdoulline, R. R. and Wade, R. C. (1997). Simulation of the diffusional association of barnase and barstar. *Biophys. J.*, **72**, 1917–29.
45. Northrup, S. H., Luton, J. A., Boles, J. O., and Reynolds, J. C. (1988). Brownian dynamics simulation of protein association. *J. Comput. Aided Mol. Des.*, **1**, 291–311.
46. Northrup, S. H., Boles, J. O., and Reynolds, J. C. (1988). Brownian dynamics of cytochrome *c* and cytochrome *c* peroxidase association. *Science*, **241**, 67–70.
47. Northrup, S. H. and Erickson, H. P. (1992). Kinetics of protein–protein associations explained by Brownian dynamics computer simulations. *Proc. Natl. Acad. Sci. USA.*, **89**, 3338–42.
48. Mauk, A. G., Mauk, M. R., Moore, G. R., and Northrup, S. H. (1995). Experimental and theoretical analysis of the interaction between cytochrome *c* and cytochrome b_5. *J. Bioenerg. Biomemb.* **27**, 311–30.
49. Salemme, F. R. (1976). An hypothetical structure for an intermolecular electron transfer complex of cytochromes *c* and b_5. *J. Mol. Biol.*, **102**, 563–8.
50. Mauk, M. R., Mauk, A. G., Weber, P. C., and Matthew, J. B. (1986). Electrostatic analysis of the interaction of cytochrome *c* with native and dimethyl ester heme substituted cytochrome b_5. *Biochemistry*, **25**, 7085–91.
51. Hazzard, J. T., McLendon, G., Cusanovich, M. A., and Tollin, G. (1988). Formation of electrostatically-stabilized complex at low ionic strength inhibits interprotein electron transfer between yeast cytochrome *c* and cytochrome *c* peroxidase. *Biochem. Biophys. Res. Commun.*, **151**, 429–34.
52. Meyer, T. E., Zhao, A. G., Cusanovich, M. A., and Tollin, G. (1993). Transient kinetics of electron transfer from a variety of c-type cytochromes to plastocyanin. *Biochemistry*, **32**, 4552–9.
53. Petersen, L. C. and Cox, R. P. (1980). The effect of complex-formation with polyanions on the redox properties of cytochrome *c*. *Biochem. J.*, **192**, 687–93.
54. Yoshimura, T., Matsushima, A., and Aki, K. (1985). Oxidation and reduction of cytochrome *c* bound to the phosphoprotein phosvitin. *Arch. Biochem. Biophys.*, **241**, 50–7.
55. Adam, G. and Delbrück, M. (1968). Reduction of dimensionality in biological diffusion processes. In *Structural chemistry and molecular biology* (ed. A. Rich and N. Davidson), pp. 198–215. W. H. Freeman, San Francisco, CA.
56. Mathews, F. S. and Durley, R. C. E. (1996). Structure of electron transfer proteins and their complexes. In *Biological electron transfer* (ed. D. S. Bendall), pp. 99–123. Bios Scientific Publ., Oxford.
57. Jentoft, J. E., Gerken, T. A., Jentoft, N., and Dearborn, D. G. (1981). [^{13}C]Methylated ribonuclease A. ^{13}C NMR studies of the interaction of lysine 41 with active site ligands. *J. Biol. Chem.*, **256**, 231–6.

58. Dick, L. R., Sherry, A. D., Newkirk, M. M., and Gray, D. M. (1988). Reductive methylation and ^{13}C NMR studies of the lysyl residues of fd gene 5 protein. Lysines 24, 46, and 69 may be involved in nucleic acid binding. *J. Biol. Chem.*, **263**, 18864–72.
59. Burch, A. M., Rigby, S. E. J., Funk, W. D., MacGillivray, R. T. A., Mauk, M. R., Mauk, A. G., and Moore, G. R. (1990). NMR characterization of surface interactions in the cytochrome b_5–cytochrome *c* complex. *Science*, **247**, 831–3.
60. Moore, G. R., Cox, M. C., Crowe, D., Osborne, M. J., Rosell, F. I., Bujons, J., Barker, P. D., Mauk, M. R., and Mauk, A. G. (1998). N^{ε}, N^{ε}-dimethyl-lysine cytochrome *c* as an NMR probe for lysine involvement in protein–protein complex formation. *Biochem. J.*, **332**, 439–49.
61. Tamburini, P. P. and Gibson, G. G. (1983). Thermodynamic studies of the protein–protein interactions between cytochrome P-450 and cytochrome b5. Evidence for a central role of the cytochrome P-450 spin state in the coupling of substrate and cytochrome b5 binding to the terminal hemoprotein. *J. Biol. Chem.*, **258**, 13444–52.
62. Erman, J. E. and Vitello, L. B. (1980). The binding of cytochrome *c* peroxidase and ferricytochrome *c*. A spectrophotometric determination of the equilibrium association constant as a function of ionic strength. *J. Biol. Chem.*, **255**, 6224–7.
63. Mauk, M. R., Reid, L. S., and Mauk, A. G. (1982). Spectrophotometric analysis of the interaction between cytochrome b_5 and cytochrome *c*. *Biochemistry*, **21**, 1843–6.
64. Michel, B. and Bosshard, H. R. (1984). Spectroscopic analysis of the interaction between cytochrome *c* and cytochrome *c* oxidase. *J. Biol. Chem.*, **259**, 10085–91.
65. Weber, C., Michel, B., and Bosshard, H. R. (1987). Spectroscopic analysis of the cytochrome *c* oxidase–cytochrome *c* complex: circular dichroism and magnetic circular dichroism measurements reveal change of cytochrome *c* heme geometry imposed by complex formation. *Proc. Natl. Acad. Sci.* USA, **84**, 6687–91.
66. Vanderkooi, J. M., Landesberg, R., Hayden, G. W., and Owen, C. S. (1977). Metal-free and metal-substituted cytochromes *c*. Use in characterization of the cytochrome *c* binding site. *Eur. J. Biochem.*, **81**, 339–47.
67. Kornblatt, J. A., Hui Bon Hoa, G., and English, A. M. (1984). Volume changes associated with cytochrome *c* oxidase–porphyrin cytochrome *c* equilibrium. *Biochemistry*, **23**, 5906–11.
68. Kornblatt, J. A., Hui Bon Hoa, G., Eltis, L., and Mauk, A. G. (1988). The effects of pressure on porphyrin *c*–cytochrome b_5 complex formation. J. *Am. Chem. Soc.*, **110**, 5909–11.
69. Simolo, K. P., McLendon, G. L., Mauk, M. R., and Mauk, A. G. (1984). Photoinduced electron transfer within a protein–protein complex formed between physiological redox partners: reduction of ferricytochrome b_5 by the hemoglobin derivative $\alpha_2^{Zn}\beta_2^{Fe(III)CN}$. *J. Am. Chem. Soc.*, **106**, 5012–13.
70. McLendon, G. L., Winkler, J. R., Nocera, D. G., Mauk, M. R., Mauk, A. G., and Gray, H. B. (1985). Quenching of zinc-substituted cytochrome *c* excited states by cytochrome b_5. *J. Am. Chem. Soc.*, **107**, 739–40.
71. Stayton, P. S., Poulos, T. L., and Sligar, S. G. (1989). Putidaredoxin competitively inhibits cytochrome b_5–cytochrome P-450$_{cam}$ association: a proposed molecular model for a cytochrome P-450$_{cam}$ electron-transfer complex. *Biochemistry*, **28**, 8201–5.
72. Nocek, J. M., Zhou, J. S., De Forest, S., Priyadarshy, S., Beratan, D. N., Onuchic, J. N., and Hoffman, B. M. (1996). Theory and practice of electron transfer within protein–protein complexes: application to the multidomain binding of cytochrome *c* by cytochrome *c* peroxidase. *Chem. Rev.*, **96**, 2459–89.
73. Fisher, M. T., White, R. E., and Sligar, S. G. (1986). Pressure dissociation of a protein–protein electron-transfer complex. *J. Am. Chem. Soc.*, **108**, 6835–7.

74. Rodgers, K. K. and Sligar, S. G. (1991). Mapping electrostatic interactions in macromolecular associations. *J. Mol. Biol.*, **221**, 1453–60.
75. Kornblatt, J. A., Kornblatt, M. J., Hoa, G. H., and Mauk, A. G. (1993). Responses of two protein–protein complexes to solvent stress: does water play a role at the interface? *Biophys. J.*, **65**, 1059–65.
76. Laskowski, M., Jr. and Finkenstadt, W. R. (1972). Study of protein–protein and of protein–ligand interactions by potentiometric methods. In *Methods in enzymology*, Vol. 26C (eds C. H. W. Hirs and S. N. Timasheff), 193–277. Academic Press, New York.
77. Mauk, M. R., Barker, P. D., and Mauk, A. G. (1991). Proton linkage of complex formation between cytochrome *c* and cytochrome b_5: electrostatic consequences of protein–protein interactions. *Biochemistry*, **30**, 9873–81.
78. Mauk, M. R., Ferrer, J. C., and Mauk, A. G. (1994). Proton linkage in formation of the cytochrome *c*–cytochrome *c* peroxidase complex: electrostatic properties of the high-and low-affinity cytochrome binding sites on the peroxidase. *Biochemistry*, **33**, 12609–14.
79. Millett, F. and Durham, B. (1996). Chemical modification of surface residues on cytochrome c. In *Cytochrome c: a multidisciplinary approach* (ed. R. A. Scott and A. G. Mauk), pp. 573–91. University Science Books, Sausilito, Ca.
80. Seiter, C. H., Margalit, R., and Perreault, R. A. (1979). The cytochrome *c* binding site on cytochrome *c* oxidase. *Biochem. Biophys. Res. Commun.*, **86**, 473–7.
81. Morand, L. Z., Frame, M. K., Colvert, K. K., Johnson, D. A., Krogmann, D. W., and Davis, D. J. (1989). Plastocyanin cytochrome *f* interaction. *Biochemistry*, **28**, 8039–47. (See erratum in *Biochemistry* (1989), **28**, 10093.)
82. Willing, A. and Howard, J. B. (1990). Cross-linking site in *Azotobacter vinelandii* complex. *J. Biol. Chem.*, **265**, 6596–9.
83. Mauk, M. R. and Mauk, A. G. (1989). Crosslinking of cytochrome *c* and cytochrome b_5 with a water-soluble carbodiimide. Reaction conditions, product analysis and critique of the technique. *Eur. J. Biochem.*, **186**, 473–86.
84. Fuller, S. D., Darley-Usmar, V. M., and Capaldi, R. A. (1981). Covalent complex between yeast cytochrome c and beef heart cytochrome *c* oxidase which is active in electron transfer. *Biochemistry*, **20**, 7046–53.
85. Bisson, R., Azzi, A., Gutweniger, H., Colonna, R., Montecucco, C., and Zanotti, A. (1978). Interaction of cytochrome *c* with cytochrome *c* oxidase. Photoaffinity labeling of beef heart cytochrome *c* oxidase with arylazido–cytochrome *c*. *J. Biol. Chem.*, **253**, 1874–80.
86. Dowe, R. J., Vitello, L. B., and Erman, J. E. (1984). Sedimentation equilibrium studies on the interaction between cytochrome *c* and cytochrome *c* peroxidase. *Arch. Biochem. Biophys.*, **232**, 566–73.
87. Holloway, P. W. and Mantsch, H. H. (1988). Infrared spectroscopic analysis of salt bridge formation between cytochrome b_5 and cytochrome *c*. *Biochemistry*, **27**, 7991–3.
88. Salamon, Z., Brown, M. F., and Tollin, G. (1999). Plasmon resonance spectroscopy: probing molecular interactions within membranes *Trends Biochem. Sci.*, **24**, 213–19.
89. Bosshard, H. R., Anni, H., and Yonetani, T. (1991). Yeast cytochrome *c* peroxidase. In *Peroxidases in chemistry and biology*, Vol. 2 (ed. J. Everse, K. Everse, and M. B. Grisham), pp. 52–83. CRC Press, Boca Raton, FL.
90. Mochan, E. and Nicholls, P. (1971). Complex-formation between cytochrome c and cytochrome c peroxidase. Equilibrium and titration studies. *Biochem. J.*, **121**, 69–82.
91. Finzel, B. C., Poulos, T. L., and Kraut, J. (1984). Crystal structure of yeast cytochrome *c* peroxidase refined at 1.7- Å resolution. *J. Biol. Chem.*, **259**, 13027–36.

92. Sivaraja, M., Goodin, D. B., Smith, M., and Hoffman, B. M. (1989). Identification by ENDOR of Trp191 as the free-radical site in cytochrome *c* peroxidase compound ES. *Science*, **245**, 738–40.
93. Pelletier, H. and Kraut, J. (1992). Crystal structure of a complex between electron transfer partners, cytochrome *c* peroxidase and cytochrome *c*. *Science*, **258**, 1748–55.
94. Poulos, T. L., Sheriff, S., and Howard, A. J. (1987). Cocrystals of yeast cytochrome *c* peroxidase and horse heart cytochrome *c*. *J. Biol. Chem.*, **262**, 13881–4.
95. Kang, C. H., Ferguson-Miller, S., and Margoliash, E. (1977). Steady state kinetics and binding of eukaryotic cytochromes *c* with yeast cytochrome *c* peroxidase. *J. Biol. Chem.*, **252**, 919–26.
96. Zhou, J. S., Tran, S. T., McLendon, G., and Hoffman, B. M. (1997). Photoinduced electron transfer between cytochrome *c* peroxidase (D37K) and Zn-substituted cytochrome *c*—probing the two-domain binding and reactivity of the peroxidase. *J. Am. Chem. Soc.*, **119**, 269–77.
97. Jeng, M. F., Englander, S. W., Pardue, K., Rogalskyj, J. S., and McLendon, G. (1994). Structural dynamics in an electron-transfer complex. *Nature Struct. Biol.*, **1**, 234–8.
98. Sukits, S. F., Erman, J. E., and Satterlee, J. D. (1997). Proton NMR assignments and magnetic axes orientations for wild-type yeast *iso*-1-ferricytochrome *c* free in solution and bound to cytochrome *c* peroxidase. *Biochemistry*, **36**, 5251–9.
99. Hahm, S., Miller, M. A., Geren, L., Kraut, J., Durham, B., and Millett, F. (1994). Reaction of horse cytochrome *c* with the radical and the oxyferryl heme in cytochrome *c* peroxidase compound I. *Biochemistry*, **33**, 1473–80.
100. Neuvo, M. R., Chu, H. H., Vitello, L. B., and Erman, J. E. (1993). Salt dependent switch in the pathway of electron transfer from cytochrome *c* to cytochrome *c* peroxidase compound I. *J. Am. Chem. Soc.*, **115**, 5873–4.
101. Miller, M. A., Liu, R.-Q., Hahm, S., Geren, L., Hibdon, S., Kraut, J., Durham, B., and Millett, F. (1994). Interaction domain for the reaction of cytochrome *c* with the radical and the oxyferryl heme in cytochrome *c* peroxidase compound I. *Biochemistry*, **33**, 8686–93.
102. Zhou, J. S., Nocek, J. M., De Van, M. L., and Hoffman, B. M. (1995). Inhibitor-enhanced electron transfer: copper cytochrome *c* as a redox-inert probe of ternary complexes. *Science*, **269**, 204–7.
103. Curry, W. B., Grabe, M. D., Kurnikov, I. V., Skourtis, S. S., Beratan, D. N., Regan, J. J., Aquino, A. J., Beroza, P., and Onuchic, J. N. (1995). Pathways, pathway tubes, pathway docking, and propagators in electron transfer proteins. *J. Bioenerg. Biomembr.*, **27**, 285–93.
104. Adman, E. T. (1991). Copper protein structures. *Adv. Protein Chem.*, **42**, 145–97.
105. Cramer, W. A., Martinez, S. E., Huang, D., Tae, G. S., Everly, R. M., Heymann, J. B., Cheng, R. H., Baker, T. S., and Smith, J. L. (1994). Structural aspects of the cytochrome b_6f complex; structure of the lumen-side domain of cytochrome *f*. *J. Bioenerget. Biomemb.*, **26**, 31–47.
106. Farver, O., Blatt, Y., and Pecht, I. (1982). Resolution of two distinct electron transfer sites on azurin. *Biochemistry*, **21**, 3556–61.
107. He, S., Mondi, S., Bendall, D. S., and Gray, J. C. (1991). The surface-exposed tyrosine residue Tyr83 of pea plastocyanin is involved in both binding and electron transfer reactions with cytochrome *f*. *EMBO J.*, **10**, 4011–16.
108. Gray, J. C. (1992). Cytochrome *f*: structure, function and biosynthesis. *Photosynth. Res.*, **34**, 359–74.

109. Martinez, S. E., Huang, D., Szczepaniak, A., Cramer, W. A., and Smith J. L. (1994). Crystal structure of chloroplast cytochrome *f* reveals a novel cytochrome fold and unexpected heme ligation. *Structure*, **2**, 95–105.
110. Qin, L. and Kostic, N. M. (1993). Importance of protein rearrangement in the electron-transfer reaction between the physiological partners cytochrome *f* and plastocyanin. *Biochemistry*, **32**, 6073–80.
111. Takabe, T. and Ishikawa, H. (1989). Kinetic studies on a cross-linked complex between plastocyanin and cytochrome *f*. *J. Biochem*. (Tokyo), **105**, 98–102.
112. Kannt, A., Young, S., and Bendall, D. S. (1996). The role of acidic residues of plastocyanin in its interaction with cytochrome *f*. *Biochim. Biophys. Acta*, **1277**, 115–26.
113. Qin, L. and Kostic, N. M. (1992). Electron-transfer reactions of cytochrome *f* with flavin semiquinones and with plastocyanin. Importance of protein–protein electrostatic interactions and of donor-acceptor coupling. *Biochemistry*, **31**, 5145–50.
114. Ullmann, G. M., Knapp, E.–W., and Kostic, N. M. (1997). Computational simulation and analysis of dynamic association between plastocyanin and cytochrome *f*. Consequences for the electron-transfer reaction. *J. Am. Chem. Soc.*, **119**, 42–52.
115. Pearson, D. C., Jr and Gross, E. L. (1998). Brownian dynamics study of the interaction between plastocyanin and cytochrome *f*. *Biophys. J.*, **75**, 2698–711.
116. Ubbink, M., Ejdeback, M., Karlsson, B. G., and Bendall, D. S. (1998). The structure of the complex of plastocyanin and cytochrome *f*, determined by paramagnetic NMR and restrained rigid-body molecular dynamics. *Structure*, **6**, 323–35.
117. Guiles, R. D., Sarma, S., DiGate, R. J., Banville, D., Basus, V. J., Kuntz, I. D., and Waskell, L. (1996). Pseudocontact shifts used in the restraint of the solution structures of electron transfer complexes. *Nature Struct. Biol.*, **3**, 333–9.
118. Ubbink, M. and Bendall, D. S. (1997). Complex of plastocyanin and cytochrome *c* characterized by NMR chemical shift analysis. *Biochemistry*, **36**, 6326–35.
119. Roberts, V. A., Freeman, H. C., Olson, A. J., Tainer, J. A., and Getzoff, E. D. (1991). Electrostatic orientation of the electron-transfer complex between plastocyanin and cytochrome *c*. *J. Biol. Chem.* **266**, 13431–41.
120. Husain, M. and Davidson, V. L. (1985). An inducible periplasmic blue copper protein from *Paracoccus denitrificans*. Purification, properties, and physiological role. *J. Biol. Chem.*, **260**, 14626–9.
121. Husain, M. and Davidson, V. L. (1986). Characterization of two inducible periplasmic c-type cytochromes from *Paracoccus denitrificans*. *J. Biol. Chem.*, **261**, 8577–80.
122. Davidson, V. L. and Kumar, M. A. (1989). Cytochrome *c*-550 mediates electron transfer from inducible periplasmic *c*-type cytochromes to the cytoplasmic membrane of *Paracoccus denitrificans*. *FEBS Lett.*, **245**, 271–3.
123. McIntire, W. S., Wemmer, D. E., Chistoserdov, A., and Lidstrom, M. E. (1991). A new cofactor in a prokaryotic enzyme: tryptophan tryptophylquinone as the redox prosthetic group in methylamine dehydrogenase. *Science*, **252**, 817–27.
124. Chen, L., Doi, M., Durley, R. C. E., Chistoserdov, A. Y., Lidstrom, M. E., Davidson, V. L., and Mathews, F. S. (1998). Refined crystal structure of methylamine dehydrogenase from *Paracoccus denitrificans* at 1.75 Å resolution. *J. Mol. Biol.*, **276**, 131–49.
125. Davidson, V. L. (1989). Steady-state kinetic analysis of the quinoprotein methylamine dehydrogenase from *Paracoccus denitrificans*. *Biochem. J.*, **261**, 107–11.
126. Bishop, G. R., Brooks, H. B., and Davidson, V. L. (1996). Evidence for a tryptophan tryptophylquinone aminosemiquinone intermediate in the physiologic reaction between methylamine dehydrogenase and amicyanin. *Biochemistry*, **35**, 8948–54.

127. Husain, M., Davidson, V. L., Gray, K. A., and Knaff, D. B. (1987). Redox properties of the quinoprotein methylamine dehydrogenase from *Paracoccus denitrificans*. *Biochemistry*, **26**, 4139–43.
128. Gray, K. A., Knaff, D. B., Husain, M., and Davidson, V. L. (1986). Measurement of the oxidation–reduction potentials of amicyanin and *c*-type cytochromes from *Paracoccus denitrificans*. *FEBS Lett.*, **207**, 239–42.
129. Cunane, L. M., Chen, Z.-W., Durley, R. C. E., and Mathews, F. S. (1996). X-ray structure of the cupredoxin amicyanin from *Paracoccus denitrificans* refined at 1.3 Å resolution. *Acta Crystallog*, **D52**, 676–86.
130. van Spanning, R. M. J., Wansell, C. W., de Boer, T., Hazelaar, M. J., Anazawa, H., Harms, N., Oltmann, L. F., and Stouthamer, A. H. (1991). Isolation and characterization of the moxJ, moxG, moxI, and moxR genes of *Paracoccus denitrificans*: inactivation of moxJ, moxG, and moxR and the resultant effect on methylotrophic growth. *J. Bacteriol.*, **173**, 6948–61.
131. Chen, L., Durley, R. C., Mathews, F. S., and Davidson, V. L. (1994). Structure of an electron transfer complex: methylamine dehydrogenase, amicyanin, and cytochrome c_{551i}. *Science*, **264**, 86–90.
132. Davidson, V. L. and Jones, L. H. (1991). Intermolecular electron transfer from quinoproteins and its relevance to biosensor technology. *Anal. Chim. Acta*, **249**, 235–40.
133. Kumar, M. A. and Davidson, V. L. (1990). Chemical cross-linking study of complex formation between methylamine dehydrogenase and amicyanin from *Paracoccus denitrificans*. *Biochemistry*, **29**, 5299–304.
134. Bishop, G. R. and Davidson, V. L. (1995). Intermolecular electron transfer from substrate-reduced methylamine dehydrogenase to amicyanin is linked to proton transfer. *Biochemistry*, **34**, 12082–6.
135. Davidson, V. L. and Jones, L. H. (1996). Electron transfer from copper to heme within the methylamine dehydrogenase–amicyanin–cytochrome c_{551i} complex. *Biochemistry*, **35**, 8120–5.
136. Chen, L., Durley, R., Poliks, B. J., Hamada, K., Chen, Z., Mathews, F. S., Davidson, V. L., Satow, Y., Huizinga, E., Vellieux, F. M. D., and Hol, W. G. J. (1992). Crystal structure of an electron-transfer complex between methylamine dehydrogenase and amicyanin. *Biochemistry*, **31**, 4959–64.
137. Regan, J. J. (1993). Pathways II Software v. 2.01, UC San Diego, CA.
138. Davidson, V. L., Jones, L. H., Graichen, M. E., Mathews, F. S., and Hosler, J. P. (1997). Factors which stabilize the methylamine dehydrogenase–amicyanin electron transfer protein complex revealed by site-directed mutagenesis. *Biochemistry*, **36**, 12733–8.
139. Merli, A., Broderson, D. E., Morini, B., Chen, Z.-W., Durley, R. C. E., Mathews, F. S., Davidson, V. L., and Rossi, G. L. (1996). Enzymatic and electron transfer activities in crystalline protein complexes. *J. Biol. Chem.*, **271**, 9177–80.
140. Rivetti, C., Mozzarelli, A., Rossi, G. L., Henry, E. R., and Eaton, W. A. (1993). Oxygen binding by single crystals of hemoglobin. *Biochemistry*, **32**, 2888–906.
141. Zhu, Z. Y. and Davidson, V. L. (1998). Redox properties of tryptophan tryptophylquinone enzymes. Correlation with structure and reactivity. *J. Biol. Chem.*, **273**, 14254–60.
142. Pochapsky, T. C., Ye, X. M., Ratnaswamy, G., and Lyons, T. A. (1994). An NMR-derived model for the solution structure of oxidized putidaredoxin, a 2-Fe, 2-S ferredoxin from *Pseudomonas*. *Biochemistry*, **33**, 6424–32.
143. Poulos, T. L., Finzel, B. C., and Howard, A. J. (1987). High resolution crystal structure of cytochrome $P450_{cam}$. *J. Mol. Biol.*, **195**, 687–700.

144. Hintz, M. J., Mock, D. M., Peterson, L. L., Tuttle, K., and Peterson, J. A. (1982). Equilibrium and kinetic studies of the interaction of cytochrome $P450_{cam}$ and putidaredoxin. *J. Biol. Chem.*, **257**, 4324–32.
145. Davies, M. D. and Sligar, S. G. (1992). Genetic variants in the putidaredoxin–cytochrome $P450_{cam}$ complex. *Biochemistry*, **31**, 11383–9.
146. Roitberg, A. E., Holden, M. J., Mayhew, M. P., Kurnikov, I. V., Beratan, D. N., and Vilker, V. L. (1998). Binding and electron transfer between putidaredoxin and cytochrome P450cam. *J. Am. Chem. Soc.*, **120**, 8927–32.
147. Pochapsky, T. C., Lyons, T. A., Kazanis, S., Arakaki, T., and Ratnaswamy, G. (1996). A structure-based model for cytochrome $P450_{cam}$–putidaredoxin interactions. *Biochimie*, **78**, 723–33.
148. Aoki, M., Ishimori, K., and Morishima, I. (1998). NMR studies of putidaredoxin: association of putidaredoxin with NADH-putidaredoxin reductase and cytochrome $P450_{cam}$. *Biochim. Biophys. Acta*, **1386**, 168–78.
149. Aoki, M., Ishimori, K., and Morishima, I. (1998). Roles of negatively charged surface residues of putidaredoxin in interactions with redox partners in $P450_{cam}$ monooxygenase system. *Biochim. Biophys. Acta*, **1386**, 157–67.
150. Holden, M., Mayhew, M., Bunk, D., Roitberg, A., and Vilker, V. (1997). Probing the interactions of putidaredoxin with redox partners in Camphor P-450 monooxygenase by mutagenesis of surface residues. *J. Biol. Chem.*, **272**, 21720–5.
151. Unno, M., Shimada, H., Toba, Y., Makino, R., and Ishimura, Y. (1996). Role of Arg 112 of cytochrome $P450_{cam}$ in the electron transfer from reduced putidaredoxin. *J. Biol. Chem.*, **271**, 17869–74.
152. Aoki, M., Ishimori, K., Fukada, H., Takahashi, K., and Morishima, I. (1998). Isothermal titration calorimetric studies on the associations of putidaredoxin to NADH-putidaredoxin reductase and $P450_{cam}$. *Biochim. Biophys. Acta*, **1384**, 180–8.

Note

1. (H), (L), (A), and (C) in atom designations refer to the heavy and light chains of MADH, amicyanin, and cytochrome *c*551i, respectively.

4 Molecular basis of integrin-dependent cell adhesion

MARTIN J. HUMPHRIES and ROBERT C. LIDDINGTON

1. Introduction

The integrins are a family of heterodimeric ($\alpha\beta$) plasma membrane proteins that bind the extracellular matrix or counter-receptors on other cells (1). They transduce bidirectional signals across the plasma membrane that are critical to many biological processes, including embryonic cell migration, wound healing, and the immune response. In pathogenic states they play an essential role in tumour metastasis and angiogenesis.

The adhesiveness of integrins' extracellular domains is allosterically controlled by binding events in their cytoplasmic domains that trigger conformational changes across the plasma membrane ('inside-out' signalling). Under some conditions it is ligand binding to the extracellular domains that triggers an intracellular signal ('outside-in' signalling). There are two mechanisms by which integrins could change the adhesiveness of cells, which are not necessarily exclusive. 'Affinity' changes caused by conformational changes in individual integrin molecules are believed to be responsible for both the initial rapid and reversible change in the affinity of platelets for damaged blood vessels and of leucocytes for sites of inflammation (2). 'Avidity' changes brought about by the synergistic effect of multiple clustered integrins on the cell surface binding to clustered or polyvalent ligands are thought to play a role in the stable attachment of cells to the extracellular matrix (ECM) (3).

Atomic-resolution structural information for integrins is so far only available for one extracellular domain, the ligand-binding A- or I-domain (4–7), and for one intracellular fragment from the atypical integrin $\alpha_6\beta_4$ (8). In contrast, the high-resolution crystal structures of several integrin ligands are now available and their integrin recognition sites have been determined. By combining the structural data of the integrin and its ligands with a broad range of biochemical and genetic data, a fascinating picture is beginning to emerge of how these large and complex molecules work.

2. Integrin structure

Integrins are heterodimers consisting of non-covalently bound α- and β-subunits. So far, there are 24 known combinations between 18 α-subunits and 8 β-subunits. Cryo-electron microscopic images reveal a globular head, 70–100 Å in diameter, containing the N-terminal ligand-binding regions of the α- and β-chains, which are connected to the membrane by a long rigid stalk containing both chains (9). The three major ligand-binding regions are the α-subunit I-domain, the seven N-terminal repeats of the α-subunit, and a conserved region of the β-subunits that appears to be a functional and structural homologue of the α-subunit I-domain. A cartoon is shown in Fig. 1.

2.1 The α-chain sevenfold repeats

At the N-terminus of the α-subunits are seven repeats of about 60 amino acids each. Two competing models exist for the three-dimensional structure of this region. A hypothetical model has been presented by Springer (10) who has predicted the fold to be a β-propeller, a cyclic structure made of seven modules or blades that can fold as a unit but not independently. In this model, the ligand contact points are located on the upper surface of the propeller. The Ca^{2+}-binding EF-hands are located on the lower surface of the domain and are far from the ligand-binding site. The I-domain (see below), which is located between repeats 2 and 3, would sit on top of the propeller where it would replace or augment the ligand-binding surface. Mutagenesis and the identification of antibody epitopes have been presented in support of this model by the same group (11) and independently by Takada's group (12).

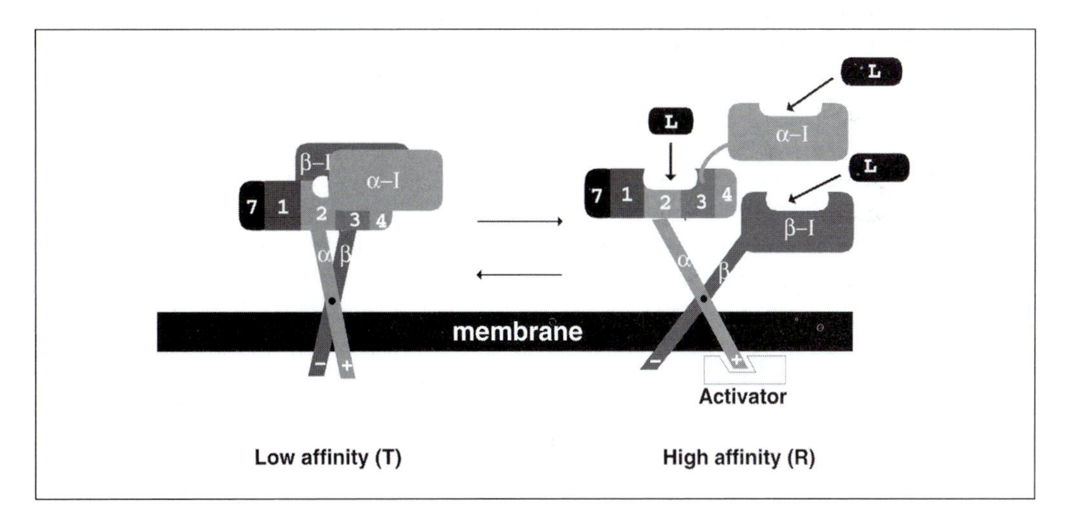

Fig. 1 Hypothetical 'scissor' model of an integrin with three ligand-binding sites (L): the sevenfold repeat in the α-subunit, the α-subunit I-domain, and the β-conserved region which may also resemble an I-domain. The integrin is in equilibrium between low- and high-affinity quaternary states (T and R). The switch to a high-affinity state involves a scissor action of the α- and β-subunits, leading to shape-shifting in the I-domains and unmasking of the ligand-binding epitope in the sevenfold repeat.

Parello's group have published biophysical data which challenge the propeller model, suggesting that the repeats more closely resemble EF-hand domains of the calmodulin class (13), consistent with earlier studies (14). Baneres *et al.* (13) expressed a recombinant fragment of α_5 corresponding to repeats 4 through 7, and showed that it binds four Ca^{2+} or Mg^{2+} ions, and binds fibronectin in a cation-dependent manner, displacing a water molecule from a cation coordination sphere (consistent with the notion of the metal ion forming a bridge between integrin and ligand). Circular dichroism (CD) spectra suggest that it contains approximately 30% α-helix. On the face of it, these biophysical data are wholly inconsistent with Springer's propeller model, since:

1. The propeller contains no helix.
2. The propeller is an all-or-nothing fold, and fragments with less than seven blades should not be stable, yet Parello's group have reported crystals of repeats 4–7 and 6–7.
3. In the propeller model, ligands bind at the face distal from the Ca^{2+}-binding sites, while Parello claims that they are directly involved in ligand binding.

A crystal structure of an appropriate fragment is eagerly awaited in order to resolve this issue.

2.2 The α-subunit I-domain

Eight α-subunits (α_1, α_2, α_{10}, α_D, α_E, α_L, α_M, and α_X) contain an additional domain ('A' or 'I') that is inserted between N-terminal repeats 2 and 3, where it plays a central role in ligand binding: thus, recombinant I-domains recapitulate many of the ligand-binding properties of the intact integrin (see, for example, refs 15–17). The first crystal structure of an α-subunit I-domain demonstrated that it adopts the 'dinucleotide-binding' or 'Rossmann' fold, with a central mostly parallel β-sheet surrounded on both sides by amphipathic α-helices (4). This fold is very common amongst intra-cellular phosphoryl transfer enzymes but had not been observed previously in an extracellular domain. In this class of fold, the functional surface of the molecule always lies at the C-terminal end of the β-sheet in a crevice formed by loops linking the β-strands to the α-helices (18). A unique, divalent-cation coordination sphere is located in the I-domain, which has been called the metal ion-dependent adhesion site or MIDAS. In the first crystal structure determined, a glutamate side chain from a neighbouring I-domain in the crystal lattice completed the octahedral coordination sphere of the metal, leading to the suggestion that the glutamate was acting as a ligand mimetic (4; Fig. 2). This was consistent with the fact that all integrin ligands possess critical aspartate or glutamate residues as a key feature of their integrin-binding motifs, and that mutation of the metal-coordinating side chains of the I-domain abolishes ligand binding in a dominant-negative fashion (e.g. ref. 19). The concept of a metal ion providing a bridge between ligand and receptor had been widely anticipated, and the crystal structure provided tantalizing but not direct proof of its existence. Very recently, the crystal structure of an authentic integrin–

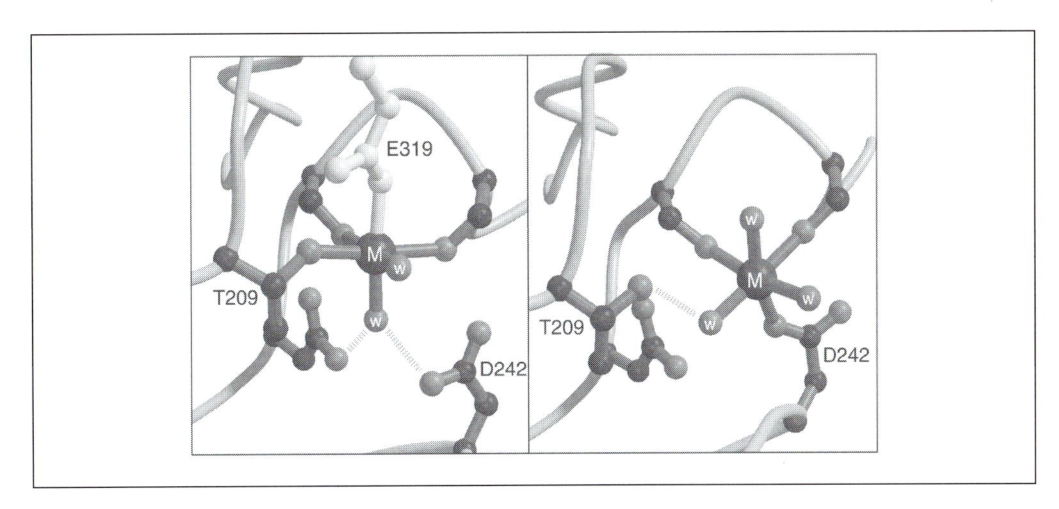

Fig. 2 Two conformations of the metal ion-dependent adhesion site (MIDAS motif; Lee *et al.* (4); protein database (pdb) code 1IDO). In the 'active' high-affinity state (left panel) a glutamic acid from a neighbouring molecule in the crystal lattice completes the octahedral coordination sphere around the metal ion; while in the inactive low-affinity state (right panel) the bond to T209 breaks, a direct bond to D242 is formed, and the ligand mimetic is replaced by a water molecule.

ligand complex has been determined, and confirms in detail the prediction of the integrin–metal–ligand bridge (20).

While the metal-coordinating residues of the MIDAS motif are highly conserved, the upper surface of the domain surrounding the MIDAS motif is highly variable among integrin I-domains, suggesting that the MIDAS face could provide the observed specificity by forming a complementary surface with ligand. Several groups have now reported the results of structure-based mutagenesis studies (21–26), and confirmed that the MIDAS face does indeed form a ligand-binding surface for both natural ligands (C3bi to α_M; intercellular adhesion molecule-1 (ICAM-1) to α_L; collagen to α_2) and the pathogenic ligands (neutrophil inhibitory factor (NIF) to α_M).

2.3 The β-subunit conserved region

Near the N-terminus of the β-subunit is a region of about 240 amino acids (102–344 in β_2) that shows the highest conservation between species and among different subunits, which has been termed the 'conserved region'. Strong support for this region of β_3 in ligand binding is provided by the observations that ligand peptides crosslink to this region (27, 28), mutations within this region abolish ligand binding, and several antibodies that inhibit ligand binding map to this region. Mutations in the corresponding region of other β-subunits similarly block ligand binding, and function-blocking as well as activating monoclonal antibodies (mAbs) directed against β_1 map to a discrete region within this fragment. Chimera studies reveal short segments within the conserved region that confer ligand specificity among β-subunits (29, 30).

The presence of an invariant DXSXS sequence required for ligand binding within this region of the β-subunits, as well as within the α-subunit I-domain, led to the suggestion that these two regions might be structurally and functionally related (31). Based on the similarity of hydropathy plots between the α-subunit I-domain and this region of the $β_3$-subunit, a hypothetical atomic model was built of the $β_3$-subunit domain that predicts that this region is likely to share many of the structural features of the I-domain (32). Mutagenesis of candidate oxygenated residues in the $β_3$-subunit predicted to be involved in cation coordination and ligand binding fully support this model. Similar results obtained with the corresponding regions of $β_2$ (33) and $β_1$ (34) further support the hypothesis that this region of the β-subunits adopts a similar, but not identical, fold to the I-domain and might engage ligand via a MIDAS-like motif.

2.4 Structure of the αβ heterodimer

The regions N- and C-terminal to the conserved region of the β-subunits are both rich in disulfides and fold independently of the α-subunit. They contain a long-range disulfide (C5–C435) and are highly resistant to proteolysis. Thus these regions may fold together into a rigid structural unit. By contrast, the β-conserved region does not fold completely in biosynthesis until after association with the α-subunit (35). Similarly, the α-subunit sevenfold repeat does not fold until after association with the β-subunit, while regions C-terminal to these repeats do fold autonomously. Furthermore, the majority of mutations in leucocyte adhesion-deficiency I, which block the association of the α- and β-chains, map to the β-conserved region (36), suggesting that the β-conserved region is required for αβ association.

Thus, the picture emerging is that the α-subunit sevenfold repeat and the β-conserved region form an intimate and extensive interface, and constitute much of the integrin 'head' (the I-domain presumably also contributes to the head in those integrins that have one), with each half of the head connected to rigid stalks. In Springer's model, the β-conserved region would pack against the β-propeller. However, it is possible to envisage alternative pairings for the two halves of the head. For example, the functional 'mini-integrin' described by Kunicki's group (37) comprises repeats 1–3 in the α-subunit and the β-conserved region; this could comprise one half of the head, while Parello's functional α-subunit repeats 4–7 (13) could comprise the other. Clearly, the tertiary structure of the head domains and their quaternary arrangement must await crystallographic analysis.

2.5 The role of divalent cations

Integrins bind multiple cations at two classes of sites: one inhibitory and the other directly involved in ligand-binding (reviewed in ref. 38). Although the concentration of divalent cations is normally strictly controlled in the plasma, it remains an open question whether integrin activation can be driven in some environments by dynamic alteration in the ratio of cation concentration, for example at sites of vascular or tissue injury or bone resorption.

The different specificities and opposite function of the two classes of binding sites have made unravelling the contributions of individual cations very difficult. However, three recent analyses have clarified the relative role in different integrins (39–41). The only site that has been fully characterized is the Mg^{2+}-binding site in the I-domain, which is required for ligand binding and which forms a bridge between integrin and ligand. This site is specific for Mg^{2+} over Ca^{2+}, but it can also bind Mn^{2+} and other transition metals of a similar size and charge. There is mounting evidence that the metal-binding site in the β-conserved region is also of this type, although there is not such a strict dependence on Mg^{2+}. For example, peptides based on the RGD sequence bind to $\alpha_{IIb}\beta_3$ equally well in the presence of Ca^{2+} or Mg^{2+}. The 'EF hands' in the N-terminal repeats are also high-affinity sites; these do not discriminate between Ca^{2+} and Mg^{2+}, but their role in ligand binding is controversial, as noted above.

A distinct class of sites is inhibitory and binds Ca^{2+}. These sites presumably lie at αβ interfaces and stabilize the 'T' state. For example, $\alpha_V\beta_3$ and $\alpha_{IIb}\beta_3$ contain a low-affinity, Ca^{2+}-specific inhibitory site and a high-affinity, 'ligand-competent' site that is not specific for Ca^{2+} vs. Mg^{2+}. Mn^{2+} binds to both sites, but Mg^{2+} only binds to the high-affinity, ligand-competent site (40). Although Mn^{2+} can certainly function as a bridge at the ligand-binding sites, the ability of Mn^{2+} to 'activate' integrins *in vitro* is somewhat puzzling. One possibility is that Mn^{2+} binds to the allosteric site but is not effective in stabilizing the T state.

A thorough analysis has been carried out by Labadia *et al.* on the I-domain integrin, $\alpha_L\beta_2$ (39). They showed that the leucocyte function-associated antigen-1 (LFA-1) contains two distinct sites (A and B). Site A binds Ca^{2+} specifically and with high affinity (K_d = 15 μM) (it also enhances the affinity of Mg^{2+} for site B (from 160 μM to 12 μM)). Ca^{2+} can also bind to site B but with low affinity (1.5 mM), competitively displacing Mg^{2+} and decreasing the binding of soluble ligand (ICAM-1). If site B is equated with the I-domain, this result confirms the prediction of Lee *et al.* (42) based on the switch of the metal-coordinating residues in the two crystal forms representing active and inactive conformations.

These results are in good agreement with those of Mould *et al.* (41), who showed that the non-I-domain integrin $\alpha_5\beta_1$ has a high-affinity and specific inhibitory Ca^{2+}-binding site (30 μM). Occupancy of this site increases the affinity of Mg^{2+} for the ligand-competent site from 1 mM to 40 μM. The similarity with the $\alpha_L\beta_2$ result suggests that the 'β-I-domain' behaves similarly to the α-I-domain.

A hypothesis of 'cation displacement' on ligand binding to integrins has been described (43). Integrins contain several metal-binding sites, as noted above, some of which stabilize the low-affinity quaternary conformation. It is therefore reasonable that ligand binding which switches the quaternary state leads to the release of cations (e.g. from site A in $\alpha_L\beta_2$). However, it has also been proposed that those metal ions directly involved in ligand binding are displaced. For example, Dickeson *et al.* (44) have shown that Tb^{3+} (which substitutes for Mg^{2+}) luminescence is quenched upon collagen binding to the recombinant α_2-I-domain, and that EDTA fails to dissociate the complex; which was interpreted as evidence for displacement of the cation upon complex formation.

The notion that a metal ion is, on the one hand, required for ligand binding to a domain, but is then displaced from the complex, makes no thermodynamic sense, and could only make sense kinetically if there was a high activation barrier to ligand binding—but there is no evidence of this. An alternative explanation, in the light of our crystal structure of an I-domain:collagen complex (which shows no evidence of cation displacement but instead a key role for the cation), is that changes in cation coordination and protein conformation are responsible for the quenching, and that the formation of the integrin–metal–ligand bridge creates a high-affinity metal-binding site that inhibits chelation by EDTA.

3. Extracellular matrix ligands for integrins

Integrins recognize many different components of extracellular matrices, including the quantitatively most common molecules, fibronectin, collagens, and laminins, and the more minor components such as osteopontin, bone sialoprotein, and thrombospondins.

3.1 Fibronectin

Fibronectin is a large, dimeric glycoprotein that exists either in a non-polymerized, soluble form in body fluids (such as plasma) or as polymeric, fibrillar assemblies in extracellular matrices (45). The distribution of fibronectin is widespread, and many of its functions centre on its ability to mediate the adhesion of a wide variety of cell types. Thus, fibronectin provides an adhesive track for cell migration during tissue morphogenesis and is a major component of granulation tissue during wound healing. The fact that most cells encounter dense assemblies of fibronectin, and may need to do so in a transient manner, implies that kinetically these interactions need special properties. The interactions are likely to be of moderate affinity and should be capable of being disrupted in a regulated manner. Indeed, current estimates of the binding affinity derived from cell-based or isolated receptor-based assays (46; Curley and Humphries, unpublished results) are consistent with a moderate affinity (8×10^{-7} M). As yet, there is little insight into the mechanisms used by cells to break their adhesive contacts with fibronectin.

The fibronectin monomer, like many matrix proteins, is modular in nature. It is made up from more than 30 polypeptide repeats which fall into three classes (type I–III repeats). At three positions, individual modules can be alternatively spliced, thereby increasing the variety of fibronectin proteins that can be generated from the single gene. The modular nature of fibronectin has permitted it to evolve a series of functionally independent binding domains that can also be isolated and separated following limited proteolysis. Using this approach, cell-binding domains were mapped to two regions of the molecule; the so-called central cell-binding domain (47) and the heparin-II/type-III connecting segment (HepII/IIICS; 48).

Progressive truncation of fragments from these domains led to the identification of cell recognition sites. A surprising, but important, strategic transition came with the discovery that it was possible to truncate to the point that the cell adhesive activity of fibronectin could be reproduced in the form of a synthetic peptide (the first such peptide being RGD from the 10th type-III repeat; 49). This finding was surprising since it suggested a binding mechanism involving the recognition of a very short, linear epitope as opposed to a multicontact interaction across an extended portion of the protein. None the less, RGD-containing peptides were able to promote adhesion directly and inhibit fibronectin-mediated adhesion. Subsequently, it was found that mutation of this site in recombinant fibronectin fragments abolished their cell adhesive activity, confirming the importance of this sequence.

3.2 The RGD motif

Following the discovery of the RGD sequence in fibronectin, many more adhesive extracellular matrix proteins were cloned and sequenced. Remarkably, many of them contained RGD sequences, and RGD peptides were found to perturb their adhesive activity, suggesting the common use of this sequence (50, 51). Definitive reductionist analyses of other extracellular matrix proteins, like those performed for fibronectin, have generally not been performed for other matrix molecules, but mutation of RGD sites in thrombospondin-1, vitronectin, and von Willebrand factor does compromise their adhesive activity. The use of RGD as a common motif implies the existence of a common mechanism of receptor binding for extracellular matrix proteins, and indeed a common family of receptors. We now know that this is the case, and that all these proteins are recognized by integrins.

Parenthetically, the discovery of the RGD cell-recognition sequence aroused enormous interest in cell adhesion within the pharmaceutical industry. Adhesion contributes to the progression of many of the most common human diseases, including inflammation (in asthma, autoimmune diseases, and atherosclerosis), neoplasia (in tumour metastasis), and infection (bacterial, viral, and parasitic infections) so that antiadhesive agents would have widespread potential utility in large markets. Since RGD peptides are so small (molecular weights of less than 500), they are virtually drugs in their own right. Thus most major pharmaceutical companies seized on the discovery to establish their own drug development programmes aimed at producing non-peptidic mimics of RGD. This work is now at an advanced stage and a number of agents are in clinical trials, initially for the prevention of thrombosis following coronary angioplasty. The widespread use of RGD by proteins, however, suggests other possible applications in the future (52, 53).

Although RGD had been established as a key cell-recognition site in fibronectin, deletion of this sequence by nested truncation of the central cell-binding domain (CCBD) only decreased the adhesive activity of recombinant proteins by about 97%, and subsequently a second functional site over 150 residues N-terminal to RGD was identified by deletion mutagenesis (54). Mutation of this site also showed a more than 96% loss of activity, indicating cooperativity between the sites and leading to the naming of this second site a 'synergy' site. The amino-acid sequence responsible for synergy activity lies in the ninth type-III repeat and contains the motif Pro–His–Ser–Arg–Asn (PHSRN), with the arginine residue being essential for activity (55). Together, these findings are consistent with both RGD and PHSRN binding simultaneously to one integrin receptor.

Although the 9th and 10th type-III repeats appear to function in concert, the first structural information on this domain came from nuclear magnetic resonance (NMR) analysis of the isolated 10th repeat (56). This module was found to exhibit a fold similar to that of an immunoglobulin domain. Subsequently, the structure of the entire CCBD of fibronectin, containing type-III repeats 7–10 was solved by X-ray crystallography (57). The 10th type-III repeat (and subsequently all other type-III repeats) was found to comprise seven β-strands arranged as a pair of opposing β-sheets. The RGD tripeptide was solvent-exposed and projected at the apex of the flexible loop between the F and G strands. Since the N- and C-termini of each repeat were at opposite ends, it was no surprise that the fragment containing repeats 7–10 had an extended, rod-like structure. Nevertheless, each repeat was a slightly different shape and each inter-repeat connection had a characteristic tilt and rotation. The synergy sequence in the 9th type-III repeat was located in the C'–E loop and, owing to an unusually small rotation between repeats 9 and 10, was presented on the same face of the fibronectin molecule as the RGD sequence. This suggests that both peptides are available to engage a single integrin receptor (see Fig. 3).

3.3 The HepII/IIICS region

The second, cell-binding domain within fibronectin comprises the HepII/IIICS region. HepII is the major heparin/proteoglycan-binding domain in the molecule; it contains three type-III repeats (the 12th–14th) and has two heparin-binding sites in the 13th and 14th repeats. The IIICS region is 120 amino acids long and is alternatively spliced from fibronectin RNA in a complex manner; it contains three segments (A–C) that give rise to five possible variants. Two $\alpha_4\beta_1$ integrin-binding sites have been pinpointed using synthetic peptides to the IIICS-A and -C; these are represented by the 25- and 20mer peptides CS1 and CS5 (48, 58). The minimal active sequence within the IIICS-C is a tetrapeptide incorporating the RGD recognition motif, but IIICS-A, which has a dominant contribution to the adhesive activity of the IIICS, has a different active sequence, LDV (59). Interestingly, RGD and LDV appear to be functionally equivalent; they are both short acidic peptides that have an absolute dependence on their D residue, both peptides are able to competitively inhibit each other's function (60), and they both bind to the same region of their respective integrin receptors (61, 62). In recent studies, a further heparin-binding site has been localized to the IIICS-B region, adding further to the complexity with which this domain interacts with its binding molecules (Whittard, Askari, and Humphries, unpublished results).

As discussed in more detail below, the LDV peptide now appears to form a second common integrin-binding peptide motif. Initial studies, based on peptide com-

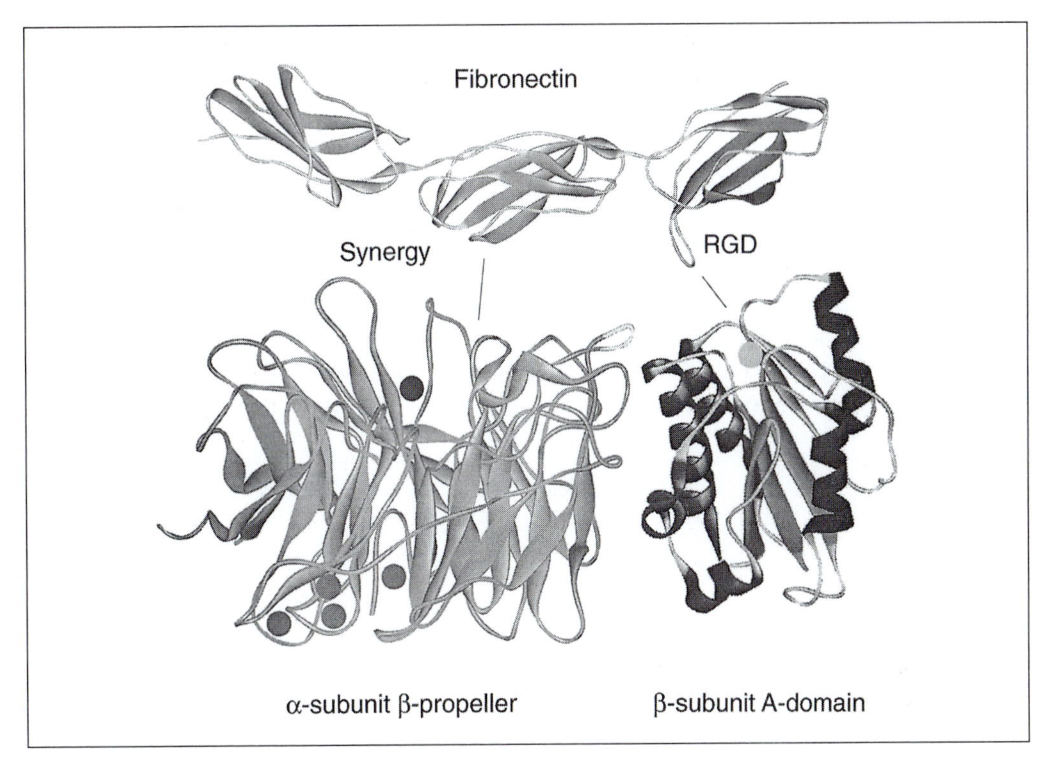

Fig. 3 A schematic cartoon of the fibronectin-bound form of an integrin. At the top is an image of the crystal structure of the 9th and 10th type-III repeats of fibronectin (pdb code 1FNF; from Leahy *et al.* (57)). At the bottom, arranged next to each other, are models for each of the ligand-binding sites in both subunits of the integrin. Current evidence suggests that the RGD motif in the type-III repeat 10 of fibronectin engages a cation in the A-domain of the β-subunit (right), while the PHSRN 'synergy' sequence in type-III repeat 9 engages a site in the α-subunit (left).

petition and mutagenesis, suggested that a second LDV-like peptide (with the sequence IDA) was located in the 14th type-III repeat of fibronectin and was responsible for the integrin-binding activity of the HepII in the absence of the IIICS sequence (63). Recently, however, the crystal structure of HepII has been solved (64) and this sequence is seen to form the linker between the 13th and 14th repeats. Inspection of the sequence of the region of fibronectin type III repeat 14 (FN14) that hydrogen-bonds to the IDAPS sequence reveals an interesting homology to the previously characterized PHSRN 'synergy' sequence in FN9. In FN14, the peptide sequence is PRARI. PHSRN relies on its arginine residue for activity, and, similarly, mutation of either arginine in PRARI caused a substantial loss of integrin-binding activity. This suggests that perturbation of IDA inhibits integrin binding because it has a knock-on structural effect on this 'synergy' sequence mimic. As yet, there is no evidence that PRARI acts as a 'synergy' sequence, for example for sites in the IIICS, but it does appear to be a bona fide integrin-binding site. The PRARI sequence in repeat 14 is conserved across all species of fibronectin. From this finding it is tempting to speculate that, like RGD and LDV peptides which form common integrin-binding motifs in adhesive ligands, PHSRN / PRARI sequences may be more commonplace than previously thought.

4. Cell-surface ligands for integrins

The tertiary structures of four immunoglobulins that employ LDV motifs have now been solved: vascular cell-adhesion molecule-1 (VCAM-1; 65) and mucosal addressin cell-adhesion molecule-1 (MAdCAM-1; 66), two endothelial proteins that are substrates for leucocyte extravasation; and intercellular cell-adhesion molecule-1 (ICAM-1; 67,68) and ICAM-2 (69), which participate in lymphocyte recirculation and leucocyte–leucocyte interactions (see Fig. 4). The major form of VCAM-1 has an

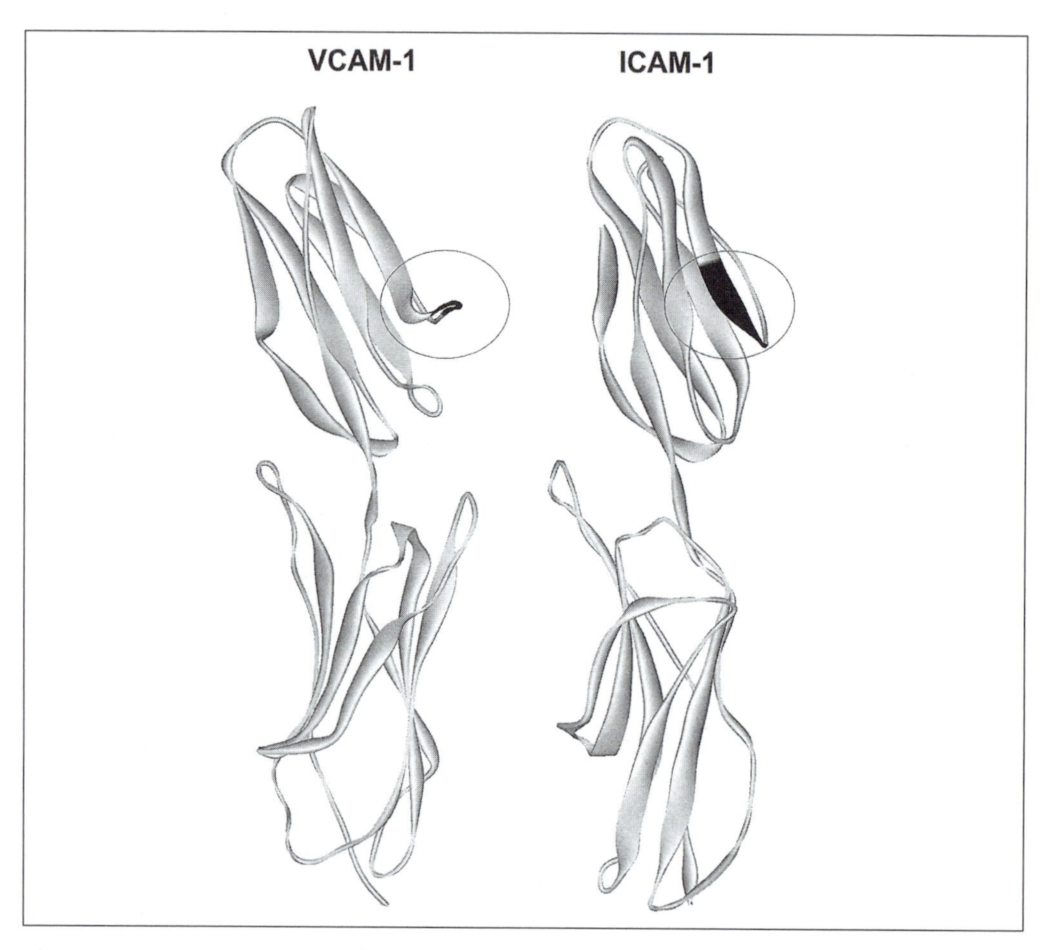

Fig. 4 Ribbon diagram of the two N-terminal immunoglobulin domains of VCAM-1 (left) and ICAM-1 (right) (from Jones *et al.* (65) and Bella *et al.* (68); pdb codes 1VCA and 1IAM, respectively). The β-strands are shown as solid ribbons. The major integrin-binding 'LDV' sites are shown in black and are encircled. In VCAM-1, the site is found in the C–D loop, while in ICAM-1, it is found at the end of the C-strand.

extracellular domain containing seven immunoglobulin domains, with domains 1–3 and 4–6 arising from a gene duplication event. The crystal structure of a fragment of VCAM-1 containing domains 1 and 2 has now been solved. Module 1 is a member of the I subset of immunoglobulins, while module 2 is a typical C2-set fold. The LDV motif responsible for binding to the integrins $\alpha_4\beta_1$ and $\alpha_4\beta_7$ is located in the solvent-exposed C–D loop of module 1, the region projected the furthest from the module, and therefore closely resembles the fibronectin RGD site (Fig. 4). VCAM-1, like fibronectin, appears to contain 'synergy' sequences that cooperate with the acidic motif. The most important accessory site is found in a long C′–E loop within domain 2.

MAdCAM-1 is a member of the immunoglobulin superfamily that is expressed on mucosal endothelium in the gut and mediates the homing of lymphocytes into these tissues. Human MAdCAM-1, but not rodent, serves as a ligand both for the integrins $\alpha_4\beta_7$ and $\alpha_4\beta_1$ (primarily the former) and for L-selectin. Integrin binding is mediated by the two N-terminal immunoglobulin domains in the protein, and selectin binding is mediated by oligosaccharides in a mucin-like domain close to the plasma membrane. The crystal structure of a fragment of MAdCAM-1 containing the two immunoglobulin domains has been solved (66). The LDV peptide motif within MAdCAM-1 (LDS) is located in the C–D loop of domain-1, just like VCAM-1, although comparison of the two structures reveals an 8-Å difference that might explain the subtle difference in their receptor specificities. Like VCAM-1, MAdCAM-1 also employs accessory sequences in the D strand of domain-2. This site is even longer than in VCAM-1, it forms a ribbon-like structure and is extremely acidic in nature.

ICAM-1 not only binds integrins but is also a receptor for a group of human rhinoviruses and for the malarial parasite *Plasmodium falciparum*. The crystal structure of the two N-terminal domains of ICAM-1 has recently been solved by two groups (67, 68). Surprisingly, the LDV motif (IET) that is responsible for binding to the integrins $\alpha_L\beta_2$ and $\alpha_M\beta_2$ differs substantially from the sites in VCAM-1 and MAdCAM-1, since the critical glutamate residue is located at the end of the C-strand rather than the C–D loop (Fig. 4). It is also on a relatively flat face of the domain rather than at the apex of a projected loop. Similar results were obtained for ICAM-2 (69). Since ICAMs bind to a different region of their integrin receptors to VCAM-1 and MAdCAM-1 (the αA-domain and putative βA-domain, respectively), this may explain the different active site requirements. A current working hypothesis is that the IET motif of ICAM-1 interacts directly with a divalent cation coordinated by loops at the surface of the αA-domain of α_L and α_M. Docking of the ICAM-1 crystal structure with that of the α_L A-domain produced a model that was consistent with this mode of interaction.

Until the solution of the crystal structure of ICAM-1 and ICAM-2, the different modes of ligand engagement by αA-domain and non-αA-domain-containing integrins was difficult to reconcile with the relatedness of their active site sequences. It is still possible that the αA-domain creates an additional rather than an alternative site for ligand binding. In this scenario, αA-domain-containing integrins might bind a dimeric ligand with simultaneous engagement of both the αA- and putative βA-domains. In this regard, there is evidence that the active form of ICAM-1 on cell surfaces is dimeric (70). Interestingly, the crystal form of ICAM-1 was a dimer, created by a hydrophobic interaction between the BED faces of the protein and with both IET sequences in the C-strand exposed for integrin binding.

Previous mutational studies had demonstrated significant differences in the binding sites on ICAM-1 for integrins, rhinoviruses, and *Plasmodium*. Docking of a rhinovirus structure with that of ICAM-1, suggested contacts with the apexes of the B–C, C–D, D–E, and F–G loops of domain-1. The structure of these sites is one of the major differences between the two ICAMs and presumably explains the selective binding of rhinoviruses to ICAM-1. Surprisingly, one of the key residues implicated in binding to *Plasmodium falciparum*-infected erythrocytes is in the dimer interface (67, 68).

5. Integrin–ligand interactions

The existence of common, receptor-binding motifs as key recognition sequences for integrins is suggestive of a common mechanism of ligand engagement. Even though each ligand is grossly different in structure, they appear to have active sites that may be hypothesized to interact with cognate binding sites on their respective receptors.

Although these studies support the idea that acidic motifs interact at a conserved site on their respective receptors, they do not identify the actual sites. However, recent approaches based on the use of antifunctional, anti-integrin mAbs have permitted a major advance by determining the topology of integrin–ligand binding. Surprisingly, kinetic analyses of the inhibition of fibronectin binding to purified $\alpha_5\beta_1$ by the anti-β_1 monoclonal mAb13 revealed an allosteric mode of action; ligand blockade of antibody binding was not outcompeted by increasing the concentration of antibody, and antibody and ligand binding were mutually exclusive (71). This suggested that destruction of the mAb13 epitope was a normal consequence of ligand binding, and therefore the antibody was probably exerting its inhibitory activity by preventing a shape change in the integrin β-subunit that was required for ligand engagement.

Subsequent analyses of the mode of action of a series of anti-β1, -5, and -4 mAbs demonstrated that they all function allosterically (61, 62). In turn, this suggests that there are sites on integrin ligands that contact both the α- and β-subunits and result in local shape changes. By examining the ability of mutant fibronectin CCBD fragments, lacking either the RGD or PHSRN sites, to block the binding of anti-α5 or -β1 mAbs to the integrin $\alpha_5\beta_1$, it was determined that RGD was responsible for β-subunit shape changes and PHSRN for the changes in the α-subunit (61). This suggests that two binding pockets exist for the same ligand (shown schematically in Fig. 3). These two sites are 35 Å apart in the crystal structure of fibronectin, suggesting that the two binding pockets are the same distance apart. Similar results were found for the $\alpha_4\beta_1$ integrin, with LDV motif sequences eliciting changes in β_1 (62). However, to date, it has not been possible to assign specific sites to the regions of ligands causing shape changes in α_4. This probably indicates a role for both sites in perturbing the α-subunit. None the less, these data now explain the use of common acidic peptides to bind to a common site on the 'shared' integrin β-subunit. This conclusion is supported by studies in which a potent peptidic LDV agent was chemically cross-linked to $\alpha_4\beta_1$ and the binding site localized to the MIDAS region of the β_1-subunit (72).

These findings also provide a partial insight into the molecular mechanisms underlying the specificity of ligand binding by different integrins. Thus the use of a particular acidic peptide motif and a particular 'synergy' site will presumably select those receptors able to bind to that ligand. In addition, the different conformations of different ligand active sites will also influence the fit that the ligand has into the receptor-binding pocket. All these findings have implications for the development of therapeutic agents based on ligand active sites: first, derivatives of either RGD or LDV may be useful for blocking the activity of many, if not all, integrins; second, since different conformations of these sequences help determine receptor specificity, constrained non-peptidic versions should exhibit receptor selectivity; third, mimics that perturb the binding of ligands to the α-subunit may also show efficacy and selectivity.

5.1 Shape-shifting in the integrin I-domain

A key question is how ligand binding by integrins triggers conformational changes that regulate signalling. The first structural views of this process came from analyses of A-domain structure. Thus, Lee *et al.* showed that the α_M I-domain adopts two different crystal forms with different tertiary conformations (42). The second crystal form did not include a ligand mimetic, and other details of the metal coordination had changed. These subtle changes in metal stereochemistry were linked to a large downward shift of the C-terminal helix (by 10 Å), which altered the shape of the ligand-binding surface and propagated changes to the opposite pole of the domain (Fig. 5). Lee *et al.* pointed out at the time the intriguing similarities with the structural changes observed in the G proteins in which, on exchanging GDP for GTP, a threonine side chain gains a direct bond to Mg^{2+}, leading to the exposure of a hydrophobic, effector-binding motif. In this model, it is the presence or absence of a ligand mimetic that determines the conformation, and it is therefore not contradicted by the lack of conformational changes observed in crystal structures of the α_L and α_M I-domains grown in the presence or absence of different metals (73, 74).

Recently, biochemical support for the model of Lee *et al.* has come from two groups. Li *et al.* have carried out mutagenesis and binding studies to show that the α_M I-domain exists in a solution as an equilibrium between high-affinity and low-affinity conformations, and that mutations predicted to destabilize the inactive conformation increase the affinity for the natural ligand C3bi but do not affect NIF-binding (24). Oxvig *et al.* (75) have mapped the epitope for an antibody that binds only to the active integrin to a region that undergoes extensive conformational changes in the two crystal forms and is widely exposed, as judged by reactivity with other antibodies.

Very recently, the crystal structure of a complex between the α_2 I-domain and a triple-helical collagen containing the peptide motif GER has been determined,

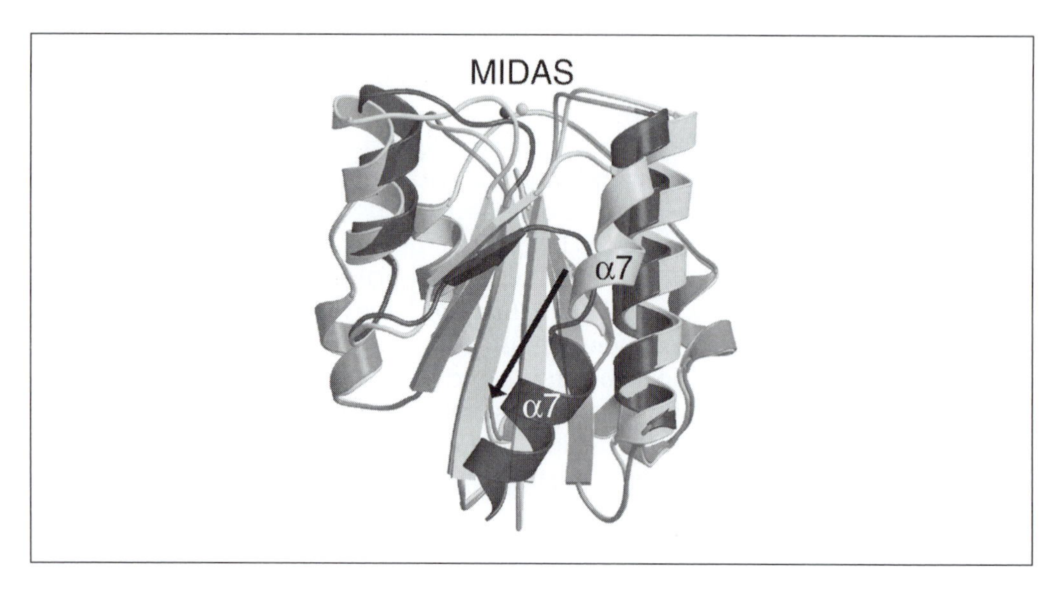

Fig. 5 'Shape-shifting' in the α_M I-domain (from Lee *et al.* (42); pdb codes: 1IDO, 1JLM). Regions with large changes are shown. Note the large movement of the C-terminal helix (α_7) which propagates structural changes from the MIDAS motif at the top of the domain to the base of the domain. 'Gain-of-function' mutations that enhance adhesiveness map to the bottom of the domain.

resolving the controversy (20). The structure confirms that the glutamic acid residue is directly involved in coordinating the metal ion and that ligand binding switches the coordination of the critical threonine residue and triggers conformational changes that closely mirror those seen in α_M (Fig. 6).

Given that the I-domain undergoes shape-shifting to create a ligand-binding surface, we must ask what is the nature of the quaternary restraints that prevent this shape-shifting from occurring in the resting integrin. The available evidence points to contacts between the lower surface of the domain (at the opposite pole from the MIDAS face) and the main body of the integrin. Thus, Zhang and Plow made $\alpha_M\beta_2/\alpha_L\beta_2$ chimeras that led to a constitutively active integrin (76). The substitutions are on the face of the I-domain distal to the MIDAS face, near to the domain termini. Recently Oxvig *et al.* (75) have confirmed this earlier study with a series of point mutants to the same region. They also pointed out that this lower surface is not recognized by a large panel of monoclonal antibodies, further supporting the role for this region in interdomain contacts. The structural link between the ligand-binding and -repressor surfaces is provided by the 10 Å movement of the C-terminal helix. At the top of this helix, a phenylalanine packs against the MIDAS motif in the inactive conformer, while the bottom of the helix forms part of the lower surface. The recent structure determination of the α_L I-domain in complex with the inhibitor lovastatin reveals that it binds in a hydrophobic pocket at the base of the domain, suggesting that it inhibits integrin function allosterically by preventing the conformational change that creates a high-affinity binding site for ICAM-1 at the top of the domain (77).

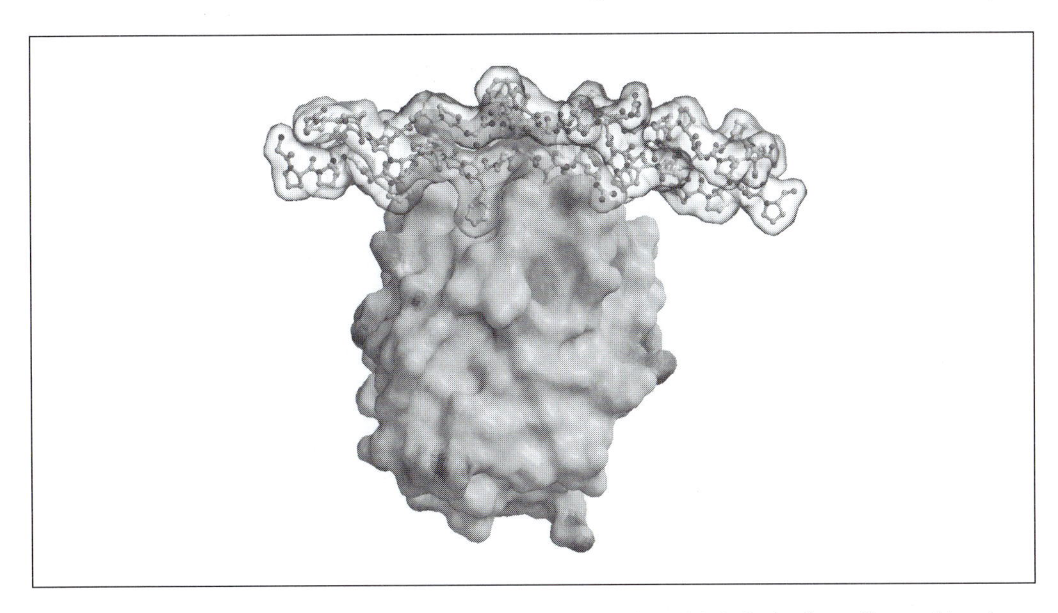

Fig. 6 Solvent-accessible surface of the α_2 I-domain in complex with a triple-helical collagen-like peptide, shown as a ball-and-stick model within a transparent solvent-accessible surface. (From Emsley *et al.* (20))

These results are highly reminiscent of the natural type-IIB mutants in the plasma protein von Willebrand factor (vWf), which lead to constitutively active ligand binding. The A domain of vWF has an identical fold, although it lacks the MIDAS motif, and the mutants lie on the analogous surface (78). A common feature here is that the A/I-domains function in the plasma and are involved in the binding of cells to substrates under conditions of high flow/shear. The active conformation, with its large movement of the C-terminal helix, is more extended or 'stretched' than the inactive conformer. Could the forces on cell contacts under high flow provide a direct mechanism of activation by 'molecular stretching'? Such bonds, called 'catch bonds', that strengthen when tensile force is applied, are theoretically possible (79).

We know much less about conformational changes in the other ligand-binding domain. The similarities between the β-conserved region and the I-domain raise the possibility that it undergoes analogous shape-shifting, but experimental data are lacking. The issue will only be fully resolved by determining crystal structures of the integrin head in both the T and R states.

6. A model for quaternary structure changes

There is abundant evidence that integrins change their conformations during ligand binding and activation (80–82). A simple model of quaternary change in integrins is a 'scissor' model (Fig. 1). In the absence of other proteins, we can imagine the integrin to be in the low-affinity 'T' quaternary state (see below) in which salt bridges and other bonds between the α- and β-cytoplasmic tails hold the blades of the scissors together. The T-state structure holds the integrin in a low-affinity conformation in which the ligand-binding surfaces are cryptic, either because they are masked sterically by other domains, or because they are inhibited by intersubunit contacts from undergoing tertiary structure changes ('shape-shifting') that would create a ligand-binding surface.

Activation entails a rotation of the two rigid subunits about the scissor pivot point, simultaneously splaying apart the cytoplasmic tails and opening the ligand-binding sites (the location of the pivot point is unknown). In principle, the integrin can be activated either by binding events in the cytoplasmic tails or the extracellular domains, for example:

(1) cytoplasmic proteins that bind preferentially to isolated α- or β-chains rather than to the αβ-dimer;

(2) phosphorylation of residues in the cytoplasmic tails that disrupts αβ association;

(3) ligand binding to the integrin head that forces the exposure of cryptic binding sites;

(4) antibody binding preferentially to any surface that is unmasked (or created by shape-shifting) in the R state;

(5) removal of calcium or other ions that stabilize the T state.

7. A two-state model of integrin activation

The emerging picture is that integrins have up to three, but typically two, distinct ligand-binding sites which are intimately linked and allosterically controlled. The simplest allosteric model with which we could try to explain the ligand binding behaviour of integrins is the 'two-state' model formulated by Monod *et al.* (83). The two-states are two distinct quaternary arrangements (T and R) for the αβ-hetero-dimer, which generates four distinct energetic states: unliganded T (Tu); liganded T (TL); unliganded R (Ru); and liganded R (RL).

The simplest way to visualize this model is in terms of two energy 'ladders' (see Fig. 7). There is a T-state ladder with narrowly spaced rungs (low affinity for ligand) and an R-state ladder with broadly spaced rungs (high ligand affinity). Tu is in equilibrium with Ru and TL with RL, and the position of equilibrium depends on the relative heights of these ladders.

It is instructive to imagine various possible scenarios:

1. The condition for the resting integrin (Fig. 7A). Here Tu is lower in energy than Ru, so that the T state is more stable in the absence of ligand. TL is lower in energy than RL, so that ligand will bind with low affinity and the integrin will remain in the T state.
2. Poised for outside-in signalling (Fig. 7B). Suppose now that that the R state gains stabilizing bonds, for example by the binding of a cytosolic protein that sequesters

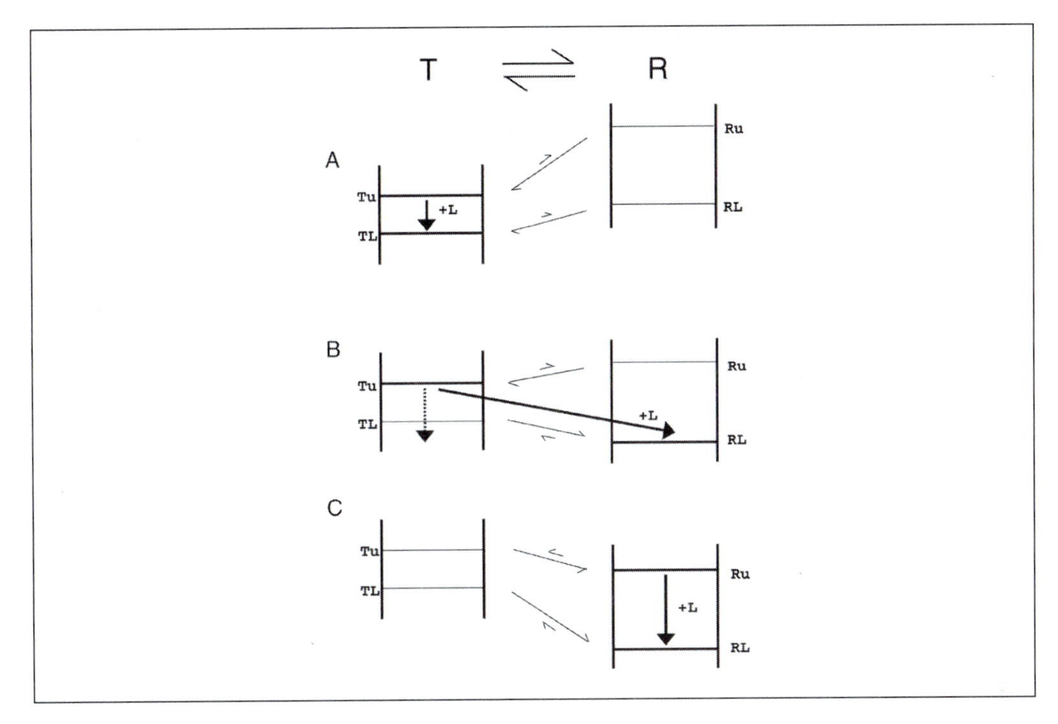

Fig. 7 A two-state model of integrin allostery, visualized as energy ladders. The integrin is in equilibrium between two quaternary conformations (T and R), which may be liganded (TL/RL) or unliganded (Tu/Ru). Tu is in equilibrium with Ru and TL with RL. The T-state has a lower affinity for ligand than the R state, so the rungs on the T ladder are closer together than the R rungs. Fig. 7A is the condition for the resting integrin. Here Tu is lower in energy than Ru, so that the T-state is more stable in the absence of ligand. TL is lower in energy than RL, so that ligand will bind with low affinity and the integrin will remain in the T-state. Suppose now that that the R-state gains stabilizing bonds, for example by the binding of a cytosolic protein that sequesters one of the tails. In this case, Tu is still lower in energy than Ru so that the integrin remains in the T-state in the absence of ligand. However, the integrin has still been 'activated' in the sense that the affinity for ligand has increased, although to an intermediate level (inside-out signalling). Furthermore, since RL is lower in energy than TL the binding of the extracellular ligand triggers the quaternary switch to the R-state (outside-in signalling). In Fig. 7C, the R-state has gained further stabilizing bonds, or the T-state has become destabilized (perhaps by the removal of extracellular Ca^{2+}). Here, Ru is lower than Tu and the integrin will remain in the high-affinity R state whether ligand is bound or not, and ligand binding will induce only local structural effects.

one of the tails. In this case, Tu is still lower in energy than Ru so that the integrin remains in the T state. However, the integrin has still been 'activated' in the sense that the affinity for ligand has increased, although to an intermediate level (inside-out signalling). Furthermore, since RL is lower in energy than TL the binding of the extracellular ligand triggers the quaternary switch to the R state (outside-in signalling).

3. In Fig. 7C, the R state has gained further stabilizing bonds, or the T state has become destabilized (perhaps by the removal of extracellular Ca^{2+}). Here, Ru is lower than Tu and the integrin will remain in the high-affinity R state whether ligand is bound or not, and ligand binding will induce only local structural effects.

It is also instructive to consider some quantitative estimates of the energies involved. The experiments by Hughes *et al.* (84) showed that the breakage of a single salt bridge between the α- and β-cytoplasmic tails was enough to destabilize the low-affinity (T) conformation. A single salt bridge is probably worth about 3 $kcal\,mol^{-1}$ and corresponds to the difference in energy between Tu and Ru. If the ligand-binding energies are in the μM range (~8 $kcal\,mol^{-1}$) for the R state and mM (~4 $kcal\,mol^{-1}$) for the T state, this generates the condition depicted in Fig. 7B, i.e. poised for outside-in signalling. Since this system relies on the quaternary organization of the T state to reduce the intrinsic ligand-binding affinity (the R state has the affinity expected of isolated domains) most mutations in the quaternary interface are likely to destabilize the T state and lead to a constitutively high-affinity integrin locked in the R state.

This simple model can provide a rationale for most natural and unnatural activators and inhibitors. In the case of two ligand-binding sites, the situation becomes more complex. There is now the possibility of cooperative ligand binding, in which the first ligand binds to site A with low affinity, allowing the second ligand to bind to site B with high affinity (Fig. 7B). With a bivalent ligand there is also the possibility of synergistic binding when the two binding sites are the correct distance apart.

It remains to be seen just how similar the integrin is to the classic allosteric ligand-binding protein, haemoglobin. In haemoglobin, salt bridges also stabilize the low-affinity T-state, and mutations which break those salt bridges increase the affinity for oxygen by shifting the equilibrium towards the quaternary R state. The quaternary change involves a rotation of two rigid dimeric subunits. The quaternary rotation is linked to subtle, but energetically critical, tertiary changes in the subunits, which increase the affinity for ligand by shape-shifting in the α-subunit and unmasking of the binding site in the β-subunit.

References

1. Hynes, R. O. (1992). Integrins: versatility, modulation, and signalling in cell adhesion. *Cell*, **69**, 11–25.
2. Diamond, M. S. and Springer, T. A. (1994). The dynamic regulation of integrin adhesiveness. *Curr. Biol.*, **4**, 506–17.
3. Bazzoni, G. and Hemler, M. E. (1998). Are changes in integrin affinity and conformation overemphasized? *Trends Biochem. Sci.*, **23**, 30–4.
4. Lee, J.-O., Rieu, P., Arnaout, M. A., and Liddington, R. C. (1995). Crystal structure of the A-domain from the α subunit of integrin CR3 (CD11b/CD18). *Cell*, **80**, 631–8.
5. Qu, A. and Leahy, D. J. (1995). Crystal structure of the I-domain from the CD11a/CD18 (LFA-1, αLβ2) integrin. *Proc. Natl. Acad. Sci. USA*, **92**, 10277–81.
6. Emsley, J., King, S., Bergelson, J., and Liddington, R. (1997). Crystal structure of the I domain from integrin α2β1. *J. Biol. Chem.*, **272**, 28512–17.
7. Nolte, M., Pepinsky, R. B., Venyaminov, S. Y., Koteliansky, V., Gotwals, P. J., and Karpusas, M. (1999). Crystal structure of the α1β1 integrin I-domain: insights into integrin I-domain function. *FEBS Lett.*, **452**, 379–85.

8. dePereda, J. M., Wiche, G., and Liddington, R. C. (1999). Crystal structure of a tandem pair of fibronectin type III domains from the cytoplasmic tail of integrin α6β4. *EMBO J.*, **18**, 4087–95.
9. Erb, E.-M., Tangemann, K., Bohrmann, B., Müller, B., and Engel, J. (1997). Integrin αIIbβ3 reconstituted into lipid bilayers is non-clustered in its activated state but clusters after fibrinogen binding. *Biochemistry*, **36**, 7395–402.
10. Springer, T. A. (1997). Folding of the N-terminal, ligand-binding region of integrin α-subunits into a β-propeller domain. *Proc. Natl. Acad. Sci. USA*, **94**, 65–72.
11. Oxvig, C. and Springer, T. A. (1998). Experimental support for a β-propeller domain in integrin α-subunits and a calcium binding site on its lower surface. *Proc. Natl. Acad. Sci. USA*, **95**, 4870–5.
12. Irie, A., Kamata, T., and Takada, Y. (1997). Mutiple loop structures critical for ligand binding of the integrin α4 subunit in the upper face of the β-propeller model. *Proc. Natl. Acad. Sci. USA*, **94**, 7198–203.
13. Baneres, J.-L., Roquet, F., Green, M., LeCalvez, H., and Parello, J. (1998). The cation-binding domain from the α subunit of integrin α5β1 is a minimal domain for fibronectin recognition. *J. Biol. Chem.*, **273**, 24744–53.
14. Gulino, D., Boudignon, C., Zhang, L., Concord, E., Rabiet, M.-J., and Marguerie, G. (1992). Ca^{2+} binding properties of the platelet glycoprotein IIb ligand interacting domain. *J. Biol. Chem.*, **267**, 1001–7.
15. Michishita, M., Videm, V., and Arnaout, M. A. (1993). A novel divalent cation-binding site in the A domain of the β2 integrin CR3 (CD11b/CD18) is essential for ligand binding. *Cell*, **72**, 857–67.
16. Kamata, T. and Takada, Y. (1994). Direct binding of collagen to the I-domain of integrin α2β1 (VLA-2, CD49b/CD29) in a divalent cation-independent manner. *J. Biol. Chem.*, **269**, 26006–10.
17. Tuckwell, D. S., Calderwood, D. A., Green, L. J., and Humphries, M. J. (1995). Integrin α2 I-domain is a binding site for collagens. *J. Cell Sci.*, **108**, 1629–37.
18. Branden, C. and Tooze, J. (1991). *Introduction to protein structure*. Garland Inc., New York.
19. Kamata, T., Wright, R., and Takada, Y. (1995). Critical threonine and aspartate residues within the I-domains of β2 integrins for interactions with intercellular adhesion molecule 1(ICAM-1) and C3bi. *J. Biol. Chem.*, **270**, 12531–5.
20. Emsley, J., Knight, C. G., Barnes, M. J., Farndale, R. W., and Liddington, R. C. (2000). Structural basis of integrin:collagen recognition. *Cell*, **101**, 47–56.
21. Huang, C. and Springer, T. A. (1995). A binding interface on the I domain of lymphocyte function associated antigen-1 (LFA-1) required for specific interaction with intercellular adhesion molecule 1 (ICAM-1). *J. Biol. Chem.*, **270**, 19008–16.
22. Rieu, P., Sugimori, T., Griffith, D. L., and Arnaout, M. A. (1996). Solvent-accessible residues on the metal ion-dependent adhesion site face of integrin CR3 mediate its binding to the neutrophil inhibitory factor. *J. Biol. Chem.*, **271**, 15858–61.
23. Zhang, L. and Plow, E. F. (1997). Identification and reconstruction of the binding site with in αMβ2 for a specific and hgh affinity ligand, NIF. *J. Biol. Chem.*, **272**, 17558–64.
24. Li, R., Rieu, P., Griffith, D. L., Scott, D., and Arnaout, M. A. (1998). Two functional states of the CD11b A-domain: correlations with key features of two Mn^{2+}-complexed crystal structures. *J. Cell Biol.*, **143**, 1523–34.
25. Zhang, L. and Plow, E. F. (1999). Amino acid sequences within the α subunit of integrin αMβ2 (Mac-1) critical for specific recognition of C3bi. *Biochemistry*, **38**, 8064–71.
26. Kamata, T., Liddington, R. C., and Takada, Y. (1999). Interaction between collagen and the

α2 I domain of integrin α2β1: critical role of conserved residues. *J. Biol. Chem.*, **274**, 32108–11.

27. Loftus, J. C., Smith, J. W., and Ginsberg, M. H. (1994). Integrin-mediated cell adhesion: the extracellular face. *J. Biol. Chem.*, **269**, 25235–8.
28. Bitan, G., Scheibler, L., Greenberg, Z., Rosenblatt, M., and Chorev, M. (1999). Mapping the αVβ3-ligand interface by photoaffinity cross-linking. *Biochemistry*, **38**, 3414–20.
29. Takagi, J., Kamata, T., Meredith, J., Puzon-McLaughlin, W., and Takada, Y. (1997). Changing ligand specificities of αVβ1 and αVβ3 integrins by swapping a short diverse sequence of the β subunit. *J. Biol. Chem.*, **272**, 19794–800.
30. Lin, E. C., Ratnikov, B. I., Tsai, P. M., Carron, C. P., Myers, D. M., Barbas, C. F. R., and Smith, J. W. (1997). Identification of a region in the integrin β3 subunit that confers ligand binding specificity. *J. Biol. Chem.*, **272**, 23912–20.
31. Bajt, M. L. and Loftus, J. C. (1994). Mutation of a ligand binding domain of β3 integrin. Integral role of oxygenated residues in αIIbβ3 receptor function. *J. Biol. Chem.*, **269**, 20913–19.
32. Tozer, E. C., Liddington, R. C., Sutcliffe, M. J., Smeeton, A. H., and Loftus, J. C. (1996). Ligand binding to integrin αIIbβ3 is dependent on a MIDAS-like domain in the β3 subunit. *J. Biol. Chem.*, **271**, 21978–84.
33. Goodman, T. G., and Bajt, M. L. (1996). Identifying the putative metal ion dependent adhesion site in the β2 (CD18) subunit required for αLβ2 and αMβ2 ligand interactions. *J. Biol. Chem.*, **271**, 23729–36.
34. Puzon-McLaughlin, W. and Takada, Y. (1996). Critical residues for ligand binding in an I domain–like structure of the integrin β1 subunit. *J. Biol. Chem.*, **271**, 20438–43.
35. Huang, C., Lu, C., and Springer, T. A. (1997). Folding of the conserved domain but not of flanking regions in the integrin β2 subunit requires association with the α-subunit. *Proc. Natl. Acad. Sci. USA*, **94**, 3156–61.
36. Bilsland, C. A. G. and Springer, T. A. (1994). Cloning and expression of the chicken CD18 cDNA. *J. Leukocyte Biol.*, **55**, 501–6.
37. McKay, B. S., Annis, D. S., Honda, S., Christie, D., and Kunicki, T. J. (1996). Molecular requirements for assembly and function of a minimized human integrin αIIbβ3. *J. Biol. Chem*, **271**, 30544–7.
38. Humphries, M. J. (1996). Integrin activation: the link between ligand binding and signal transduction. *Curr. Opin. Cell Biol.* **8**, 632–40.
39. Labadia, M. E., Jeanfavre, D. D., Caviness, G. O., and Morelock, M. M. (1998). Molecular recognition of the interaction between leukocyte function-associated antigen-1 and soluble ICAM-1 by divalent metal cations. *J. Immunol.*, **161**, 836–42.
40. Hu, D. D., Barbas III, C. F., and Smith, J. W. (1996). An allosteric Ca^{2+} binding site on the β3 integrins that regulates the dissociation rate for RGD ligands. *J. Biol. Chem.*, **271**, 21745–51.
41. Mould, A. P., Akiyama, S. K., and Humphries, M. J. (1995). Regulation of integrin α5β1–fibronectin interactions by divalent cations. *J. Biol. Chem.*, **270**, 26270–7.
42. Lee, J.-O., Bankston, L. A., Arnaout, M. A., and Liddington, R. C. (1995). Two conformations of the integrin A-domain (I-domain): a pathway for activation? *Structure*, **3**, 1333–40.
43. D'Souza, S. E., Haas, T. A., Piotrowicz, R. S., Byers-Ward, V., McGrath, D. E., Soule, H. R., Ciernieswski, C., Plow, E. F., and Smith, J. W. (1994). Ligand and cation binding are dual functions of a discrete segment of the integrin β3 subunit: cation displacement is involved in ligand binding. *Cell*, **79**, 659–67.

44. Dickeson, S. K., Bhattacharyya-Pakrasi, M., Mathis, N. L., Schlesinger, P. H., and Santoro, S. A. (1998). Ligand binding results in divalent cation displacement from the α2β1 integrin I domain: evidence from terbium luminescence spectrsocopy. *Biochemistry*, **37**, 11280–8.
45. Hynes, R. O. (1990). *Fibronectins*. Springer-Verlag, New York.
46. Akiyama, S. K. and Yamada, K. M. (1985). The interaction of plasma fibronectin with fibroblastic cells in suspension. *J. Biol. Chem.*, **260**, 4492–500.
47. Pierschbacher, M. D., Hayman, E. G., and Ruoslahti, E. (1981). Location of the cell-attachment site in fibronectin with monoclonal antibodies and proteolytic fragments of the molecule. *Cell*, **26**, 259–67.
48. Humphries, M. J., Akiyama, S. K., Komoriya, A., Olden, K., and Yamada, K. M. (1986). Identification of an alternatively spliced site in human plasma fibronectin that mediates cell type-specific adhesion. *J. Cell Biol.*, **103**, 2637–47.
49. Pierschbacher, M. D. and Ruoslahti, E. (1984). The cell attachment activity of fibronectin can be duplicated by small synthetic fragments of the molecule. *Nature*, **309**, 30–3.
50. Humphries, M. J. (1990). The molecular basis and specificity of integrin–ligand interactions. *J. Cell Sci.*, **97**, 585–92.
51. Yamada, K. M. (1991). Adhesive recognition sequences. *J.Biol.Chem.*, **266**, 12809–12.
52. Coller, B. S. (1997). Platelet GPIIb/IIIa antagonists: the first anti-integrin receptor therapeutics. *J. Clin. Invest.*, **99**, 1467–71.
53. Lobb, R. R. and Hemler, M. E. (1994). The pathophysiologic role of α4 integrins in vivo. *J. Clin. Invest.*, **94**, 1722–8.
54. Obara, M., Kang, M. S., and Yamada, K. M. (1988). Site-directed mutagenesis of the cell-binding domain of human fibronectin: separable, synergistic sites mediate adhesive function. *Cell*, **53**, 649–57.
55. Aota, S., Nomizu, M., and Yamada, K. M. (1994). The short amino acid sequence Pro–His–Ser–Arg–Asn in human fibronectin enhances cell-adhesive function. *J. Biol. Chem.*, **269**, 24756–61.
56. Main, A. L., Harvey, T. S., Baron, M., Boyd, J., and Campbell, I. D. (1992). The three-dimensional structure of the tenth type III module of fibronectin: an insight into RGD-mediated interactions. *Cell*, **71**, 671–8.
57. Leahy, D. J., Aukhil, I., and Erickson, H. P. (1996). 2.0 Å crystal structure of a four-domain segment of human fibronectin encompassing the RGD loop and synergy region. *Cell*, **84**, 155–64.
58. Humphries, M. J., Komoriya, A., Akiyama, S. K., Olden, K., and Yamada, K. M. (1987). Identification of two distinct regions of the type III connecting segment of human plasma fibronectin that promote cell type-specific adhesion. *J. Biol. Chem.*, **262**, 6886–92.
59. Komoriya, A., Green, L. J., Mervic, M., Yamada, S. S., Yamada, K. M., and Humphries, M. J. (1991). The minimal essential sequence for a major cell-type specific adhesion site (CS1) within the alternatively spliced IIICS domain of fibronectin is Leu–Asp–Val. *J. Biol. Chem.*, **266**, 15075–9.
60. Mould, A. P., Komoriya, A., Yamada, K. M., and Humphries, M. J. (1991). The CS5 peptide is a second site in the IIICS region of fibronectin recognized by the integrin α4β1. Inhibition of α4β1 function by RGD peptide homologues. *J. Biol. Chem.*, **266**, 3579–85.
61. Mould, A. P., Askari, J. A., Aota, S., Yamada, K. M., Irie, A., Takada, Y., Mardon, H. J., and Humphries, M. J. (1997). Defining the topology of integrin α5β1-fibronectin interactions using inhibitory anti-α5 and anti-β1 monoclonal antibodies: evidence that the synergy sequence of fibronectin is recognized by the amino-terminal repeats of the α5 subunit. *J. Biol. Chem.*, **272**, 17283–92.

62. Newham, P., Craig, S. E., Clark, K., Mould, A. P., and Humphries, M. J. (1998). Analysis of ligand-induced and ligand-attenuated epitopes on the leukocyte integrin α4β1≷ VCAM-1, mucosal addressin cell adhesion molecule-1, and fibronectin induce distinct conformational changes. *J. Immunol.*, **160**, 4508–17.
63. Mould, A. P. and Humphries, M. J. (1991). Identification of a novel recognition sequence for the integrin α4β1 in the COOH-terminal heparin-binding domain of fibronectin. *EMBO J.*, **10**, 4089–95.
64. Sharma, A., Askari, J. A., Humphries, M. J., Jones, E. Y., and Stuart, D. S. (1999). Crystal structure of a heparin and integrin-binding segment of human fibronectin. *EMBO J.*, **18**, 1468–79.
65. Jones, E. Y., Harlos, K., Bottomley, M. J., Robinson, R. C., Driscoll, P. C., Edwards, R. M., Clements, J. M., Dudgeon, T. J., and Stuart, D. I. (1995). Crystal structure of an integrin-binding fragment of vascular cell adhesion molecule-1 at 1.8Å resolution. *Nature*, **373**, 539–44.
66. Tan, K., Casasnovas, J. M., Liu, J. H., Briskin, M. J., Springer, T. A., and Wang, J. H. (1998). The structure of immunoglobulin superfamily domains 1 and 2 of MAdCAM-1 reveals novel features important for integrin recognition. *Structure*, **6**, 793–801.
67. Casasnovas, J. M., Stehle, T., Liu, J., Wang, J., and Springer, T. A. (1998). A dimeric crystal structure for the N-terminal two domains of intercellular adhesion molecule-1. *Proc. Natl. Acad. Sci. USA.*, **95**, 4134–9.
68. Bella, J., Kolatkar, P. R., Marlor, C. W., Greve, J. M., and Rossmann, M. G. (1998). The structure of the two amino-terminal domains of human ICAM-1 suggests how it functions as a rhinovirus receptor and as an LFA-1 integrin ligand. *Proc. Natl. Acad. Sci. USA*, **95**, 4140–5.
69. Casasnovas, J. M., Springer, T. A., Liu, J., Harrison, S. C., and Wang, J. (1997). Crystal structure of ICAM-2 reveals a distinctive integrin recognition surface. *Nature*, **387**, 312–15.
70. Miller, J., Knorr, R., Ferrone, M., Houdei, R., Carron, C., and Dustin M. L. (1995). Intercellular adhesion molecule-1 dimerization and its consequences for adhesion mediated by lymphocyte function associated-1. *J. Exp. Med.*, **182**, 1231–41.
71. Mould, A. P., Akiyama, S. K., and Humphries, M. J. (1996). The inhibitory anti-β1 integrin monoclonal antibody 13 recognizes an epitope that is attenuated by ligand occupancy: evidence for allosteric inhibition of integrin function. *J. Biol. Chem.*, **271**, 20365–74.
72. Chen, L. L., Lobb, R. R., Cuervo, J. H., Lin, K., Adams, S. P., and Pepinsky, R. B. (1998). Identification of ligand binding sites on integrin α4β1 through chemical crosslinking. *Biochemistry*, **37**, 8743–53.
73. Qu, A. and Leahy, D. J. (1996). The role of the divalent cation in the structure of the I domain from the CD11a/CD18 integrin. *Structure*, **4**, 931–42.
74. Baldwin, E. T., Sarver, R. W., Bryant, G. L., Curry, K. A., Fairbanks, M. B., Finzel, B. C., Garlick, R. L., Heinrikson, R. L., Horton, N. C., Kelley, L. L.C., Mildner, A. M., Moon, J. B., Mott, J. E., Mutchler, V. T., Tomich, C. S.C., Watenpaugh, K. D., and Wiley, V. H. (1998). Cation binding to the integrin CD11b I domain and activation model assessment. *Structure*, **6**, 923–35.
75. Oxvig, C., Lu, C., and Springer, T. A. (1999). Conformational changes in tertiary structure near the ligand binding site of an integrin I domain. *Proc. Natl. Acad. Sci. USA*, **96**, 2215–20.
76. Zhang, L. and Plow, E. F. (1996). A discrete site modulates activation of I domains. *J. Biol. Chem.*, **271**, 29953–7.

77. Kallen, J., Welzenbach, K., Ramage, P., Geyl, D., Kriwacki, R., Legge, G., Cottens, S., Weitz-Schmidt, G., and Hommel, U. (1999). Structural basis for LFA-1 inhibition upon lovastatin binding to the CD11a I-domain. *J. Mol. Biol.*, **292**, 1–9.
78. Emsley, J., Cruz, M., Handin, R., and Liddington, R. (1998). Crystal structure of the von Willebrand Factor A1 domain and implications for the binding of glycoprotein Ib. *J. Biol. Chem.*, **273**, 10396–401.
79. Dembo, M., Torney, D. C., Saxman, K., and Hammer, D. (1988). The reaction-limited kinetics of membrane-to-surface adhesion and detachment. *Proc. R. Soc. London*, B**234**, 55–83.
80. Frelinger, A., Du, X. P., Plow, E. F., and Ginsberg, M. H. (1991). Monoclonal antibodies to ligand-occupied conformers of integrin αIIbβ3 alter receptor affinity, specificity, and function. *J. Biol. Chem.*, **266**, 17106–11.
81. Sims, P. J., Ginsberg, M. H., Plow, E. F., and Shattil, S. J. (1991). Effect of platelet activation on the conformation of the plasma membrane glycoprotein IIb–IIIa complex. *J. Biol. Chem.*, **266**, 7345–52.
82. Kouns, W. C., Hadvary, P., Haering, P., and Steiner, B. (1992). Conformational modulations of purified glycoprotein GPIIb–IIIa allows proteolytic generation of active fragments from either active or inactive GPIIb–IIIa. *J. Biol. Chem.*, **267**, 18844–51.
83. Monod, J., Wyman, J., and Changeux, J.-P. (1963). Allosteric systems and cellular control systems. *J. Mol. Biol.*, **6**, 306.
84. Hughes, P., Diaz-Gonzalez, F., Leong, L., Wu, C., McDonald, J., Shattil, S. J., and Ginsberg, M. H. (1996). Breaking the integrin hinge. *J. Biol. Chem.*, **271**, 6571–4.

5 Structure and energetics of anti-lysozyme antibodies

BRADFORD C. BRADEN and ROBERTO J. POLJAK

1. Introduction

The formation of specific, stable protein–protein complexes is essential to biological processes. Since such complexes are fundamental for biological function, an understanding of their structure can lead to important clues about structure–function relationships in general. Such an understanding requires detailed characterization of the three-dimensional structure of the proteins as well as the physicochemical basis of interatomic interactions at the points of intermolecular association, i.e. the interface.

Advances in the techniques of protein structure determination, site-directed mutagenesis for probing amino-acid substitutions, and methods for investigating the kinetic and thermodynamic behaviour of protein–protein association have greatly enhanced our understanding of the underlying forces involved in protein–protein recognition and complex formation.

These techniques have, in the past few years, been applied to the problems of understanding the structure and energetics of antibody binding to protein antigens, the most common and diversified antigens encountered by the immune system. The primary paradigm of antibody–antigen interaction is that the three-dimensional structure of the six hypervariable loops, or complementary-determining regions (CDR), of an antibody molecule recognize and bind a specific antigenic surface which is also determined by the three-dimensional structure of the antigen. Antibody–protein antigen interactions occur over large (>650 Å^2) sterically and electrostatically complementary areas. That is, apolar patches of the antigen surface are recognized and interact with apolar patches on the surface of the antigen-binding site of an antibody; in turn, polar atoms interact with oppositely charged atoms in the antibody, and proton donors and acceptors are involved in hydrogen bonds. Although crystallographic studies of antibody–antigen interactions have revealed the stereochemistry of the intermolecular contacts and complementarity in several systems, the complexity of these contacts and limitations in our current knowledge of molecular reactions preclude a deeper understanding of the intermolecular forces governing their association. For example, on the basis of the X-ray crystal structures alone, it is still quite difficult to evaluate the individual contribution of amino acids or atoms to the free-energy changes that stabilize the complexes.

Going hand-in-hand with the complementary nature of the antibody–antigen interaction is the notion that the antibody should bind its antigen with a high affinity. Thus, the nature of the interactions involved in the formation of antibody–antigen complexes should also be reflected in the thermodynamics of the binding interactions. In this regard, a prime objective is the elucidation of the structural features of the antibody and antigen that contribute to the enthalpic (hydrogen bonds, salt bridges, dispersion forces) and entropic (solvent and conformational changes) components of the free energy of complex formation. With the application of site-directed mutagenesis, calorimetric measurements, and plasmon resonance techniques the contribution of individual residues to the energetics of complex formation can be explored, although, as will be outlined in this chapter, this is still difficult. It is indeed a convincing result when observations from three-dimensional crystal structure determinations can explain the physicochemical differences between the reactions of wild-type and mutant antibody–antigen complexes. On the other hand, results from X-ray crystal structure determinations do not inherently disclose the entropic components to free energy such as changes in the rotational, translational, and vibrational freedoms of the system or effects arising from solvent entropy. None the less, in a few instances where the crystal structures of free and complexed antibodies and antigens are known to high resolution some entropic information can be inferred, particularly in the cases of conformational changes in antibody and antigen during complex formation and the structure of water in and around the antibody–antigen interface.

The antibody molecule has been repeatedly presented as a model of biological adaptation: structure, function, and specificity being built upon a carefully conserved framework structure. Thus the genetic control and three-dimensional structure of antibody molecules are the key to adaptive responses to varied antigenic challenges. This diversity, conjoined to a constant overall molecular structure, propels the study of antibody–antigen interaction into the generalization of protein association.

However, not all the problems posed by antibody function and specificity have been worked out. In fact, the search is only beginning for a detailed understanding of the structural basis of specificity and the physicochemical characterization of antibody action at the molecular level. There is a continuing search for a molecular definition of immunogenicity and antigenicity, for new uses of antibody molecules in regulating immune responses, in mimicking foreign antigens, and in delivering therapeutic agents, etc. These applied aspects of antibody action, as well as their relevance to protein association, have attracted much attention in the last few years.

Although the understanding of those intermolecular interactions that stabilize an antibody–antigen complex arise from the study of the forces which stabilize protein folding, the study of antibody–antigen interaction can likewise contribute to the comprehension of protein–protein recognition. The search for a detailed understanding of the structural basis of specificity and the physicochemical characterization of antibody action at the molecular level are only beginning to unravel the fundamental questions relating to the recognition process. These questions are:

- Is productive binding mediated by a distinct subset of combining site residues or are more complex cooperative interactions involving both contacting and non-contacting residues responsible for the affinity of antigen for antibody?
- What is the contribution of individual residues in the antibody–antigen complex to the free energy of complex formation?
- What are the relative contributions of hydrophobicity, surface complementarity, hydrogen bonding, and ionic and van der Waals interactions to the energetics and mechanism of binding?
- To what extent do individual interatomic interactions depend on the local environment?
- What is the role of solvent in the antibody–antigen interface? Is water simply displaced upon the binding of antibody and antigen or does the antibody recognize a solvated antigen trapping water molecules in the interface? If solvent is trapped is this a productive or non-productive interaction?
- How does the antigen-binding site accommodate the structure of the antigen? Is antibody–antigen binding simply a lock-and-key mechanisms, or are subtle or even complex atomic rearrangements necessary for antigen binding?

In this review we shall report on the progress that has been made in answering these important questions. The antibody–antigen system we shall discuss is the mouse anti-hen, egg-white lysozyme antibody D1.3. We shall also discuss other mouse anti-lysozyme antibodies as well, particularly HyHEL-5 and D44.1, that have significantly added to the understanding of the structure and energetics of antibody–protein antigen interaction.

2. Historical perspective of antibody–antigen binding

The literature concerning the three-dimensional structure of antibodies has been well reviewed (1–3), so we will only present a short historical perspective to define the structure terms. The earliest crystal structures of Fab fragments, L chains, and complete immunoglobulin molecules (Ig) confirmed Porter and Edelman's proposals of a homodimer composed of two identical polypeptide chains spanning $\sim$450 amino acids (heavy or H-chain) and two identical polypeptide chains of about 250 amino acids (light or L-chain). The Ig molecule can be separated by papain digestion into three fragments; two Fab (antigen-binding fragments) and an Fc (historically, the crystallizable fragment). The Fab fragments themselves are heterodimers of the light and heavy chains, each composed of an amino-acid sequence variable domain (V_H or V_L) and an amino-acid sequence constant domain (C_H1 or C_L). The antigen-binding site of the Ig molecule is situated at the amino-terminal ends of the H and L chains and antigen specificity is determined by 'hypervariable' sequences in the V_H and V_L domains (the Fv moiety). The H chains continue to the carboxy-terminal of the immunoglobulin molecule through a 'hinge' segment and two additional constant domains (C_H2 and C_H3; i.e. the Fc fragment). While the Fab fragment of the Ig molecule is the antigen-binding constituent, the Fc fragment is the anchoring site for proteins of the complement system and for receptors on different effector cells.

Dimerization of the Ig molecule results from covalent (disulfide) bridges between the two C_H2 domains and from non-covalent interactions between constant domains C_H2 and C_H3. Linking the Fab and Fc regions of each half of the Ig molecule is a stretch of 10 or more amino acids, rich in proline and cysteine residues and known as the 'hinge region'; this allows for intersegmental flexibility of the Ig molecule, in the sense of allowing freedom in the relative orientations of the Fabs and the Fc fragments. This flexibility of antibodies is important for their roles in effector function and in facilitating the binding to equivalent antigenic determinants when these are randomly oriented in space.

Structural studies of intact antibodies allow comparison of the relative disposition of the Fab and Fc regions, thus demonstrating the flexibility allowed by the hinge region. The three-dimensional structure of a human myeloma immunoglobulin, Mcg (IgG1, λ), has been determined by X-ray diffraction techniques to a resolution of 6.5 Å (4). In this IgG, as in the previously reported IgG Dob (5), a deletion of the hinge region brings the Fab and Fc regions closer together and confers a rigidity not found on normal IgGs. The molecule has an exact twofold axis of symmetry, which can be described as T-shaped, although the angle between the Fab arms is 170° instead of the 180° value observed in IgG Dob. Thus, the absence of a hinge region evidently restricts the relative intersegmental mobility of Fab and Fc. Mobility had been observed before, even in the crystalline state, in human IgG1 Kol (6) for which the Fc region was disordered and no electron density could be observed to model it. A second example of intersegmental mobility can be added, that of IgG2 (κ) Zie, a cryoglobulin with four inter-H chain disulfide bonds in its hinge region. This IgG2 and its Fab crystallize isomorphically and give similar X-ray intensity distributions, indicating that the Fc assumes multiple orientations as if in motion relative to the Fab arms of the molecule (7).

More recently, McPherson and colleagues (8) have described the X-ray crystal structure of an intact IgG κ molecule at 3.2-Å resolution. An interesting observation of this molecule is that, although the two Fabs are related by a 179.7° rotation, a translation of about 9 Å is required to superpose these segments. The overall shape of this immunoglobulin is also roughly T-shaped. Thus, the conformational range for the disposition of antibody segments is quite large.

X-ray crystal structures of Fab and Fv antibody fragments have formed the basis of the understanding of antibody specificity for antigen at an atomic level. All the antibody domains, as defined above, have the same general tertiary structure, i.e. the so-called 'immunoglobulin fold' (9). This structure is composed of two antiparallel beta-sheets covalently linked by a disulfide bridge. The antigen-binding fragment Fab, composed of the four domains V_H, V_L, C_H1, and C_L, has been the subject of numerous X-ray crystallographic structure determinations (reviewed in refs 1, 2). Each immunoglobulin domain is paired with a second (i.e. V_H–V_L, C_H1–C_L), with the components of each of these pairs related by a pseudo-twofold rotation axis. Each antibody domain is joined to a subsequent domain (i.e. V_H–C_H1, V_L–C_L) by a short segment of polypeptide. This sequence of extended polypeptide, as in the example of the 'hinge' region, permits intersegmental flexibility of the Fab and, as such, allows for differing relative orientations of the V_H, V_L (i.e. the Fv fragment) and the C_H1 and C_L domains. The relative disposition of the variable and constant domains (the so-called elbow angle) of an Fab is defined as the angle between the pseudo-twofold rotation axes of the Fv and C_H1–C_L domains. Thus, an elbow angle of 180° specifies that the pseudo-twofold axes of the Fv and C_H1–C_L domains are co-linear and angles greater or less than 180° specify a Fab which is asymmetrical. Well before the vast majority of Fab crystal structures had been determined, Poljak *et al.* (10) and Huber *et al.* (11) had proposed that a change in the elbow angle, upon immunoglobulin binding of antigen, was important in the signalling mechanism which initiates secondary or effector functions. However, as pointed out in a review by Wilson and Stanfield (12) of the known elbow angles in X-ray crystal structures of free and liganded Fabs, there is no compelling structural evidence to support an allosteric mechanism for signalling that the antibody has bound antigen. Calorimetric analysis of antigen binding to intact IgG molecules, Fab, and Fv fragments also fails to identify an allosteric effect upon antigen binding (13).

As the Fv domain is approximately half the molecular weight of a Fab, single crystals of Fv antibody fragments generally diffract to high resolution. As a consequence, the large number of X-ray observations allows the refinement of molecular models to high accuracy in atomic position and stereochemical parameters. Therefore, structural studies of Fv fragments, both free and complexed with antigen, are of great value. The structure of the Fv fragment consists of the two immunoglobulin variable domains (V_H and V_L) that, as stated previously, are related by a pseudo-twofold rotation axis. There are six hypervariable segments, or complementarity-determining-regions (CDRs), three each from the V_H and V_L domains, that are formed from polypeptide loops which connect the beta sheets of the immunoglobulin domain. The conformation of the CDR loops are generally determined by the length of the loop, the distance between the invariant (framework) residues which anchor the loop, and the primary sequence of amino acids. By comparing the sequences and lengths of CDR loops of known structure, Chothia *et al.* (14) discovered that each CDR loop, with the exception of the third hypervariable loop of the heavy chain (CDR H-3), generally conform to one of a few structural possibilities (i.e. canonical models). While there may be an extremely large repertoire of primary sequences of hypervariable loops, there appears to be some limit to the number of tertiary structures into which the backbone polypeptide chain at each loop can fold. This limits the overall structural design, but it still provides for structural diversity at the level of the amino-acid side chain. As yet, no canonical models have been suggested for CDR H-3, owing to its variability in length; however, the local conformation of the 'stem' region can be predicted from sequence studies (15). Several recent structure determinations of antibody fragments, however, have illustrated the difficulty in predicting CDR conformation from sequence data alone with sufficient accuracy to permit modelling of binding interactions (16–19). This difficulty is especially acute in cases where local structure may be influenced by long-range interactions or effects from crystal packing.

3. Energetics of protein folding and association

The classic papers by Anson (20) and Anfinsen (21), recognizing that the stability of a folded protein is a thermodynamically driven process, garnered implications for protein–protein associations. It has long been proposed that a major contribution to the free energy of protein folding and protein–protein complexation arises from the transfer of apolar atoms from an aqueous to a close-packed hydrophobic environment, with a corresponding reduction in solvent-accessible surface area and a favourable increase in solvent entropy (i.e. the 'hydrophobic effect' (22–24)). This concept was widely suggested as the primary mode of antibody–antigen binding. As will be presented below, this view of the energetics of antibody–antigen binding is being challenged. For example, it has been argued that the hydrophobic effect alone cannot account for the observed changes in entropy and enthalpy in a protein–protein reaction which is enthalpically driven (25).

Calorimetric analysis of several antibody–protein antigen interactions demonstrate that complex formation is an enthalpically driven process with varying degrees of entropic stabilization, or even destabilization. X-ray crystallographic structure determination of these antibody–antigen systems have revealed important polar and charged interactions, and even buried solvent water molecules, which play an important role in antibody–antigen recognition.

How best then to describe the energetics of antibody–antigen interaction? Unfortunately, the natures of the forces, which stabilize protein folding, and protein–protein association are still highly debated. These forces have been variously interpreted as arising from the withdrawal of the apolar protein surface from contacts with solvent, i.e. the hydrophobic effect, or from electrostatic interactions, mainly hydrogen bonds and van der Waals (dispersion) interactions. In the following sections we will briefly describe the current views of the nature of these forces and how they have been applied to the understanding of antibody–antigen recognition.

3.1 Apolar interactions

Conveniently for the structural immunologist, much of the literature dealing with the contribution of apolar interactions to the free energy of protein folding or protein–protein association arrive at similar quantitative conclusions. Unfortunately, however, similar to the value of these contributions, interpretations as to their thermodynamic origin are in dispute. The main assumptions behind most of the experimental approaches to the understanding of protein folding and association are that each interacting atom makes some contribution, either stabilizing or destabilizing, to the overall free energy, and that these interactions can be modelled from smaller molecular systems such as phase transfer, free-energy changes of small molecules. This concept was first proposed by Kauzmann (22) and further elaborated by Tanford (26). It assumes that the change in the solvent-accessible surface area of the model compound, upon the change of phase, is proportional to the thermodynamic contribution to the stability of the intermolecular interactions. Thus, the energetic consequence of a cavity in the interior of a protein, or at the interface in a protein complex, is a loss proportional to that in interaction surface area or volume. Based on the transfer free energy of hydrocarbons, Chothia (23) proposed that the mean hydrophobic energy of carbon–carbon interactions was about 25 cal mol^{-1} Å^{-2} of buried surface area. Similar conclusions as to the nature of the hydrophobic effect have been reached based on the high-density packing of protein interiors (27), using a statistical mechanical model of protein stability (28), using atom-based solvation parameters and accessible surface areas (29), and using atom-based theoretical calculations (30).

Honig and co-workers (31, 32) arrive at a slightly larger value for the apolar driving forces in protein-folding and protein-binding interactions. These studies suggest that the free energy of solvation depends on the curvature of the apolar surface. Consequently, models of protein stability based upon the thermodynamic parameters of small convex molecules must be corrected by the effect of surface curvature on the water-accessible surface area. For a generally large planar surface of interaction, as has been noted for several antibody–protein antigen interactions (1, 2), a value of 50 cal mol^{-1} Å^{-2} for the hydrophobic interactions was proposed.

In a now classic series of experiments, Matthews and co-workers probed the effect of interior cavity-creating mutants on the stability of bacteriophage T4 lysozyme (reviewed in ref. 33). While the concomitant decrease in protein stability depended largely on the degree of protein relaxation around the subsequent cavity, they found an approximate linear relationship between the final cavity volume and the decrease in protein stability. In terms of the decrease in protein stability the cavity-creating mutants contributed to a free energy cost of 20 cal mol^{-1} Å^{-2} of lost contact surface area or 25 cal mol^{-1} Å^{-3} of cavity volume. Cavity-creating mutants of the dimeric 4-α-helical protein ROP (34), barnase (35) and staphylococci nuclease (36) suggest that the hydrophobic effect may be somewhat larger.

Refuting the classical notion that protein–protein association is driven by the hydrophobic effect, Privalov and co-workers (reviewed in ref. 37) make the point that proteins denature upon heating. This suggests that enthalpic factors, arising from close-packing dispersion interactions, not entropic factors, contribute to protein stability. Foremost in this explanation of the energetics of protein stability is that aromatic groups are not hydrophobic but exhibit a thermodynamically favourable hydration. Separating the effects of aliphatic and aromatic groups, Privalov contends that the contribution to protein stability by buried aliphatic groups is about 40 cal mol^{-1} Å^{-2} and 30 cal mol^{-1} Å^{-2} for aromatic groups.

3.2 Polar interactions

As many proteins are soluble at physiological ionic strength and pH, a large number of surface residues are hydrophilic in character. Since antigenic determinants lie on the surface of antigens, antibodies must be able to recognize and bind these polar constituents of the antigen surface. The polar interactions which may arise in the formation of an antibody–antigen complex are hydrogen bonds, occasional salt bridges, and, in a few instances, weak polar interactions to aromatic groups. The hydrogen bond is such an important potential that less than 2% of main-chain NH and CO moieties in a sample of high-resolution X-ray crystal structures exhibit no compensating interactions (38). Creighton (39), in fact, classifies the hydrogen bond as the dominant force stabilizing protein associations.

The energy of an intermolecular hydrogen bond ranges from 1.5 to 1.8 kcal mol^{-1} when the bonding pairs are uncharged and up to 6 kcal mol^{-1} when the bonding atoms are charged (40, 41). Possibly even more significant, is that uncompensated proton donors or acceptors can be quite destabilizing to complex formation, with unsatisfied buried hydrogen-bond donors or acceptors destabilizing the protein up to 3 kcal mol^{-1} (42). In the context of antibody–antigen interactions, therefore, the formation of hydrogen bonds in the molecular interface is not only stabilizing but also required when the antigenic epitope contains polar atoms. The apparent large breadth of values for the energy of a hydrogen bond arises from several factors. The strength of a hydrogen bond is dependent not only on the distance and angles between the proton and the bonding atoms but also on the dielectric environment. Thus a hydrogen bond in a low dielectric medium (such as the interior of a protein) is energetically quite different from that of a hydrogen bond in a high dielectric environment such as solvent water.

Solvent water also interacts with proton donors and acceptors forming hydrogen bonds. The strength of a water-mediated hydrogen bond is comparable to other types of hydrogen bonds (43), and the importance of water-mediated hydrogen bonds has been shown in a number of protein systems (44–46). Kornblatt *et al.* (47) demonstrated that about 12 water molecules may by essential for the formation of the cytochrome *c*–cytochrome *c* oxidase complex (see Chapter 3). Likewise, Dzingeleski and Wolfenden (48) have demonstrated that hydration of the active site of adenosine deaminase is required for hydrolytic activity of the enzyme.

As a final word on electrostatic interactions, it should be added that many analyses of the energetics of protein folding and association suggest that hydrogen bonds contribute little to protein stability. Although recognizing the requirement to electrostatically compensate proton donors and acceptors, this view suggests that hydrogen bonds are only important in the fine specificity of protein associations (23, 28). Moreover, several theoretical studies have suggested that the formation of hydrogen bonds and salt bridges may be destabilizing to complex formation. These studies, employing solutions of the classical Poisson–Boltzmann treatment of electrostatics, indicate that there is a cost to free energy when interacting pairs of polar atoms must be desolvated prior to the formation of hydrogen bonds or salt bridges (49, 50). The destabilizing consequence of these electrostatic interactions arises because the loss of favourable solvent interactions is not completely recovered by the resulting electrostatic attractions. While these results seem to have some important application in antibody–antigen interactions, suggesting that hydrophobic interactions alone are the source of complex stabilization, desolvated polar interactions are destabilizing only with respect to apolar interactions. It should also be kept in mind that there is a large penalty imposed by uncompensated polar groups, therefore, salt bridges and hydrogen bonds are still quite important, especially in determining specificity (50).

3.3 Entropy considerations and conformational changes

In addition to solvent entropy effects, arising from the hydrophobic effect, solute molecules have internal degrees of freedom: molecular rotations, translations, vibrations, and rotations about bonds, which contribute to the entropy of the system. Janin and colleagues have reviewed the theory and calculation of the magnitude of these components to system entropy (51; see Chapter 1), and have presented estimates of these terms to the association of HyHEL-5 and lysozyme (52). The change in system entropy for the HyHEL-5/HEL reaction arising from reduction of the translational and vibrational freedoms of the reacting proteins is in the order of -15 kcal mol^{-1} (at 300 K), while the reduction of side-chain mobility and atomic vibrations is even more destabilizing to complex formation (-50 kcal mol^{-1} at 300 K). Other estimates for the entropic terms in bimolecular systems have led to different results and suggest that the above estimates are high (53).

While the 'lock-and-key' mechanism of antibody–antigen association is still a valid metaphor, considering the specificity of binding, high-resolution crystal structures of both free and antigen-bound antibody fragments, coupled with the above estimates of system entropy, indicate that some molecular motion, as a product of the mutual accommodating of antibody and antigen, is essential to complex formation.

One final word on the entropy of macromolecular assemblies is the phenomenon of entropy/enthalpy compensation (32). Water molecules in a protein–protein interaction can either give up a hydrogen bond, at the cost of enthalpy and increase in entropy, or retain hydrogen bonds at a cost of entropy. Often, as we shall see, a mutation in a protein–protein interface will result in a decrease in the enthalpy of reaction with a corresponding increase in entropy. This solvent effect is difficult to perceive in X-ray crystal structures where a definitive view of solvent structure is unavailable. None the less, entropy/enthalpy compensation has been used to account for some of the energetics of antibody–antigen interactions even though the structural evidence is lacking.

Considering the above arguments one is essentially left with two views of protein association. The first representation of protein association is the one in which the hydrophobic effect is the primary source for protein stability and that polar interactions are, at best, only necessary for fine specificity. The second view is that polar and van der Waals interactions drive protein–protein associations with favourable (negative) enthalpies, in addition to providing specificity. Specific protein associations exist that exemplify and substantiate each of these views.

3.4 Morphology and topology of protein–protein associations

Although not immediately related to the quantification of the energetics of protein association, Larsen *et al.* (54) have reported on the major structural features of protein–protein association. Surveying 136 homodimeric proteins listed in the Brookhaven Protein Data Bank, this study concluded that protein–protein interfaces generally exhibit a mixture of apolar and polar interactions scattered over the binding surface. Only one-third of the proteins examined had a discernible hydrophobic core at the interface, surrounded by an annulus of polar interactions.

4. The structure and thermodynamics of the anti-lysozyme antibody D1.3

Antigenic determinants recognized by monoclonal anti-lysozyme antibodies raised in A/J mice (55, 56) and BALB/c mice (57–59) provided evidence that the entire surface of a protein molecule is potentially antigenic, giving rise to many antibodies that partially overlap their recognition surfaces.

The first antibody–protein antigen system employed to answer the questions concerning the structure and thermodynamics of antibody–antigen association was the anti-hen egg-white lysozyme antibody D1.3 (IgG1, κ). The structural interpretation of this antibody–antigen system was initially carried out with the Fab–lysozyme complex at 6-Å resolution (60), but was quickly extended to 2.8-Å resolution (61) and finally refined to 2.5 Å (62). These X-ray structure determinations established that the interaction between the antibody and antigen occurred over a large area with approximate dimensions of 20 × 30 Å. The sum of the solvent-inaccessible buried surface areas of HEL and D1.3 after complex formation was found to be about 1290 $Å^2$. The antibody combining site appeared as a rather flat surface with small protuberances and invaginations, complementary to the surface of the antigen. The antigenic determinant recognized by D1.3 was composed of two non-contiguous stretches of the lysozyme polypeptide chain, comprising residues 18–27 and residues 117 to the C-terminal residue 129. All six CDRs appeared to interact with the antigen, and, in all, 13 antibody residues made close contacts to 12 antigen residues. Most of the contacts were made by the heavy chain and in particular by CDR H3. Fischmann *et al.* (62) described an antibody–antigen interface which featured 20 hydrogen bonds and 44 interatomic van der Waals contacts.

Given the resolution at which the crystal structure of the complex was determined, no large-scale deviations between the complexed lysozyme and that of the free HEL in its tetragonal form (63) were evident. A least-squares fit of the lysozyme Cα atoms in the Fab D1.3–HEL complex with the Cα atoms of the HEL gave a root-mean-square (rms.) difference of 0.48 Å.

Although only 30 well-ordered water molecules were located in the 2.5-Å Fab D1.3–HEL crystal structure, seven of the solvent molecules were integral to the antibody–antigen interface. These water molecules were seen to participate in the hydrogen-bonding network bridging the antibody and antigen. Several other water molecules filled gaps between the buried surfaces of the antibody and antigen, contributing to the topological complementarity of the interface.

Coincident with the refinement of the 2.5-Å Fab D1.3–HEL crystal structure, the Fv fragment of D1.3 was bacterially expressed and crystals of the free Fv and Fv–HEL complex were produced (64). The X-ray crystal structures of the free Fv and Fv–HEL complex were solved by molecular replacement techniques using the 2.5-Å Fab D1.3–HEL crystal structure model and refined to 1.9-Å and 2.4-Å resolution, respectively (65). Eventually, Fv and Fv–HEL crystal data to 1.8 Å were obtained and the crystal structures refined to that resolution (66). The structure of the bacterially expressed Fv D1.3 were closely superposed with the V_H and V_L domains in the Fab D1.3–HEL complex, giving an rms difference in Cα positions of 0.39 Å. In particular, a very close correspondence was noted in the position and orientation of the antigen-contacting residues. However, exemplifying the importance of high-resolution crystal structures, the Fv D1.3–HEL complex featured a slightly different number of interatomic contacts as compared to the 2.5-Å Fab D1.3–HEL structure. A listing of the intermolecular contacts between Fv D1.3 and HEL is given in Table 1 and a view of the Fv D1.3–HEL interface is shown in Fig. 1.

The high-resolution Fv D1.3 crystal structures made possible a more detailed analysis of the effect of antigen binding on the antibody structure. Superposition of the Cα coordinates of the free Fv D1.3 V_L domain with the antigen-bound Fv D1.3 V_L domain gave an rms deviation of 0.37 Å, comparable to the estimated error in atomic positions. A similar observation was made on superimposing the V_H domains, indicating that the antigen binding did not induce major conformational changes in the individual V domains. This was borne out on comparing individual main-chain and side-chain conformations of both variable domains in the Fab–HEL, Fv free, and Fv–HEL crystal structures.

The effect of antigen binding on the relative arrangement of the variable domains was also examined by Bhat *et al.* (65). When the V_L domains of the Fv D1.3 and Fv D1.3–HEL structures were superimposed, a small relative displacement of the V_H domains was evident with an rms displacement of 0.52 Å. This systematic displacement of V_H relative to V_L moves the V_H domain closer to HEL, such that the antibody–antigen contact distances decrease by a mean value of 0.72 Å. Alternatively, when the V_H domains were superimposed, the V_L domain contact distances to the antigen decreased by a mean value of 0.53 Å. Although small but stereochemically significant movements of antibody side chains upon complex formation have been substantiated (65), it can be concluded that, in general terms, the conformations of side chains of the antibody combining site in its free form and in its complex with antigen are fairly close. This situation can be said to resemble a 'lock-and-key' fit of antigen and antibody.

The crystal structure of D1.3 complexed with turkey egg-white lysozyme (TEL) demonstrated a novel conformational change in the antigen combining site in order to accommodate the antigen (19). TEL and HEL differ in their amino-acid sequences at seven positions. However, only one residue in the antibody–antigen interface (HEL Gln121; TEL His121) is variant between the two lysozymes. HEL and TEL

Table 1 Intermolecular contacts in the FvD1.3–HEL crystal structure

D1.3	HEL	D1.3	HEL	D1.3	HEL
L1 30 Nδ1	**129 Oxt**	H1 31 C	117 Cα	H3 100 Cyy	24Cβ
L1 Tyr 32[1]	**121 Nε2**	H1 31 O	116 Cγ		24N
L2 49 OH	22 Cα		117 Cα		24 Oγ
L2 50 Cε2	18 Cγ	H1 32 Cζ	116Cε		27 Nδ2
	19 Cβ	H2 52 Cβ	117 O	**H3 100δ1**	**24 N**
L2 50 ζc	18 Cγ	H2 52 Cε2	119 Cβ		**22 O**
	18 Oδ1	H2 52 Cε3	118 C		23 C
	18 Oδ2		119 Cα		23 Cα
L2 50 OH	**18 Oδ2**		119 N	**H3 100δ2**	**24 Oγ**
	18 Oδ1	H2 52 Cζ3	118 C		**27 Nδ2**
	18 Cγ		119 N		24 Cβ
L2 50 Cδ2	19 Cβ		119 Cα	H3 101 Cδ2	121 Cβ
	19 Cγ		119 Cβ	H3 101 Cε2	121 Cβ
	19 Nδ2	H2 52 Cζ2	119 Cβ	H3 101 Cε1	120 Cγ2
L2 53 Oγ1	**19 Nδ2**		119 Cγ		119 Cα
L3 91 O	**121 Nε2**	H2 52 Cη2	119 Cα	**H3 101 OH**	**119 Oδ1**
	121 Cδ		119 Cβ		**120 N**
L3 92 Cε3	121 Cδ		119 Cγ		**121 N**
	121 Nε2		119 Oδ1		119 C
L3 92 Cζ3	121Cδ	**H2 53 N**	**117 O**		121 Cβ
L3 92 Cη2	125 Cα	H2 53 Cα	117 O		119 Cα
	125 Cγ	H2 54 Oδ2	118 Cβ	H3 101 Cζ	121 Cβ
L3 92 Cζ2	125 Cγ		118 Cγ2		119 Oδ1
	125 Cδ	**H3 99 Nη1**	**102 O**	**H3 102 Nη2**	**22 O**
L3 93 N	**121 Oε1**		102 C		
L3 93 Cα	121 Oε1				
L3 93 Cβ	125 Cδ				
	125 Nε				

Intermolecular contacts were defined by atom pair distances less than or equal to: C–C 4.1, C–N 3.8, C–O 3.7, N–N 3.4, N–o 3.4, O–O 3.3. Hydrogen bond pairs are in bold type.
[1] Hydrogen bond interaction to the phenyl ring of Tyr 32

interactions to the D1.3 main-chain atoms of CDR L3 show an exquisite electrostatic complementarity (Fig. 2). The Oε1 and Nε2 atoms of the HEL residue Gln121 form hydrogen bonds to D1.3 V_L backbone atoms Ser93 N and Phe91 O. In the D1.3–TEL crystal structure a peptide conformational change at Trp92–Ser93 exposes the carbonyl oxygen of Trp92, ensuring electrostatic complementarity and a hydrogen bond with TEL atom His121 Nε2. Thus, a subtle conformational change in D1.3 CDR L3 allows for the binding of two antigens with differing electrostatic properties in the region facing CDR L3. The association rate of the D1.3–HEL reaction is approximately 100-fold faster than for the D1.3–TEL reaction. This is probably a consequence of the observed peptide flip, suggesting that the more frequent conformation of D1.3 CDR L3 residues Trp92 and Ser93 is that found in the D1.3–HEL crystal structures of the Fv D1.3–HEL complex and the free Fv D1.3.

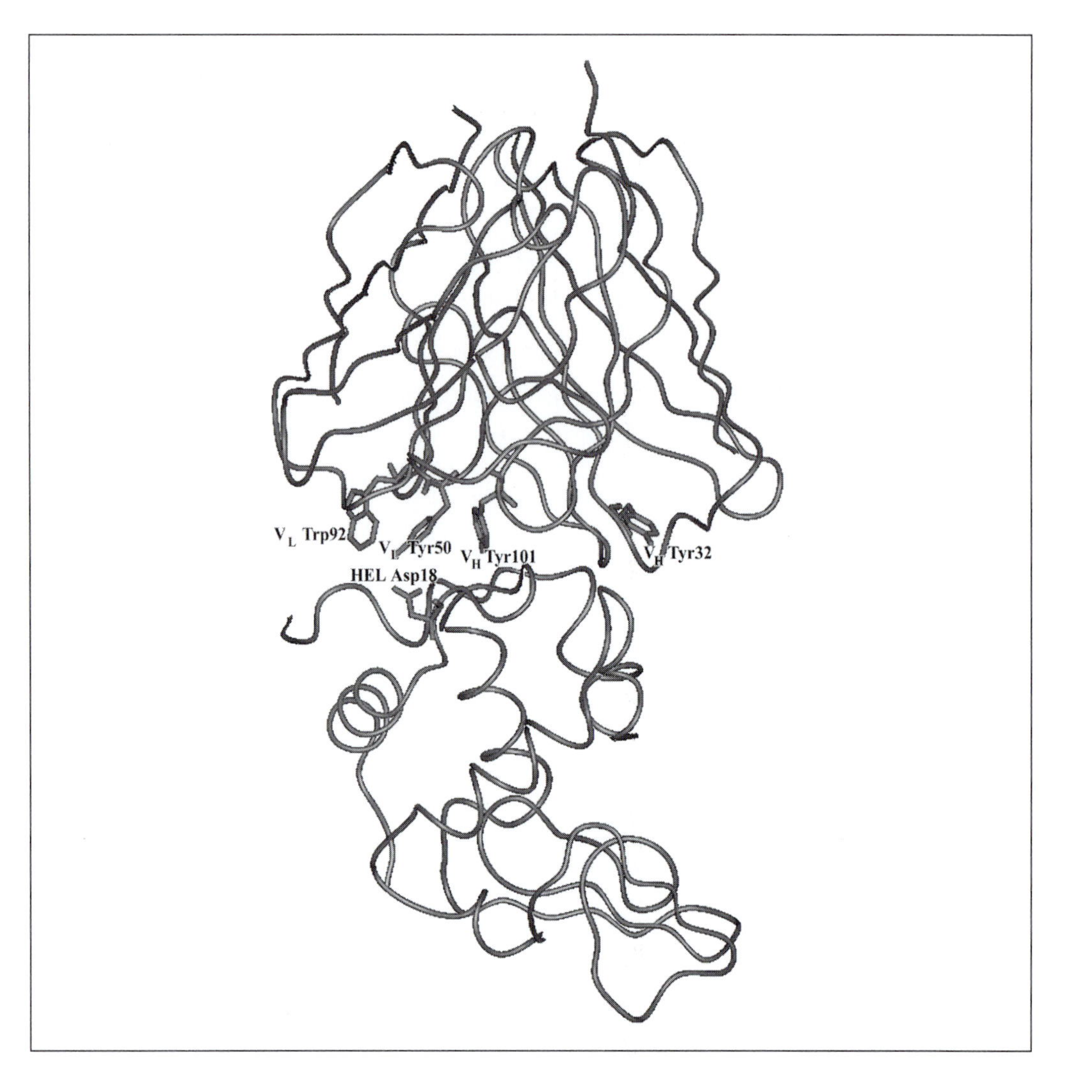

Fig. 1 View of the D1.3–HEL interface. The V_H chain, located at the upper right of the figure, the V_L chain at the upper left, and HEL below are represented by thin ribbon Cα traces. The five residues studied by site-directed mutation and X-ray crystallographic structure analysis are labelled.

4.1 Hydration and complementarity

Based on the crystallographic studies to 1.8-Å resolution of the free Fv and Fv–HEL complex, Bhat *et al.* (66) reported on the solvent structure of the complex and, particularly, on the water structure integral to the antibody–antigen interface. These water molecules formed an extensive hydrogen bond network, and having low temperature factors comparable to those of adjacent protein atoms, were considered an integral part of the complex. More than 20 water molecules were located in the free Fv combining site, and about 50 in the complex in the region of the antibody–antigen interface (Fig. 3). Based on this observation and the thermodynamics of the D1.3–HEL reaction, indicating favourable enthalpic and unfavourable entropic components to the free energy of the reaction (see Table 2), Bhat *et al.* (66) proposed that several water molecules from the bulk solvent are incorporated in the antibody–antigen interface.

Hydration has been demonstrated in many three-dimensional structures deter-

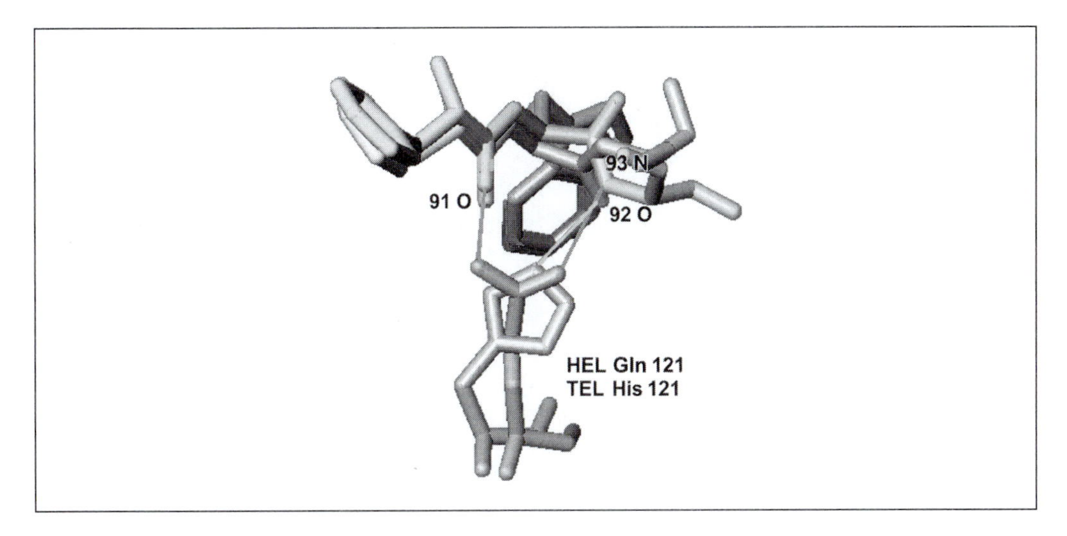

Fig. 2 View of the hydrogen bonding to lysozyme residue 121. D1.3–HEL and D1.3–TEL complexes. HEL Gln121 is involved in three hydrogen bonds to the antibody, to main-chain atoms Ser93 N and Phe91 O and the phenyl ring of Tyr32 (not shown). TEL His121 makes only one hydrogen bond to the antibody, His Nε2–Trp92 O, which is made possible by a change in peptide torsion angles for D1.3 residues Trp92 and Ser93.

Table 2 Thermodynamic Parameters of the Binding of Antibodies to Lysozyme[a]

Ab	Temp. °C	ΔH kcal mol^{-1}	K_a M^{-1}	ΔG kcal mol^{-1}	$T\Delta S$ kcal mol^{-1}
IgG D1.3	24.2	−21.7			
Fab D1.3	24.0	−21.5	2.7×10^8	−11.5	−10.0
Fv D1.3	24.5	−21.5d	2.3×10^8	−11.5[d]	−10.0[d]
IgG D44.1	22.7	−9.4	1.0×10^8	−9.4	0.0
Fab D44.1	24.2	−10.4	1.4×10^7	−9.7	−0.7
IgG F9.13.7	24.2	−10.2	3.4×10^8	−11.6	+1.4
Fab F9.13.7	23.9	−11.1	7.2×10^8	−12.1	+1.0
IgG F10.6.6	20.7	−16.0			
Fab F10.6.6	24.2	−18.3			
IgG D11.15	24.0	−18.9	1.1×10^8	−10.9	−8.0
Fab D11.15	24.9	−19.0	0.8×10^8	−10.8	−8.2
HyHEL-5	25.0	−22.6[b]	4.0×10^{10}[c]	−16.9[b]	−5.7[b]

[a] Data from Schwarz *et al.* (13)
[b] Data from Hibbits *et al.* (108)
[c] Data from Benjamin *et al.* (86)
[d] Data from Ysern *et al.* (97)

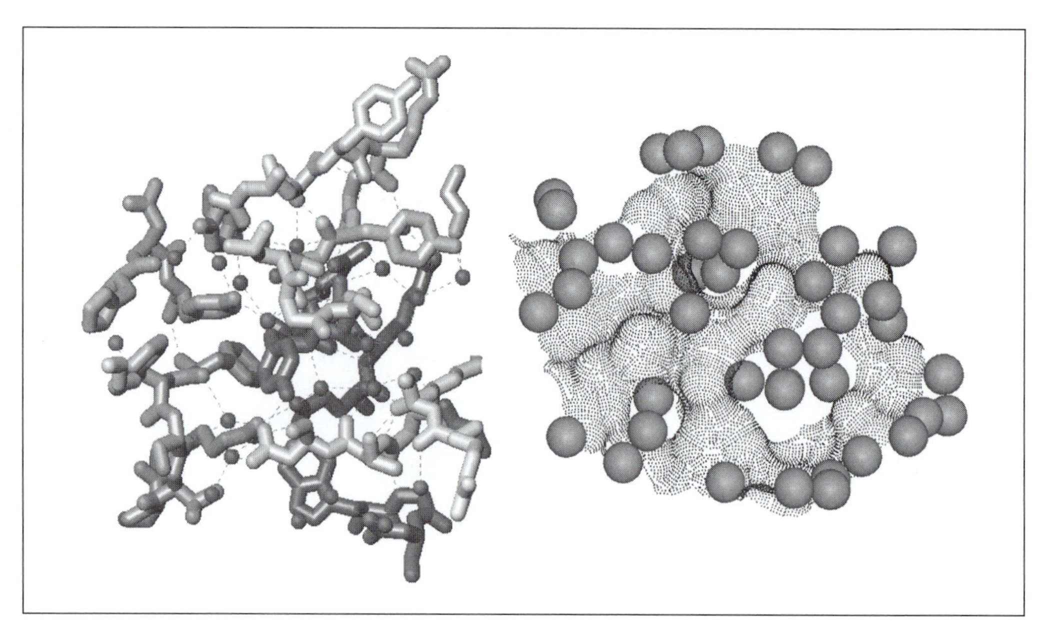

Fig. 3 Water molecules in contact with the Fv D1.3–HEL buried surface. Including the 25 bridging water molecules (left), nearly 50 solvent sites are in contact with the buried surface defined by the D1.3–HEL interface (right). Some of these water molecules fill internal cavities, further stabilizing the complex.

mined by X-ray diffraction (reviewed in ref. 67). Nevertheless, there has been a long-standing debate about the reproducibility, and indeed the structural relevance, of solvent molecules modelled by X-ray diffraction techniques (68–70). Zhang and Matthews (71) have presented a detailed accounting of consensus water molecules in 18 structures from 10 crystal forms of phage T4 lysozyme. Although no solvent sites were found to be common to all 18 lysozyme structures, 69% of the 40 most frequently occupied solvent sites were conserved. Furthermore, water molecules have been identified in the adhesion interface of the cell-adhesion protein cadherin (72) and in the interface of the barnase–barstar complex (73; see Chapter 9).

Braden *et al.* (74) examined the solvent structure of the D1.3–HEL complex in the context of the conservation of solvent sites between crystal structures of the wild-type D1.3–HEL, six mutant Fv D1.3–HEL structures, and the crystal structures of the free Fv D1.3 and five free lysozymes. From solvent sites common to the wild-type and mutant Fv–HEL complexes, 48 water molecules were conserved in the D1.3–HEL interface. Of these, 25 are bound directly or through other water molecules to both antibody and antigen, and thus comprise a large component of the hydrogen bond network linking the antibody and antigen (Fig. 3). The solvation of the D1.3–HEL interface again demonstrates the high degree of complementarity in an antibody–antigen interface. Not only do the interface waters contribute hydrogen bonds to the stability of the complex, but also the mere presence of the solvent molecules precludes the existence of destabilizing cavities and uncompensated polar atoms. The unique availability of a number of antibody–antigen structures and of free antibody

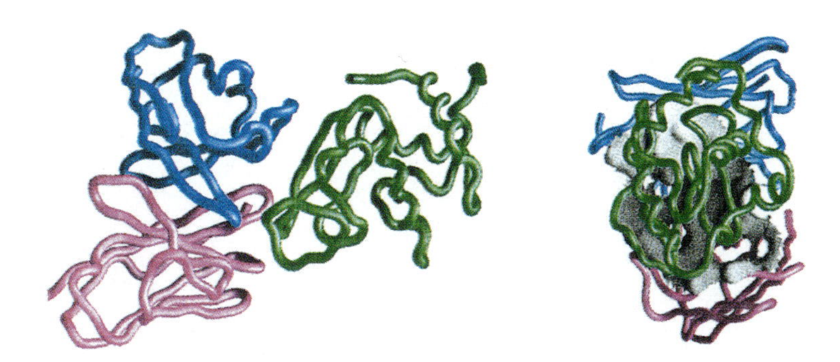

Plate 1 An antigen–antibody complex (PDB file 3HFL) (23). Two orthogonal views of hen egg lysozyme (green) associated to monoclonal antibody HyHEL5. Of the antibody, only the variable domains of the heavy (purple) and light (blue) chains are shown. Left: the lysozyme epitope comprises a loop and a β-strand; it forms a flat surface that covers the combining site of the antibody. Right: the surface of the combining site in contact with the antigen is viewed through the lysozyme backbone. This plate and the ones that follow are made with GRASP (117).

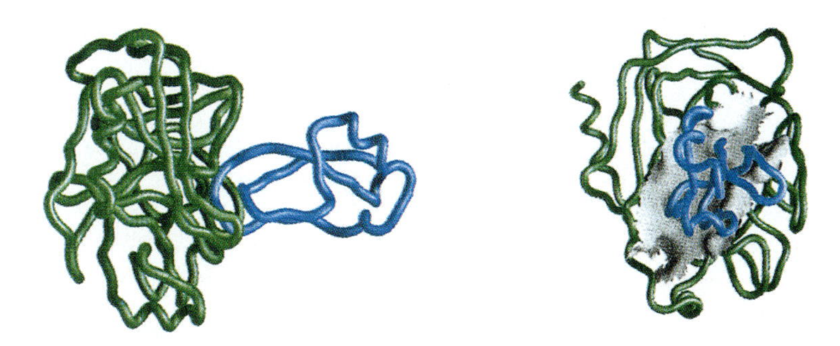

Plate 2 A protease–inhibitor complex (PDB file 2PTC) (24). Two orthogonal views of the pancreatic trypsin inhibitor (PTI, cyan) binding to trypsin (green). Left: the tip of the inhibitor penetrates the active site of the enzyme. Right: the surface of the active site in contact with the inhibitor is viewed through the PTI backbone.

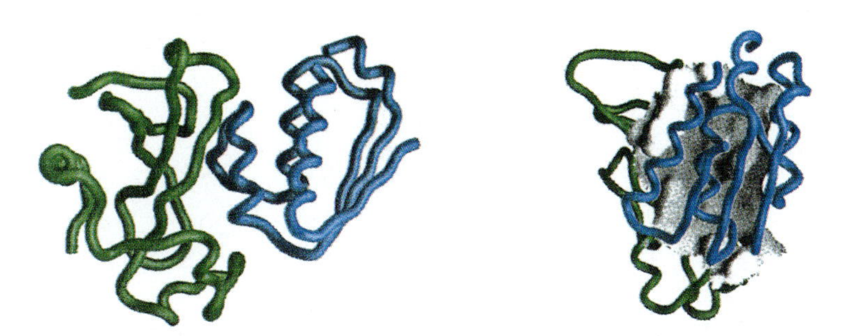

Plate 3 The barnase–barstar complex (PDB file 1BRS) (26). Two orthogonal views of barnase (green), a small bacterial ribonuclease, with bound barstar (blue), a natural inhibitor produced by the same bacterium. Left: a region of barstar comprising an α-helix and a loop cover the active site of the ribonuclease. Right: the surface of the active site in contact with the inhibitor is viewed through the barstar backbone.

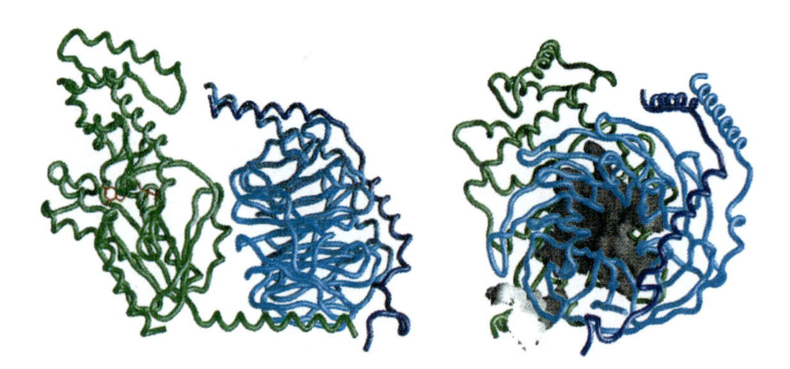

Plate 4 The transducin G_α–$G_{\beta\gamma}$ complex (PDB file 1GOT) (27). Two orthogonal views of the heterotrimeric G-protein with the G_α chain in light blue, bound GDP in red, the G_β chain in green, and the G_γ chain in dark blue. Left: on G_α, the interaction with $G_{\beta\gamma}$ involves a long N-terminal α-helix on the bottom, and two segments of polypeptide chain called Switch I and Switch II at the centre of the complex. G_β is a protein with a characteristic β-propellor fold where loops connecting the β-strands interact with the two Switch segments of G_α. The G_γ chain makes no contact with G_α. Right: the surface of G_α in contact with G_β is viewed through the β-propellor. The N-terminal α-helix on the bottom left and the Switch I–Switch II region forms two distinct patches within the interface. In free G_α, the N-terminal segment is disordered and the Switch segments adopt a different conformation.

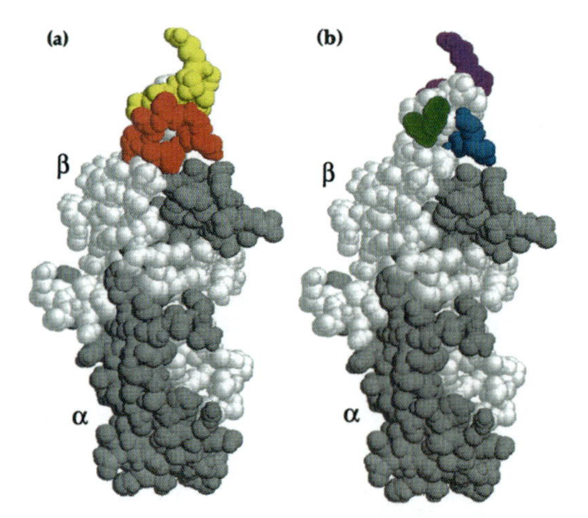

Plate 5 CPK diagram of human chorionic gonadotropin (hCG). The two subunits (α and β) are labelled and coloured using two shades of grey. (a) The two residue segments on subunit-β (residues 17–27 in red and 70–78 in yellow) indicated as structural epitopes from PATCH predictions. (b) Three residues identified by mutagenesis experiments that affect the binding of monoclonal antibodies (see Table 2 for coding).

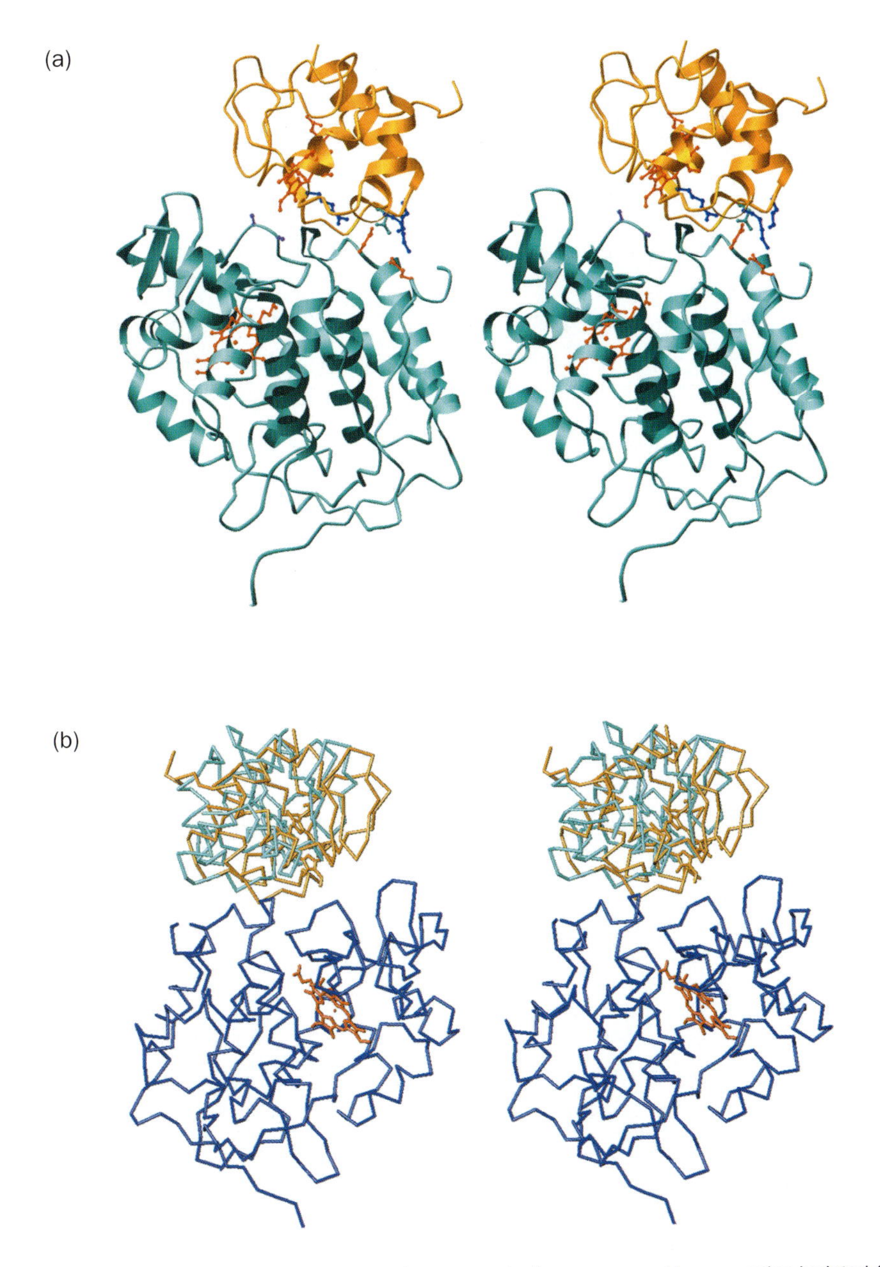

Plate 6 (a) Stereo ribbon diagram of the cytochrome *c*–cytochrome *c* peroxidase complex isolated from bakers yeast (2PCC). The polypeptide chain of C*c*P is shown in light blue and that of CY is in gold while both haem groups are in red. Within the interface, acidic side chains and a carbonyl on C*c*P are in red, while basic side chains and a neutral hydrophilic side chain (Asn) on CY are shown in dark blue and light blue, respectively. Ala193 and Ala194 of C*c*P that interact with the haem of CY are violet. (b) Comparison of the C^{α} backbones of yeast cytochrome *c* (gold) (2PCC) and horse-heart cytochrome *c* (light blue) (2PCB) in complex with yeast C*c*P (dark blue). The haem of C*c*P is red, whereas those of yeast and horse-cytochrome *c* are gold and light blue, respectively.

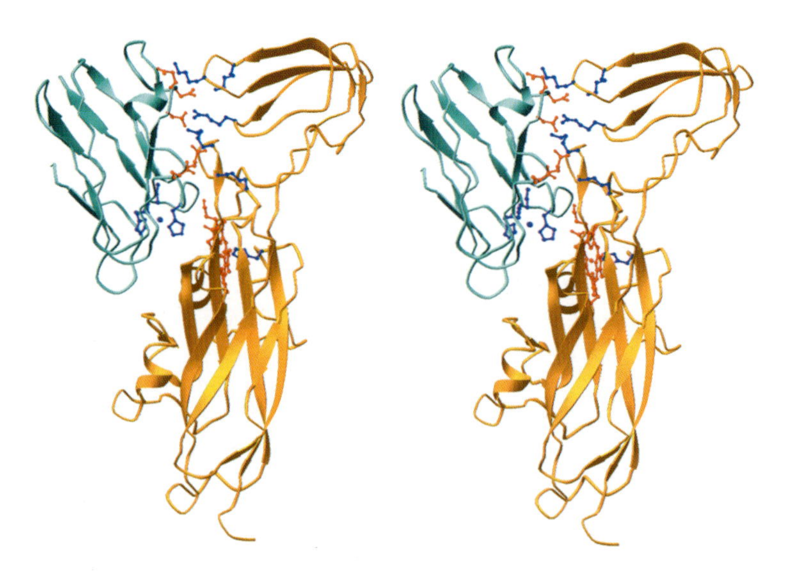

Plate 7 Stereo ribbon diagram of cytochrome *f* (gold chain tracing) and plastocyanin (light-blue chain tracing) (2PCF). Five basic side chains (dark blue) on cytochrome *f* that interacts with five acid side chains (red) of plastocyanin are indicated. The copper is purple and the haem group is red; the ligands to the copper and iron, respectively, are purple.

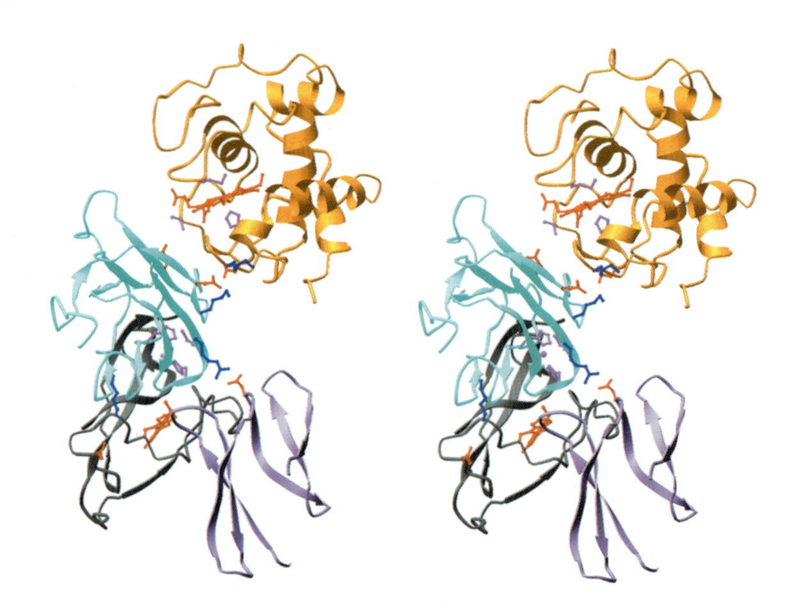

Plate 8 Close-up view of portions of the MADH H- and L-subunits, shown in violet and grey, respectively. In addition, the haem and copper ligands are violet, while the surface acidic and basic residues that interact at the interface are red and dark blue, respectively. A surface threonine residue on the cytochrome that forms a hydrogen bond to an aspartic acid on amicyanin is shown in violet.

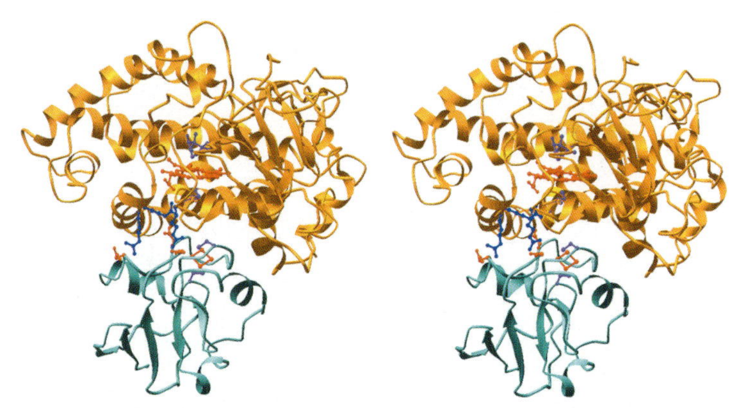

Plate 9 Stereo ribbon diagram of the cytochrome P450–putidaredoxin model complex (coordinates supplied by T. Pochapsky, personal communication). The chain tracings of the P450 and the putidaredoxin are in gold and light blue, respectively, and their 2Fe-2S and haem prosthetic groups are shown in red. The acidic groups on putidaredoxin (red) that interact with the basic side chains on P450 (dark blue) are also shown, and the cysteine haem ligand and camphor substrate of P450 above the haem are presented in violet.

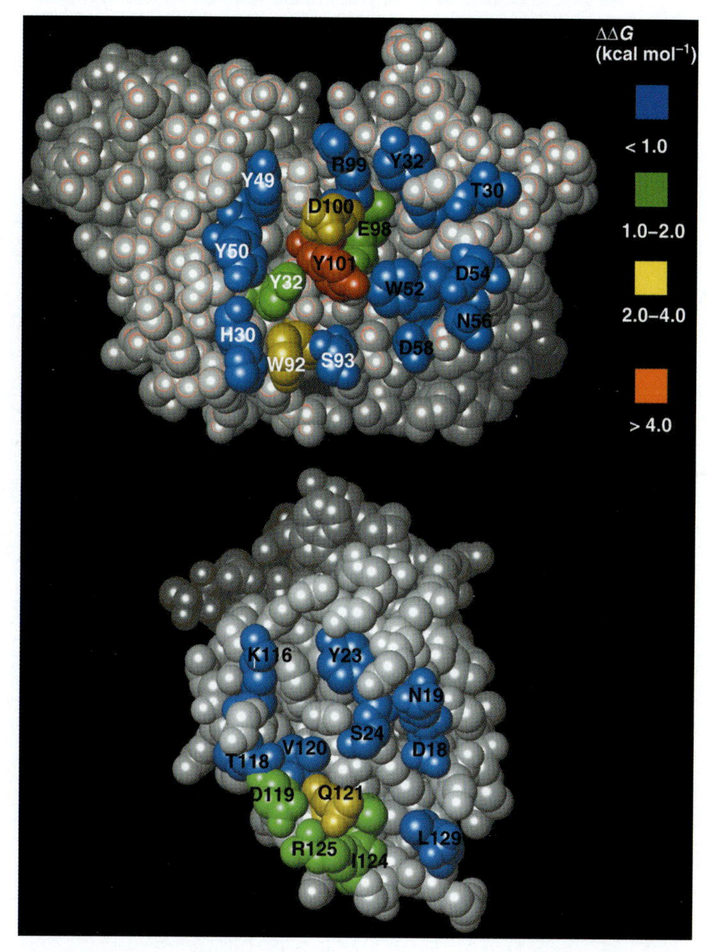

Plate 10 Space-filling model of the surface of D1.3 (above) in contact with HEL and of the surface of HEL (below) in contact with D1.3. Residues are colour-coded according to the loss of binding free energy upon alanine substitution: red, > 4 kcal mol^{-1}; yellow, 2–4 kcal mol^{-1}; green, 1–2 kcal mol^{-1}; blue 1 kcal mol^{-1}. V_L residues are labelled in white and V_H residues in black.

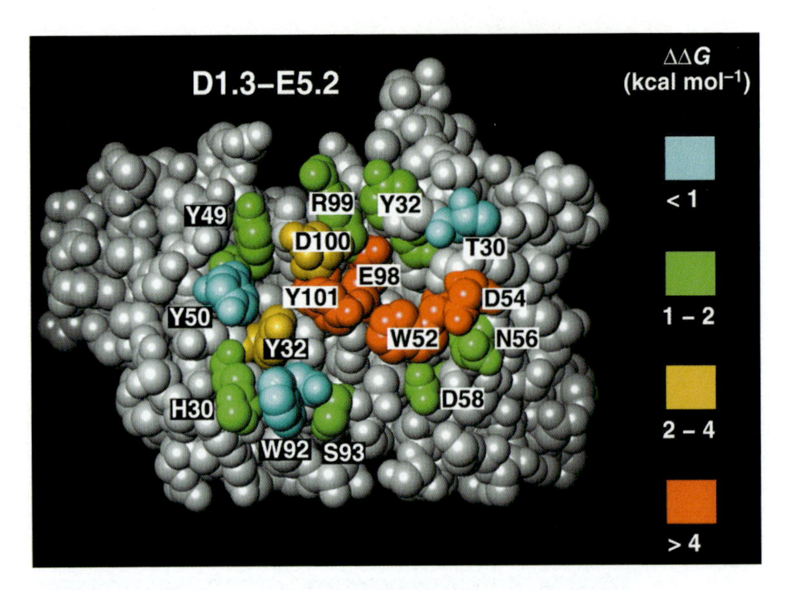

Plate 11 Space-filling model of the surface of D1.3 in contact with the anti-idiotopic antibody E5.2. Residues are colour-coded according to the loss of binding free energy upon alanine substitution: red, > 4 kcal mol^{-1}; yellow, 2–4 kcal mol^{-1}; green, 1–2 kcal mol^{-1}; blue 1 kcal mol^{-1}.

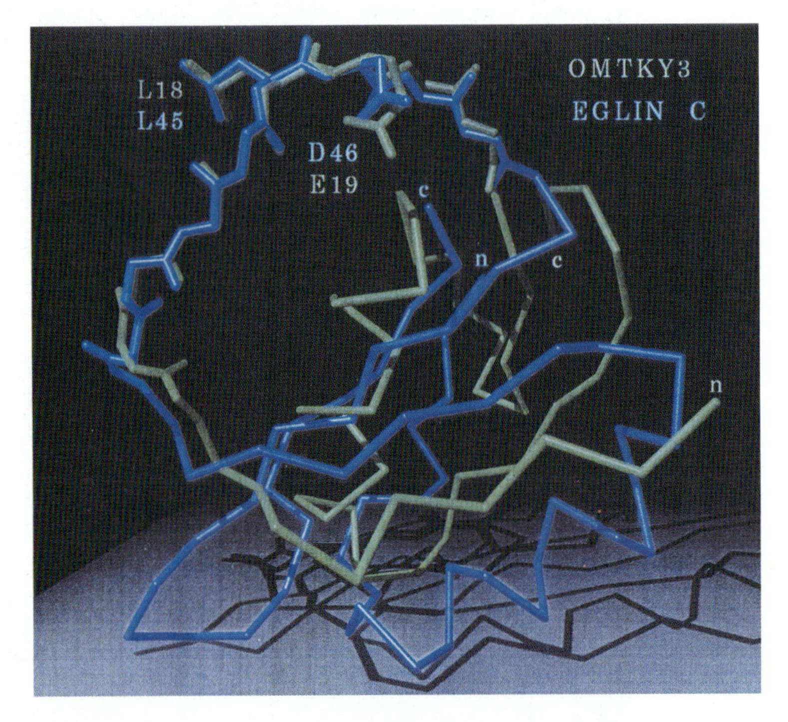

Plate 12 Superimposition of the backbone of two standard mechanism inhibitors: OMTKY3 (green) and eglin c (blue). The backbone atoms from P_6 to P_3' (reactive-site loop, residues 13–21 for OMTKY3 and 40–48 for eglin c) and the side-chain atoms for P_1 (L18, L45) and P_1' (E19, D46) are shown (42). The rest of the structure of the two inhibitors is shown as a Cα trace.

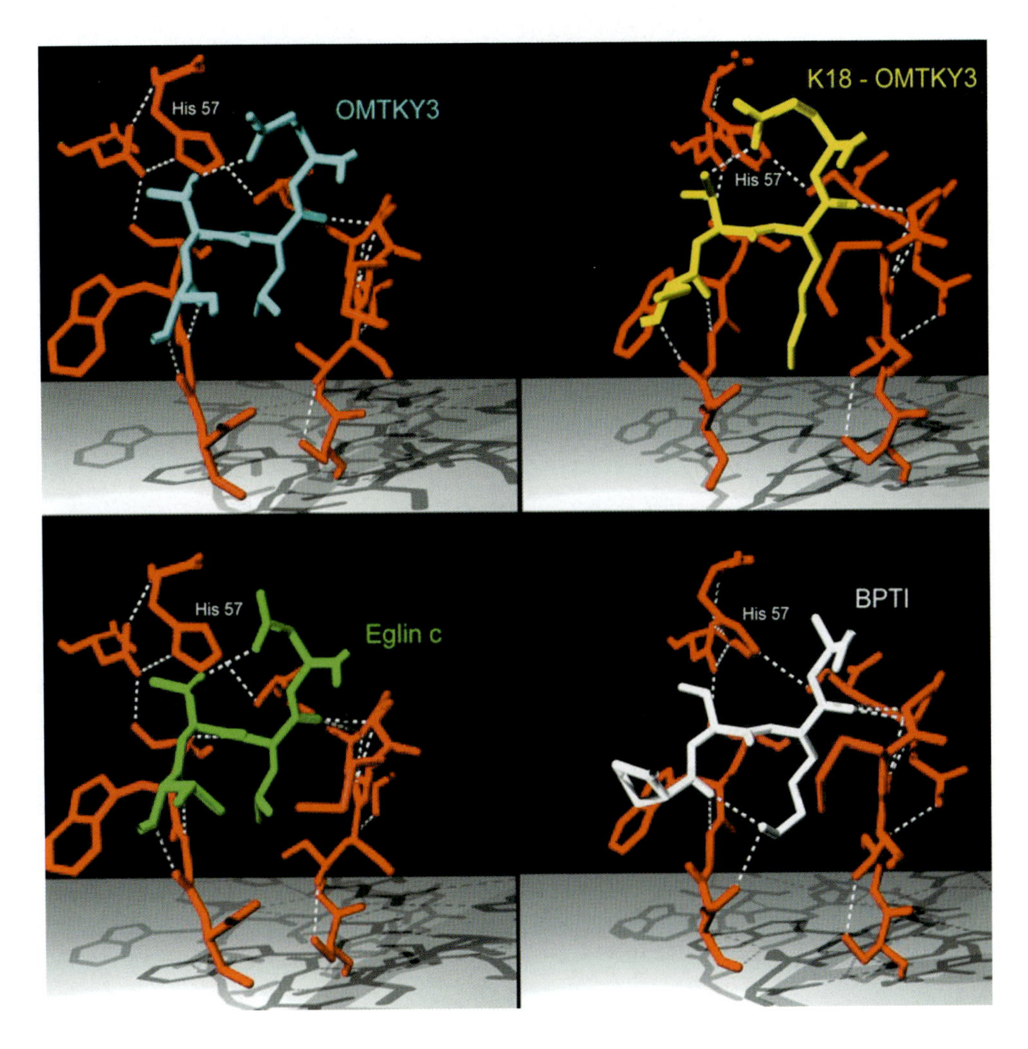

Plate 13 Interaction of the P_1 residue of four inhibitor variants with the S_1 pocket of CHYM (in red). Top left, portion of OMTKY3 including P_1 Leu18 in blue, 1CHO. Bottom left, portion of eglin c including P_1 Leu45 in green, 1ACB. Note the similarity in the two P_1 Leu conformations (42). Top right, portion of P_1 Lys18 OMTKY3 in yellow (Ding, Qasim, Laskowski, Jr, James, unpublished data), 1HJA. Bottom right, portion of BPTI including P_1 Lys15 in white, 1CBW. Note the dissimilarity in the P_1 Lys conformation (106).

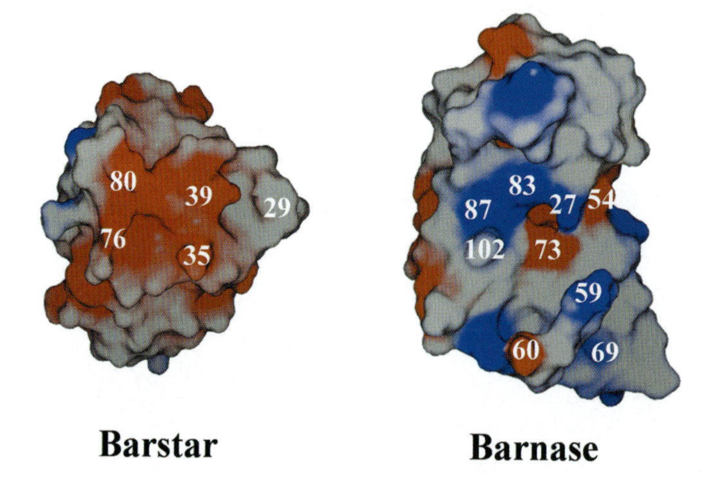

Plate 14 Electrostatic potentials of the interfacial regions of barnase and barstar. Potentials were calculated using the SPOCK program (90), assuming a uniform dielectric constant of 80 for solvent and 2 for the protein interior. The molecules have been rotated to show the charged surfaces that form the protein–protein interaction. Selected residues have been identified in each of the proteins and are discussed in the text. (Adapted from Buckle *et al.* (45).)

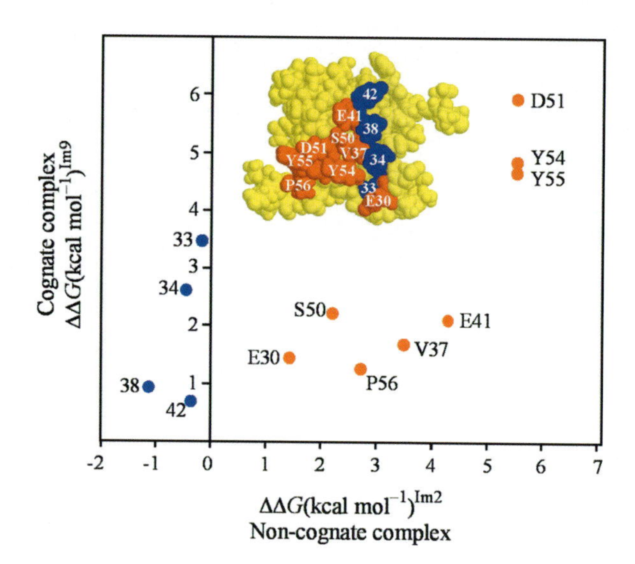

Plate 15 Comparative alanine scan of the E9 DNase binding site in two immunity proteins supports a dual-recognition model of protein–protein interaction specificity. The figure shows $\Delta\Delta G$binding data for 12 Im9 mutants plotted against the equivalent data for Im2 mutants binding the E9 DNase in a non-cognate complex. The insert identifies the locations of the residues in each protein. The 12 residues account for most of the E9 DNase binding energy in both proteins. Residues labelled red are conserved between Im2 and Im9, and these dominate DNase binding energy, while those labelled blue are variable and define the specificity of the protein–protein interaction by modulating this binding energy through positive, negative, or neutral effects. (Adapted from Li *et al.* (86).)

and free antigen structures permitted an extensive analysis of the solvent structure of the antibody combining site and antigen epitope before and after complex formation. During the formation of the complex, 15 water molecules are displaced and 5 water molecules are added. Of the 15 'removed' waters, 9 would have been within 1 Å of a proton donor or acceptor in the interface. As such, these nine waters may be removed from the free proteins as a result of complex formation, while the solvent site itself retains hydrogen bond interactions to protein atoms. The free antibody and antigen contribute the remaining interface water molecules (about 30). Even with the desolvation of 15 water molecules, interface waters contribute a net gain of 10 hydrogen bonds to the stability of the D1.3–HEL complex.

However, even at the resolution of 1.8 Å attained for the Fv D1.3–HEL complex, a crystallographic analysis cannot give a complete picture of all the water molecules that participate in the formation and stabilization of the antibody–antigen complex. This is particularly true for water molecules that are present in the outer hydration shells of proteins. In addition, the crystals of the Fv D1.3–HEL complex, as well as those of the free reactants, were obtained under conditions of osmotic stress and may not necessarily reflect the total hydration in solution. To further test this structural model and to assess the role of water molecules in the formation of the D1.3–HEL complex, Goldbaum *et al.* (75) studied the thermodynamics of association under conditions that reduce water activity. Following the examples of other systems and the experiments performed by other laboratories (45, 47, 48), Goldbaum *et al.* used different co-solutes to increase osmotic pressures and measured—by titration calorimetry, ultracentrifugation, and by plasmon resonance (BIAcore^{TM} (76)) techniques—association constants and the enthalpy of reactions under different water activities. For this purpose, they used the mAb D1.3, the bacterially expressed Fv D1.3, as well as a site-directed mutant of Fv D1.3, V_L Tyr32→Ser. The Y32S mutant has a lower equilibrium association constant (in the micromolar range) than the wild-type Fv D1.3, making it easier to follow the titration of antibody combining sites at room temperature (21–25 °C) by calorimetry and the dissociation of the complex by sedimentation studies. The addition of ethanol, methanol, or dioxane resulted in a decrease of the association constant of Fv D1.3 and the Fv D1.3 Y32S mutant with lysozyme. The values of the association constants, plotted against the logarithm of water activity give a linear plot, as shown in Fig. 4. The three co-solutes gave comparable results. The slope of the best (least squares) lines that can be traced in Fig. 4 would indicate that an additional 13.5 water molecules are bound upon complex formation between D1.3 and HEL. The co-solutes affect the association reactions by competing with the proteins for the available water. Solvent viscosity cannot explain the observed results, since, for example, the viscosity of methanol solutions increases and then decreases in the range of concentrations used in these experiments. Thus, it was concluded that water plays an essential role in the stability of the D1.3–HEL complex, as a decrease in water concentration results in lower association constants. The estimate of the number of water molecules that are bound upon complex formation, obtained from the slopes of the curves of association constants vs. the logarithm of water activity, may not strictly correlate with that observed in the crystal structures due to several factors. For instance:

1. The crystals are obtained at concentrations of 15–20% ((w/v) polyethylene glycol, i.e. under conditions of decreased water activity.
2. The resolution of the structure determination was 1.8 Å, which may not allow for a complete enumeration of bound water molecules.
3. The water molecule positions observed in the crystallographic studies do not have full occupancies, so that evaluation of the moles of water bound per mole of complex is not straightforward.

However, within these limitations, the results reported support the conclusions previously derived from the crystallographic analysis and the calorimetric studies.

The D1.3–HEL complex has also been analysed as to the shape complementarity (77) and polar–apolar surface complementarity (78). Lawrence and Colman (77) define a shape statistic that measures the correlation of buried surface area to shape complementarity. Analysis of antibody–protein antigen interactions suggested that these interfaces are somewhat poorer in shape complementarity as compared to other protein–protein associations. In addition, this analysis showed that the D1.3–HEL interface has a shape complementarity similar to other antibody–protein antigens. However, Lawrence and Colman used the 2.5-Å structure of D1.3–HEL and did not include the water molecules buried in the interface. It remains to be determined whether this analysis, including bound water molecules, will identify the antibody–antigen interface with greater complementarity.

Braden *et al.* (78) devised a method to quantitate the complementarity of the apolar and polar surfaces of the complexes D1.3–E5.2 (idiotope–anti-idiotope) and D1.3–HEL

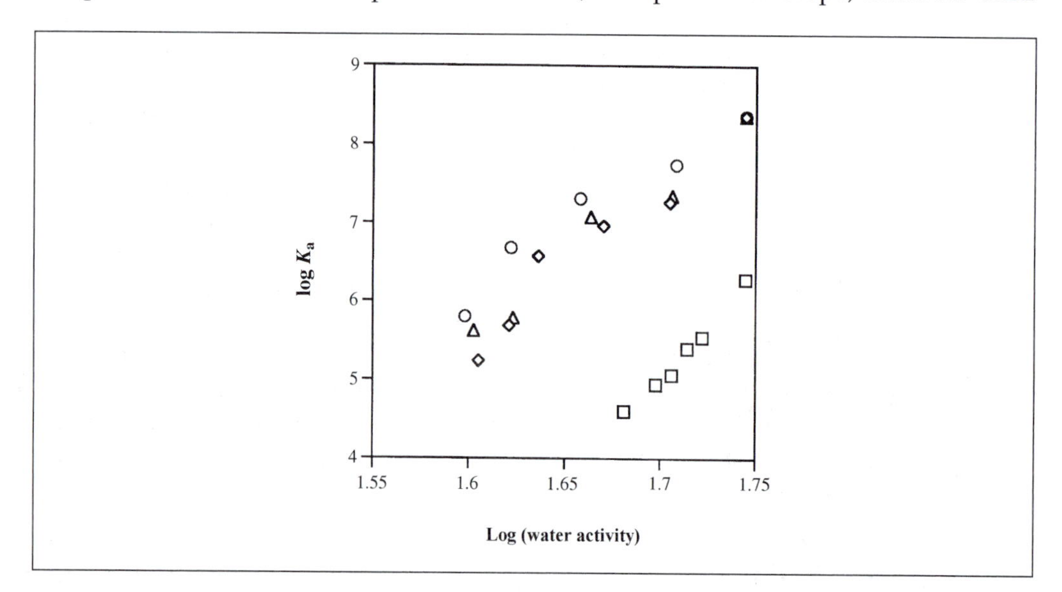

Fig. 4 Plot of the association constants of Fv D1.3 and the mutant Y32S against water activity. The K_a values for the reaction of Fv D1.3 with HEL in methanol (○), ethanol (◇), and dioxane (△) were obtained by the BIAcore™ technique. The values of K_a for the Y32S mutant were measured by sedimentation equilibrium at different concentrations of dioxane (□).

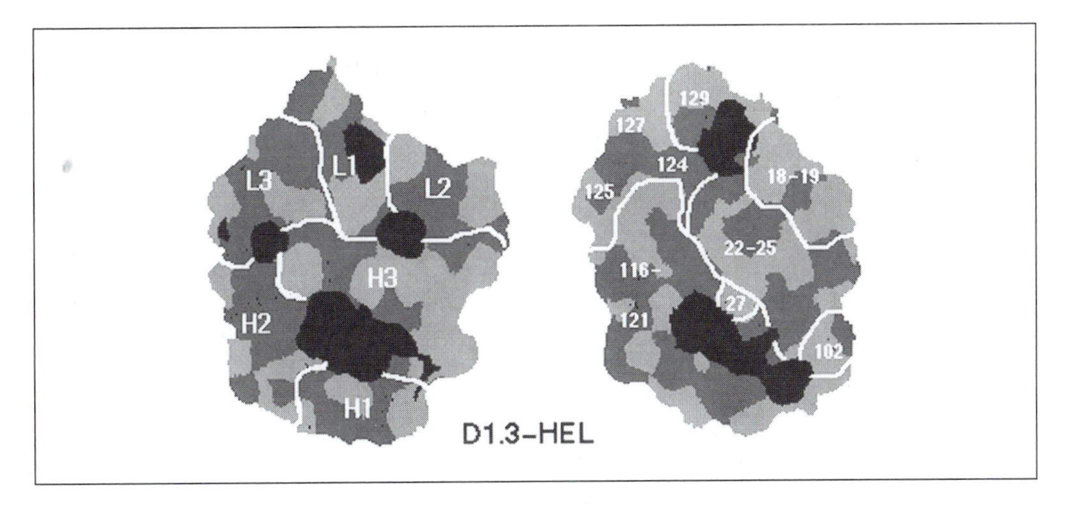

Fig. 5 Buried surface representation of the D1.3–HEL complex. The hydrophilic surfaces (defined by nitrogen and oxygen atoms) are shaded light and the hydrophobic surfaces (defined by carbon atoms) are shaded dark. Black represents solvated cavities in the molecular interface. The surfaces for the complex are separated by a translation, such that the D1.3 molecule would be behind the page and the HEL molecule would be in front. Mapped on to each surface are the CDR loop designations and the HEL residue numbers of the atoms that form the buried surfaces.

(antigen–antibody) surfaces, analysed as the percentage of surface points on the D1.3 buried surface whose nearest neighbours on the HEL and E5.2 buried surfaces are of the same apolar/polar character. From this analysis, the polar/apolar similarity of the two D1.3 buried surfaces in the two complexes is 0.78; i.e. more than three-quarters of the D1.3 surface points in the complexes with HEL and E5.2 are of the same polar/apolar character. The complementarity of the D1.3–HEL and D1.3–E5.2 buried surfaces are 0.58 and 0.55, respectively. Thus, only slightly more than half of the surface points of D1.3 are opposed by points on the HEL and E5.2 surfaces that are of the same polar/apolar character. As demonstrated by a majority of the proteins in the Larsen *et al.* survey (54), the D1.3–HEL interface is a somewhat random mixture of polar–apolar surface characteristics (Fig. 5). The polar complementarity of the D1.3–HEL interface is 0.75, whereas the apolar complementarity is only 0.30. Therefore, this interface demonstrates the geometric requirements for polar and apolar interactions. Hydrogen bonds require a stereochemical disposition much stricter than apolar interactions. Thus it is not surprising that the interface complementarity of polar surfaces in the antibody–antigen interface is much greater than that of the apolar surfaces.

4.2 Thermodynamic analysis of antibody–lysozyme reactions

The energetics of antibody–protein antigen interactions and the recognition of antigenic determinants by antibodies has been investigated by thermodynamic analysis of antibody–antigen reactions and by mutation of epitope and combining-site amino-acid residues (13, 79–85). In particular to the anti-lysozyme model system, Benjamin *et al.* (86) and Schwarz *et al.* (13) have analysed the thermodynamics of antibody–antigen binding of several anti-lysozyme antibodies. Table 2 summarizes the thermodynamics of binding reactions displayed by fragments of the anti-lysozyme antibodies D1.3, D11.15, D44.1, F9.13.7, F10.6.6 and HyHEL-5. It is particularly noteworthy that these reactions are exothermic, enthalpy-driven, with some opposition from a negative entropy contribution. These properties, in conjunction with the X-ray crystal structures, forced the conclusion that the main forces stabilizing antibody–lysozyme complexes arise from hydrogen bonds and van der Waals interactions. It is also crucial, especially regarding the significance of the Fab and Fv crystal structures, that the binding constants and thermodynamics for the corresponding Fab and Fv fragments are essentially the same as the binding parameters for the parent IgG. This observation implied that, upon antigen binding, there are no conformational changes in the C domains of the Fab and IgG that affect the thermodynamics of the reaction. As discussed above, conformational changes had been considered to be the trigger for effector functions of antibody molecules such as complement binding and activation. However, the lack of any differences between the HEL-binding thermodynamics of the IgG and Fv and Fab fragments rules out an allosteric mechanism which transmits conformational signals from one binding site to the other binding site of the IgG or to the Fc domains of the IgG via the light and heavy chains and the hinge region (13).

Ito *et al.* (81), Hawkins *et al.* (82), and Dall'Acqua *et al.* (84, 85) have presented mutational analysis of the D1.3–lysozyme interaction. In these studies, a small number of residues appear to dominate the energetics of reaction, suggesting a 'functional' or 'energetic' epitope as a subset of all the interaction residues. However, the contribution of main-chain to main-chain interactions to the energetics of antibody–antigen association cannot be described, as these interactions cannot be modified by site-directed mutagenesis. In the case of D1.3–HEL there are three hydrogen bonds between the antibody and antigen which arise from interactions with D1.3 main-chain atoms. Two of these hydrogen bonds, V_L 93 N–HEL Gln121 Oε1 and V_L 91 O–HEL Gln121 Nε2 are discussed below. However, without serious alteration to the secondary structure of the protein, a third such hydrogen bond between the amide nitrogen of V_H Gly53 and the carbonyl atom from HEL Gly117 cannot be investigated by mutagenesis studies.

Table 3 summarizes the effect of single-mutant experiments on the D1.3–HEL system. The data presented therein and graphically in Plate 10 demonstrate the change in the overall free energy of the antibody–antigen interaction via alanine-scanning mutagenesis of the D1.3–HEL interface. Association constants and relative free-energy changes (85) reveal that residues in V_L CDR1 and V_H CDR3 contribute more to binding than residues in the remaining CDRs. By far, the greatest reductions in affinity ($\Delta G_{mutant} - \Delta G_{wild\ type} > 2.5$ kcal mol^{-1}) occurred on substituting three residues: V_L Trp92, V_H Asp100, and V_H Tyr101. Significant effects (1–2 kcal mol^{-1}) were also seen for substitutions at positions V_L Tyr32, V_H Tyr32, and V_H Glu98, even though the latter is not involved in direct contacts with HEL. Mutations at seven

Table 3 Association Constants and Relative Free Energy Changes for Fv D1.3 Single Mutant-HEL Complexes

D1.3	K_A (M^{-1})	$\Delta\Delta G$ (kcal mol^{-1})	No. intermolecular contacts made by wild type residue
Wild Type	$8.0 \pm 2.4 \times 10^7$		
V_LH30A[a]	$1.2 \pm 0.2 \times 10^7$	0.8	1
V_LY32A[b]	$4.4 \pm 1.3 \times 10^6$	1.7	1
V_LY49A[b]	$1.3 \pm 0.2 \times 10^7$	0.8	1
V_LY50A[b]	$3.3 \pm 1.1 \times 10^7$	0.5	11
V_LS93A[a]	$2.8 \pm 0.4 \times 10^7$	0.3	4
V_LW92A[b]	$2.8 \pm 0.8 \times 10^5$	3.3	7
V_HT30A[a]	$5.5 \pm 0.8 \times 10^7$	0.1	0
V_HY32A[b]	$1.2 \pm 0.3 \times 10^7$	1.1	1
V_HW52A[b]	$1.7 \pm 0.5 \times 10^7$	0.9	15
V_HN56A[a]	$3.7 \pm 0.6 \times 10^7$	0.2	0
V_HD54A[b]	$1.5 \pm 0.4 \times 10^7$	1.0	2
V_HD58A[a]	$7.1 \pm 1.1 \times 10^7$	−0.2	0
V_HE98A[a]	$7.1 \pm 1.1 \times 10^6$	1.1	0
V_HR99A[a]	$5.9 \pm 0.9 \times 10^7$	0.1	2
V_HD100A[b]	$5.6 \pm 0.6 \times 10^5$	2.9	11
V_HY101A[b]	NB	>4	12
V_HY101F[b]	$5.5 \pm 1.1 \times 10^6$	1.6	

[a]From Dall' Acqua *et al.* (84)
[b]From Dall' Acqua *et al.* (85)
NB, no binding.

other contact positions (V_L His30, V_L Tyr49, V_LTyr50, V_L Ser93, V_H Trp52, V_H Asp54, and V_H Arg99) had little or no effect (i.e. <1 kcal mol^{-1}). Therefore, by this analysis, the energetics of binding of HEL by D1.3 appears to be largely mediated by only 5 of the 16 residues tested. This is similar to findings reported in the cases of a humanized anti-p185[HER2] antibody (80), of human growth hormone binding to its receptor (87, 88) and g the immunity protein Im9 binding the DNase domain of colicin E9 (Chapter 9).

Because of the large number of interatomic interactions in the antibody–antigen contacting region, it is generally not feasible to attribute the significance of each and every intermolecular interaction between an antibody and antigen to the overall physicochemical behaviour of the binding reaction. The 'group additivity' assumption of modelling protein associations fails here, in that the sum total of all the interactions, as shown in Table 3, cannot describe the overall energetics of the antibody–antigen association. While residues of the 'energetic' epitope or combining site may dominate the free energy of complex formation, the contribution by the remaining contacting atoms in the antibody–antigen interface cannot be ignored. Furthermore, while only five residues in the D1.3 paratope appear to contribute to complex stabilization, a sixth hydrogen bond is equally important—that between the amide nitrogen of V_H Gly53 and the carbonyl atom from HEL Gly117. In total, these remaining interactions could contribute significantly to the free energy and, more importantly, may provide the local environment required by the 'energetic' residues, such as the screening of polar interactions from the solvent continuum.

In contrast to the interaction of D1.3 and HEL, a mutational analysis of D1.3 binding to the anti-D1.3 antibody E5.2 demonstrates that many of the contacting residues play a significant role in the energetics of ligand binding (84). The results of this analysis are graphically presented in Plate 11. Although the anti-idiotypic antibody E5.2 has been shown to mimic the antigen HEL, providing several identical or equivalent contacts to D1.3 (78, 89), the mutational analysis showed that two D1.3 V_H CDR2 residues, Trp52 and Asp54 and a CDR3 residue, Glu98, contribute significantly to the D1.3–E5.3 interaction ($\Delta\Delta G > 4$ kcal mol^{-1}) while seemingly contributing little to the energetics of the D1.3–HEL interactions ($\Delta\Delta G$ changes of 0.9, 1.0, and 1.1 kcal mol^{-1}, respectively). Such a large energetic surface was also noted in the interaction of barnase–barstar (90). In the D1.3–E5.2 and barnase–barstar interfaces, the relative strengths of intermolecular interactions were, in most cases, broadly consistent with expectations based on the crystal structures.

Alanine scanning mutagenesis of HEL epitope residues in contact with D1.3 were used to further characterize the interface between D1.3 and HEL (85). A total of 12 non-glycine residues of HEL were individually mutated to alanine, and the affinities of the mutants for D1.3 determined using surface plasmon resonance detection (BIACore™). Only four positions showed a significant decrease in binding (i.e. $\Delta\Delta G = \Delta G_{mutant} - \Delta G_{wild\ type} > 1$ kcal mol^{-1}); with HEL residue Gln121, which makes hydrogen bonds to main-chain D1.3 atoms, proving the most critical for binding ($\Delta\Delta G = 2.9$ kcal mol^{-1}). This is similar to the D1.3 side of the interface where only 5 out of 17 contact residues contribute significantly to binding HEL. Thus, in both the D1.3 and HEL sides of the interface, only small subsets of total contacting residues appear to account for a large portion of the binding energy.

The residues of HEL most important for binding D1.3 (Asp119, Gln121, Ile124, and Arg125) form a contiguous patch located at the periphery of the surface contacted by the antibody (Plate 10). In the three-dimensional structure of the complex, Gln121 penetrates a hydrophobic pocket where it is surrounded by the side chains of V_L Tyr21, V_L Trp92, and V_H Tyr101. A comparison of the D1.3 and HEL surfaces at the interface reveals that hot-spot residues on the D1.3 side of the interface generally correspond to hot-spot positions on the HEL side (Plate 10). For example, HEL residues Gln121 ($\Delta\Delta G = 2.9$ kcal mol^{-1}) and Arg125 (1.8 kcal mol^{-1}) contact D1.3 residues V_L Trp92 (3.3 kcal mol^{-1}), with HEL Gln121 making additional contacts to V_L Tyr32 (1.7 kcal mol^{-1}) and V_H Tyr101 (>4 kcal mol^{-1}) and to D1.3 main-chain atoms V_L 93 N and V_L 91 O. Likewise, apparently less-important D1.3 and HEL residues tend to be juxtaposed in the antibody–antigen interface: HEL Asp18 ($\Delta\Delta G = 0.3$ kcal mol^{-1}) and Thr118 (0.8 kcal mol^{-1}) interact with D1.3 V_L Tyr50 (0.5 kcal mol^{-1}) and V_H Trp52 (0.9 kcal mol^{-1}). The results from the alanine scans demonstrate that the D1.3–HEL interface is remarkably tolerant to mutations which, on the basis of the three-dimensional structure of the wild-type complex, might be expected to have pronounced effects on affinity.

A modelling analysis of the D1.3–HEL interface by Covell and Wallqvist (91), based on pairwise surface preferences, predicted that the largest changes in binding free energy should arise from mutation of HEL residues Asn19 and Gln121. These calculations further suggested that three glycine residues at HEL positions 22, 102, and 117, all of which make hydrogen bonds to D1.3, also contribute significantly to binding. While the energetic contribution of glycine residues cannot be evaluated by alanine mutation, Gln121 is indeed functionally the most important residue for HEL binding to D1.3. However, contrary to prediction, HEL residue Asn19 makes essentially no net contribution to complex stabilization (85). This is surprising since Asn19 makes one short hydrogen bond (V_L Thr53 Oγ1–HEL Asn19 Nδ2; 2.8 Å), as well as a number of van der Waals contacts with D1.3 (see Table 1).

To further dissect the energetics of specific interactions in the D1.3–HEL interface, double-mutant cycles (92–94; see also Chapters 1 and 9) were carried out by Dall'Acqua *et al.* (85) to measure the coupling of 14 amino-acid pairs. Of the 14 pairs tested, 10 have interacting side chains as judged from the crystal structure, two do not form direct contacts but are in proximity (4 Å), and two are far apart (6–15 Å). Surprisingly, only three of the 10 residue pairs in direct contact in the crystal structure show coupling energies of greater than 1 kcal mol^{-1}: V_LTyr32–Gln121 (2.0 kcal mol^{-1}), V_LTrp92–Gln121 (2.7 kcal mol^{-1}), and V_LTrp92–Arg125 (1.7 kcal mol^{-1}). None of the remaining seven residue pairs have coupling energies exceeding the estimated experimental error of ±0.3 kcal mol^{-1}: V_L Tyr50–Asp18 (–0.4 kcal mol^{-1}), V_H Tyr32–Lys116 (0.2 kcal mol^{-1}), V_H Trp52–Asp119 (–0.3 kcal mol^{-1}), V_H Asp100–Ser24 (0.3 kcal mol^{-1}), V_H Tyr101–Asp119 (–0.1 kcal mol^{-1}), and V_H Tyr101–Val120 (0.0 kcal mol^{-1}). In fact, in terms of their coupling energies, these residue pairs are indistinguishable from the four pairs that do not form direct contacts: V_L Tyr32–Ile124 ($\Delta\Delta G_{int}$ = 0.0 kcal mol^{-1}), V_L Tyr50–Asp119 (0.3 kcal mol^{-1}), V_L Trp92–Ile124 (0.7 kcal mol^{-1}), and V_L Trp92–Leu129 (0.2 kcal mol^{-1}). These findings are in marked contrast to those for the D1.3-E5.2 (95) and barnase–barstar (90) complexes in which nearly all residues within 4 Å of each other showed significant coupling (>1 kcal mol^{-1}).

The interaction of Gln121 with V_L Trp92 had the largest coupling energy ($\Delta\Delta G_{int}$ = 2.7 kcal mol^{-1}). Since this interaction involves only three van der Waals contacts, whose strength has been estimated at about 0.5 kcal mol^{-1} per contact, this implies that this $\Delta\Delta G_{int}$ almost certainly includes a significant contribution from the hydrophobic effect arising from burial of the tryptophan side chain. Similar considerations may apply to the interaction of V_L Trp92 with Arg125 ($\Delta\Delta G_{int}$ = 1.7 kcal mol^{-1}).

The only hydrogen-bonded residues in the antibody–antigen interface to show significant coupling are HEL Gln121 and V_L Tyr32 ($\Delta\Delta G_{int}$ = 2.0 kcal mol^{-1}). This interaction is a hydrogen bond between the Nε2 group of Gln121 acting as a hydrogen-bond donor and the centre of the phenyl ring of V_L Tyr32 acting as a hydrogen-bond acceptor. It has been calculated that hydrogen bonds with aromatic rings as proton acceptors could contribute up to 3 kcal mol^{-1} of stabilizing free energy to molecular associations (96), which is in good agreement with the measured value of 2.0 kcal mol^{-1}. None of the other hydrogen bonds examined make significant net contributions to complex stabilization. These include hydrogen bonds which, on the basis of donor-acceptor distance and relative orientation of the interacting groups, would be predicted to be strong, as well as others expected to be weak.

4.3 Correlation of X-ray crystal structures of mutant D1.3–HEL with thermodynamics

Thus far, five of the mutant D1.3 antibody fragments have been analysed by X-ray crystallographic methods.

An important correlation between the extent of buried surface area and the thermodynamics and equilibrium binding constants of an antibody–antigen complex has been demonstrated by Ysern *et al.* (97). The mutant, Fv D1.3 V_L Trp92→Asp, results in a 1000-fold reduction in the equilibrium association constant (K_a) compared to the wild-type D1.3 reaction with HEL (from ~10^8 M^{-1} to ~10^5 M^{-1}). V_L CDR3 residue Trp92 makes extensive van der Waals contacts with HEL residues Gln121, Arg125, and Ile124. In the mutant Fv D.13–HEL crystal structure, a rearrangement of water molecules at the V_L–HEL interface is observed such that two water molecules partially fill in the void created by replacement of the bulky tryptophan with the smaller aspartate side chain (Fig. 6). The buried surface area thus lost, compared to that of the wild type, is about 150 Å^2. This net loss in buried surface area, paired with a 3.8-kcal mol^{-1} less-favourable binding enthalpy (97), corresponds to a hydrophobic energy loss of 28 cal mol^{-1} Å^{-2}, in agreement with several of the estimates of the contribution made by hydrophobic forces to the free energy of complex stabilization cited above (23, 29, 31). Thus, analysis of the site-directed mutant structure has

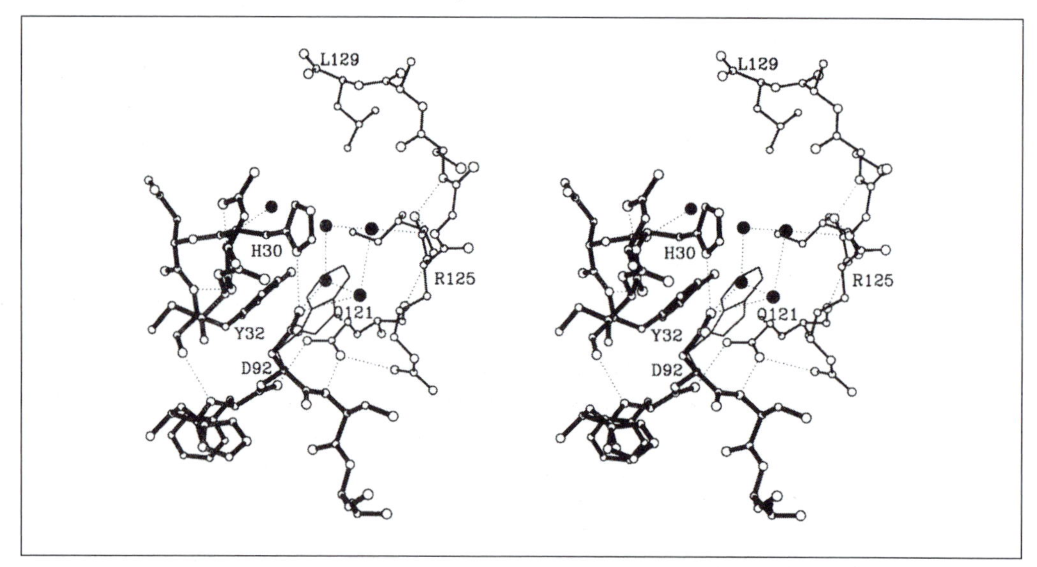

Fig. 6 Stereo view of the Fv D1.3–HEL W92D mutant complex in the area of the mutated residue. The wild-type V_L Trp92 is superimposed (in light trace) with the structure of the mutant Asp92. HEL residues are shown in lighter trace. HEL residues Gln121, Arg125, and Leu129, as well as VL residues His30, Tyr32, and Asp92, are labelled. Hydrogen bonds are shown by broken lines, water molecules as dark circles.

shown the importance of close-packed van der Waals interactions, and has demonstrated the structural basis for the reduced association of the mutant Fv D1.3–HEL complex.

Fields *et al.* (98) have determined the X-ray crystal structures of three mutant D1.3 Fv fragments in complex with HEL. These mutants, V_L Tyr50→Ser, V_H Tyr32→Ala, and V_H Tyr101→Phe, were designed to disrupt hydrogen bonds in the wild-type D1.3–HEL interaction. However, the resulting changes in binding and free energy were rather small, making correlation of the structural changes and energetics difficult.

The binding constant of the D1.3 V_L Tyr50→Ser mutant to HEL is 2.6×10^7 M^{-1}, 10-fold less that that of the wild-type complex. This affinity loss corresponds to a $\Delta\Delta G$ of 1.57 kcal mol^{-1} in which a $\Delta\Delta H$ of 2.1 kcal mol^{-1} enthalpy loss is partially compensated by a Δ ($T\Delta S$) of 0.6 kcal mol^{-1}. Tyr50 of the antibody V_L domain makes one direct hydrogen bond to an aspartate side chain of the lysozyme antigen (Fv D1.3 Tyr50 OH–HEL Asp Oδ2, length 2.7 Å) and also participates in the interfacial water network. When V_L Tyr50 is mutated to Ser, a two-solvent molecule bridge to the HEL side-chain Asp18 (Fig. 7A) replaces this direct hydrogen bond. These two water molecules occupy some of the volume that was taken up by the V_L Tyr50 side chain in the wild-type structure. Several neighbouring water molecules are in positions that correspond closely to those of the wild-type structure, while others are not observed in the mutant crystal structure, probably because of the lower resolution of the mutant crystal structure compared to the wild-type structure. Conformational changes in the vicinity of the mutation are largest at HEL side chains Asp18 and Asn19, those residues that are in closest contact to V_L Tyr50 of Fv D1.3 in the wild-type structure. The largest difference in atomic positions (1.1 Å) occurs at Asp18 Oδ2. In addition, the temperature factors of these side chains increase when V_L Tyr50 is replaced by serine. As well as losing a direct hydrogen bond to the lysozyme, several van der Waals contacts are also lost, which should contribute to the observed unfavourable enthalpy change. Geometric criteria, namely the hydrogen-bond, donor-acceptor distance and the relative orientation of the interacting groups, would classify the V_L Y50S hydrogen bond with HEL as 'strong'. However, the $\Delta\Delta H$ for this reaction compared to the wild-type is only 2.1 kcal mol^{-1}. Furthermore, as shown in Table 1, several van der Waals contacts are lost along with the hydrogen bond when V_L Tyr50 is substituted by serine. Based on the Ysern *et al.* analysis of the contribution of van der Waals contacts by V_L Trp92 (97), the lost contacts of Tyr50 could provide about an additional 1.9 kcal mol^{-1}. Tyr50 is also involved in hydrogen-bonding interactions with the interface solvent network. Thus, Tyr50 interacts with HEL through a direct hydrogen bond, several van der Waals contacts, and a solvent network and would therefore appear to be energetically important. There are several ways to explain the apparent small difference in enthalpy between wild-type and V_L Y50S:

1. The new bridging solvent in the Y50S complex provides most of the stabilizing enthalpy.
2. The Tyr50 hydrogen bond and/or van der Waals contribute much less to the enthalpy of the wild-type complex than assumed.

3. The solvation of residue V_L Tyr50 in the D1.3 wild-type and Y50S mutant, in solution, differ significantly.
4. Or a combination of these.

The side chain of V_H Tyr32 makes a long, potential hydrogen bond (3.5 Å) with the side chain of lysozyme residue Lys116 and, like V_L Tyr50, is involved in a solvent

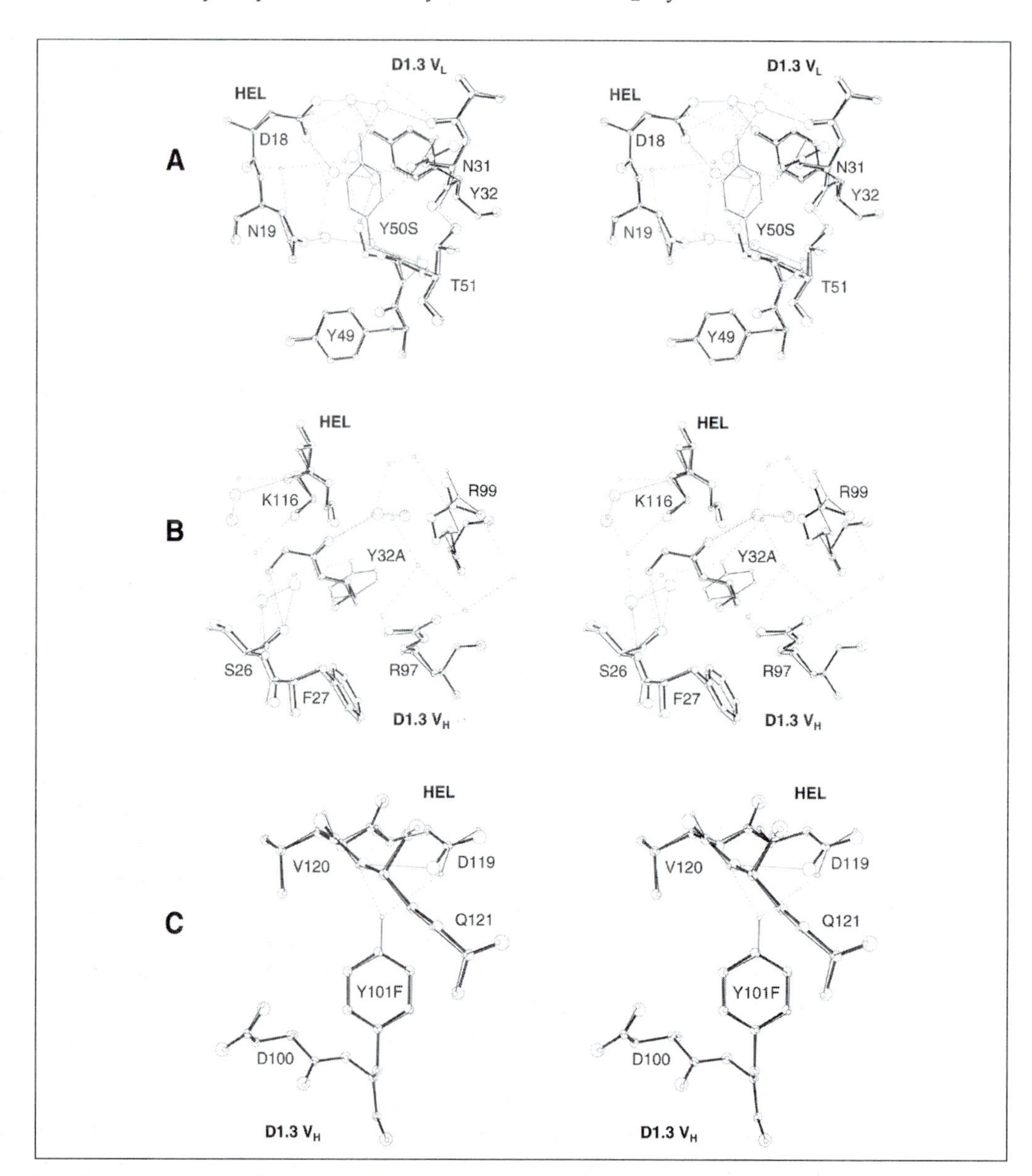

Fig. 7 Stereo diagrams of the mutant Fv D1.3 structures (thick bonds, large atoms, continuous hydrogen bonds) superposed on the wild-type structure (thin bonds, small atoms, dashed hydrogen bonds. (A) V_L Y50S; (B) V_H Y32A; and (C) V_H Y101F.

network that bridges antigen and antibody. Mutation to alanine disrupts the solvent network but does not affect the conformation or temperature factors of HEL Lys116 (Fig. 7B). The V_H Y32A mutant binds lysozyme with only about a fourfold lower affinity than that of the wild-type Fv. In this case the large ΔH (4.3 kcal mol^{-1}) is almost fully compensated for by the entropic term resulting in the small decrease in affinity. As for the V_L Y50S mutant, a rearrangement in the local solvent structure is clear. However, in contrast to V_L Y50S and W93D, only one van der Waals contact is lost along with the long hydrogen bond. There are two side chains in the vicinity of the mutation whose mobilities are affected by the V_H Y32A mutation. When Tyr32 is truncated to alanine, V_H Arg99 loses stabilizing contacts with V_H Tyr32 and becomes more mobile, as indicated by an increase in the mean temperature factor of side-chain atoms from 28 to 40 Å^2. A conformational change in V_H Arg99, resulting in a maximum atomic displacement of 1.1 Å from the wild-type structure, accompanies the temperature factor increase. In contrast, the mean temperature factor of a neighbouring side chain, V_H Phe27, decreases (from 16 to 9 Å^2), while the shifts in atomic positions for this side chain are smaller (<0.5 Å).

The V_H Tyr32 OH–HEL Lys116 Nξ separation of 3.5 Å would classify this interaction as a relatively weak hydrogen bond compared to the V_L Tyr50 interaction. In contrast to the situation for V_L Y50S, the $\Delta\Delta H$ for the reaction of V_H Y32A with HEL, compared to wild-type, is large (4.3 kcal mol^{-1}), a value which may be expected to correspond to a strong hydrogen bond. In this case, however, Tyr32 is also involved in a bridging solvent network, which is completely lost when tyrosine is replaced by alanine and is not compensated for by new bridging-solvent interactions. A new water molecule that occupies some of the space taken up by the tyrosine ring in the wild-type structure is too distant from HEL to act as a bridge. Thus, a possible explanation for the ΔH, despite the apparent weak hydrogen bond, is that the existing solvent network in the wild-type complex makes a sizeable contribution to ΔH. Thus, analysis of the V_H T32A and V_H Y50S complexes with HEL supports the view that localized bridging solvent can make a significant enthalpic contribution.

Tyr101 of D1.3 V_H is situated near the centre of the Fv–HEL interface and makes two hydrogen bonds with HEL: V_H Tyr101 OH–HEL Asp119 Oδ1, length 2.7 Å and V_H Tyr101 OH–HEL Gln121 N, length 3.0 Å. The effective substitution of the tyrosine hydroxyl group by a hydrogen atom to give the phenylalanine mutant results in small shifts in atomic positions (Fig. 7C). The largest atomic displacement (0.5 Å) in the interface compared to the wild-type structure occurs at HEL Oδ1, an atom involved in one of the above-mentioned hydrogen bonds. In contrast to the other mutants, there are no increases in temperature factors for atoms in the vicinity of the mutation. The calorimetric data indicate that the unfavourable enthalpy change resulting from the loss of two interface hydrogen bonds is strongly compensated by a favourable entropy change. This complex exhibits the largest $\Delta\Delta H$, 7.4 kcal mol^{-1}, and the largest Δ ($T\Delta S$), 4.8 kcal mol^{-1}, of the mutant D1.3–HEL complexes. It is important to note that there is no space for a water molecule to insert into the interface when the OH group of Tyr101 is replaced by a hydrogen. Significantly, the structure shows that the side chain of HEL residue Asp119 is not left with its hydrogen-bonding potential unsatisfied in the mutant. Instead, HEL Asp119 forms an intramolecular hydrogen bond with HEL Gln121 N; length 3.2 Å. In the wild-type structure, the corresponding distance is 3.8 Å, which is too long to be a hydrogen bond.

Compared to the previous two mutants, an analysis of V_H Y101F appears straightforward since there is no solvent network directly associated with the hydroxyl group of V_H Tyr101 in the complex interface. Truncation to phenylalanine results in an unfavourable $\Delta\Delta H$ of 7.4 kcal mol^{-1}. In this case it is clear that two hydrogen bonds are lost. Other structural differences compared to wild-type are small, and HEL Asp119 is not left with its hydrogen-bonding potential unsatisfied. Thus, it is probably valid to assume that most of $\Delta\Delta H$ component arises from the loss of the two hydrogen bonds. The resulting estimate of the contribution of each hydrogen bond is therefore ~4 kcal mol^{-1}. Since hydrogen bonds are essentially electrostatic interactions, the strength of a hydrogen bond depends mostly on the donor–acceptor distance, the magnitude of the partial charges, and the local dielectric constant. With regard to the dielectric constant, it is pertinent that hydrogen bonds between V_H Tyr101 and HEL are buried in the interface of the complex, while those from V_L Tyr50 and V_H Tyr32 are on the periphery. Thus, it is possible that the hydrogen bonds from V_H Tyr101 are elevated in strength compared to those from V_L Tyr50 and V_H Tyr32, due to a lower dielectric constant in the centre of the interface compared to the periphery. Calculation of such an effect, however, is likely to be difficult since the D1.3–HEL interface has several polar residues and is solvated.

Dall'Acqua *et al.* (85) have examined the crystal structure of a mutant HEL–D1.3 complex, D1.3–HEL Asp18→Ala. This mutation, again, shows little affect in the overall free energy of reaction, $\Delta\Delta G$ 0.3 kcal mol^{-1}. The major effect of the substitution of aspartic acid by alanine is a rearrangement of the solvent structure, such that three additional water molecules are incorporated in the interface at the site of the mutation. These bound waters form several hydrogen bonds with protein atoms and with other water molecules, some of which mimic hydrogen bonds made by HEL Asp18. In the wild-type structure one of the carbonyl atoms of HEL Asp18 forms two hydrogen bonds with two water molecules, one of which in turn is hydrogen-bonded to the hydroxyl group of V_L Tyr32. In the mutant complex, a water molecule serves as a substitute for Asp18 by partially occupying some of the volume taken up by the side chain and by making hydrogen bonds to other water molecules making up the interface water network. In addition, the same water molecule forms a hydrogen bond with the main-chain nitrogen of HEL Leu25, which further anchors it in the interface. The other carbonyl oxygen of HEL Asp18 forms a direct hydrogen bond with the hydroxyl of V_L Tyr50 in the wild-type structure. In the mutant, V_L Tyr50 makes hydrogen bonds with two waters that are also positioned to fill the cavity created by the mutation and form part of the rearranged solvent network bridging antibody and antigen. Thus, the loss of complementarity in the D1.3–HEL interface resulting from the replacement of HEL Asp18 by alanine is compensated for by the stable inclusion of additional water molecules and by local rearrangements in solvent structure, rather than by adjustments in the conformation of the protein.

This structure and its physical interpretation are in marked contrast to the mutant

Fv D1.3 W92D interaction with the lysozyme. In the W92D mutant the reordered solvent water network only provides one hydrogen bond to the antigen, and the loss of apolar surface area dominates the energetics of the reaction.

5. Anti-lysozyme antibodies HyHEL-5 and D44.1

While it is evident that the ideal antibody–antigen interaction should occur with complementary surfaces, the analysis of surface areas becomes more complex when comparing different antibody–antigen pairs, even when the antigen and indeed the antigenic epitope are the same. For example, murine anti-HEL antibodies D44.1 (BALB/c IgG1 κ; (59)) and HyHEL-5 (BALB/c IgG1 κ; (58)) both recognize the same epitope on the lysozyme, but they have different equilibrium binding constants ($\sim 10^{10}$ M^{-1} for HyHEL-5 (86); $\sim 10^{7}$ M^{-1} for D44.1 (99)). Structural similarities between the D44.1–HEL and HyHEL-5–HEL complexes include three salt links between V_H glutamates (35 and 50) and HEL arginines (45 and 68).

The importance of the salt link in the HyHEL-5–HEL association was made clear by a two-orders of magnitude reduction in the association constant with bob-white quail lysozyme (BWQEL) and a HEL Arg68→Lys mutation which mimics the BWQEL epitope (100). The crystal structure of the Fab HyHEL-5–HEL complex contains 16 hydrogen bonds and 3 salt bridges (100, 101). Semi-empirical energy calculations of the free energy of binding have argued that these glutamate and arginine residues are significant contributors to complex stability (102). The crystal structure of a site-directed mutant of HEL in which Arg 68 is replaced with lysine, as in the bobwhite quail protein, has been reported (100). Not surprising, since arginine and lysine side chains have very similar solvent-accessible surfaces, the buried surface area of the HyHEL-5–mutant HEL and wild-type HEL complexes are nearly identical.

The significant difference between the wild-type HyHEL-5–HEL complex and the mutant HEL complex occurs in the vicinity of the HEL Arg68→Lys mutation. A single water molecule is inserted which compensates for the loss of one of the wild-type NH–Oε salt bridges; however, there is an overall net loss of one hydrogen bond between antibody and antigen. The 1000-fold decrease in binding affinity for the mutant HEL compared to the wild-type HEL corresponds to a free energy difference of 4 kcal mol^{-1}. Since the lost hydrogen bond is charged and therefore contributes up to 6 kcal mol^{-1} to the stability of the wild-type complex, the net loss of one hydrogen bond explains the reduced binding of HyHEL-5 and the mutant HEL (100).

The HyHEL-5–HEL interaction buries about 1500 $Å^2$ of the solvent-accessible antibody and antigen surfaces (100, 101), whereas the total surface area of the D44.1–HEL interface is somewhat less ($\sim$ 1250 $Å^2$; (103)). In addition, the D44.1–HEL surfaces in one region are separated by up to 3 Å as a result of an unsolvated buried cavity. As noted in the case of the Fv D1.3 V_L Trp92→Asp mutation, buried surfaces that are not in van der Waals contact will contribute less to the stability of the complex. Can the smaller interaction area and the existence of significant gaps in the antibody–antigen interface explain the reduced association of D44.1 with HEL as compared to HyHEL-5? In this case it is not quite as clear. True, while the difference in buried surface areas (~ 250 Å^2) is quite consistent with the 1000-fold difference in binding constants, there may be differences in the nature of the electrostatic interactions, particularly in the relative strengths of the salt bridges due to different solvent accessibility.

Gibas *et al.* (104) analysed the HyHEL-5 and D44.1 interactions to HEL using continuum electrostatic methods to model the pH-dependent energetics of association. This study suggests that subtle differences in the geometry of the salt-link configurations can account for the observed differences in affinity for HEL. The combining site of HyHEL-5 and HEL can be seen to be surrounded by an annulus of hydrophobic residues, which form contacts that shield the polar residues in the combining site from bulk solvent; this is not true at the interface of D44.1 and HEL. This insulation by apolar residues has implications both for van der Waals interactions and for electrostatic interactions at the antibody–lysozyme interface, in that it alters the effective dielectric constant in the interface (104).

The Fab HyHEL-5–HEL and Fab D44.1–HEL structures were analysed at medium resolution, 2.8 Å and 2.5 Å, respectively. Experience dictates that these resolutions are not ideal for a full disclosure of solvent water molecules bound to the proteins. Where as many as 50 water molecules are bound at the Fv D1.3–HEL interface (66, 74), the 2.5 Å Fab D44.1–HEL structure revealed only three buried solvent water molecules which bridged the antibody and antigen (103). This is not to say that one would expect to find as many buried water molecules in the high-resolution structure of D44.1 as compared to D1.3. On the contrary, Goldbaum *et al.* (75) found, in the analysis of antibody–antigen binding under conditions of reduced water concentration, that D44.1 apparently buries few water molecules upon complex formation.

6. Conclusions

Considering all the structural, kinetic, thermodynamic, and mutational studies of the D1.3–HEL interaction what have we learned? The structural analysis of D1.3 in complex with HEL reveals a structure of exquisite complementarity, in large part due to the inclusion of solvent water molecules. There are no destabilizing voids or gaps in the interface, nor are there any uncompensated hydrogen bond donors or acceptors. Only minor variations between the structure of the free and antigen-complexed antibody point to a binding mechanism, whereby little of the free energy of complex formation is wasted by a decrease in entropy due to the loss of rotational, translational, and side-chain degrees of freedom. However, analysis of the shape complementarity of antibody–antigen interfaces, albeit without the inclusion of bound water molecules, indicates that these complexes have a somewhat poorer complementarity compared to other protein–protein interactions.

The D1.3–HEL interaction features many van der Waals and polar interactions, including a large network of hydrogen-bonded solvent molecules bridging the interface. The energetics of D1.3 binding to HEL are dominated by only three contact residues whose side-chain contributions can be probed by site-directed mutagenesis. These residues, one apolar and two polar, form a central core to the interaction and are surrounded by residues whose apparent contributions to complex stabilization are much less pronounced, although clearly required to provide the environment for the dominant interactions. On the other hand, the interaction of D1.3 with the anti-idiotypic antibody E5.2 features an interface where most of the contacting residues play a significant role in ligand binding. Thus, the same protein may recognize different ligands in ways that are structurally similar yet energetically distinct. This is in agreement with the work of Tulip *et al.* (105), who found that the anti-neuraminidase antibodies NC10 and NC41 recognize the same protein surface through different key residues.

Antibody D1.3 is designed to recognize a solvated antigen. Of the 25 well-ordered bridging water molecules in the antibody–antigen interface, 20 are found in the crystal structures of the free proteins. Upon complex formation, 15 water molecules are displaced, some of which are substituted by polar protein interactions. Moreover, a physicochemical analysis of the binding of D1.3 with HEL has demonstrated that ligand binding is severely inhibited upon replacing water with an organic solvent. Even more telling, some amino-acid substitutions in the D1.3–HEL interface fully or partially compensate for the loss in antibody–antigen contacts by incorporating or rearranging the solvent network, providing new stabilizing interatomic interactions.

The crystal structures of the free and HEL-complexed Fv fragment of D1.3 have been determined at high resolution (1.8 Å) allowing the detection of small main-chain and side-chain movements. A small relative displacement of the V_H and V_L domains, by reference to their positions in the free Fv, bring contacting residues closer to the antigen. Moreover, a distinct decrease in the mobility of V_H CDR3 upon binding of antigen points to one of the origins of the unfavourable entropic components to the free energy of the reaction. Similar results were obtained on comparison of the free and HEL-complexed Fab D44.1. In this example there are also marked reductions in the mobility of the antigen-contacting residues and, in fact, the entire Fv moiety exhibits reduced mobility upon complex formation. As in the case of the D1.3–HEL complex, the V domains of D44.1 show a change in relative disposition upon complex formation as a result of small perturbations in the positions of the CDR when bound to antigen. On the other hand, the reaction of D1.3 with TEL, having only one amino-acid difference in the epitope compared to HEL, leads to a conformational change in the peptide configuration of V_L CDR3 resulting in the conservation of polar interactions to the antibody, an excellent example of electrically compensating conformational change.

The 'lock-and-key' model for antibody–antigen association must be modulated by these descriptions of conformational change. It has been observed in several antibody–hapten systems that maturation of immune responses leads to an increase in the association constants of induced antibodies (106). Furthermore, it has been proposed that one of the mechanisms of this maturation is the recruitment of new clones of cells producing antibodies with higher association rates (107). Undoubtedly, adequate high association rates and equilibrium constants may constitute an essential feature of the secondary immune response. These features will obviously be better achieved with antibodies that are highly complementary to their specific antigens.

From a consideration of thermodynamic and kinetic parameters, and comparisons of free and antigen-bound, three-dimensional structures, a complementarity with minimal conformational changes, maximal van der Waals contacts, and electrostatic compensation of contacting atoms, will result in the most stable association.

There is a startling amount of information regarding the structure of antibody–antigen complexes being made available on what seems to be a daily basis. Therefore, we must temper our understanding of the nature of antibody–antigen interactions with the thought that the anti-lysozyme antibodies may not necessarily reflect antibody–antigen systems in general, but only in a specific case. Moreover, as we have seen, even the interactions to a specific antibody by cross-reacting antigens and antigen mimics fail to be classified by a single mechanism. As more antibody–antigen systems become available through structural and thermodynamic study, most likely we shall see antibodies for what they are: a protein engineering system developed by nature for the generation of a virtually unlimited repertoire of complementary molecular surfaces, and, as such, exhibiting a wide range of structural and energetic phenomena.

References

1. Padlan, E. A. (1994). Anatomy of the antibody molecule. *Mol. Immun.*, **31**, 169.
2. Padlan, E. A. (1997). X-ray crystallography of antibodies. *Adv. Protein Chem.*, **50**, 57.
3. Edmundson, A. B., Guddat, L. W., Rosauer, R. A., Andersen, K. N., Shan, L., and Fan, Z.-C. (1996). Three-dimensional aspects of IgG structure and function. In *The antibodies*, Vol. 1 (ed. M. Zanetti and J. D. Capra), p. 41. Harwood Academic, Amsterdam.
4. Rajan, S. S., Ely, K. A., Abola, E. E., Wood, M. K., Colman, P. M., Athay, R. J., and Edmundson A. B. (1983). Three-dimensional structure of the Mcg IgG1 immunoglobulin. *Mol. Immun.*, **20**, 787.
5. Silverton, E. W., Navia, M. A., and Davies, D. R. (1977). Three-dimensional structure of an intact human immunoglobulin. *Proc. Natl. Acad. Sci. USA*, **74**, 5140.
6. Marquart, M., Deisenhofer, J., Huber, R., and Palm, W. (1980). Crystallographic refinement and atomic models of the intact immunoglobulin molecule Kol and its antigen-binding fragment at 3.0 Å and 1.9 Å resolution. *J. Mol. Biol.*, **141**, 369.
7. Ely, K. R., Colman, P. M., Abola, E. E., Hess, A. C., Peabody, D. S., Parr, D. M., Connell, G. E., Laschinger, C. A., and Edmundson, A. B. (1978). Mobile Fc region in the Zie IgG2 cyroglobulin: comparison of crystals of the $F(ab')_2$ fragment and the intact immunoglobulin. *Biochemistry*, **17**, 820.
8. Harris, L. J., Skaletsky, E., and McPherson, A. (1998). Crystallographic structure of an intact IgG1 monoclonal antibody. *J. Mol. Biol.*, **275**, 861.
9. Poljak, R. J., Amzel, L. M., Avey, H. P., Chen, B. L., and Phizackerly, R. P. (1973). Three-dimensional structure of the Fab′ fragment of a human immunoglobulin at 2.8 Å resolution. *Proc. Natl. Acad. Sci. USA* **70**, 3305.
10. Poljak, R. J., Amzel, L. M., Chen, B. L., Phizackerley, R. P., and Saul, F. (1974). The three-dimensional structure of the Fab′ fragment of a human myeloma immunoglobulin at 2.0 Å resolution. *Proc. Natl. Acad. Sci. USA*, **71**, 3440.
11. Huber, R., Deisenhofer, J., Colman, P. M., Matsushima, M., and Palm, W. (1976). Crystallographic structure studies of an IgG molecule and an Fc fragment. *Nature*, **264**, 415.

12. Wilson, I. A. and Stanfield, R. L. (1994). Antibody–antigen interactions: new structures and new conformational changes. *Curr. Opin. Struct. Biol.*, **4**, 857.
13. Schwarz, F. P., Tello, D., Goldbaum, F. A., Mariuzza, R. A., and Poljak, R. J. (1995). Thermodynamics of antigen-antibody binding using specific anti-lysozyme antibodies. *Eur. J. Biochem.*, **228**, 388.
14. Chothia, C., Lesk, A. M., Tramontano, A., Levitt, M., Smith-Gill, S. J., Air, G., Sheriff, S., Padlan, E. A., Davies, D.,Tulip, W. R., Colman, P. M., Spinelli, S., Alzari, P. M., and Poljak, R. J. (1989). Conformations of immunoglobulin hypervariable regions. *Nature*, **342**, 877.
15. Al-Lazinkani, B., Lesk, A. M., and Chothia, C. (1997). Standard conformations for the canonical structures of immunoglobins. *J. Mol. Biol.*, **273**, 927.
16. Eigenbrot, C., Randal, M., Presta, L., Carter, P., and Kossiakoff, A. A. (1993). X-ray structures of the antigen-binding domains from three variants of humanized anti-p185^{HER} antibody 4D5 and comparison with molecular docking. *J. Mol. Biol.*, **229**, 969.
17. Steipe, B., Plucktun, A., and Huber, R. (1992). Refined crystal structure of a recombinant immunoglobulin domain and a complementary-determining region grafted mutant. *J. Mol. Biol.*, **225**, 739.
18. Bajorath, J. and Sheriff, S. (1996). Comparison of an antibody model with X-ray structure: the variable fragment of BR96. *Proteins: Struct., Funct., Genet.*, **24**, 152.
19. Braden, B. C., Fields, B. A., Ysern, X., Goldbaum, F. A., Dall'Acqua, W., Schwarz, F. P., Poljak, R. J., and Mariuzza, R. A. (1996). Crystal structure of the complex of the variable domain of antibody D1.3 and turkey egg white lysozyme: a novel conformational change in antibody CDR-3 selects for antigen. *J. Mol. Biol.*, **257**, 889.
20. Anson, M. L. (1945). Protein denaturation and the properties of protein groups. *Adv. Protein Chem.*, **2**, 361.
21. Anfinsen, C. B. (1956). The limited digestion of ribonuclease with pepsin. *J. Biol. Chem.*, **221**, 405.
22. Kauzman, W. (1959). Some factors in the interpretation of protein denaturation. *Adv. Protein Chem.*, **14**, 1.
23. Chothia, C. (1974). Hydrophobic bonding and accessible surface area in proteins. *Nature*, **248**, 338.
24. Chothia, C. and Janin, J. (1975). Principles of protein–protein recognition. *Nature*, **256**, 705.
25. Ross, P. D. and Subramanian, S. (1968). Thermodynamics of protein associations: forces contributing to stability. *Biochemistry*, **20**, 3096.
26. Tanford, C. (1968). Protein denaturation. *Adv. Protein Chem.*, **23**, 121.
27. Richards, F. M. (1977). Areas, volumes, packing and protein structure. *Annu. Rev. Biophys. Bioeng.*, **6**, 151.
28. Dill, K. A. (1990). Dominant forces in protein folding. *Biochemistry*, **29**, 7133.
29. Eisenberg, D. and McLachlan, A. D. (1986). Solvation energy in protein folding and binding. *Nature*, **319**, 199.
30. Lazaridis, T., Archontis, G., and Karplus, M. (1995). Enthalpic contribution to protein stability: insights from atom-based calculations and statistical mechanics. *Adv. Protein Chem.*, **47**, 231.
31. Nicholls, A., Sharp, K. A., and Honig B. (1991). Protein folding and association: insights from the interfacial and thermodynamic properties of hydrocarbons. *Proteins: Struct. Funct. Genet.*, **11**, 281.
32. Sharp, K. A., Nicholls, A., Fine, R. M., and Honig, B. (1991). Reconciling the magnitude of the microscopic and macroscopic hydrophobic effects. *Science*, **252**, 106.

33. Matthews, B. W. (1995). Studies on protein stability with T4 lysozyme. *Adv. Protein Chem.*, **46**, 249.
34. Steif, C., Hinz, H.-J., and Cesareni, G. (1995). Effects of cavity-creating mutations on conformational stability and structure of the dimeric 4-α-helical protein ROP: thermal unfolding studies. *Proteins: Struct., Funct., Genet.*, **23**, 83.
35. Kellis, J. T., Jr, Nyberg, K., and Fersht, A. R. (1989). Energetics of complementary side-chain packing in a protein hydrophobic core. *Biochemistry*, **28**, 4914.
36. Shortle, D., Stites, W. E., and Meeker, A. K. (1990). Contributions of the large hydrophobic amino acids to the stability of staphylococci nuclease. *Biochemistry*, **29**, 8033.
37. Makhatadze, G. I. and Privalov, P. L. (1995). Energetics of protein structure. *Adv. Protein Chem.*, **47**, 307.
38. McDonald, I. K. and Thornton, J. M. (1994). Satisfying hydrogen bonding potential in proteins. *J. Mol. Biol.*, **238**, 777.
39. Creighton, T. (1991). Stability of folded conformations. *Curr. Opin. Struct. Biol.*, **1**, 5.
40. Fersht, A. R., Shi, J.-P., Knill-Jones, J., Lowe, D. M., Wilkinson, A. T., Blow, D. M., Brick, P., Carter, P., Ware, M. M. Y., and Winter, G. (1985). Hydrogen bonding and biological specificity analysed by protein engineering. *Nature*, **314**, 235.
41. Fersht, A. R. (1987). The hydrogen bond in molecular recognition. *TIBS*, **12**, 301.
42. Blaber, M., Zhang, X.-J., and Matthews, B. W. (1993). Structural basis of α-helix propensity. *Science*, **260**, 1637.
43. Del Bene, J. E. (1975). Molecular orbital theory of the hydrogen bond. XII. Amide hydrogen bonding in formamide–water and formamide–formaldehyde systems. *J. Chem. Phys.*, **62**, 1961.
44. Rand, R. P. (1992). Raising water to new heights. *Science*, **256**, 618.
45. Colombo, M. F., Rau, D. C., and Parsgian, V. A. (1992). Protein solvation in allosteric regulation: a water effect on hemoglobin. *Science*, **256**, 655.
46. Douzou, P. (1994). Osmotic regulation of gene action. *Proc. Natl. Acad. Sci. USA*, **91**, 1657.
47. Kornblatt, J. A., Kornblatt, M. J., Hui Bon Hoa, G., and Mauk, A. G. (1993). Responses of two protein–protein complexes to solvent stress: does water play a role at the interface? *Biophys. J.*, **65**, 1059.
48. Dzingeleski, G. D. and Wolfenden, R. (1993). Hypersensitivity of an enzyme reaction to solvent water. *Biochemistry*, **32**, 9143.
49. Yang, A. S., Sharp, K. A., and Honig, B. (1992). Analysis of the heat capacity dependence of protein folding. *J. Mol. Biol.*, **227**, 889.
50. Hendsch, Z. S. and Tidor, B. (1994). Do salt bridges stabilize proteins? A continuum electrostatic analysis. *Protein Sci.*, **3**, 211.
51. Finkelstein, A. V. and Janin, J. (1989). The price of lost freedom: entropy of bimolecular complex formation. *Protein Eng.*, **3**, 1.
52. Janin, J. (1995). Elusive affinities. *Proteins: Struct., Funct., Genet.*, **21**, 30.
53. Murphy, K. P., Xie, D., Thompson, K. S., Amzel, L. M., and Freire, E. (1994). Entropy in biological binding processes: estimation of translational entropy loss. *Proteins:Struct., Funct., Genet.*, **18**, 63.
54. Larsen, T. A., Olson, A. J., and Goodsell, D. S. (1998). Morphology of protein–protein interfaces. *Structure*, **6**, 421.
55. Miller, A., Ch'ng, L.-K., Benjamin, C., Sercarz, E. E., Brodeur, P., and Riblet, R. (1983). Detailed analysis of the public idiotype of anti-hen egg-white lysozyme antibodies. *Ann. NY Acad. Sci.*, **418**, 104.
56. Metzger, D. W., Ch'ng, L.-K., Miller, A., and Sercarz, E. E. (1984). Sharing of an idiotypic

marker by monoclonal antibodies specific for distinct regions of hen lysozyme. *Nature*, **287**, 540.

57. Kobayashi, T., Fujio, H., Kondo, K., Dohi, Y., Hirayama, A., Takagaki, Y., Kosaki, G., and Amano, T. (1982). A monoclonal antibody specific for a distinct region of hen egg-white lysozyme. *Mol. Immun.*, **19**, 619.
58. Smith-Gill, S. J., Wilson, A. C., Potter, M., Prager, E. M., Feldmann, R. J., and Mainhart, C. R. (1982). Mapping the antigenic epitope for a monoclonal antibody against lysozyme. *J. Immun.*, **128**, 314.
59. Harper, M., Lema, F., Boulot, G., and Poljak, R. J. (1987). Antigen specificity and cross-reactivity of monoclonal anti-lysozyme antibodies. *Mol. Immunol.*, **24**, 97.
60. Amit, A. G., Mariuzza, R. A., Phillips, S. E. V., and Poljak, R. J. (1985). Three-dimensional structure of an antigen–antibody complex at 6 Å resolution. *Nature*, **313**, 156.
61. Amit, A. G., Mariuzza, R. A., Phillips, S. E. V., and Poljak, R. J. (1986). Three-dimensional structure of an antigen–antibody complex at 2.8 Å resolution. *Science*, **233**, 747.
62. Fischmann, T. O., Bentley, G. A., Bhat, T. N., Boulot, G., Mariuzza, R. A., Phillips, S. E. V., Tello, D., and Poljak, R. J. (1991). Crystallographic refinement of the three-dimensional structure of the FabD1.3–lysozyme complex at 2.5 Å resolution. *J. Biol. Chem.*, **266**, 12915.
63. Blake, C. C. F., Koenig, D. F., Mair, G. A., North, A. C. T., Phillips, D. C., and Sarma, V. R. (1965). Structure of hen egg-white lysozyme. A three-dimensional Fourier synthesis at 2 Å resolution. *Nature*, **206**, 757.
64. Boulot, G., Eiselé, J.-L., Bentley, G. A., Bhat, T. N., Ward, E. S., Winter, G., and Poljak, R. J. (1990). Crystallization and preliminary X-ray diffraction study of the bacterially expressed Fv from the monoclonal anti-lysozyme antibody D1.3 and its complex with the antigen, lysozyme. *J. Mol. Biol.*, **213**, 617.
65. Bhat, T. N., Bentley, G. A., Fischmann, T. O., Boulot, G., and Poljak, R. J. (1990). Small rearrangements in structures of Fv and Fab fragments of antibody D1.3 on antigen binding. *Nature*, **347**, 483.
66. Bhat, T. N., Bentley, G. A., Boulot, G., Greene, M. I., Tello, D., Dall'Acqua, W., Souchon, H., Schwarz, F. P., Mariuzza, R. A., and Poljak, R. J. (1994). Bound water molecules and conformational stabilization help mediate an antigen–antibody association. *Proc. Natl. Acad. Sci., USA*, **91**, 1089.
67. Westhof, E. (1993). *Water and biological macromolecules*. CRC Press, Boca Raton, FL.
68. Wlodawer, A., Nachman, J., and Gilliland, G. L. (1987). Structure of form III crystals of bovine pancreatic trypsin inhibitor. *J. Mol. Biol.*, **198**, 469.
69. Otting, G., Liepnish, E., and Wuthrich, K. (1991). Protein hydration in aqueous solution. *Science*, **254**, 974.
70. Levitt, M. and Park, B. H. (1993). Water: now you see it, now you don't. *Structure*, **1**, 223.
71. Zhang, X.-J. and Matthews, B. W. (1994). Conservation of solvent-binding sites in 10 crystal forms of T4 lysozyme. *Protein Sci.*, **3**, 1031.
72. Shapiro, L., Fannon, A. M., Kwong, P. D., Thompson, A., Lehmann, M. S., Grubel, G., Legrand, J.-F., Als-Nielsen, J., Colman, D. R., and Hendrickson, W. A. (1995). Structural basis of cell–cell adhesion by cadherins. *Nature*, **374**, 327.
73. Guillet, V., Lapthorn, A., Harley, R. W., and Mauguen, Y. (1993). Recognition between a bacterial ribonuclease, barnase and its natural inhibitor, barstar. *Structure*, **1**, 165.
74. Braden, B. C., Fields, B. A., and Poljak, R. J. (1995). Conservation of water molecules in an antibody–antigen interaction. *J. Mol. Recogn.* **8**, 317.
75. Goldbaum, F. A., Schwarz, F. P., Eisenstein, E., Cauerhff, A., Mariuzza, R. A., and Poljak, R. J. (1996). The effect of water activity on the association constants and the enthalpy

of reaction between lysozyme and the specific antibodies D1.3 and D44.1. *J. Mol. Recogn.*, **9**, 6.

76. Karlsson, R., Michaelsson, A., and Mattsson, L. J. (1991). Kinetic analysis of monoclonal antibody–antigen interactions with a new biosensor based analytical system. *Immunol. Meth.*, **145**, 229.
77. Lawrence, M. C. and Colman, P. M. (1993). Shape complementarity at protein/protein interfaces. *J. Mol. Biol.*, **234**, 946.
78. Braden, B. C., Fields, B. A., Ysern, X., Dall'Acqua, W., Goldbaum, F. A., Poljak, R. J., and Mariuzza, R. A. (1996). Crystal structure of an Fv–Fv idiotope–antiidiotope complex at 1.9 Å resolution. *J. Mol. Biol.*, **264**, 137.
79. Jin, L., Fendly, B., and Wells, J. A. (1992). High resolution functional analysis of antibody–antigen interactions. *J. Mol. Biol.*, **226**, 851.
80. Kelley, R. F. and O'Connell, M. P. (1993). Thermodynamic analysis of an antibody functional epitope. *Biochemistry*, **32**, 6828.
81. Ito, W., Iba, Y., and Kurosawa, Y. (1993). Effects of substitutions of closely related amino acids at the contact surface in an antigen-antibody complex on the thermodynamic parameters. *J. Biol. Chem.*, **268**, 16639.
82. Hawkins, R. E., Russell, S. J., Baier, M., and Winter, G. (1993). The contribution of contact and non-contact residues of antibody in the affinity of binding to antigen: the interaction of mutant D1.3 antibodies with lysozyme. *J. Mol. Biol.*, **234**, 958.
83. Kelley, R. F., O'Connell, M. P., Carter, P., Presta, L., Eigenbrot, C., Covarrubias, M., Snedecor, B., Bourell, J. H., and Vetterlein, D. (1992). Antigen binding thermodynamics and antiproliferative effects of chimeric and humanized anti-p185^{HER2} antibody Fab fragments. *Biochemistry*, **31**, 5435.
84. Dall'Acqua, W., Goldman, E. R., Eisenstein, E., and Mariuzza, R. A. (1996). A mutational analysis of the binding of two different proteins to the same antibody. *Biochemistry*, **35**, 9667.
85. Dall'Acqua, W., Goldman, E. R., Lin, W., Teng, C., Tsuchiya, D., Li, H., Ysern, X., Braden, B. C., Li, Y., Smith-Gill, S. J., and Mariuzza, R. A. (1998). A mutational analysis of binding interactions in an antigen–antibody protein–protein complex. *Biochemistry*, **37**, 7981.
86. Benjamin, D. C., Williams, D. C., Jr., Smith-Gill, S. J., and Rule, G. S. (1992). Long-range changes in a protein antigen due to antigen–antibody interaction. *Biochemistry*, **31**, 9539.
87. Cunningham, B. C. and Wells, J. A. (1993). Comparison of a structural and functional epitope. *J. Mol. Biol.*, **234**, 554.
88. Clackson, T. and Wells, J. A. (1995). A hot spot of binding energy in a hormone-receptor interface. *Science*, **267**, 383.
89. Fields, B. A., Goldbaum, F. A., Ysern, X., Poljak, R. J., and Mariuzza, R. A. (1995). Molecular basis of antigen mimicry by an anti-idiotope. *Nature*, **374**, 739.
90. Schreiber, G. and Fersht, A. R. (1995). Energetics of protein–protein interaction: analysis of the barnase–barstar interface by single mutant and double mutant cycles. *J. Mol. Biol.*, **248**, 478.
91. Covell, D. G. and Wallqvist A. (1997). Analysis of protein–protein interaction and the effects of amino acid mutations on their energetics. The importance of water molecules in the binding epitope. *J. Mol. Biol.*, **269**, 281.
92. Carter, P. J., Winter, G., Wilkinson, A. J., and Fersht, A. R. (1984). The use of double mutants to detect structural changes in the active site of tyrosyl-tRNA synthetase. *Cell*, **38**, 835.
93. Ackers, G. K. and Smith, F. R. (1985). Effects of site-specific amino acid modification on protein interaction and biological function. *Annu. Rev. Biochem.*, **54**, 597.

94. Horovitz, A. (1987). Non-additivity in protein–protein interactions. *J. Mol. Biol.*, **196**, 733.
95. Goldman, E. R., Dall'Acqua, W., Braden, B. C., and Mariuzza, R. A. (1997). Analysis of binding interactions in an idiotope–antiidiotope protein–protein complex by double mutant cycles. *Biochemistry*, **36**, 49.
96. Levitt, M. and Perutz, M. F. (1988). Aromatic rings act as hydrogen bond acceptors. *J. Mol. Biol.*, **201**, 751.
97. Ysern, X., Fields, B. A., Bhat, T. N., Goldbaum, F. A., Dall'Acqua, W., Schwarz, F. P., Poljak, R. J., and Mariuzza, R. A. (1994). Solvent rearrangement in an antigen–antibody interface introduced by site-directed mutagenesis of the antibody combining site. *J. Mol. Biol.*, **238**, 496.
98. Fields, B. A., Goldbaum, F. A., DallÁcqua, W., Malchodi, E. L., Cauerhff, A., Schwarz, F. P., Ysern, X., Poljak, R. J., and Mariuzza, R. A. (1996). Hydrogen bonding and solvent structure in an antigen–antibody interface. Crystal structures and thermodynamic characterization of three Fv mutants complexed with lysozyme. *Biochemistry*, **35**, 15494.
99. Tello, D., Goldbaum, F. A., Mariuzza, R. A., Ysern, X., Schwarz, F. P., and Poljak, R. J. (1993). Three-dimensional structure and thermodynamics of antigen binding by anti-lysozyme antibodies. *Biochem. Soc. Trans.*, **21**, 943.
100. Chacko, S., Silverton, E., Kam-Morgan, L., Smith-Gill, S. J., Cohen, G., and Davies, D. (1995). Structure of an antibody–lysozyme complex: unexpected effect of a conservative mutation. *J. Mol. Biol.*, **245**, 261.
101. Sheriff, S., Silverton, E. W., Padlan, E. A., Cohen, G. H., Smith-Gill, S. J., Finzel, B. C., and Davies, D. R. (1987). Three-dimensional structure of an antibody–antigen complex. *Proc. Natl. Acad. Sci. USA*, **84**, 8075.
102. Novotny, J., Bruccoleri, R. E., and Saul, F. A. (1989). On the attribution of binding energy in antigen–antibody complexes McPC 603, D1.3, and HyHEL-5. *Biochemistry*, **28**, 4735.
103. Braden, B. C., Souchon, H., Eiselé, J.-L., Bentley, G. A., Bhat, T. N., Navaza, J., and Poljak, R. J. (1994). Three-dimensional structures of the free and the antigen-complexed Fab from monoclonal anti-lysozyme antibody D44.1. *J. Mol. Biol.*, **243**, 767.
104. Gibas, C. J., Subramaniam, S., McCammon, J. A., Braden, B. C., and Poljak, R. J. (1997). pH dependence of antibody / lysozyme complexation. *Biochemistry*, **36**, 15599.
105. Tulip, W. R., Harley, V. R., Webster, R. G., and Novotny, J. (1994). N9 neuraminidase complexes with antibodies NC41 and NC10: empirical free energy calculations capture specificity trends observed with mutant binding data. *Biochemistry*, **33**, 7986.
106. Eisen, H. N. and Siskind, G. W. (1964). Variations in affinities of antibodies during the immune response. *Biochemistry*, **3**, 996.
107. Foote, J. and Milstein, C. (1991). Kinetic maturation of an immune response. *Nature*, **352**, 530.
108. Hibbits, K. A., Gill, D. S., and Wilson, R. C. (1994). Isothermal titration calorimetric study of the association of hen egg lysozyme and the anti-lysozyme antibody HyHEL-5. *Biochemistry*, **33**, 3584.

6 Proteins of the major histocompatibility complex and their interactions with T-cell receptors

TIM R. DAFFORN and ARTHUR M. LESK

1. Introduction

Surgical patients, if not immunosuppressed by drugs, will reject transplanted organs—unless the donor is an identical twin—because the transplant is recognized as being foreign. The immunological distinction between 'self' and 'non-self' resides in the proteins of the Major Histocompatibility Complex (MHC) and their interaction with T-cell receptors. MHC proteins bind intracellularly produced peptides and present them on cell surfaces. The triggering event in alerting the immune system to the presence of a foreign protein is the recognition, by a T-cell receptor, of a complex between an MHC protein and a peptide derived from the foreign protein.

MHC proteins fall into two classes, with related structure and function. The two classes function in parallel, to produce different immune responses to intracellular and extracellular pathogens, respectively (Fig. 1). Class I MHC molecules appear on the surfaces of most cells of the body, and present peptides derived from proteins degraded in the cytosol. These peptide–MHC complexes alert the body to intracellular pathogens. They interact with cytotoxic T cells, and direct the immune response to the presenting cell and those in its vicinity. They can also form complexes with super-antigens. Class II MHC molecules appear on the surfaces of specialized cells of the immune system: B lymphocytes and antigen-presenting macrophages. They present oligopeptides derived from exogenous antigens (which have been endocytosed and chopped into peptides). These peptide–MHC complexes interact with helper T cells, mediating the proliferation of cells synthesizing antibodies that circulate in the blood and the activation of macrophages.

In addition to triggering immune responses in mature individuals, MHC–peptide complexes are also involved in the removal of self-complementary T cells in the thymus during development.

Each individual in vertebrate species expresses a set of MHC proteins selected from a diverse genetic repertoire in species. In humans, the MHC complex is a set of linked genes on chromosome 6 (1). The system is highly polymorphic, with 50–150 alleles per locus, showing greater sequence variation than most polymorphic proteins. Each of us produces six class I molecules and a somewhat higher complement of class II (2, 3). Each MHC protein must therefore be able to bind many peptides, if about 30 MHC proteins are to present the large number of possible antigens.

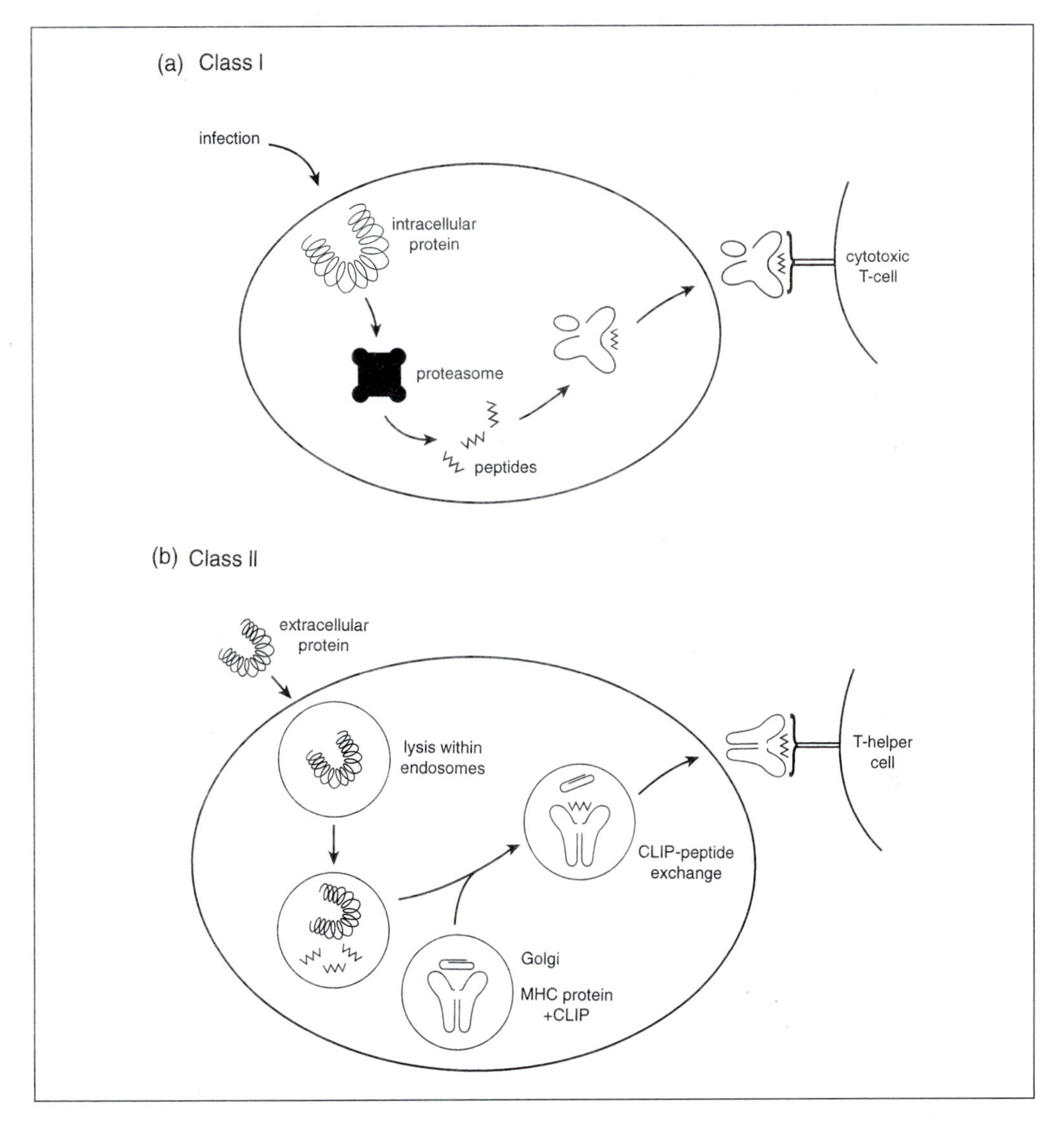

Fig. 1 (a) Class I and (b) class II MHC molecules participate in two parallel systems to trigger the immune response to foreign proteins originating inside and outside cells. Peptides derived from foreign proteins are loaded intracellularly and transported to the surface where they are presented to T cells. The representation of the class II invariant peptide (CLIP) is an icon, not a drawing of its structure.

2. MHC system

2.1 Structures of MHC proteins

MHC domains are modular proteins that contain several characteristic domains. These include peptide-binding domains with folds unique to the MHC system plus a related Fc receptor (4, 5) and the HFE protein (6), MHC proteins also contain immunoglobulin-like domains.

Class I MHC proteins contain two polypeptide chains (Fig. 2). The longer chain (44 kDa in humans; 47 kDa in mice) has a modular domain structure, inherited from the exons of the gene, of the form α_1–α_2–α_3–, a short hydrophobic membrane-spanning segment, and a 30-residue cytoplasmic tail. The α-domains are approximately 90 residues in length. The second chain is β_2-microglobulin, a non-polymorphic structure (constant in the species), the gene for which is unlinked from the MHC complex (Fig. 3).

The α_1- and α_2-domains of class I MHC proteins have a common fold, and interact to form a symmetrical combined structure. They bind peptides between them in a groove, created by two long curved α-helices (Fig. 4a). The variability in the amino-acid sequence, i.e. the polymorphism, of MHC proteins is high in the regions that surround the groove, to create variety in specificity. The α_3-domain and β_2-microglobulin, which do not interact with bound peptides, are double β-sheet proteins with topologies in the immunoglobulin superfamily.Their sheet structure is characteristic of the C1-type of immunoglobulin domain (7). Each has the disulfide bridge, and the tryptophan packed against it, common to immunoglobulin domains (8).

The α_1- and α_2-domains of class I MHC proteins, and the α_1- and β_1-domains of class II, have a common folding pattern, specific to the MHC. Each has a four-stranded

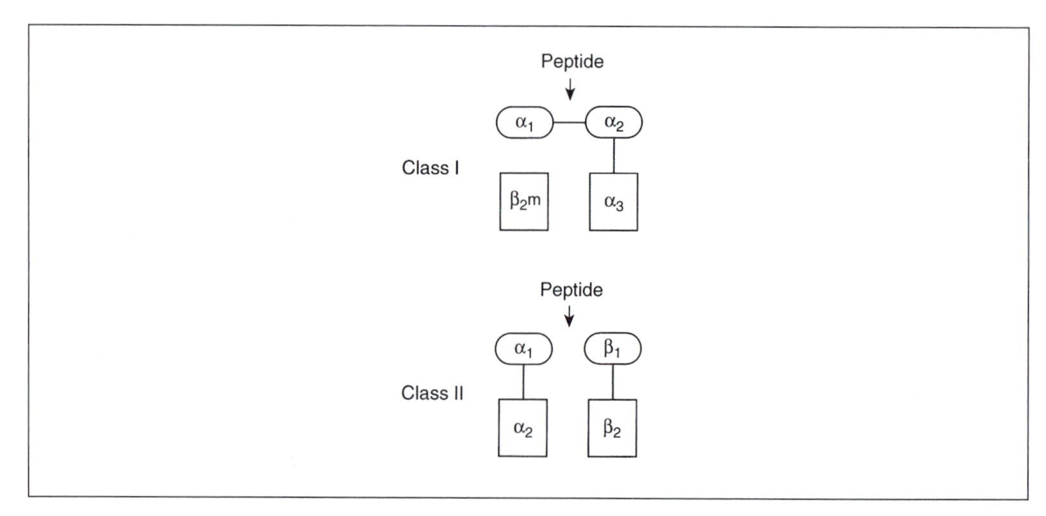

Fig. 2 Domain structure of MHC proteins. MHC proteins are modular proteins containing two peptide-binding domains (round cartouche) and two immunoglobulin-like domains (rectangles). Peptides bind in a groove created by the α_2/α_3-domains in class I MHC proteins and by the α_1/β_1-domains in class II MHC proteins.

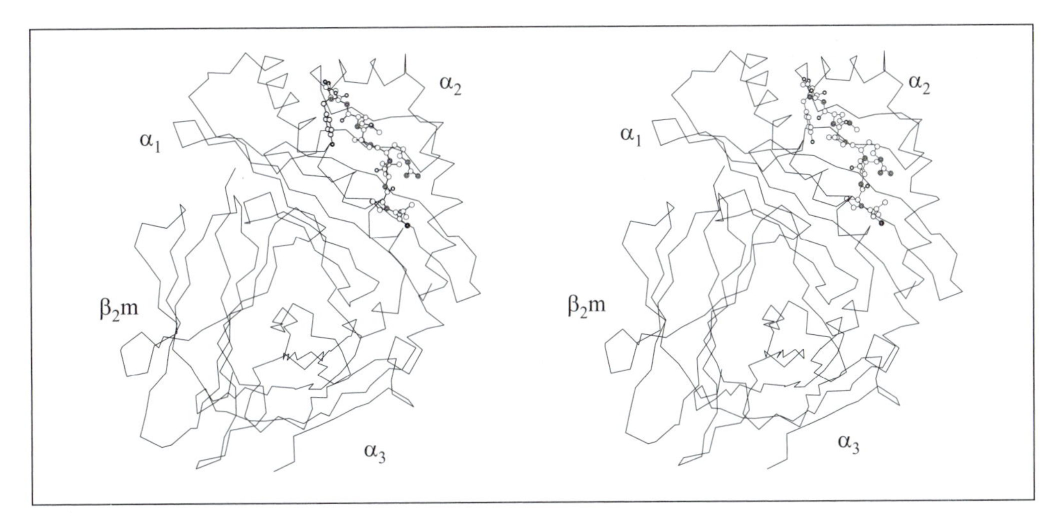

Fig. 3 A stereo representation of the structure of a class I MHC protein, B3501, binding the peptide VPLRPMTY from the nef protein of HIV-1 (14).

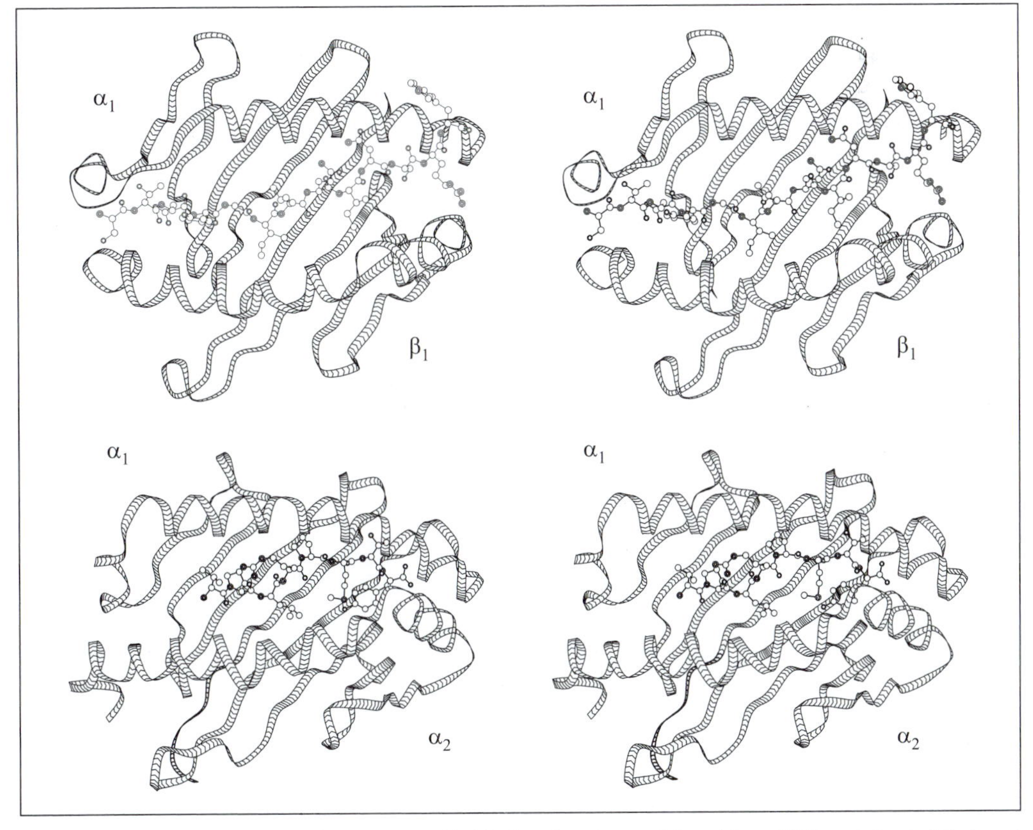

Fig. 4 A stereo representation of the two peptide-binding domains and ligand from MHC proteins in (a) class I (B3501, see Fig. 3), (b) class II (I-Ak, see Fig. 5).

β-sheet at the N-terminus, followed by a short bridging helix and a C-terminal helix that lies across the β-sheet. The strands from each domain form an eight-stranded β-sheet. The two C-terminal helices form the sides of the peptide-binding cleft. Each has a pronounced curvature. Because of the very rough twofold symmetry of the α_1–α_2 unit, the long C-terminal helices run antiparallel. Peptides bind in an orientation parallel to that of the α_1-helix.

Class II MHC proteins contain an α-chain (34 kDa) containing two domains, $\alpha_1 + \alpha_2$, a homologous β-chain (29 kDa) containing two domains, $\beta_1 + \beta_2$, and an invariant chain I1 (31 kDa) (Fig. 5). The α_1- and β_1-domains pack together to make a structure similar to that formed by the α_2- and α_3-domains of class I MHC proteins, and bind peptides in a similar mode (Fig. 4b). Figure 6 shows a superposition of the peptide-binding domains from class I and class II MHC proteins, each containing a peptide ligand. The binding domains have a similar structure, and the peptides occupy a similar region of space and orientation relative to the MHC protein. Note that in the class I molecules, the cleft is pinched off at both ends: at the left by the unwinding of two turns at the N-terminus of the helix in the α_1-domain, resulting from a deletion in this region in class I molecules relative to class II; and at the right by side chains. The side chains that bind the chain termini tend to be conserved; in class I these include tyrosines at positions 7, 59, 159, and 171 at the N-terminus of the peptide, and Tyr84, Lys146, and Trp147 at its C-terminus. In class II the cleft is also closed at the right, but open at the left.

Class I molecules can bind peptides of limited length—from about 8–11 residues, but most commonly 8–9. Two factors impose the limits: the closure of the cleft at either

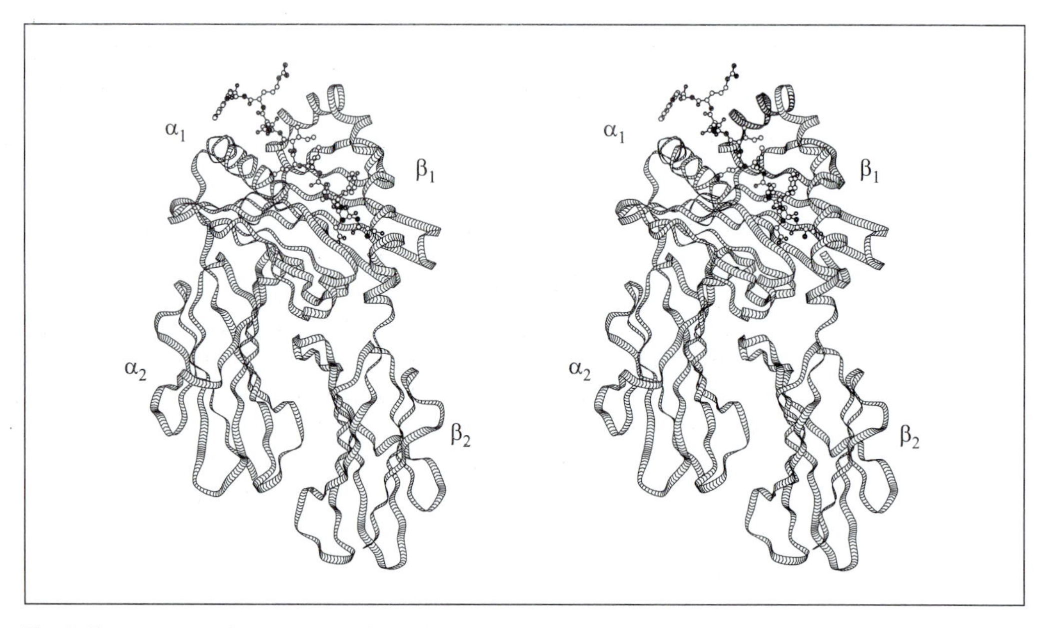

Fig. 5 The structure of a class II MHC protein, I-Ak, binding the peptide STDYGILQINSRW from hen egg-white lysozyme shown in stereo (47).

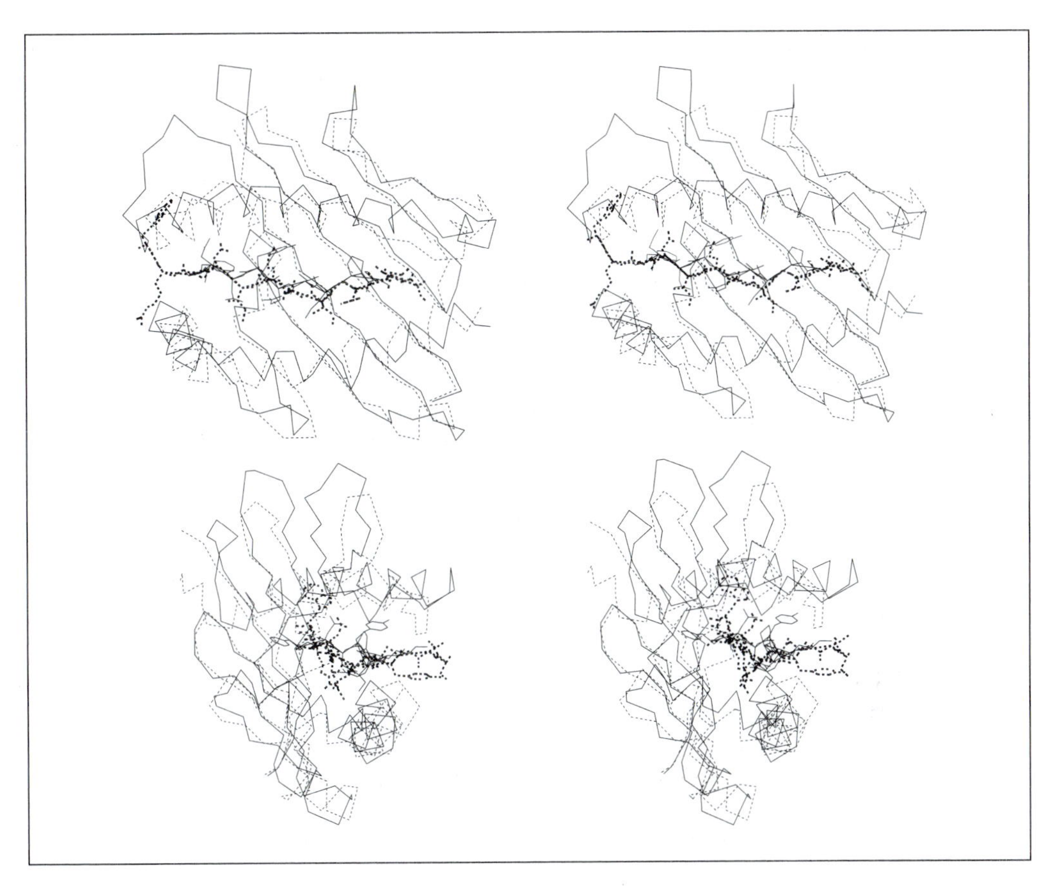

Fig. 6 Superposition of peptide-binding domains and ligand from class I (B3501) (pdb code 1AIO) (solid lines) and class II (I-Ak) MHC (pdb code1IAK) (broken lines) proteins, showing the overall similarity in structure. The ligands occupy similar regions of molecular space, in the groove between the two helices. Front and side views.

end, and a salt bridge to the C-terminal carboxyl group of the peptide. The cleft will accommodate nine-residue peptides in a nearly extended conformation; longer peptides can bulge out or zig-zag (9), or their C-termini can extend out beyond the end of the pocket (10) The α_2- and β_2-domains of class II MHC proteins, like the α_3- and β_2-microglobulin domains of class I, resemble immunoglobulin constant domains.

We note that the nomenclature is somewhat unfortunate. The two peptide-binding domains are called α_1 and α_2 in class I, and α_1 and β_1 in class II. In class I molecules, domain-α_3 is *not* homologous to the other two α-domains. In class II, domains-α_1 and -β_1, and -α_2 and -β_2, respectively, are homologous to each other.

2.2 Specificity of the MHC system

Two aspects of MHC specificity are: (1) the self–foreign distinction, which depends on T-cell scrutiny of MHC–peptide complexes; and (2) the selection of different types of immune response to different categories of threats—extracellular vs. intracellular—which is accomplished by directing peptides derived from extracellular and intracellular sources through the class I and class II systems, to form complexes recognized by different types of T cells.

Each MHC protein has broad specificity, binding many peptides including those of self and non-self origin. Cell surfaces contain large numbers of MHC–peptide complexes, among which those binding foreign peptides are a small minority. T-cell receptors, in contrast, have narrow specificity, and pick out the complexes containing foreign peptides.

T cells 'see' both the MHC and peptide components of the complex they recognize. The variability in MHC proteins extends to the residues that interact with T cells but do not interact with the bound peptides. Therefore, two MHC proteins binding the same peptide could be recognized by different T cells.

The parallel functions of class I and class II MHC proteins select different immune responses to extracellular and intracellular pathogens. Challenge by internally synthesized foreign proteins, as in virus-infected cells, leads to peptide cleavage by cytosolic proteasomes. They encounter class I MHC proteins in the endoplasmic reticulum (ER), from where the peptide complexes are moved to the cell surface. Challenge by external foreign proteins leads to cleavage in endosomes. Class II MHC proteins start out in the ER associated with the Invariant (Ii) chain, in a complex in which the peptide-binding site is blocked (to suppress flooding the system with self-derived peptides and to prevent picking up peptide fragments from internally synthesized proteins which are to be bound to the class I MHC molecules). The complex moves to specialized organelles where it is exposed to peptide fragments of imported proteins. Cleavage of the I chain leaves the CLIP (class II invariant chain peptide) in the binding site. A molecule related to the MHC proteins—HLA-DM in humans and H-2M in mice—catalyses peptide exchange, removing the CLIP and enhancing the rate of binding. Peptide binding stabilizes the α-chain–β-chain dimers, which are brought to the cell surface.

2.3 Peptide binding

Crystal structures of both class I and class II MHC proteins, with and without bound peptides, reveal:

(1) the nature of the peptide–MHC protein interactions;

(2) the mechanism of the broad specificity;

(3) the nature of the surface presented to T-cell receptors (11).

The complexes of the class I MHC protein HLA-B5301 with two nonapeptides show the conformation and interactions of the ligand (12). Figure 7 shows the shape of the cleft from the complex of HLA-B5301 with peptide ls6 (KPIVQYDNF) from the malarial parasite *Plasmodium falciparum*. The cleft is a deep groove, with a hump in

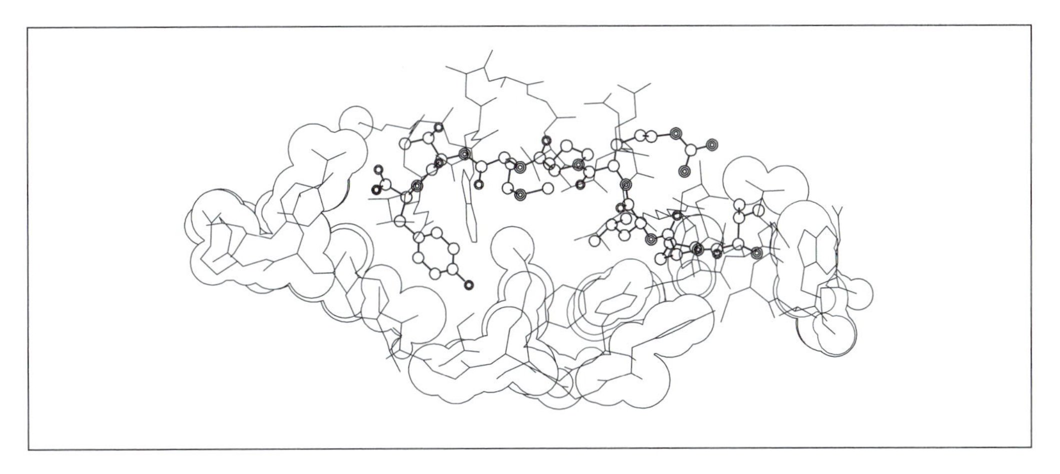

Fig. 7 Slice through the cleft of the complex between class I MHC protein B5301 binding peptide Is6 from malarial parasite (1A1O). View looking across the interface.

the middle, and pockets at either end which receive inward-pointing side chains from near the chain termini of the peptide. It is interesting that the middle part of the floor of the groove is lined by polar side chains: Tyr9, Asn70, Tyr74, and Arg97. The peptide is in an extended conformation, the conformational angles of all residues being in the β-region of the Ramachandran plot. The main-chain residues (N, Cα, C, O) of all but two central peptide residues make contact with the MHC protein, and every side chain of the peptide makes contact with the MHC protein. Upon forming the complex, the peptide loses 80% of its accessible surface area (ASA) (13). (The ASA of the peptide residues in the complex is 300 $Å^2$. The ASA of the peptide residues, removed from the complex but calculated with the same structure that appears in the complex, is 1425 $Å^2$.) The complex buries 1918 $Å^2$, relative to its separated components. Side chains Pro2, Ile3, Gln5, and Phe9 are almost completely buried.

Interactions between the peptide and the MHC protein can be divided into those that involve the main chain of the peptide—and are therefore sequence-independent—and those that involve the side chains—and are sequence-dependent. The former explain the affinity and the broad specificity of the interaction; the latter explain the selectivity.

The terminal residues of the peptide in the Is6–HLA-B53 complex are anchored by networks of hydrogen bonds (Table 1) (Lys is the N-terminal residue and Phe the C-terminal residue): the residues from the MHC protein that make these hydrogen bonds are highly conserved. Residues 3 and 4 of the peptide also make backbone hydrogen bonds to the MHC protein. All these hydrogen bonds are peptide-sequence independent.

Two of the buried peptide residues, Pro2 and Phe9, lie in crevices at the base of the interhelix cleft. The other two, Ile3 and Gln5, are buried between the α_2-domain C-terminal helix and the peptide itself. Figure 8 shows two sets of slices through the

Table 1 The hydrogen bonds formed between the MHC and bound peptide in the complex of LS6 (sequence KPIVQYDNF) and HLA-B53 (PDB entry 1A10)

Peptide Residue	MHC protein residue
Lys1 N	OH 7Tyr
Lys1 N	OH 171Tyr
Lys1 O	OH 159Tyr
1Phe N	OD1 77Asn
1Phe O	NZ 146Lys
1Phe OXT	OG1 143 Thr

complex: (a) at a relative high 'altitude' in the cleft; and (b) nearer the base of the cleft. In (b) it can be seen that the side chains Pro2 and Phe9 project through the lowest slice, to be buried in a pocket in the MHC protein. In (a) the side chains of Ile3 and Gln5 are not pointing down into the MHC protein but are nevertheless buried between it and the peptide backbone. The HLA-B53 system reveals how the MHC protein can bind to alternative peptides. It has been solved by binding both the Is6 peptide (KPIVQYDNF) from *P. falciparum* and also with a peptide from the gag protein of HIV-2 (TPYDINQML). Some of the buried residues have mutated; but how does the structure accommodate this?

Smith *et al.* (12, 14) have analysed the binding of different peptides to the same MHC (HLA-B53) and also compared it to a closely related one (HLA-B5301). The sequences are IVQY in Is6 and YDIN in the HIV-2 peptide. The first and third residues point into the protein. The Y in the HIV-2 peptide is much larger than the I in Is6. The I in the HIV-2 peptide is hydrophobic, whereas the Q in Is6 is polar. The molecule accommodates these changes by: (1) changing the position of a buried arginine (R97); and (2) by reconfiguring the water structure in the cleft.

Another degree of freedom used by the class I protein HLA-B3501 (PDB code 1a1n) is 'induced fit': conformational changes in the MHC protein accommodate changes in sequence in the ligand (14). A comparison of the closely related MHC proteins HLA-B53 and HLA-B3501 shows that sequence-changes between the two MHC proteins alter the pocket that binds the C-terminal aromatic residue, and changes the position of the C-terminal region of the backbone. The position of the C-terminal residue in B3501 would clash sterically with the position of the short helix in B53, as would residue Asp7 in the peptide. Second, HLA-B53 binds the nonamer KPIVQYDNF, whereas HLA-B5301 binds the octomer VPLRPMTY. The P in the second position of these peptides is locked into a crevice at one end of the groove; the C-terminal residue (Y or F) is locked into a crevice at the other. To fit in an extra residue, something has to give. Guo *et al.* (15) observed that when peptides of different length bind to HLA-Aw68, the conservation of the binding of the termini requires that the longer peptides buckle in the middle.

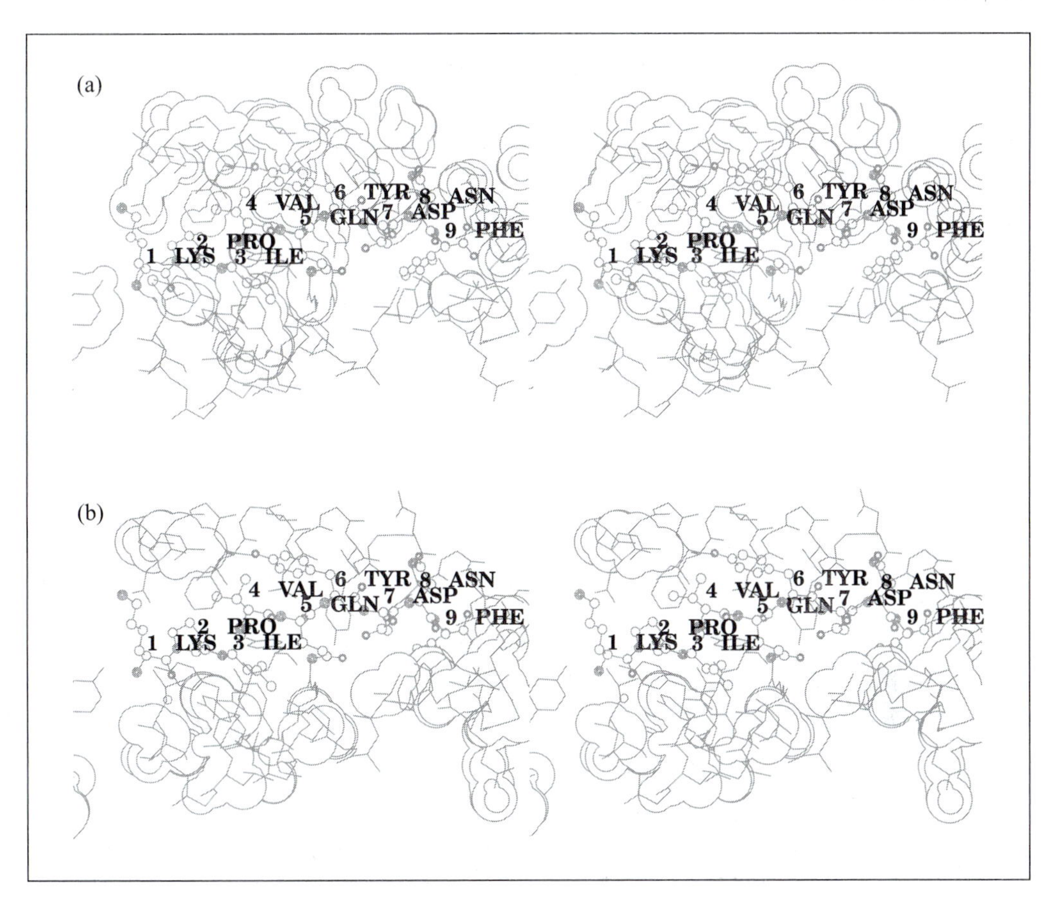

Fig. 8 Slices through the cleft of the complex between class I MHC protein B5301 binding peptide Is6 from malarial parasite (1A10). Stereoviews looking down on to the interface. (a) Nearer the top of the cleft (oblique view). (b) Fairly low down within the cleft. Note that the Pro and Phe protrude through these slices to the bottom of the cleft. Note also the side chains buried between the MHC protein and the peptide.

The reader may ask why the question of ligand-induced conformational changes is addressed by comparison of different peptides binding to the same, or very similar, MHC molecules, rather than by comparing an MHC molecule in the free and ligated states. Peptide binding is important for stabilizing the structure of MHC proteins, and it has not been possible to crystallize the unligated state (see ref. 16).

2.4 Binding of peptides to class II MHC proteins

Fremont *et al.* (17) have solved the structure of the class II MHC protein I-Ak binding the peptide STDYGILQINSRW at high resolution (1.9 Å) (see Fig. 4b). The peptide lies in the cleft with both its ends protruding. As the groove is pinched off at the right, the ligand turns up, out of the cleft. There is a difference between class I and class II in the interactions between peptide side chains and the binding groove. In class I the terminal side chains of the ligand sit in pockets at the bottom of the groove, and the central side chains do not point down into the groove but at right angles to it. In the class II I-Ak, the central side chains are buried in pockets in the base of the groove.

Upon forming the complex the peptide loses over 60% of its accessible surface area. (The ASA of the peptide residues in the complex is 782 Å^2. The ASA of the peptide residues, removed from the complex but calculated with the same structure that appears in the complex, is 2144 Å^2.) The complex buries 2334 Å^2, relative to its separated components. Side chains Asp3, Ile6, Gln8, Ile9, and Ser11 are almost completely buried.

The peptide ligand makes numerous hydrogen bonds to the MHC protein. In Table 2 the hydrogen bonds are classified according to the participation of main-chain and side-chain atoms from peptide and MHC protein. Therefore the first two columns, involving peptide main-chain atoms, are peptide-sequence independent and account for the broad specificity. Of the 13 peptide residues, 9 make main-chain hydrogen bonds, mostly but not exclusively with MHC protein side chains. The two hydrogen bonds forming the interaction between residues Ser1–Asp3 of the peptide with Arg53 of the MHC protein are like those of a parallel β-sheet. Table 3 summarizes the interactions of the peptide with I-Ak.

Table 2 The hydrogen bonds between peptide STDYGILQINSRW from hen egg white lysozome and class II MHC protein I-Ak (PDB code 1IAK)

Peptide mainchain MHC mainchain	**Peptide mainchain MHC sidechain**	**Peptide sidechain MHC mainchain**	**Peptide sidechain MHC sidechain**
1 O 54 N			
		1 OG 279 O	
	2 O 275 NE2		
3 N 54 O			
			3 OD1 53 NH2
			3 OD1 280 OG1
			3 OD2 53 NE
	4 N 276 OD1		
	4 O 276 ND2		
6 N 9 O			
	7 N 268 OE2		
	8 N 63 OD1		
		8 NE2 63 O	
		8 NE2 67 N	
			8 OE1 205 ND1
	9 N 226 OH		
	9 O 70 ND2		
	11 N 70 OD1		
	11 O 69 NE2		
			11 OG 253 OD1

Table 3 A summary of the environments surrounding residues of peptide bound to class II MHC protein I-Ak (1IAK) (see Table 2)

Residue	
Ser1	On surface
Thr2	On surface
Asp3	Buried in a cavity, double salt bridge with sidechain of Arg52
Tyr 4	Sidechain on surface
Gly 5	Two aromatic residues (Tyr22, Phe24) and backbone of Tyr9 and Gly 9A occupy space where a sidechain of peptide residue 5 would be.
Leu 6	On surface
Gln 7	Hydrogen bonds to O of N62 and H9. Sidechain packs in pocket formed by the sidechains of T65 and F11
Ile8	Buried in a pocket formed by sidechains of W61 adn Y47, covered by sidechain of Y67
Asn10	On the surface. Forms hydrogen bond to T65
Ser11	Sidechain sits in polar pocket formed by sidechains of D57 and R76
Arg12	On the surface
Trp13	On the surface; sidechain slots into pocket formed by E71 and K75

2.5 Conclusions about peptide binding

1. Class I and class II molecules bind peptides in ways that are generally similar but different in detail. The most obvious difference is the interaction of the terminal residues of peptide ligands with class I molecules in a way that limits the length of peptides that can be bound.
2. MHC proteins achieve the goal of broad specificity with fairly tight binding by numerous contacts with the main-chain of the peptide, the contacts being peptide-sequence independent, and by conformational changes that allow tuning of the binding site to several peptides.

3. The T-cell receptor and its interactions with peptide–MHC complexes

The ability of mammalian cells to present peptides derived from self and pathogenic material on to the cell surface highlights both infected and cancerous cells to the immune system. The first stage of the immune response involves T lymphocytes and circulating antibodies. The T lymphocyte (or T cell) is able to respond either by destroying the infected cell (in the case of a cytotoxic T cell or CD8+ T cell) or proliferating and activating B lymphocytes and macrophages (the helper T cell or CD4+ T cell) (18).

In this section we discuss the interactions involving those proteins on the T-cell surface involved in the recognition of MHC molecules and part of the subsequent 'activation' of the cell.

3.1 The T-cell receptor (TCR)

The TCR comprises a disulfide-linked heterodimer of α- (40–50 kDa) and β- (35–47 kDa) subunits. Each subunit consists of a single polypeptide chain folded into two immunoglobulin-like domains containing approximately 110 amino acids (Fig. 9). The TCR is anchored to the membrane by a transmembrane region at its C-terminus and is associated with the γ-, δ-, ε-, and χ-subunits of the auxiliary protein CD3 to form the TCR complex. The N-terminus of the TCR contains areas of high sequence variability, which are required to allow the receptor to bind to the enormous repertoire of possible MHC–peptide complexes generated throughout life.

The genetic organization of this region is homologous to that observed for the variable domains of immunoglobulins (19, 20). The domain is constructed by the rearrangement of a library of *V* and *J* gene segments for the α-chain and *V, D,* and *J* segments for the β-chain. Sequence analysis of these gene segments shows areas of particularly high variability within the *V* segments, which are very similar to the complementarity-determining regions (CDRs) of immunoglobulins. There are two sets of three of these CDRs present in the TCR. CDR3 (6–13 amino acids in length) has a larger variation than either CDR1 (6 amino acids) or CDR2 (8–12 amino acids) due to somatic rearrangement involving the *V, D,* and *J* segments of the β-chain and the activity of deoxynucleotide transferase. It is estimated that this gives this particular loop higher variability than that normally observed in immunoglobulins.

In 1996, the three-dimensional structure of the TCR was determined in both its

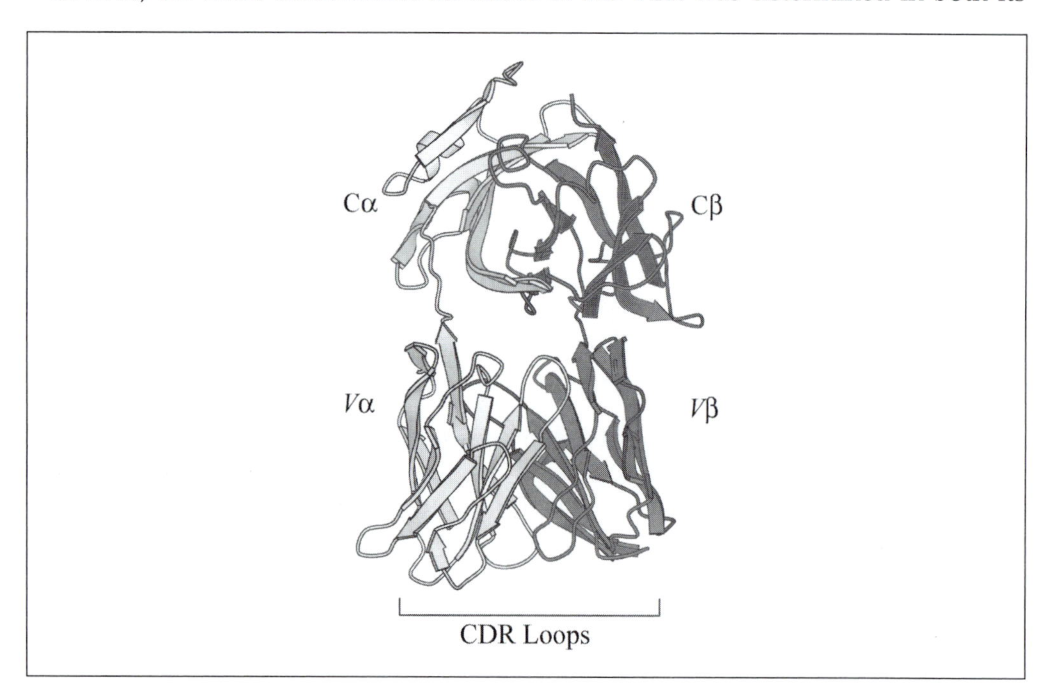

Fig. 9 The structure of the 2C T-cell receptor (2CKB) showing the two subunit structure (α in dark grey, and β in light grey). Diagram produced using MOLSCRIPT (48).

uncomplexed form (21, 22), in complex with an MHC (23–25) and in complex with other auxiliary proteins (CD8, superantigen) (26, 27). For the first time, this has allowed us to see in great detail not only the structure of the TCR but also the interactions between the T cell and antigen-presenting cell.

As suggested by the earlier analysis of the sequence of the TCRs, the structure is immunoglobulin-like, each subunit consisting of a constant and variable domain. The two subunits are twisted about another (see Fig. 9) with interactions being made between the Vα and Vβ, and the Cα- and Cβ-domains. The V and C domains within each subunit are linked to one another by an elbow region formed by the FG loop which, like that in immunoglobulins, affords some flexibility between the domains. It is notable that for the relatively small number of structures available the elbow angle inhabits a more limited set (140°–150°) than that observed in immunoglobulins (127°–224°).

Unlike the interactions observed between the heavy and light chains of immunoglobulins, the TCR shows little symmetry in the interactions between the C domains. The Cα- and Cβ-domains dimerize along an interface made up of β-strands A, B, E, and D, two from each subunit. This allows extensive sheet–sheet packing to take place, so producing an interface of 2375–2400 $Å^2$ (the largest such interface in the TCR). More detailed examination of the interface in the available structures shows it to be made up of mainly hydrophilic interactions, with a skew towards acidic residues in the Cα- and basic residues in the Cβ-domains.

The Vα and Vβ domains of the two subunits combine together at the other end of the protein to form a surface made out of six loops representing the CDR3s of each subunit (Fig. 10). As can be seen, the loops representing the two CDR3s (the most highly variable of the CDRs) are situated between both CDR1s (CDR1α and CDR1β from the α- and β-subunits, respectively) and CDR2s (CDR2α and CDR2β from the α- and β-subunits, respectively).

In the structures so far determined a difference has been observed as to the fold of the Vα domain compared to the canonical immunoglobulin fold (7, 28). This involves the switch of a polypeptide strand between two of the β-sheets so that the Vα C″ strand is hydrogen-bonded to the D strand instead of the C′ strand (Fig. 11). This alteration in the canonical fold removes a surface protrusion from the Vα domain and has been hypothesized to allow T-cell activation via αβ-TCR dimerization (22). Further careful examination of the Vα and Vβ domains and the interface between them shows a smaller interface (typically 1350 $Å^2$), which has a completely different character to that of the Cα and Cβ interface. The interface is made up of mainly hydrophobic contacts, in contrast to the mainly ionic ones seen between Cα and Cβ. This difference between the two interfaces has led Wang and co-workers (29) to propose that the C-domain interface forms a rigid structure, while the V-domain interface has inherent flexibility due to the less spatially defined interactions formed by the hydrophobic residues. Indeed, work by Li and co-workers (30), presenting the structure of a Vα-homodimer (with a mutated CDR3 region), shows a 14° rotation when compared to the equivalent wild-type structure (22). It has been further postulated that the V-domain interface may be implicated in some kind of conformational change induced upon MHC–peptide binding.

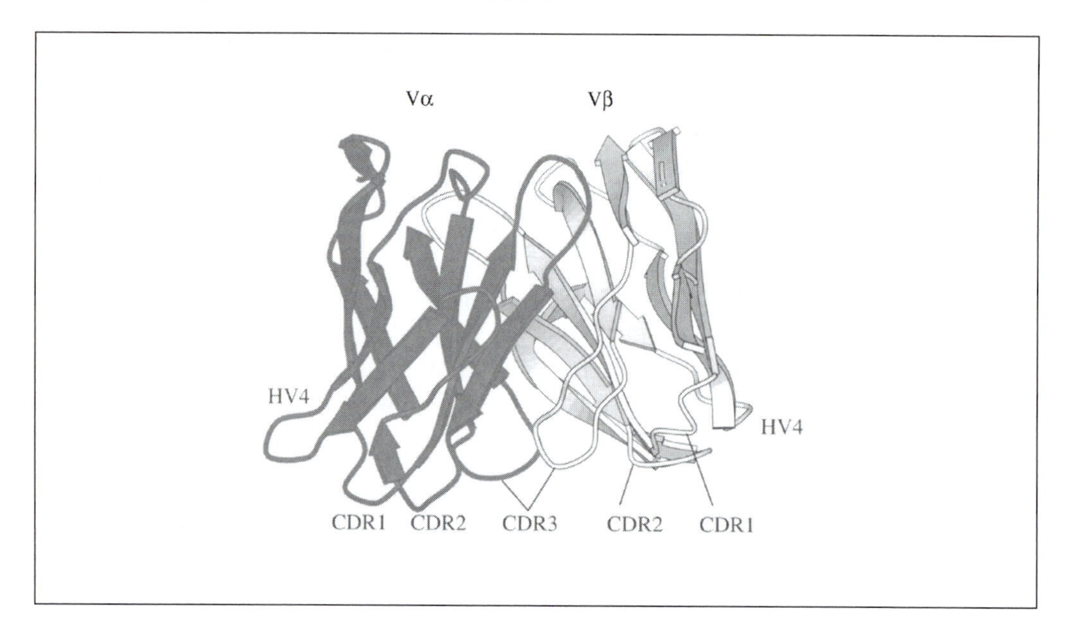

Fig. 10 The orientation of the CDR loops and HV4 region within the TCR V domains (2CKB). Diagram produced using MOLSCRIPT (48).

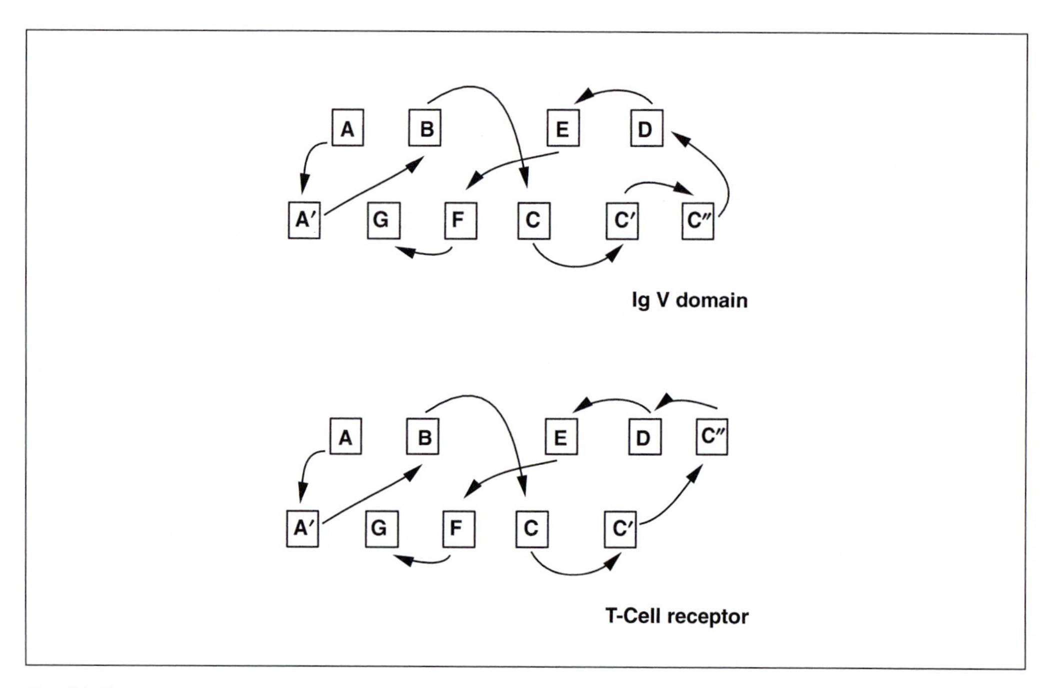

Fig. 11 The strand topology of a TCR compared to that of the immunoglobin fold.

3.2 The interaction of the TCR with MHCs and peptides

The major function of TCRs is to act as receptors for peptides presented by MHCs on the surface of target cells. The receptor must then transduce the formation of a complex into an intracellular signal that affects the T cell. Extensive experimental work has been carried out investigating these interactions at a cellular level, with some interesting results. It has been discovered that peptide-loaded MHCs not only can act as agonists to T-cell activity via the TCR (and so activate the T cell to become cell lytic or enhance B-cell activity), but they can also exert an antagonistic response when loaded with different peptide–MHC complexes (31, 32). It has also been found that those MHC–peptide complexes that behave as antagonists typically have much lower binding affinities, as a result of higher off-rates than those measured for agonist complexes. This may suggest a kinetic route for the differential activities.

3.3 The orientation of the TCR with respect to the MHC

The first crystal structure showing the complex between an MHC peptide complex and a TCR was published by Garcia and co-workers in 1996 (23). This structure showed at 2.6-Å resolution the 2C TCR and at 3.4-Å resolution the complex of the 2C TCR with its cognate MHC–peptide pair (H-2K^b/dEVB). Thus, for the first time, the position of the TCR with respect to the peptide in the binding groove of the MHC could be observed. Greater detail of the complex was then shown later in 1996 by a structure from Garboczi and co-workers (24). They had solved the structure of the A6 TCR bound to the MHC–A2/Tax complex at a resolution of 2.6 Å.

Both structures showed the V domains of the TCR sitting on top of the α_1- and α_2-helices of the MHC with the peptide in between (Fig. 12a). Intriguingly, in both complexes, the TCR is slightly skewed compared to the long axis of the peptide-binding slot. This diagonal mode of binding was proposed as a common feature amongst TCR–MHC-peptide structures by Garbozi and co-workers (24). Since then two further structures have been solved (one at 6.0-Å resolution (33) and another at 2.5 Å (34)), these show a similar arrangement between the MHC and TCR.

Explanation for the conservation of this binding motif is thought to come from the observation that there are two 'high points' at the centre of the α_1- and α_2-helices of the MHC (25). These limit the possible combinations of orientations between the MHC and TCR, allowing for only a skewed interaction. As these high points seem to be an inherent feature of the MHC fold (caused by the twist of the β-sheet underlying the helices (Fig. 12b)), it is likely that the motif is conserved. Although the skew of the complex seems to be constant amongst the currently available structures, Teng and co-workers (33) have noted deviations in the positions of TCR and MHC about other axes (Table 4).

These differences have been denoted as 'tilt' and 'shift' (33). The 'tilt' being the rotation of the TCR about an axis in the plane of the MHC β-sheet perpendicular to the strands; the 'shift' being any movement of the TCR laterally along the peptide-binding grove of the MHC. Analysis of the available structures shows large

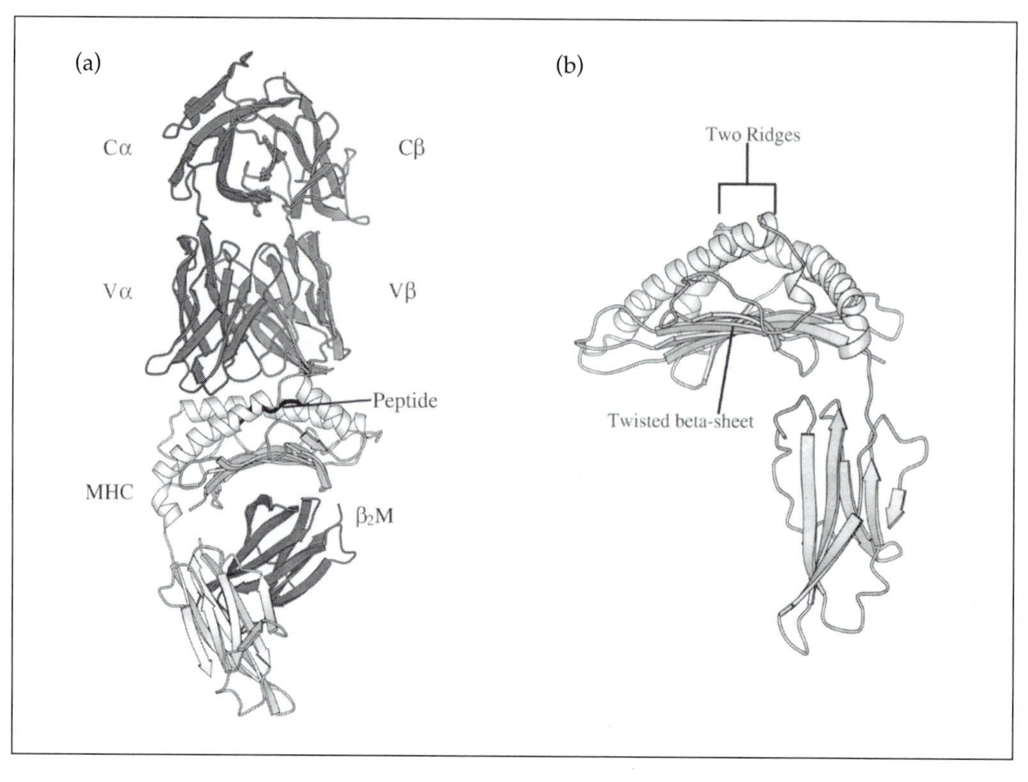

Fig. 12 (a)The structure of the complex between an HLA (H-2K^b), a peptide (dEV8), and a TCR (2C). The auxiliary protein β_2-microglobulin is also shown. (b) The structure of HLA (H-2K^b) showing the 'kinked' helices making up the peptide-binding groove, and the underlying 'twisted' β-sheet. Diagram produced using MOLSCRIPT (48).

deviations in both tilts (21°) and shifts (9.2 Å) which seem to correlate well with the length of the CDR3 loop.

3.4 Structure of the CDR loops in the MHC–TCR complex

By analysing the available structures, a closer examination of the contributions of the CDRs on the TCR to the binding interface shows that, generally, the CDR1 and CDR3 loops lie along the antigen-binding cleft (Fig. 13). The Vβ CDR1 and CDR3 interact directly with the carboxy-terminal half of the peptide, whereas the Vα CDR1 and CDR3 contact the amino-terminal. The CDR2 loops of both subunits lie outside the cleft making interactions with the MHC. In the case of the structure of the A6/HLA–A2 complex (the complex with the largest 'tilt' angle), the CDR1 and CDR2 loops from the β-subunit are lifted almost off the surface. This is demonstrated by the small percentage of the interface surface area made up by the two loops when compared to the 2C/H-2K^b structure (Table 5). This is proposed to be the result of a long CDR3 in the same subunit which 'levers' the β-subunit from the MHC.

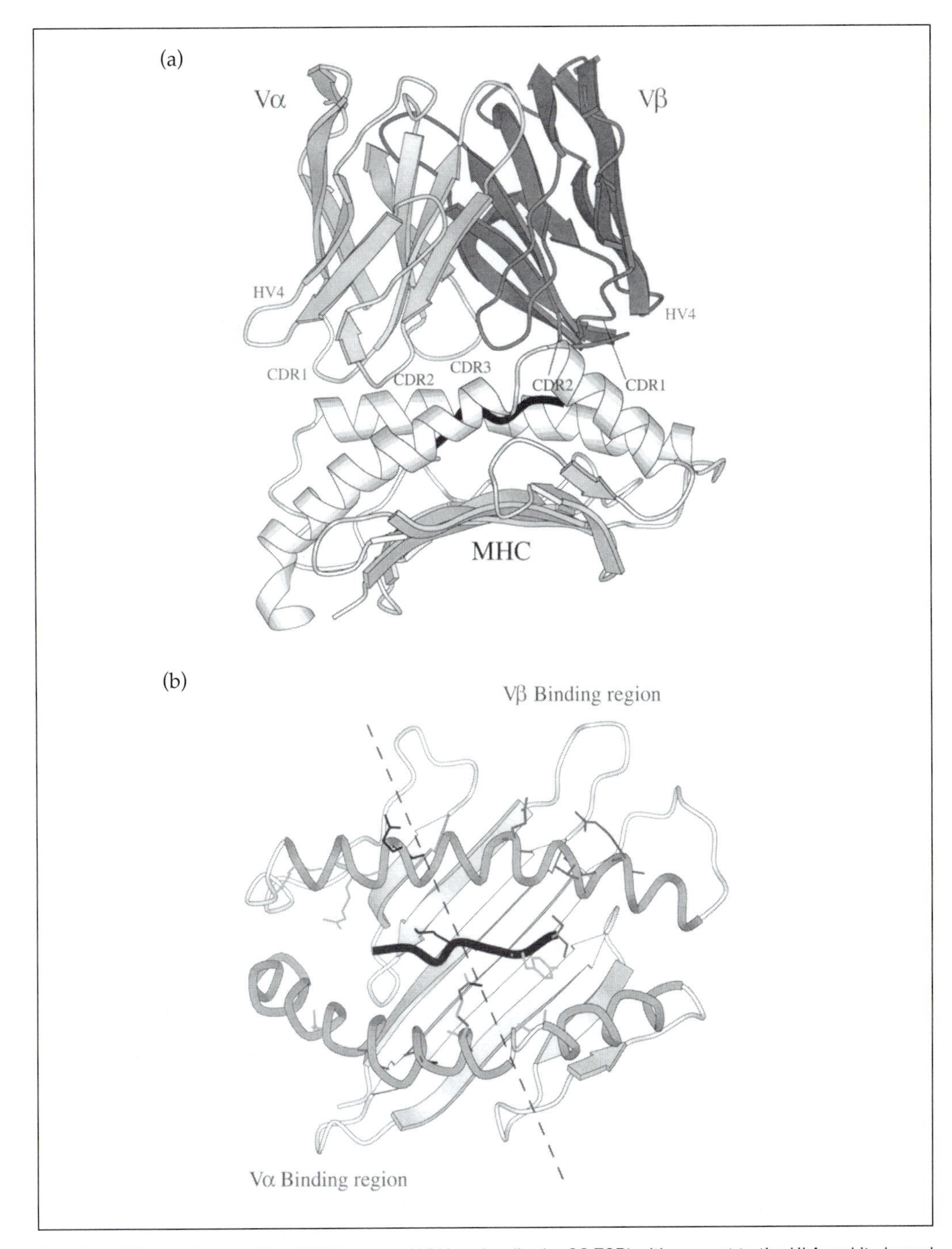

Fig. 13 (a) The positions of the CDR loops and HV4 region (in the 2C TCR) with respect to the HLA and its bound peptide. (b) View from above the HLA looking down into the peptide-binding groove showing the regions contacted by the TCR. Black, CDR3 contacts; light grey, CDR2 contacts; dark grey, CDR1 contacts. Diagram produced using MOLSCRIPT (48).

Table 4 Summary of geometric parameters which define the relative orientation between TCRs and MHCs as determined by [31] for 4 TCR complex structures. These values are calculated from two separate structures which make up the asymmetric unit of the crystal structure.

Complexes TCR type	Peptide	MHC type	CDR3 Length (Residues)	Twist (°)	Tilt (°)	Shift (Å)
A6	Tax	MHC-A2	11	−4.5	−16.0	9.2
2C	dEV8	H-2K^b	6	10.0	5.5	2.8
N15*	VSV8*	H-2K^b	9	1.0	0.0	0.2
N15*	VSV8*	H-2K^b	9	0.0	0.0	0.0

Table 5 The surface areas of contact for the CDR loops in the TCR 2C/H-2K^b/dEV8 complex

CDR	Percentage buried surface in the TCR-MHC complex	
	A6/HLA-A2-Tax	2C/H-2K^b-dEV8
1α	24	24
2α	10	13
3α	24	15
1β	2	18
2β	1	16
3β	33	10

3.5 Energetic contribution of the CDR loops to MHC–TCR complex formation

In the years between the solving the first MHC structure (6) and that of the first TCR, immunologists have tried to unravel the balance made by the TCR between binding to the relatively conserved areas of the MHC and the essentially random-bound peptide. Several approaches using site-directed mutagenesis of both MHCs and TCRs produced intriguing results, which, when viewed in the context of the complex structures, give a clear account of the forces holding the complex together.

In particular, Manning and co-workers (35) demonstrated an approach to unravelling the energetics of the TCR–MHC interaction which we will discuss here. Work also of note in this area is that by Wang *et al.* (29) and Chang *et al.* (36) who studied the interaction of TCRs with a peptide from the vesicular stomatitis virus bound to a MHC. Returning to the work carried out by Manning and co-workers, they have presented a heroic analysis of the individual contributions made to the binding of the QL9 peptide to H2-L^d, and then to the TCR 2C. To draw a complete picture of the contribution of all residues within the binding region, 47 residues within the CDR regions and 15 within a second region denoted HV4 were individually mutated to alanine. The HV4 region was selected, as it, like the CDRs, shows significant sequence variation (37). Despite the quite drastic change in sequence which occurred in some regions, only nine of the above proteins failed to express and the rest were assessed for functional binding against the peptide–MHC complex. Looking at the results by protein region (Table 6) it becomes clear that the majority of the binding energy originates from the CDR 1 and 2 loops. Indeed the ranking of energies for each region is:

$$\text{CDR1}\alpha > \text{CDR1}\beta > \text{CDR2}\alpha = \text{CDR2}\beta > \text{CDR3}\alpha = \text{CDR3}\beta.$$

This suggests that in the case of the QL9/L^d pairing, areas of recognition reside primarily in the α_1- and α_2-helices of the MHC, the areas making contact with the CDR1 and CDR2 loops.

Examining the individual residues responsible for the binding energy, a quartet of tyrosines (49α, 50α, 48β, 50β) can be seen to contribute almost 6 kcal mol^{-1}. Two of these residues, 50α and 50β, interact directly with the MHC in the analogous 2C/dEV8/K^b (Fig. 14). However 49α and 48β show no direct interaction with the MHC, and instead are thought to provide a scaffold over which the CDR loops lie.

T29α also forms an important part of the TCR–MHC interface, whereas residue N31β contributes by interacting directly with the P6 residue of the dEV8 peptide. Collation of additional data for all the alanine mutants made by Manning and co-workers (35) allows an overall assignment of the contributions made by the peptide and the MHC residues to the binding energy. In the case of this system, 37% of the energy comes from the peptide and 63% from the MHC. Thus it can be concluded that the TCRs primarily recognize the body of the MHC using the CDR1 and CDR2 loops, while a smaller contribution comes from the CDR3 peptide interaction. However, Manning and co-workers (35) point out that such a division may be slightly

Table 6 Summary of the contribution to HLA-L^d/QL9/TCT 2C binding of residues within the CDR loops. For simplicity only interactions exceeding 0.6 kcal mol^{-1} are shown. The data concerning the energetics of binding are taken from [35]. The data concerning surface areas of interaction is calculated using the structure of HLA-K^b/dEV8/TCR 2C

CDR Loop	TCR Residue	ΔΔG:QL9/L^d (kcal/mol ± SD)	Contact Surface Area (Å^2)
CDR1α	Y24	0.98 ± 0.10	0
	T29	2.05 ± 0.23	45.98
	Y31	1.33 ± 0.24	76.08
CDR2α	Y49	> 1.48	1.26
	Y50	> 1.48	71.7
CDR3α	F100	0.76 ± 0.12	0
	L104	1.42 ± 0.05	0
CDR1β	N31	> 1.48	11.64
CDR2β	Y48	> 1.48	0
	Y50	> 1.48	88.68

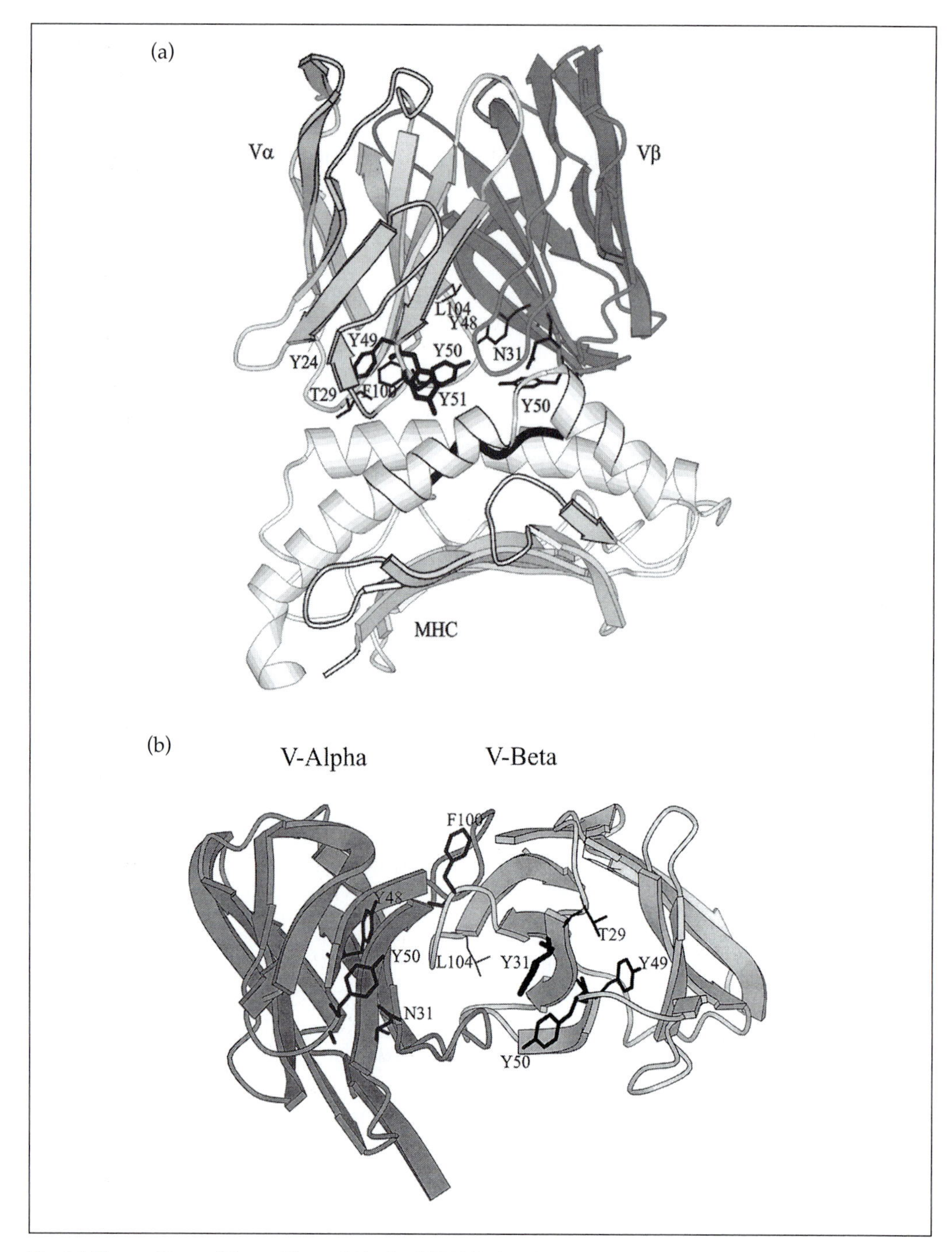

Fig. 14 The positions of the residues within the TCR (2C) thought to be important to its interaction with the HLA-peptide/peptide complex. (a) A view looking across the peptide-binding cleft of the MHC. (b) A view of the MHC binding region of the TCR taken from the perspective of the MHC peptide-binding region. Diagram produced using MOLSCRIPT (48).

atypical in this case as TCR 2C has been selected for alloreactivity, and as such has slightly shorter CDR3 loops than seen in foreign peptide/self specific TCRs.

3.6 Signalling and CD4 and CD8 co-receptors

The binding of the MHC–peptide complex to a TCR on the surface of a T cell is not the final stage in the functioning of the TCR. Complex formation must still be transduced into a signal which triggers the T cell to react appropriately. Prior to the solution of the TCR–MHC complex structure this transduction had been thought to occur by one of two methods: (1) The formation of the complex induces a conformational change in the TCR, which is transmitted to the intracellular segment of the protein and activates cell signalling pathways; (2) formation of a TCR–MHC complex between the T cell and its target is proposed not to be a lone event, and, in fact, many such conjunctions occur on the surfaces between the two cells. This leads to oligomerizations of the TCR complexes to produce a macromolecular structure, which somehow initiates the appropriate cell signalling pathways.

Structural comparisons between the TCR in the complex TCR–MHC structure and the unliganded TCR show little conformational change within the constant domains of the TCR (the region connecting the active V domains to the intracellular domains) (Fig. 15) (25). This provides strong evidence against proposal 1.

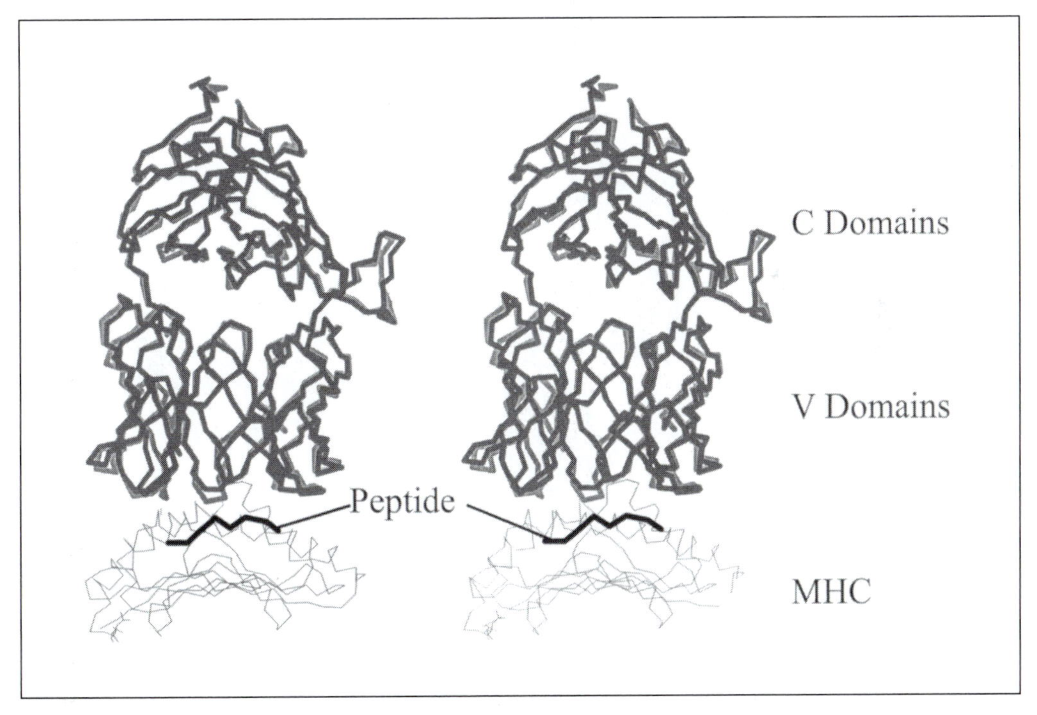

Fig. 15 A stereo overlay (C-α atom trace shown) of unliganded TCR 2C (thick black) and TCR 2C bound to the H-$2K^b$/dEV complex (thick grey).

This leaves the second proposal as perhaps the more likely, and in support of which studies of the co-receptors (CD4 and CD8) of the TCR have been implemented (for a review see ref. 38). These two cell-surface glycoproteins are expressed individually on the two types of T cell which also express the TCR. Both proteins bind to areas of the MHC class I and II proteins separate from the peptide-binding groove, allowing a complex to be formed that contains the MHC, TCR, and CD4 or CD8 (26, 39–42). Both proteins are known to associate with the T cell-specific intracellular protein tyrosine kinase $p56^{lck}$ (Lck) (43, 44) which, it is thought, may be one of the initiators for the signalling complex formation. Indeed, transgenic mice that lack the co-receptors show a severely impaired differentiation of their corresponding T-cell lines. Structures of both receptors (Fig. 16) show them both to be made up of immunoglobulin related domains. CD4 is a single polypeptide folded into four separate domains with unique strand organizations between domains D1 and 2, and between domains D3 and 4 (45). Intriguingly, CD4 has been shown to dimerize, thus lending weight to the signal transduction mechanism by macromolecular association (proposal 2 above). CD8 is made up of a disulfide-linked dimer of α- and β-polypeptides, each containing immunoglobulin-like N-terminal domains. CD8 is linked to its transmembrane domain by an extended polypeptide region high in O-linked glycosylation sites. Analysis of the CD8–MHC–A2 structure (26) (Fig. 17) shows that the α-domain makes contacts with the MHC in such a way as not to disrupt its peptide-binding region. This allows the possibility of a quaternary complex to form

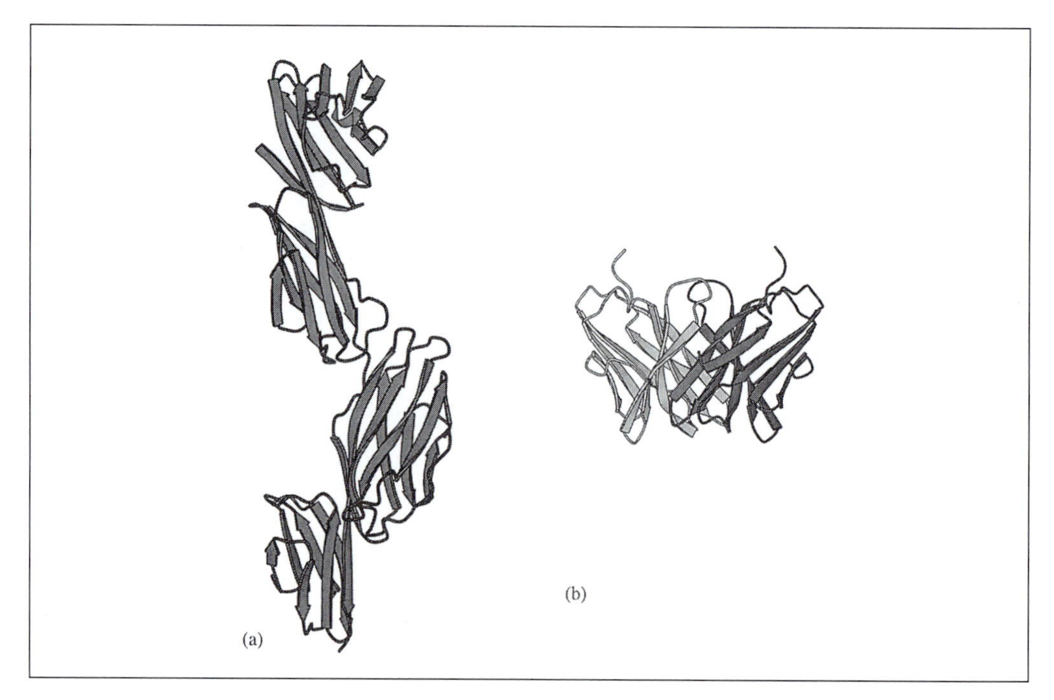

Fig. 16 The structures of CD4 (left) and CD8 (right), co-receptors associated with the TCR. Diagram produced using MOLSCRIPT(48).

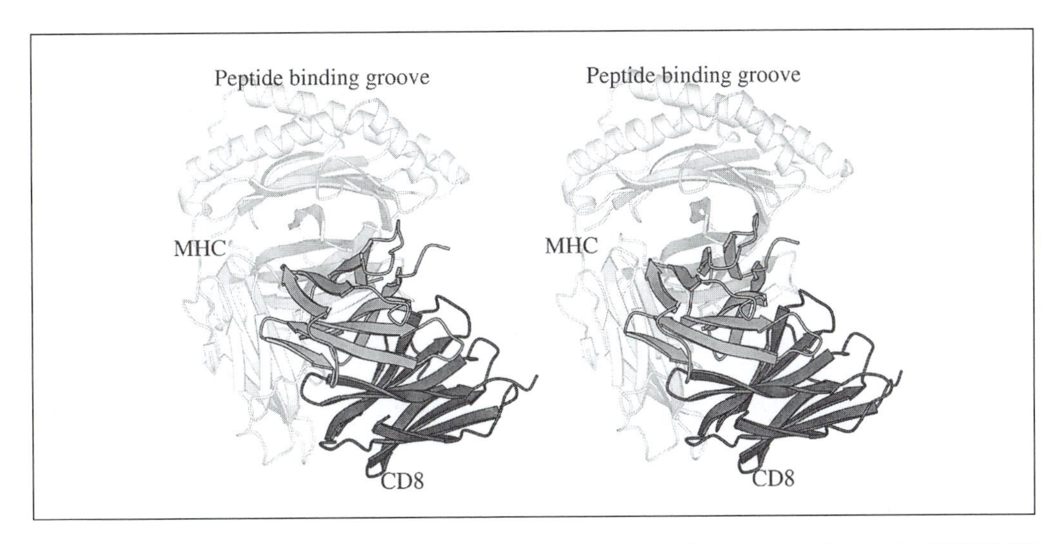

Fig. 17 Stereo picture of the complex formed when CD8 binds to an MHC. Diagram produced using MOLSCRIPT (48).

containing the MHC, CD8, TCR, and β_2-microglobulin. The observation that only one β-domain binds to the MHC has led to the speculation that the 'free' β-domains are able to oligomerize in a similar way to that suggested for CD4. This again would favour transduction mechanism 2.

4. Summary

The MHC proteins, originally discovered through the investigation of transplantation rejection, are now recognized as mediating the immunological 'self/non-self' distinction. They underlie individual variation in sensitivity to particular antigens, including susceptibility to disease. MHC proteins control immune responses by presenting peptides on cell surfaces. Recognition of a peptide–MHC complex by a T cell identifies the peptide as a fragment of a foreign protein. Groups of these complexes then coalesce on the cell surface and interact with co-receptors, activating the T cell and hence the immune system.

References

1. Campbell, R. D. and Trowsdale, J. (1993). Map of the human MHC. *Immunol. Today*, **14**, 349–52.
2. Klein, J., Takahata, N., and Ayala, F. J. (1993). MHC polymorphism and human origins. *Sci. Am.*, **269**, 46–51.
3. Iminishi, T., Wakisaka, A., and Gojobori, T. (1992). Genetic relationships among various human populations indicated by MHC polymorphisms. In *HLA* 1991, Vol. 1 (ed. K. Tsugi, M. Aizawa, and T. Sasazuki), p. 627. Oxford University Press, Oxford.

4. Simister, N. E. (1993). IgG Fc receptors that resemble class I major histocompatibility complex antigens. *Biochem. Soc. Trans.*, **21**, 973–6.
5. Burmeister, W. P., Huber, A. H., and Bjorkman, P. J. (1994). Crystal structure of the complex of rat neonatal Fc receptor with Fc. *Nature*, **372**, 379–83.
6. Lebron, J. A., Bennett, M. J., Vaughn, D. E., Chirino, A. J., Snow, P. M., Mintier, G. A., Feder, J. N., and Bjorkman, P. J. (1998). Crystal structure of the hemochromatosis protein HFE and characterization of its interaction with transferrin receptor. *Cell*, **93**, 111–23.
7. Harpaz, Y. and Chothia, C. (1994). Many of the immunoglobin superfamily domains in cell adhesion molecules and surface receptors belong to a new structural set which is close to that containing variable loops. *J. Mol. Biol.*, **238**, 528–39.
8. Lesk, A. M. and Chothia, C. (1982).The evolution of proteins formed by β-sheets. *J. Mol. Biol.*, **160**, 325–42.
9. Madden, D. R., Garboczi, D. N., and Wiley, D. C. (1993). The antigenic identity of peptide–MHC complexes: a comparison of the conformations of five viral peptides presented by HLA-A2. *Cell*, **75**, 693–708.
10. Collins, E. J., Garboczi, D. N., and Wiley, D. C. (1994). Three-dimensional structure of a peptide extending from one end of a class I MHC binding site. *Nature*, **371**, 626–9.
11. Batalia, M. A. and Collins, E. J. (1997). Peptide binding by class I and class II MHC molecules. *Biopolymers*, **43**, 281–302.
12. Smith, K. J, Reid, S. W., Stuart, D. I., McMichael, A. J., Jones, E. Y., and Bell, J. I. (1996). An altered position of the α-2 helix of MHC class I is revealed by the crystal structure of HLA-B*3501. *Immunity*, **4**, 203-13.
13. Lee, B. K. and Richards, F. M. (1971). The interpretation of protein structures: estimation of static accessibility. *J. Mol. Biol.*, **55**, 379–400.
14. Smith, K. J, Reid, S. W., Harlos, K., McMichael, A. J., Stuart, D. I., Bell, J. I., and Jones, E. Y (1996). Bound water structure and polymorphic amino acids act together to allow the binding of different peptides to MHC class I HLA-B53. *Immunity*, **4**, 215–28.
15. Guo, H.-C., Jardetzky, T. S., Garrett, T. P. J., Lane, W. S., Strominger, J. L., and Wiley, D. C. (1992). Different length peptides bind to HLA-Aw68 similarly at their ends but bulge out in the middle. *Nature*, **360**, 364–6.
16. Bouvier, M. and Wiley, D. C. (1998). Structural characterisation of a soluble and partially folded class I major histocompatibility heavy chain/beta 2m heterodimer. *Nature Struct. Biol.*, **5**, 377–84.
17. Fremont, D. H., Hendrickson, W. A., Marrack, P., and Kappler, J. (1996). Structures of an MHC class II molecule with covalently bound single peptides. *Science*, **272**, 1001–4.
18. Swain, S. (1983). T cell subsets and the recognition of MHC class. *Immunol. Rev.*, **74**, 129–42.
19. Davis, M. M. and Bjorkman, P. J. (1988). T cell antigen receptor genes and T cell recognition. *Nature*, **334**, 395–402.
20. Chothia, C., Boswell, D. R., and Lesk, A. M. (1988). The outline structure of the T-cell αβ receptor. *EMBO J.*, **7**, 3745.
21. Housset, D., Mazza, G., Gregoire, C., Piras, C., Malissen, B., and Fontecilla-Camps, J. C. (1997). The three-dimensional structure of the T-cell antigen receptor VαVβ heterodimer reveals a novel arrangement of the Vβ domain. *EMBO J.*, **16**, 4205–16.
22. Fields, B. A., Ober, B., Malchiodi, E. L., Lebedeva, M. I., Brandem, B. C., Ysern, X., Kim, J., Shao, X., Ward, E. S., and Mariuzza, R. A. (1995). Crystal structure of the Vα domain of a T-cell receptor. *Science*, **270**, 1821–4.
23. Garcia, K. C., Degano, M., Stanfield, R. L., Brunmark, A., Jackson, M. R., Peterson, P. A.,

Teyton, L., and Wilson, I. A. (1996). An $\alpha\beta$ T cell receptor structure at 2.5 angstroms and its orientation in the TCR–MHC complex. *Science*, **274**, 209–19.

24. Garboczi, D. N., Ghosh, P., Utz, U., Fan, Q. R., Biddison, W. E., and Wiley, D. C. (1996). Structure of the complex between human T-cell receptor, viral peptide and HLA-A2. *Nature*, **384**, 134–41.
25. Garcia, K. C., Degarno, M., Pease, L. R., Huang, M., Peterson, P. A., Teyton, L., and Wilson, I. A. (1998). Structural basis of plasticity in T cell receptor recognition of a self peptide-MHC antigen. *Science*, **279**, 1166–72.
26. Gao, G. F., Tormo, J., Gerth, U. C., Wyer, J. R., McMichael, A. J., Stuart, D. I., Bell, J. I., Jones, E. Y., and Jakobsen, B. Y. (1997). Crystal structure of the complex between human CD8a(a) and HLA-A2. *Nature*, **387**, 630–4.
27. Fields, B. A., Malchiodi, E. L., Hongmin, L., Ysern, X., Stauffacher, C. V., Schlievert, P. M., Karjalainen, K., and Mariuzza, R. A. (1996). Crystal structure of a T-cell receptor β-chain complexed with a superantigen. *Nature*, **384**, 188–92.
28. Bork, P., Holm, L., and Sander, C. (1994). The immunoglobulin fold: structural classification, sequence patterns and common core. *J. Mol. Biol.*, **242**, 309–20.
29. Wang, F., Ono, T., Kalergis, A. M., Zhang, W., Delorenzo, T. P., Lim, K., and Nathenson, S. G. (1998). On defining the rules for interactions between the T cell receptor and its ligand: a critical role for a specific amino acid residue of the T-cell receptor β chain. *Proc. Natl. Acad. Sci. USA*, **95**, 5217–22.
30. Li, H., Lebedeva, M. I., Ward, E. S., and Mariuzza, R. A. (1997). Dual conformations of a T cell receptor V α homodimer: implications for variability in Vα Vβ domain association. *J. Mol. Biol.*, **269**, 385–94.
31. Lyons, D. S., Lieberman, S. A., Hampl, J., Boniface, J. J., Chien, Y., Berg, L. J., and Davis, M. M. (1996). A TCR binds to antagonist ligands with lower affinities and faster dissociation rates than to agonists. *Immunity*, **5**, 53–61.
32. Alam, S. M., Travers, P. J., Wung, J. L., Nasholds, W., Redpath, S., Jameson, S. C., and Gascoigne, N. R. J. (1996). TCR affinity and thymocyte positive selection. *Nature*, **381**, 616–20.
33. Teng, M., Smoylar, A., Tse, A. G. D., Lui, J., Hussey, R. E., Nathenson, S. G., Chang, H., Reinherz, E. L., and Wang, J. (1998). Identification of a common docking topology with substantial variation among different TCR–peptide–MHC complexes. *Curr. Biol.*, **8**, 409–12.
34. Ding, Y., Smith, K. J., Garboczi, D. N., Utz, U., Biddison, W. E., and Wiley, D. C. (1998). Two human T cell receptors bind in a similar diagonal mode to the HLA-A2/Tax peptide complex using different TCR amino acids. *Immunity*, **8**, 403–11.
35. Manning, T. C., Schlueter, C. J., Brodnicki, T. C., Parke, E. A., Speir, J. A., Garcia, C., Teyton, L., Wilson, I. A., and Kranz, D. M. (1998). Alanine scanning mutagenesis of an $\alpha\beta$ T cell receptor: mapping the energy of antigen recognition. *Immunity*, **8**, 418–25.
36. Chang, H., Smolyar, A., Spoerl, R., Witte, T., Yao, Y., Goyarts, E. C., Nathenson, S. G., and Reinherz, E. L. (1997). Topology of the T cell receptor-peptide/class I MHC interaction defined by charge reversal complementation and functional analysis. *J. Mol. Biol.*, **271**, 278–93.
37. Pattern, P., Yokota, T., Rothbard, J., Chien, Y., Arai, K., and Davis, M. M. (1984). Structure, expression and divergence of T-cell receptor β-chain variable regions. *Nature*, **312**, 40–6.
38. Zamoyska, R. (1998). CD4 and CD8: modulators of T-cell receptor recognition of antigen and immune responses? *Curr. Opin. Immunol.*, **10**, 82–7.

39. Cammarota, G., Schierle, A., Takacs, B., Doran, D., Knorr, R., Bannworth, W., Guardiola, J., and Sinigaglia, F. (1992). Identification of a CD4 binding site on the β2 domain of HLA-DR molecules. *Nature*, **356**, 796–8.
40. Konig, R., Huang, L., and Germain, R. (1992). MHC class II interaction with CD4 mediated by a region analogous to the MHC class I binding site for CD8. *Nature*, **356**, 796–8.
41. Norment, A. M., Salter, R. D., Parham, P., Engelhard, V. H., and Littman, V. H. (1988). Cell–cell adhesion mediated by CD8 and MHC class I molecules. *Nature*, **336**, 79–81.
42. Salter, R. D., Benjamin, H., Wesley, P. K., Buxton, S. E., Garrett, T. P. J., Clayberger, C., Krensky, A. M., Norment, A. M., Littman, D. R., and Parham, P. (1990). A binding site for the T-cell co-receptor CD8 on the α3 domain of HLA-A2. *Nature*, **345**, 41–6.
43. Veilette, A., Bookman, M. A., Horak, E. M., and Bolen, J. B. (1988). The CD4 and CD8 T-cell surface antigens are associated with the internal membrane tyrosine-protein kinase $p56^{lck}$. *Cell*, **55**, 301–8.
44. Rudd, C., Trevillyan, J., Dasgupta, J., Wong, L., and Schlossman, S. (1988). The CD4 receptor is complexed in detergent lysates to a protein-tyrosine kinase (pp58) from human T lymphocytes. *Proc. Natl. Acad. Sci. USA*, **85**, 5190–4.
45. Wu, H., Kwong, P. D., and Hendrickson, W. A. (1997). Dimeric association and segmental variability in the structure of human CD4. *Nature*, **387**, 617–20.
46. Wilson, I. A. and Garcia, K. C. (1997). T-cell receptor structure and TCR complexes. *Curr. Opin. Struct. Biol.*, **7**, 839–48.
47. Fremont, D. H., Monnaie, D., Nelson, C. A., Hendrickson, W. A., and Unanue, E. R. (1998). Crystal structure of I-Ak in complex with a dominant epitope of lysozyme. *Immunity*, **8**, 305–17.
48. Kraulis, P. J. (1991). Molscript—a program to produce both detailed and schematic plots of protein structure. *J. Appl. Crystallog.*, **24**, 946–50.

7 Protein–protein interactions in eukaryotic signal transduction

MARKO HYVÖNEN, JAKE BEGUN, and TOM BLUNDELL

1. Introduction

Correct transmission of extracellular signals is vital both for unicellular and multicellular organisms. Stimulation of cells from outside can trigger complicated cascades of signal transduction which result in cellular responses such as growth, differentiation, and movement. These signals can be transduced both by small molecules, such as cyclic AMP, Ca^{2+}, or inositol triphosphate ($Ins(1,4,5)P_3$) and by networks of interacting proteins (1, 2). Many interactions have their main effect by bringing two proteins together and allowing a reaction to occur; thus dimerization of receptor tyrosyl kinases allows each subunit to phosphorylate the other and so activate signal transduction pathways. Quite often this is mediated by an adaptor molecule or module that serves to bring two other components together. Thus, the presence of an SH3 domain in a protein allows it to recognize a proline-rich sequence and form a polyproline structure, so bringing the proteins containing these modules into close juxtaposition.

The interactions may involve the recognition of either continuous or discontinuous protein epitopes. When epitopes are discontinuous the proteins usually have a stable tertiary structure, although conformational changes may be induced during interaction. When the epitopes are continuous there is no necessity for a tertiary structure and very often the binding conformation is induced during the process of interaction. Continuous epitopes usually adopt a regular motif or secondary structure, and we shall see that α-helices, β-strands, polyproline helices, and β-turns all feature in the continuous epitopes that provide ligands for adaptors, and that some of these conformations may be preferred in solution.

In many cases, protein interactions transduce signals through allostery, by inducing a conformational change or stabilizing an existing conformation, causing a change of activity at another site. This is a characteristic of molecules that activate G proteins, for example the activation of Gα by G-protein coupled receptors; such proteins act as molecular switches. Cell-cycle regulators (cyclins and cyclin-dependent kinase (Cdk) inhibitors) also employ a similar mechanism to control cyclin-dependent kinases. Efficient signal transduction must maintain fidelity and decrease noise while amplifying the signal. The cellular cytoplasm is, in fact, quite a dense soup of proteins and other molecules, so how is this achieved? The solution is to make signal transduction dependent on multicomponent complexes. If the binary interactions are weak, then chance collisions would not cause noise. On the other hand, multicomponent complexes can take advantage of cooperativity, so that the weak interactions in binary complexes are replaced by much stronger and more specific interactions in the higher complex. We shall see that this strategy is adopted by some adaptor domains which often exist in series on a protein, for example the Grb2 adaptor which contains two SH3 domains. But it is also becoming apparent that most signalling complexes have many components and are often much more complex than was earlier assumed.

During the transduction process signal termination is as important as signal initiation, otherwise the sensitivity of the system would be lost. Failure to terminate can have severe consequences, as demonstrated by the oncogenic mutants of signalling proteins (2). 'Switching off' can be achieved in a number of different ways. First, interacting surfaces must be stable both as a component of, and apart from, the complex. Thus, they are rarely completely hydrophobic, unlike the interfaces of permanent complexes that are retained during the lifetime of the protein. They contain a mixture of ion pairs, hydrogen bonds, water-mediated interactions, and hydrophobic interactions. The off-rates may be increased by intrinsic processes, like the hydrolysis of GTP in activated Gα-GTP, which leads to an inactive Gα-GDP that dissociates from downstream components of the pathway, such as adenylate cyclases. This process can be accelerated by the formation of higher order complexes, for example through binding GTPase activating proteins (GAPs) that increase the rate of hydrolysis of GTP by direct participation in the catalytic process. Alternatively, the off-rates can be increased by the allosteric binding of small regulator molecules or larger proteins, or by covalent modification of the complex by an additional enzyme.

In cases of covalent modification the cell provides an array of enzymes that can reverse the reaction, many of which will also be regulated by signalling pathways; phosphorylation is reversed by phosphatases and second messengers are degraded by specific enzymes. Thus all steps of the signal transduction process are kept in control by a yin and yang mechanism.

In this chapter we are interested in the protein–protein interactions that occur in intracellular signalling. We will begin by reviewing interactions in adaptor systems that normally involve continuous epitopes (see Table 1). We will then discuss two major superfamilies of globular proteins that are widely involved in signal transduction: the G proteins, which occur in smaller G-protein switches like Ras and as part of the heterotrimeric G proteins; and the cyclin-dependent kinases that are involved in the cell cycle (see Table 2). Finally, we will ask what generalizations, if any, can be made about the nature of the interacting proteins in signalling, the surface areas buried in the complexes, the types of residues involved, the conformational changes induced, and the complexity of the oligomeric systems involved.

Table 1 List of X-ray and NMR structures of signalling domains in complex with ligand peptides. For the SH2 and SH3 domains mainly the X-ray crystallographic structures have been listed for brevity. SPR = Surface plasmon resonance, fluor. = fluoresence spectroscopy

PDB	Protein	Resol.	Ligand	Length	Affinity (method)	Ref.
SH2 domains						
1LCJ	Lck	1.8 Å	middle-T antigen	11	1 nM	(4)
1LKK	Lck	1.0 Å	middle-T antigen	4	140 nM	(5)
1LKL	Lck	1.8 Å	middle-T antigen	4	1.54 μM	(5)
1LCK	Lck SH3-SH2	2.5 Å	Lck tail peptide	10		(6)
1SPS	v-Src	2.7 Å	middle-T antigen	11		(7)
1SHA	v-Src	1.5 Å	PDGF receptor	5		(8)
1SHB	v-Src	2.0 Å	EGF receptor	5		(8)
1SHD	c-Src	1.5 Å		5	50 nM (ITC)	(9)
–	PI3-kinase	2.0 Å	c-Kit	11	0.9 μM	(10, 11)
–	PI3-kinase	2.4 Å	PDGFR	11	1.1 μM	(10, 11)
1TZE	Grb2	2.0 Å	BCR-Abl	7	(IC_{50} = 150 nM)	(12)
1BMB	Grb2	1.8 Å	BCR-Abl	9	(IC_{50} = 150 nM)	(137)
1AYA	SH-PTP2 N-term	2.05 Å	PDGF receptor	11		(13)
1AYB	SH-PTP2 N-term	3.0 Å	IRS-1	12		(13)
1AYC	SH-PTP2 N-term	2.3 Å	PDGF receptor	11		(13)
–	ZAP-70	1.9 Å	ITAM ζ	19		(14)
–	SH-PTP2	2.1 Å	PDGF receptor	11		(15)
SH3 domains						
1ABO	Abl	2.0 Å	3BP-1	10	34 μM (fluor.)	(16, 17)
1CKA	c-Crk (N-term)	1.5 Å	C3G	10	1.9 μM (fluor.)	(18)
1CKB	c-Crk (N-term)	1.9 Å	Sos peptide	10	5.2 μM (fluor.)	(18)
1FYN	Fyn	2.3 Å	3BP-2	11	34 μM (fluor.)	(16, 17)
1BBZ	Abl	1.6 Å	p41	10	1.5 μM (fluor.)	(19, 20)
1SEM	SEM-5 (C-term)	2.0 Å	Sos peptide	9	ca. 37 μM (fluor.)	(21)
1EFN	Fyn SH3 (R96I)	2.5 Å	Nef protein		380 nM (SPR)	(22, 23)
1AVZ	Fyn SH3	3.05 Å	Nef protein		> 20 μM (SPR)	(23, 24)
PTB domains						
–	IRS-1	1.8 Å	Insulin receptor	9	(ID_{50} = 6.5 μM)	(25)
1IRS	IRS-1	NMR	IL-4 receptor	11	6 μM (NMR)	(26)
1SHC	Shc	NMR	Trka Receptor	12	53 nM (NMR)	(27)
1AQC	X11	2.3 Å	APP	10	4.56 μM (SPR)	(28)
1X11	X11	2.5 Å	APP	14	320 nM (SPR)	(28)
PDZ domains						
1BE9	PSD-95	1.82 Å	PDZ-3 peptide	9		(29)
2PDZ	syntrophin	NMR	VGSVC	7		(30)
Calmodulin						
1CDM	Calmodulin	2.2 Å	SmMLCK	25	1 nM	(31)
1CDL	Calmodulin	2.0 Å	CaMKII	20	1 nM	(32)
14-3-3 proteins						
–	14-3-3 ζ	2.6 Å	middle-T antigen	11	730 nM (SPR)	(33)

Table 2 List of X-ray structures of signalling protein complexes.

PDB	Resol	Proteins	Other ligands	Ref.
Ras and related proteins				
1CIY	2.2 Å	Rap1A, RafRBD	GMPPNP, Mg^{2+}	(34)
1GUA	2.0 Å	Rap1A (E30D, K31E), RafRBD	GMPPNP, Mg^{2+}	(35)
1WQ1	2.5 Å	Ras, RasGAP	GDP, AlF_4, Mg^{2+}	(36)
1LFD	2.1 Å	Ras, RalGDS	GMPPNP, Mg^{2+}	(37)
1RRP	2.96 Å	Ran, RanBD1	GMPPNP, Mg^{2+}	(38)
1ZBD	2.6 Å	Rab3A, Rabphilin-3A	GTP, Mg^{2+}	(39)
1BKD	2.8 Å	Ras, Sos-GEF		(40)
–	2.8 Å	Arf1, Sec7		(41)
1AM4	2.7 Å	Cdc24Hs, RhoGAP	GMPPNP, Mg^{2+}	(42)
1TX4	1.65 Å	Rho, RhoGAP	GDP, AlF_4, Mg^{2+}	(43)
Trimeric G-proteins				
1AGR	2.8 Å	Giα, RGS4	GDP, AlF_4	(44)
1AZS	2.3 Å	Gsα, Adenylyl cyclase	GTPγS	(45)
1GOT	2.0 Å	Gαβγ	GDP	(46)
1GG2	2.4 Å	Gαβγ	GDP	(47)
1GP2	2.3 Å	Gα (G203A)βγ	GDP	(47)
1TBG	2.1 Å	Gβγ		(48)
2TRC	2.4 Å	Gβγ, phosducin		(49)
1AOR	2.8 Å	Gβγ, phosducin		(50)
Cyclin dependent kinases				
1FIN	2.3 Å	Cdk2, cyclinA	ATP	(51)
1JST	2.6 Å	Cdk2 (phosphor.), cyclinA	ATP	(52)
1BUH	2.6 Å	Cdk2, CksHs1		(53)
1JSU	2.3 Å	Cdk2, cyclinA, p27(Kip)		(54)
1BI7	2.8 Å	Cdk6, $p19^{INK4d}$		(55)
1BI8	3.4 Å	Cdk6, $p16^{INK4a}$		(55)
1BLX	1.9 Å	Cdk6, $p19^{INK4d}$		(56)
Others				
1YCQ	2.3 Å	p53, MDM2		(57)
1YCS	2.2 Å	p53, 53BP2		(58)
1GUX		Retinoblastoma, HPV E7		(59)

2. Interactions with linear epitopes

Intracellular signalling proteins in eukaryotes are often modular in structure, with different domains for different purposes (3). There are catalytic domains, such as kinases and lipases, membrane-targeting domains (pleckstrin homology (PH) and Cys2 and C2 domains), as well as domains that complex proteins by recognizing linear epitopes on their targets. The latter group includes the well-characterized SH2 and SH3 domains as well as the more recently identified PTB, PDZ, and WW domains. A common feature of these domains is that they can function independently of the rest of the protein. They can occupy different positions and they can be found in multiple copies in a single protein. In some cases they can also modulate the activity or function of other domains in the same molecule. Multiple domains can also introduce cooperativity to complex formation by forming several independent interactions with their targets. These small, independent protein domains have allowed nature to use similar activation mechanisms in different pathways by shuffling these domains between different proteins. It allows also cross-talk between different pathways and creates functional redundancy in signalling.

A great deal of structural data has emerged during the last 5–10 years on signalling domains, and their interactions with ligand peptides are rather well understood (Table 1). In the following discussion domain–ligand interactions of isolated domains are described for different classes.

2.1 SH2 domains

The Src homology domains-2 (SH2) and -3 (SH3) were among the first signalling modules described. They were identified as regions outside the catalytic kinase domain (the SH1 domain) of cytoplasmic protein tyrosine kinases (PTKs) by sequence homology with other signalling proteins (60, 61). These domains are also the hallmark of the Src-superfamily of PTKs (62). The SH2 domains are typically found in proteins involved in growth-factor signalling such as PTKs, phospholipase C (PLC) isoforms, and adaptor proteins like Grb2 (63). No SH2 domains have been found in yeasts or plants, which is in correspondence with the lack of tyrosine phosphorylation in these organisms.

The structure of the SH2 domain consists of a central β-sheet sandwiched between two α-helices (8). The N- and C-termini are very close to each other, a feature believed to facilitate the introduction of these domains into new proteins without disturbing the existing fold of the host protein.

SH2 domains interact with phosphorylated tyrosine residues (pY) in their target proteins. Dissociation constants for SH2–peptide interactions are in the micromolar to nanomolar range (see Table 1). The SH2 domains are strict in their requirement for the phosphotyrosine, and unphosphorylated peptides bind SH2 domains very weakly, if at all. The different SH2 domains distinguish between different ligand peptides by the amino-acid sequence surrounding the phosphotyrosine residue (11). In particular, the sequence that is carboxy-terminal to the phosphotyrosine is important for specificity (64, 65). Typically, the SH2 domains show greatest selectivity for residues +1 and +3 (the phosphotyrosine position is assigned as 0 and other residues are numbered in relation to it, i.e. +3 refers to the third residue C-terminal to pY). Although several peptide residues interact with the domain and affect the specificity, they seem to contribute little to the overall binding energy.

All the SH2 domains bind the phosphotyrosyl residue in a specific pocket, with a strictly invariant arginine βB5 (R154 in Fig. 1) interacting with the phosphate group of the phosphotyrosine and the loop between β-strands B and C closing over the phosphate (Fig. 1) (4, 7, 8, 12, 13). In the crystal structures of the Src and Lck SH2

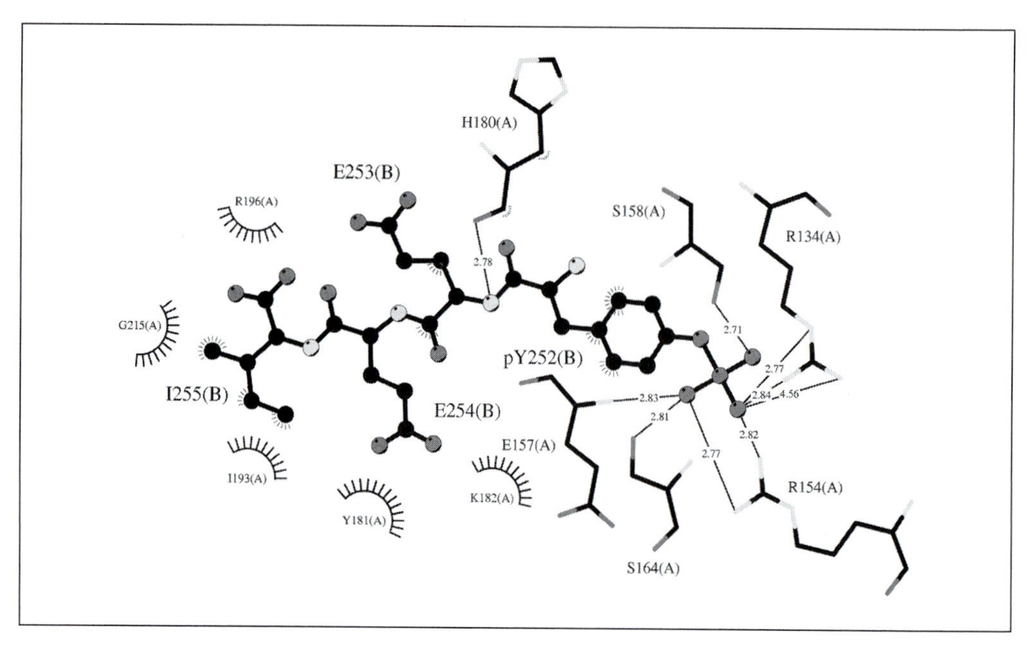

Fig. 1 Ligand–domain interactions between the Lck SH2 domain and pYEEI peptide (PDB:1LKK). Peptide atoms are shown as ball-and-stick models and interacting residues from the SH2 domain as sticks. Hydrogen bonds are drawn as dashed lines and hydrophobic interactions are indicated by spoked arcs. The plots show the extensive hydrogen bonding of the phosphotyrosine residue and hydrophobic interactions between peptide residue +3 (I255) and the SH2 domain. The figure was created using program LIGPLOT (67).

domains, the peptide lies in extended conformation perpendicular to the central β-sheet of the domain with a deep pocket serving as a docking site for a hydrophobic residue in position +3 of the peptide (4, 7). The two SH2 domains of PI3-kinase preferentially bind peptides with a methionine in position +3 (64). In the crystal structure of a ligand-bound SH2 domain g PI3-kinase, rotation of a tyrosine residue opens up a well-defined pocket for the methionine (10).

The Syp SH2 domain binds phosphopeptides in the same orientation as Src, Lck, and PI3-kinase, but, rather than having a specific pocket for residue +3, it binds the peptide in a continuous groove which allows a hydrophobic residue in the position +1 (13). These results are in accordance with the different specificities of these domains for the +1 position (64).

The structure of the SH2 domain of Grb2 in complex with a phosphopeptide has revealed an alternative binding mode, in which the peptide adopts a β-turn conformation and residue +2 interacts specifically with the domain (12). Tryptophan 121 in the EF loop blocks the binding groove for the peptide and forces the β-turn conformation. The corresponding residue is threonine in Src and serine in Lck. Asparagine in position +2 of the peptide points towards the domain forming multiple hydrogen bonds, explaining the preference for this residue (65, 66).

2.2 SH3 domains

The SH3 domains contain around 60 amino acids and are characterized by several highly conserved aromatic residues. They are more universal than SH2 domains and are found in yeasts, invertebrates, and vertebrates (68, 69). SH3 domains consist of a five-stranded, antiparallel β-barrel (70–72). The most distinctive feature of this domain is the clustering of conserved aromatic residues on one side. This hydrophobic surface has proved to be the interaction site for the SH3-domain ligands.

Deletion analysis and site-directed mutagenesis of the first SH3-domain binding proteins, 3BP1 and 3BP2, identified a 10-amino acid, proline-rich sequence as the binding site for SH3 domains (73, 74). A similar recognition motif has since been confirmed for several different SH3 domains. These peptides can adopt left-handed, polyproline type-II (PPII) helical conformations in solution and they bind to SH3 domains with typically micromolar affinities (17).

As previously predicted, the proline-rich peptides bind the SH3 domain in a PPII conformation, which allows the side chains of the peptide to intercalate between the aromatic residues on the domain (16). The proline residues in the ligand peptide seem to be important for two reasons: first, they stabilize the PPII conformation of the peptide, which is required for the correct positioning of the side chains; and, second, their pyrolidone rings provide hydrophobic surfaces for the interaction. PPII helices contain three residues per turn: in SH3–peptide complexes two residues are in contact with the domain, the third points away, and the next two residues are again in the interface. The main-chain carbonyl groups do not form intramolecular hydrogen bonds, as they do in α-helices, and are therefore free to accept hydrogen bonds. In the SH3–peptide complexes three semi-conserved intermolecular hydrogen bonds to the peptide carbonyls can be identified, these involve the side chains of a tryptophan, asparagine, and a tyrosine in the SH3 domain (see Fig. 2). Thus the two conserved

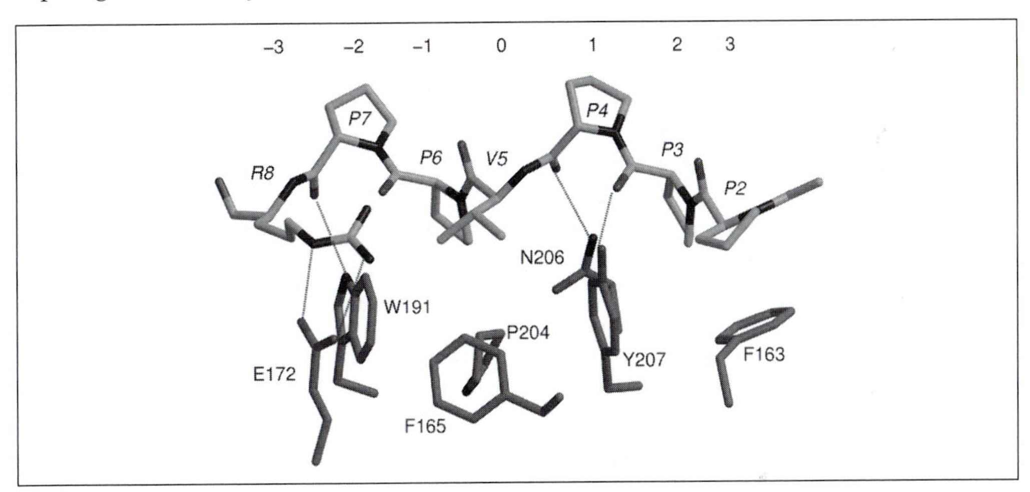

Fig. 2 SEM-5 (PDB:1SEM) SH3 domain in complex with a peptide. The peptide (above) and interacting residues of the SH3 domain are shown as stick models. Intramolecular hydrogen bonds, including the ones to the peptide carbonyls, are indicated by dashed lines. Numbering on the top refers to the binding pockets on the domain (69).

aromatic residues have a dual role: to form part of the hydrophobic surface for the peptide side chains, and to coordinate the main-chain carbonyls. The interactions between the proline-rich peptide and SH3 domain are not, however, very specific—a feature that partly explains the ability of an SH3 domain to interact with several different peptides, and vice versa, with similar affinities.

The PPII helix is a pseudosymmetrical structure that allows binding of the ligand peptides to SH3 domains in two opposite orientations. The SH3 domain of SEM-5, a *Drosophila melanogaster* orthologue of Grb2, binds the proline-rich peptide of Son of sevenless (Sos) in an opposite orientation to that of the Abl and Fyn SH3 domains (16, 21). The peptide–domain interactions are very similar in both peptide orientations (including the hydrogen bonding to the main-chain carbonyls of the peptide), with the residues outside the proline-rich core of the peptide determining the orientation of the binding (*R8* in Fig. 2) (21). Nuclear magnetic resonance (NMR) studies of the c-Src SH3 domain in complex with two different peptides demonstrate how the same domain can bind similar ligands in different orientations (75). The orientation of the surface peptide is determined by an ionic interaction between an arginine in the peptide and an aspartate in the SH3 domain. If the arginine precedes the proline-rich region, the peptide binds in the so-called plus-orientation; if the arginine follows the region, the peptide binds in the minus-orientation. This mechanism seems to be general for many of the SH3 domains. The RT (Arg-Thr) loop of the SH3 domain, which connects the first two β-strands, seems to play a more general role in determining the specificity of the interactions, as demonstrated by the complex structure of the Abl-SH3 domain complexed with a designed, high-affinity peptide (19). In this structure, a tyrosine residue in the N-terminal part of the peptide is hydrogen-bonded by two residues in the RT loop. The following serine residue functions as a hinge between the N-terminal part and the PPII helical C-terminus, enabling the optimal positioning of both parts on the SH3 domain. It seems that the PPII helix provides the common motif for the peptide–SH3 interactions, and additional contacts outside this region affect the specificity of the interaction.

The structure of the Nef protein from the human immunodeficiency virus (HIV) in complex with the R96L mutant of the Fyn SH3 domain demonstrates how interactions outside the PPII helix can contribute to the specificity and affinity of the interaction (22). Binding of the proline-rich tail of Nef by the SH3 domain resembles the binding of an isolated peptide, but the specificity and higher affinity of the interaction (over 300-fold difference in binding to full-length Nef vs. proline-rich peptide of Nef) is governed by interactions between other parts of the molecules. The RT loop of the SH3 domain, in particular, interacts very specifically with the Nef protein, increasing the interface surface area from the 750 $Å^2$ buried by the polyproline helix to 1300 $Å^2$ in total.

2.3 Phosphotyrosine-binding domains

The phosphotyrosine-binding (PTB, also called PI or SAIN) domains represent a second class of domains that specifically recognize phosphorylated tyrosines in their target proteins. The PTB domain was first discovered in the N-terminus of an adaptor molecule, Shc, with a capacity to interact with tyrosine-phosphorylated proteins (76–79). All PTB domains have a core of a seven-stranded β-sandwich capped by a C-terminal α-helix, similar to PH domains. The Shc PTB domain has an insertion in the first loop containing two additional β-strands and an α-helix (27).

The substrate specificity of PTB domains differs from that of SH2 domains. While the SH2 domains distinguish between different peptides from the sequence C-terminal to the phosphotyrosine, the PTB domains differentiate the peptides by their N-terminal sequences. The Shc PTB domain recognizes a consensus motif, NPXpY, in the target proteins c-ErbB2, epidermal growth-factor (EGF), and interleukin-4 (IL-4) receptors (78, 80). The IRS-1 and Shc PTB domains recognize slightly different phosphotyrosine peptides, which vary in positions -1, -5, and -6 with respect to phosphotyrosine (81). Dissociation constants of PTB domains for peptides derived from their biological binding partners can be as low as 20 nM (82).

The phosphopeptides bind to the PTB domains between the C-terminal α-helix and the adjacent β-strand 5 by hydrogen-bonding with their N-termini to this β-strand in an antiparallel orientation (Fig. 3) (25–27). The NPXpY phosphopeptides have β-turn conformations in solution and the same conformation is retained upon binding to PTB domains (81). The conserved asparagine in the recognition motif plays a structural role by forming an intrapeptide hydrogen bond with its carboxamide oxygen, with the backbone amide of residue-1, stabilizing the β-turn. The phosphotyrosine residue is coordinated both in Shc and IRS-1 PTB domains by positively charged amino acids, which, however, are not equivalent between the domains. Arginine 67 in Shc PTB derives from the long insertion missing from IRS-1; also, the residues in β-strand 5 and loop 6–7 (numbering corresponds to IRS-1) are different.

The specificity of PTB domains for particular peptides is affected by the spacing of hydrophobic residues in the amino-terminus of the ligand. An aliphatic leucine in position -5 binds to a hydrophobic pocket in the Shc PTB domain, whereas IRS-1 has a similar hydrophobic interaction with residues in positions -6 and -8 of the peptide.

In contrast to SH2 domains, tyrosine phosphorylation is not required for all PTB domains to interact with their ligands. The Shc PTB domain interacts with the NPLH peptide sequence in the protein–tyrosine phosphatase PEST, but with a 10-fold lower affinity compared to the phosphorylated EGF receptor peptide (83). Neuron-specific X11 and FE65 proteins are able to bind unphosphorylated peptides from β-amyloid precursor protein (β-APP) with their PTB domains (84). The peptides are similar in sequence to IRS-1 and Shc ligands and carry the NPXY motif, although the tyrosine is not essential for the binding. The structures of X11 PTB in complex with 10- and 14-residue peptides show very similar domain–peptide interactions in the core region as in the other PTB domain–peptide complexes (28). The peptide hydrogen bonds to the β-strand 5 of the domain and adopts a β-turn conformation. C-terminal to the tyrosine residue, the peptide adopts a 3_{10} helical conformation and two phenylalanines bind to a hydrophobic patch on the domain formed by the C-terminal α-helix. These additional contacts increase the affinity for the peptide and mutation of either of the two phenylalanines results in a 10-fold reduction in affinity.

2.4 PDZ domains

The PDZ domains (named after the three proteins in which it was first found: PSD-95, DlgA, and ZO1—also called the DHR domains) are one of the more recently identified signalling domains (85, 86). They are approximately 100 amino acids in size and are found both in signalling and cytoskeletal proteins. The crystal structure of an unliganded PDZ domain of Dlg revealed a five-stranded β-barrel capped at both ends by short α-helices (87). The structure resembles PH and PTB domains in the orientation of the β-strands and the C-terminal α-helix, but it has completely different connectivity (Fig. 3). The ligands for some of the PDZ domains were identified by yeast two-hybrid screening and found to be the C-terminal tails of transmembrane proteins carrying a T/SXV motif (88, 89). PDZ domains from different proteins can also interact with each other to form larger protein complexes (90).

The crystal structure of the PSD-95 PDZ domain complexed with a nine-residue peptide TKNYKQTSV yielded the first glimpse of the basis of the specificity of PDZ–peptide interactions (29). The peptide binds on the surface of the domain forming an additional β-strand to the pre-existing β-sheet of the domain, reminiscent of the PTB-domain–peptide interactions (see Fig. 3). The carboxy-terminal residue of the peptide is inserted into a small cavity formed by main-chain amides and a side chain of an arginine residue in the conserved loop β1–β2. The terminal carboxylate is coordinated by these residues and the side chain of the C-terminal valine points into a hydrophobic cavity, thereby explaining the specificity for this residue. A smaller residue would not fill the space as completely, whereas a larger side chain could not be accommodated. Side chains of the glutamine (residue-3 from the C-terminus) and threonine (residue-2) are in contact with the peptide forming multiple hydrogen bonds. Residue-1 (serine) does not make specific side-chain contacts with the domain, in agreement with its poor conservation between ligand peptides.

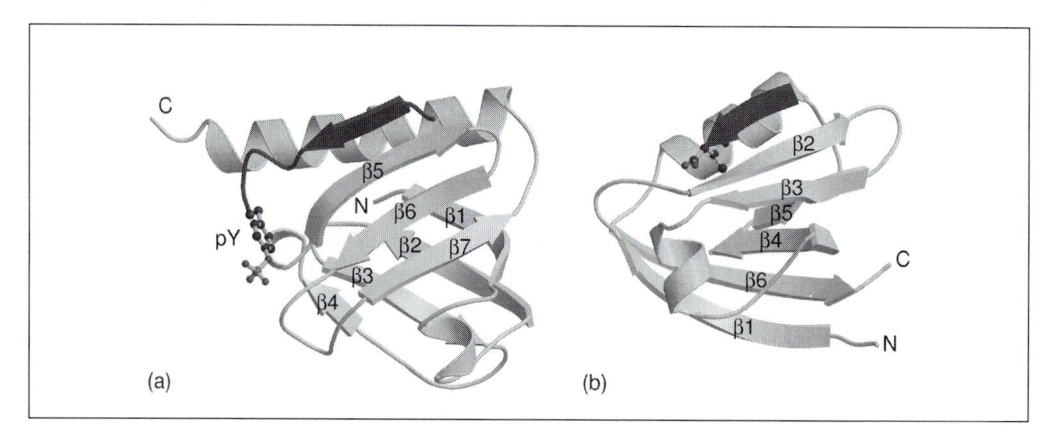

Fig. 3 Ribbon diagrams of PTB and PDZ domains in complex with their ligand peptides. (a) The IRS-1 PTB domain is complexed with phosphopeptide. The additional β-strand formed by the peptide is shown in darker grey and phosphotyrosine as a ball-and-stick model. (b) Similar view of the PSD-95 PDZ domain with ligand peptide (1BE9). The C-terminal valine is shown as a ball-and-stick model.

The PDZ domain of syntrophin shows a very different specificity towards its ligands, favouring positive residues in position -4 and glutamate in -3. In the NMR structure of syntrophin PDZ in complex with a peptide from a voltage-dependent sodium channel, these residues form ionic interactions with the domain (30). Serine in position -2 points towards the domain, hydrogen-bonding to it—and, similar to PSD-95, residue-1 makes only non-specific contacts.

2.5 Calmodulin

Calmodulin (CaM) is a ubiquitous calcium-binding protein that regulates several enzymes such as smooth muscle, myosin light-chain kinase (smMLCK) and Ca^{2+}/ CaM-dependent protein kinase (CaMK). It is the prototypical member of a larger Ca^{2+}-binding protein family that binds the metal ion using a so-called EF hand motif. Calmodulin contains two of these small α-helical domains connected by a long linker. Calmodulins bind, in a calcium-dependent fashion, a number of different peptides which contain a large proportion of hydrophobic residues and form amphipathic α-helices (91). Binding of Ca^{2+} to EF hands induces the rearrangement of the α-helices and exposure of hydrophobic surfaces. CaM can then interact with its target peptides by wrapping the amphipathic α-helices between the two EF hands. This is feasible because of the unwinding and bending of the long connecting α-helix between the EF hands (see Fig. 4) (31). Wrapping of the ligand between the EF hands results in a large interface between the ligand and CaM, over 2000 Å^2. Most of the interactions are hydrophobic, but the orientation of the peptide seems to be determined by ionic interactions between positive residues in the peptide and a patch of negatively charged residues in the N-terminal EF hand.

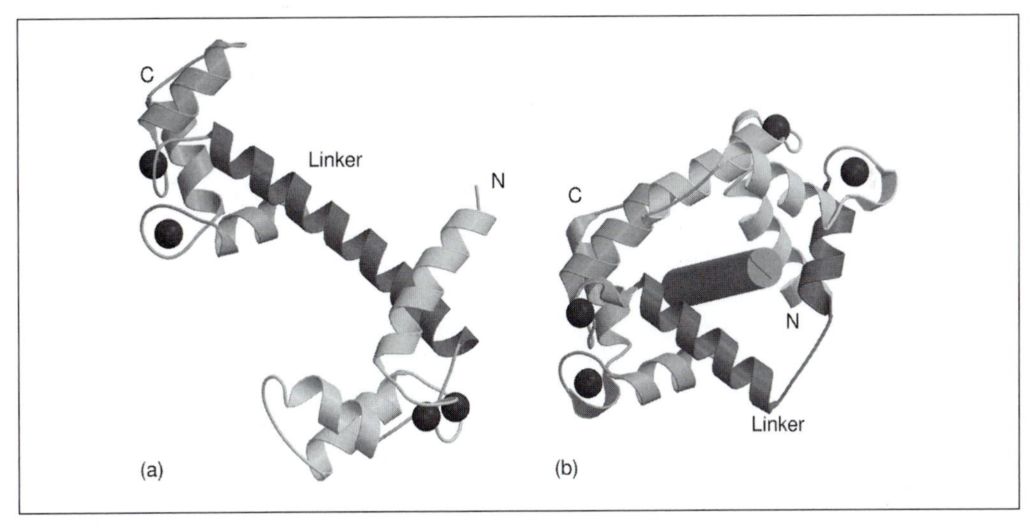

Fig. 4 Calmodulin (a) without (PDB:1CLL) and (b) with (1CDL) a ligand peptide. The N- and C-terminal EF hands and the bound calcium atoms are shown in different shades of grey. The C-terminal EF hands are aligned in the two structures. The ligand peptide in (b) is shown as a grey cylinder.

The ability of CaM to interact with different peptides is facilitated by several factors, as became evident by comparing the crystal structures of CaM complexed with different peptides (32). The ligand peptides form a large number of van der Waals contacts, which as such are relatively non-specific, and in CaM, a number of hydrophobic contacts are made by methionines, the side chains of which are flexible and can adopt different conformations. In addition, the orientation of the two EF-hand domains relative to each other is not fixed, and can change to accommodate different peptides.

2.6 Cooperativity in protein–peptide interactions

The intermediate-to-low affinity of the domains towards their ligand peptides discussed above can, *in vivo*, be compensated for either by additional contacts outside the linear peptide sequences, as seen in the SH3–Nef complex, or alternatively by forming multiple interactions between domains in one protein and epitopes in another. Sequence, orientation, and distance of the different interaction sites with respect to each other can influence the specificity and affinity of complexation.

The structure of the tandem SH2 domains from ZAP-70 protein tyrosine kinase in complex with a doubly phosphorylated peptide from the ζ-subunit of the T-cell receptor has unveiled an interesting mechanism, by which two SH2 domains can cooperate to achieve high specificity and affinity interactions with their ligands (14). The two SH2 domains are connected by a 65-residue linker, which, by forming a coiled-coil, brings the domains in contact thus creating a long, continuous binding site for the peptide. Remarkably, the binding pocket for the second phosphotyrosine is formed by the interface of the two SH2 domains. Both SH2 domains have a hydrophobic cavity for the +3 residues, similar to the SH2 domains of Lck and Src. The peptide makes a large number of contacts to the two SH2 domains, burying a total of 1300 $Å^2$. There are a large number of direct hydrogen bonds and bridging water molecules between the peptide and the SH2 domains. The relative orientation of the two SH2 domains and the distance between the phosphotyrosine pockets enables a highly selective interaction with the natural ligand.

The structure of the tandem SH2 domains of protein tyrosine phosphatase, SH–PTP2, shows a different interdomain arrangement, with two phosphopeptides binding individual domains in roughly antiparallel orientations (15). The SH2 domains are fixed in their relative positions by a disulfide linkage and are predicted to restrict both the length and conformation of the natural biphosphorylated ligand peptides. In the crystal structure of the inactive SH–PTP2 the two SH2 domains have very different orientations relative to each other, and the previously formed disulfide bond is broken (92). The N-terminal SH2 (N-SH2) domain blocks the active site of the phosphatase, and as a consequence the phosphopeptide-binding site is disrupted. The C-terminal SH2 domain is speculated to function as an anchor, attaching SH–PTP2 to a target peptide and thus increasing the local concentration of the second phosphotyrosine for the N-terminal SH2 domain. Binding of the N-SH2 to a phosphopeptide induces a conformational change in the domain, which disrupts the inhibitory interaction with the active site and results in the activation of the phosphatase. A double-phosphorylated peptide can activate SH–PTP2 10-fold compared to a monophosphorylated one, supporting the cooperative role of the SH2 domains in the activation of this enzyme.

A similar example of a cooperative interaction of the SH2 and SH3 domains in regulating enzymatic activity has been observed in the crystal structures of Src and Hck PTKs (93–95). These structures describe the kinases in the inactive (or closed) state, caused by intramolecular interactions of the SH2 and SH3 domains. The structures show that the SH2 domain interacts with the inhibitory phosphotyrosine at the C-terminus of the kinase (Y527 in Src), as was previously known. The SH3 domain binds the linker between the SH2 and kinase domains and sandwiches it in a PPII conformation against the upper lobe of the kinase (see Fig. 5). Both the SH2 and SH3 interactions are required for inactivation of the kinase, as deletion of the SH3 domain or mutagenesis of the phosphorylated tyrosine residue can separately activate the enzyme (96).

14–3–3s are a small class of proteins involved in several intracellular signalling pathways, and serve as another example of multidentate binding to linear epitopes (97). They recognize peptides carrying phosphoserines in the RSXpSXP sequence in several proteins, including various protein kinase-C isoforms and the Ras effector Raf. 14–3–3s form either homo- or heterodimers and are α-helical in structure (98, 99). The 14–3–3ζ homodimer binds two peptides in antiparallel orientation to each other

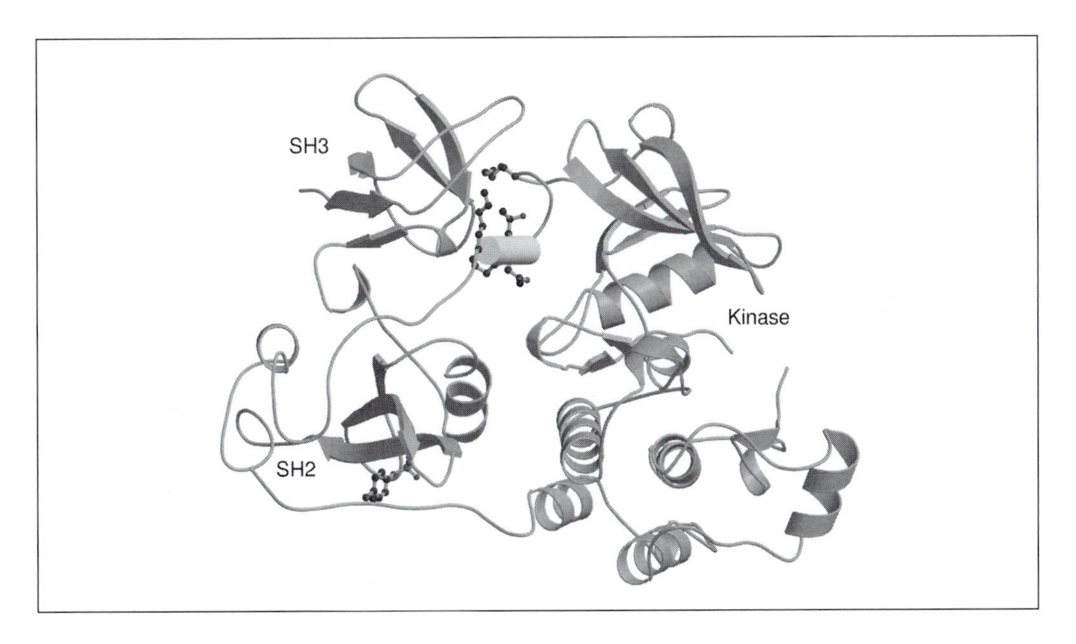

Fig. 5 Src tyrosine kinase in inactive conformation (PDB:1FMK). The part of the linker connecting the SH2 and kinase domains sandwiched in PPII conformation between the SH3 and kinase domains is shown as a cylinder, with residues in contact with the SH3 domain as ball-and-stick models. The autoinhibitory phosphotyrosine 527 bound to the SH2 domain is also shown.

in long grooves lined with conserved residues, and specifically recognizes the phosphorylated serine residue (33). Several positively charged side chains are grouped together to coordinate the phosphate. *Cis*-conformation of the proline +2 allows additional hydrogen-bonding to take place between the peptide and the protein and directs the C-terminus of the peptide away from the binding site. The opposite direction of the two peptides in the 14–3–3 dimer and the close proximity of the binding grooves would allow a continuous, double-phosphorylated peptide to bind to the protein thereby forming a high-affinity complex. Binding studies using peptides with two phosphorylated binding sequences have shown a more than 30-fold increase in affinity compared to singly phosphorylated peptides. This again illustrates the effect of cooperative interactions in increasing the affinity of a signalling protein towards its ligands. Binding to a single polypeptide in a bidentate fashion is speculated to play a role in regulating the activity of target proteins (97). With the possibility to form heterodimers of different 14–3–3 isoforms, the range of ligands that can be bound increases significantly.

3. G proteins

G proteins are a large superfamily of GTP hydrolysing enzymes that share a common evolutionary origin and whose functions are regulated by the nature of the bound nucleotide—GDP or GTP. Members of this superfamily include the well-characterized *ras* ($p21^{ras}$) oncogene, the α-subunit of the trimeric G proteins, elongation factor Tu (EF-Tu) which is involved in protein translation, and many others. Ras and its close relatives are the smallest G proteins (often referred to as the small GTPases) and comprise the canonical and common core of the G-protein fold. This approximately 200-residue core contains a seven-stranded β-sheet surrounded by α-helices (100, 101). The nucleotide-binding site is formed by loops connecting the β-sheet to the flanking α-helices. These loops are conserved among the G proteins and provide a sequence fingerprint for the family.

G proteins function by shuttling between two forms: GTP- or GDP-bound. They are normally activated upon binding GTP, and the conformational changes that occur as a result of this allow interactions with other molecules and the propagation of signals to take place. The G proteins do not, however, function on their own, and there are several regulatory / accessory proteins for each of the G proteins. Some of these affect the exchange of the bound nucleotide, either enhancing (guanine nucleotide-exchange factors—GEFs) or inhibiting it (guanine nucleotide-dissociation inhibitors—GDIs), others enhance the often low intrinsic catalytic activity of the G proteins and are referred to as GTPase-activating proteins (GAPs). This control of the GTP–GDP cycle is at the heart of G-protein function; many of the steps in this cycle for Ras have been characterized structurally and will be discussed below. Interactions of the Gα-subunit with other components of the heterotrimeric G proteins as well as downstream effectors are also discussed below.

Before discussing the interactions of different G proteins with other macromolecules, it is essential to briefly describe the effects that GTP binding has on the

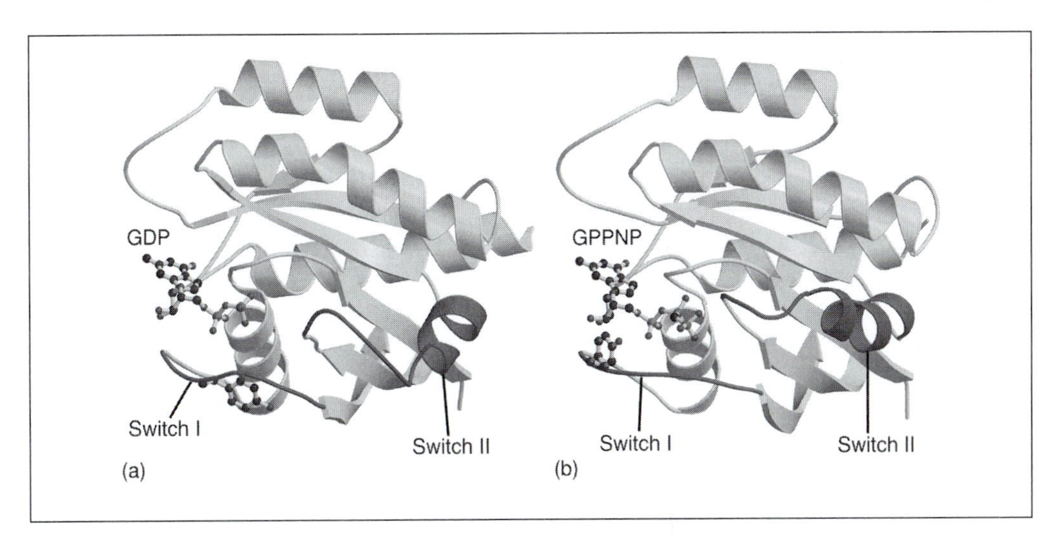

Fig. 6 Switch regions of Ras. (a) The GDP-bound form of Ras (PDB:4Q21) is shown, with the Switch-I and -II regions in dark grey, and bound nucleotide as ball-and-stick models. (b) The same view of the GPPNP–Ras (5P21) illustrates the structural changes in these regions when the bound nucleotide is GTP. Tyrosine 32 is shown to demonstrate the large changes in the Switch-I region, which are otherwise less evident in this representation.

structure of G proteins (see ref. 102 for a more detailed review). Both Ras and Gα have been structurally characterized in the GTP-bound form (or more accurately with the GTP analogues, GMPPNP or GTPγS). The minimal G-protein fold of Ras can serve as the prototypical example for the other G proteins. The overall fold of Ras is not affected by GTP binding, and the structural changes are restricted to two regions of the molecule referred to as the Switch I and Switch II regions (Fig. 6). The X-ray and NMR structures of active Ras–GMPPNP (a non-hydrolysable GTP analogue) and inactive Ras–GDP showed how the flexible Switch I and II regions of Ras are sensitive to the identity of the bound nucleotide, and how they adopt a more rigid conformation in the Ras–GMPPNP complex (103–105). This conformational change creates novel binding surfaces for downstream effector molecules which bind selectively to Ras–GTP. Effectors bound to Ras become associated with the plasma membrane, and may be activated by Ras itself or other activating agents embedded in the plasma membrane (for a review see ref. 106).

3.1 Small G proteins

3.1.1 Ras guanine nucleotide-exchange factors—GEFs

Exchange of GTP for GDP in Ras-like proteins is very slow, despite the 10-fold higher concentration of GTP in the cytoplasm. The exchange of the nucleotide, and hence activation of G proteins, can be greatly accelerated by a family of proteins called guanine nucleotide-exchange factors (GEFs). The exchange of GDP for GTP is the first step in the various signal-transduction cascades that the Ras-family of proteins regulate and is under the control of GEFs, which in turn are regulated by transmembrane receptors (for a review see refs 107, 108).

The Son of sevenless (Sos) protein is the GEF for Ras. It activates Ras after being translocated to the plasma membrane by an adaptor protein, Grb2 (109). The mechanism of the GEF function in Sos protein was elucidated with the 2.8-Å resolution structure of Sos bound to Ras in the absence of nucleotides or Mg^{2+} (40).

When Ras is complexed with the GEF region of Sos, Ras adopts a structural conformation distal from the interacting surface that is very similar to that seen in Ras–GTP, although the region of Ras that is in contact with Sos is very distorted (Fig. 7a). Sos is a predominantly α-helical protein with two domains: a C-terminal domain that interacts with Ras; and an N-terminal domain, which does not interact with Ras and seems to serve a purely structural role. The C-terminal domain consists of tightly packed α-helices with a protruding α-hairpin, which seems to be crucial for activity.

Sos interacts with the phosphate binding P-loop, and the Switch regions I and II and α-helix 3. The interface surface is very large (3600 $Å^2$) and consists of a core of hydrophobic van der Waals interactions surrounded by polar residues that form hydrogen-bonding networks. Sos inserts its helical hairpin near the active site, displacing the Switch I region and breaking the extensive network of direct and water-mediated interactions with the nucleotide found in intact Ras. This insertion does not block the ribose-binding site, but merely disrupts the binding potential of the region. Two residues from αH of the α-hairpin physically occupy the binding sites of Mg^{2+} and the α-phosphate, interacting with corresponding residues in Ras, and block the association of GDP with Ras. The tight grip of Sos on the Switch II region, which forms important interactions with the γ-phosphate of GTP, results in a

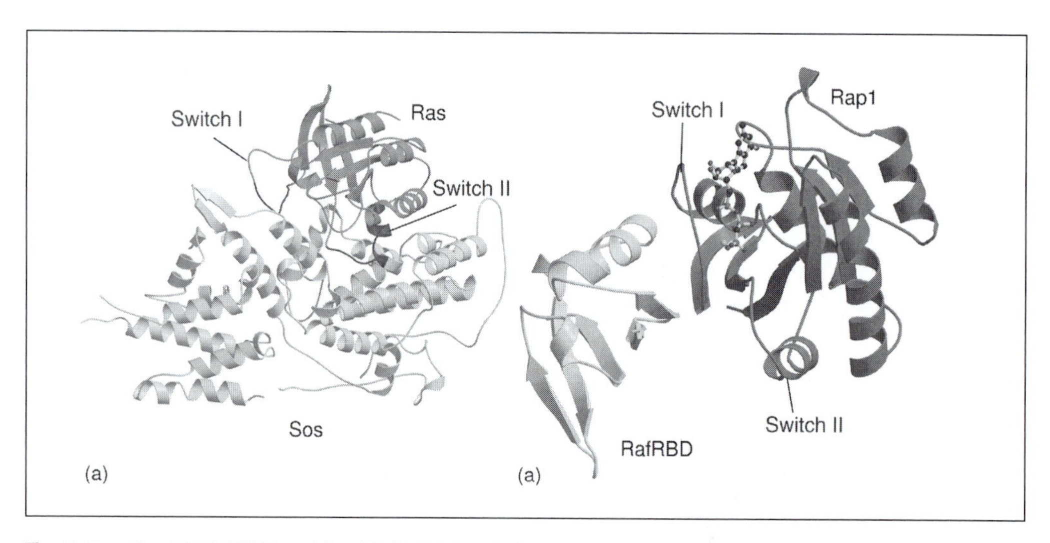

Fig. 7 Ras–Sos (PDB:1BRD) and Rap1A–RafRBD (1GUA) complexes. (a) A ribbon diagram of the Ras–Sos complex showing the greatly distorted Switch I and II regions. (b) Ribbon representation of Rap1A (E30D,K31E) and RafRBD indicating the continuous β-sheet between the two proteins.

restructuring of the peptide backbone and culminates in the occlusion of Mg^{2+} from the binding site and the removal of favourable interactions with the γ-phosphate. Other interactions between Ras and Sos do not impose notable structural changes on Ras. Small changes observed on the P-loop, for example, reflect the absence of the nucleotide and are not caused by Sos directly.

Based on the structure, the authors propose a mechanism whereby the association of Sos with Ras–GDP (or Ras–GTP) results in displacement of the nucleotide (40). In turn, Sos is displaced by the incoming nucleotide (the nucleotide would first associate through its ribose ring with the intact portion of the Ras-binding site), which causes changes in the Switch I and Switch II regions that result in dissociation of the Ras–Sos complex. Sec7, a GEF for ADP ribosylation factor (Arf1), induces similar conformational changes in the phosphate-binding region of Arf1, and an analogous mechanism for nucleotide exchange has been proposed (41).

3.1.2 Downstream effector interactions

Once Ras is loaded with GTP, it is capable of transmitting signals further. This is accomplished by interactions with various proteins called effectors, which are defined as proteins that have a higher affinity for Ras–GTP than Ras–GDP (110). This includes a great number of proteins involved in the downstream propagation of Ras signalling and also regulators of Ras. Early mutagenesis studies of H-Ras showed that the region encompassing residues 32–40, the effector region, was crucial for biological activity (111, 112). This region corresponds closely to the Switch I region (residues 30–37), which was identified as being conformationally sensitive to the state of the bound nucleotide and provided evidence that the 'on-state' of Ras was being transmitted to effector molecules through the conformational changes in this region.

The structure of the complex between the Ras-binding domain (RBD) of c-Raf1, a downstream effector of Ras, and Rap1A–GMPPNP, a Ras-family protein, confirmed the mutagenesis results and provided more details of the interaction (34). c-Raf1 is one of the initial kinases in the cytoplasmic mitogen-activated protein kinase (MAPK) cascade which can be stimulated by Ras activation. Rap1A is a small GTPase in the Ras family, sharing 50% sequence similarity with Ras and being identical in the effector region. It interacts with the same effectors as Ras but acts antagonistically; most of the Ras/Rap chimeras that have been constructed are biologically active (113, 114). The structure shows that Rap1A in complex with RafRBD is structurally very similar to Ras (overall rms 0.88 Å) with the largest deviations occurring in the protein–protein interface. The topology of RafRBD is very similar to the ubiquitin α/β roll fold, with a central five-stranded mixed β-sheet having an interrupted α-helix and two additional small 3_{10}-helices. The interface region between Rap1A and the c-Raf1 RBD is mainly hydrophilic and is centred around the formation of a continuous β-sheet, linking the β-sheets of Rap and RafRBD (Fig. 7b). The C-terminal portion of the α-helix of RafRBD also makes contact with the Switch I region β-strand. The structure provides explanations for many of the observed mutations that interfere with Ras-effector binding. For example, the well-studied Ras mutation

D38A reduces the affinity 72-fold and even a conservative D38E mutation abolishes the biological activity (115, 116). In the structure, D38 makes two direct hydrogen bonds to the hydroxyl of T68 and the side chain of R89 from RafRBD, water-mediated hydrogen bonds to V69 and R89 from RafRBD, and to the hydroxyl of T35 of Rap, which is a ligand for Mg^{2+}.

The structure of the Rap–RafRBD complex has also allowed the origins of the specificity and affinity of the Ras/Rap–Raf complex to be probed. A mutational study of Raf binding to Ras showed that the residues observed in the protein interface were crucial to the measured affinities, as well as to the biological activity of Ras. The study also pinpointed particular residues in Raf, namely Q66, K84, and R89, that provided the majority of the binding energy (117).

Ras and Rap have nearly a 70-fold difference in affinity towards RafRBD, and this is believed to account for their antagonistic effects *in vivo* (118). Based on the structure of the Rap1A–RafRBD complex, and the sequence differences between Ras and Rap1A, mutant Rap1A proteins were engineered that mimicked the effector region of Ras and allowed the structures of the mutant Rap1A proteins complexed to RafRBD to be solved (35). The single- and double-Rap1A mutants (K31E and E30D, K31E) showed RafRBD affinities very close to Ras. This affinity seems to be due mainly to the K31E mutation, which generates a tight salt bridge with K84 in RafRBD, and also by a number of new hydrophilic interactions (i.e. 89 close contacts in the mutant complex versus 54 in the native) created by minor structural changes. In general, the mutations resulted in a tighter complex, indicated by lower B-factors in the interface region and a higher diffracting power (118).

There is no direct contact between the RafRBD and the bound nucleotide, indicating that information is transmitted to the effector through purely allosteric mechanisms, although it is not clear from the structure how GTP hydrolysis would affect the interface. Presumably the conformational shift in the Switch I region disturbs the interface geometry and reduces the affinity significantly enough to dissociate the complex.

Other known complexes of small GTPases and their effectors have demonstrated similar dependence of the interactions to the identity of the bound nucleotide. Complexes of Ras with RalGDS, Rab3a with Rabphilin, and Ran with RanBD all demonstrate interactions in the Switch region which can only be formed with a GTP-bound G protein (38, 39, 119).

3.1.3 GTPase activating proteins, GAPs

To switch off the signalling by Ras, the GTP bound to the molecule has to be hydrolysed to GDP. The intrinsic rate of GTP hydrolysis by members of the Ras family is very low (the Ras turnover rate is 0.03 min^{-1}), but it can be enhanced by GTPase-activating proteins by up to five orders of magnitude (120). This is a similar phenomenon to the activation of Gα-subunits in heterotrimeric G proteins, although the mechanism of activation is very different. There was some debate about whether the GAPs for the Ras-family operated by stabilizing the enzyme in an active state, or whether they actually contributed residues to the active site. Evidence for the latter hypothesis was provided by the observation that Ras could not bind GDP–AlF_4^-, a transition-state analogue, without the presence of RasGAP (121). The two GAP-complex structures available confirm this hypothesis, illustrate the basis for this interaction, and identify the residues involved in the reaction mechanism.

The first structure to be elucidated was the 2.5-Å X-ray structure of the complex between a catalytic fragment of p120GAP, termed GAP334, and Ras–GDP in the presence of aluminium fluoride (36). GAP-334 is a purely helical protein consisting of two domains, the larger of which is conserved among RasGAPs. Activated Ras sits in a shallow groove on the surface of the catalytic domain of GAP-334 with a large buried surface area (3145 Å^2), interacting with GAP-334 through its Switch regions I and II, its phosphate-binding P-loop, and possibly the α-helix 3 (Fig. 8a). The interface consists of weak van der Waals interactions and a number of polar interactions, which account for the majority of the binding energy as deduced by the effect of ionic strength on affinity (122).

The structure shows that the activity of RasGAP is due mainly to the insertion of R789 into the active site of Ras and to the stabilizing influence of R903. The orientation of the R789 side chain places the charged guanidinium group near the aluminium fluoride ion. In Ras–GTP, this would be in an excellent location to stabilize the negative charge that would develop in the transition state on the γ-phosphate. R903 appears to maintain the conformation of the finger loop and to form hydrogen bonds with other residues on Ras in the active site including the Q61, a residue implicated in oncogenic Ras (see Fig. 8b).

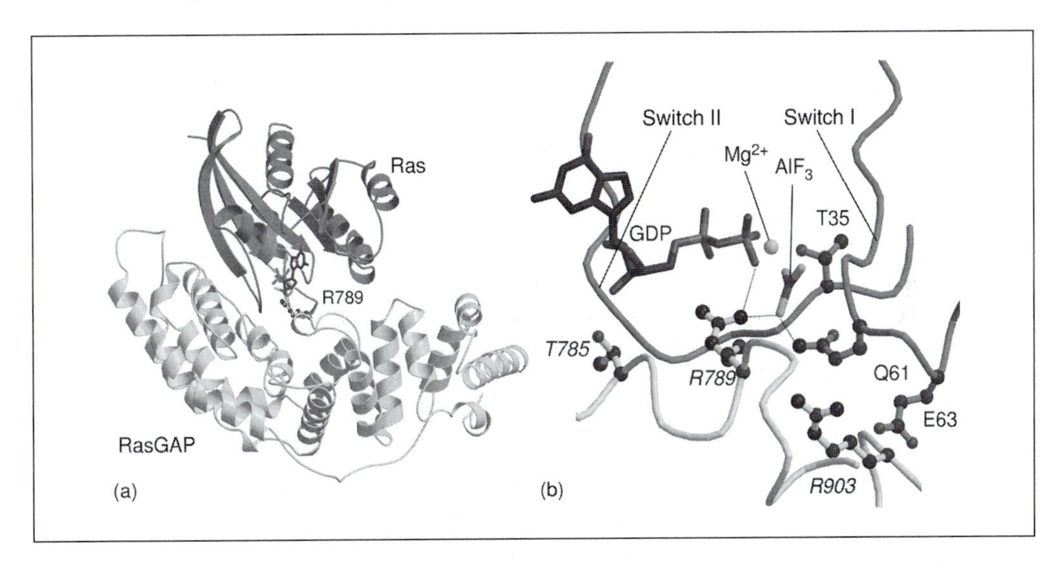

Fig. 8 Ras–RasGAP complex (PDB:1WQ1). (a) A ribbon representation of the Ras–RasGAP complex showing how Ras sits on RasGAP. Bound nucleotide, AlF_3, and arginine 789 of RasGAP are shown as ball-and-stick models. Although $A1F_4^-$ was included in the crystallization conditions the trigonal $A1F_3$ was observed in the structure (36). (b) A closer view of the active site of Ras with GDP and AlF_3 shown as stick models and protein side-chains as ball-and-stick models. The RasGAP residues are labelled in italics. Hydrogen bonds formed by R789 of GAP are indicated with dashed lines.

Structures of the related small G-proteins Cdc42 and RhoA in complex with p50rhoGAP have revealed similarities in the mechanism by which these proteins and Ras are turned off. p50rhoGAP is, like RasGAP, an α-helical protein, but shares no structural or sequence homology with it. In the complex with GMPPNP-loaded Cdc42Hs, the p50rhoGAP interacts with the G protein in a shallow groove burying ca. 2000 Å^2 of accessible surface (42). The low resolution of the structure did not allow an accurate description of the catalytic mechanism for the GAP, and the 1.65-Å structure of a RhoA–p50rhoGAP with GDP and AlF_4^- explained why this may be the case. In this structure, RhoA has rotated by 20° relative to p50rhoGAP and the interface surface area has increased by 50% to nearly 3000 Å^2 (43). As with the Ras–rasGAP complex, the p50rho–GAP stabilizes the transition state of GTP hydrolysis by providing residues to the catalytic site. It seems that the initial interaction of G protein and its GAP results in the suboptimal positioning of the two molecules, so that subsequent rearrangements are needed to gain a fully functional complex.

3.2 Heterotrimeric G proteins

G protein-coupled receptors (GPCRs) are involved in processes ranging from vision and taste to embryogenesis and development in mammals, although the downstream targets and effects vary greatly from one receptor to another, and from one cell type to another. The intracellular effectors of GPCRs are the so-called heterotrimeric G proteins containing two tightly associated subunits, Gβ and Gγ, and the GTP-binding subunit Gα. Stimulation of the extracellular portion of GPCRs causes a conformational change in the cytoplasmic face of the GPCR and induces the exchange of GDP for GTP in the associated heterotrimeric G protein. The Gα-GTP monomer and the Gβγ heterodimer then dissociate and each moiety interacts with downstream targets. Gα-subunits have an intrinsic GTPase activity, hydrolysing bound GTP and allowing it to revert to its inactive Gα-GDP form, which is able to reunite with Gβγ-subunits, reassociating with the cytoplasmic membrane surface, and ready to interact with activated GPCRs (for a review see ref. 123)

3.2.1 Subunit interaction in the heterotrimeric G proteins

The high-resolution structures of two heterotrimeric G proteins have greatly helped our understanding of the interactions between the subunits and the putative interactions with the intracellular faces of GPCRs (47, 124). The β-subunit of the heterotrimeric structures adopts a seven-bladed, β-propeller conformation with an N-terminal, α-helical extension. The Switch I and II regions of the Gα-subunit make extensive contact with the base (narrower end) of the β-propeller, and the long, amino-terminal α-helix (residues 2–34) interacts with the outer surface of the first and seventh blades of the β-propeller (see Fig. 9a). The Gγ-subunit binds to the different surfaces of the β-propeller, forming a coiled-coil interaction with the N-terminal α-helix of Gβ, and interacts with loops on the fifth and sixth blades of the β-propeller (47, 124) (see Fig. 9b).

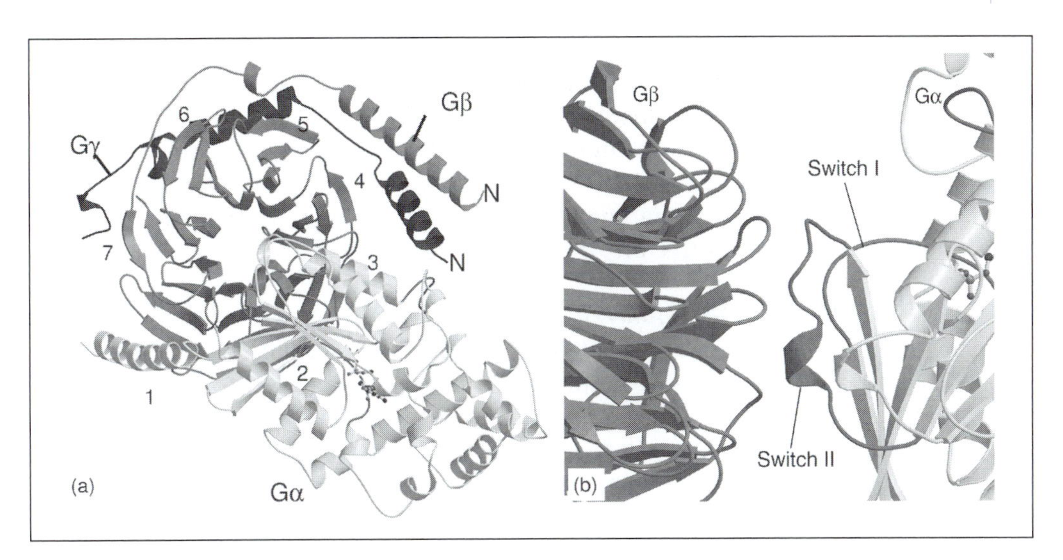

Fig. 9 Heterotrimeric G-protein complex (PDB:1GOT). (a) A top view of the complex with subunits-α, -β, and -γ shown in different shades of grey. The unique α-helical domain of Gα is on the lower right corner of the figure. (b) A close-up from the side showing the Switch I and II regions contacting the β-subunit.

In vivo, the trimeric G proteins are membrane-bound by N-terminal myristoylation of the Gα-subunit and C-terminal prenylation of the Gγ-subunit. In the crystal structures, which were determined without lipid modifications, the N-terminus of Gα is juxtaposed with the C-terminus of Gγ, and sits against a positive patch on the surface of the Gγ-subunit, providing information on the orientation of this complex with respect to the cell membrane (for a review see ref. 125). The mutual association of the proteins on the inner face of the membrane facilitates their interactions and thus increases their affinities.

3.2.2 Gα interaction with adenylate cyclase

The Gα-subunits can be classified based on their ability to either stimulate (Gsα) or inhibit (Giα) a range of downstream targets such as adenylyl cyclase. Adenylyl cyclase is a transmembrane protein with two large cytoplasmic domains that catalyse the formation of cyclic-AMP from ATP. The structure of the catalytic domains of adenylyl cyclase in a complex with Gsα–GTPγS and the activating plant terpenoid forskolin, elucidated by Tesmer and co-workers (45), helps to illuminate the likely mechanism by which Gsα-subunits can activate their targets.

In this structure the two cytosolic domains of adenylyl cyclase, C1 and C2, form a heterodimer. The dorsal surface, which would connect to the transmembrane helices of intact adenylyl cyclase, and the nearby Gsα N-terminus, which is often palmitoylated *in vivo*, must face the membrane. Thus, the shallow groove running the length of the ventral surface that contains the binding sites for the substrate, forskolin, and Gsα would face the cytoplasm. The Gsα-binding site is distal to the substrate-binding site and interacts with the Switch II α-helix as well as with residues in the Switch I and the α3–β5 loop of Gsα (see Fig 10a). Although the structure of the heterodimer without Gsα is unknown, it is believed to be similar to the structure of the C1 homodimer, and inferences have been drawn from the relative orientations of the subunits. It appears likely that the binding of Gsα changes the relative orientations of the subunits by approximately 7°, bringing residues important for catalysis into the correct orientation to stabilize the active site. Thus Gsα activates adenylyl cyclase, and possibly other effectors, through allosteric effects.

3.2.3 Interactions with regulators

An important aspect of G-protein signalling is the regulation of the pathway. Gα, like other G proteins, is self-regulated by its intrinsic GTPase activity. This intrinsic activity is very low (for example, k_{cat} for the GTPase of Ras is in the order of 2–5 min^{-1}), but it can be increased by more than 50-fold by a family of proteins known as regulators of G-protein signalling (RGS). They appear to have little or no affinity for Gα-GDP, moderate affinity for Gα-GTP, and high affinity for Gα-GDP–AlF_4 (a transition state analogue); supporting the observation that they enhance the catalytic activity of Gα-subunits by transition-state stabilization, rather than nucleotide dissociation. In addition, some RGS proteins appear to compete for effector-binding sites, for example RGS4 can block the activation of phospholipase-C β1 by Gqα–GTPγS. The best-characterized RGS is RGS4, which interacts with high affinity to Giα-GDP–AlF_4 ($K_d < 100$ nM).

The structure of the RGS4/Giα-GDP–AlF_4 complex has been solved to a resolution of 2.8 Å by Tesmer and co-workers (44). In this structure RGS4 is an α-helical protein consisting of two subdomains: a 'terminal' subdomain containing both N- and C-termini; and a larger 'bundle' subdomain consisting of a classical, right-handed, four-helix bundle. The terminal subdomain interacts with the Switch I region of Giα, whereas the bundle subdomain interacts with the Switch II and Switch III regions

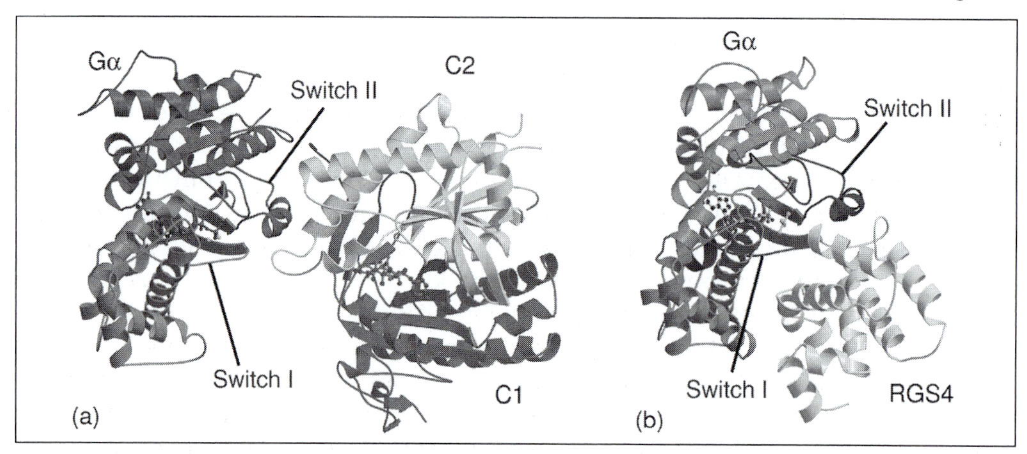

Fig. 10 Gα in complex with (a) adenylate cyclase (PDB:1AZS) and (b) RGS4 (1AGR). In both (a) and (b) Gα is shown in medium grey with darker Switch regions I, II, and III. (a) The C1 and C2-subunits of adenylate cyclase are shown in darker and lighter grey, respectively. Forskolin bound between these subunits is also shown, as well as the nucleotides in the Gα-subunits. In (b) the RGS4 molecule is shown in lighter grey.

(see Fig. 10b). Since these three regions are known to be conformationally sensitive to the identity of the nucleotide in the active site, this is consistent with observations that RGS4 has no affinity for Giα-GDP. Because these regions are also known to be important in Gα-effector interactions, this explains why RGS binding can interfere in downstream interactions. The interface is characterized by hydrophilic interactions and an overall surface charge complementarity. The structure confirms that RGS4 activates Giα by stabilizing the transition state of GTP hydrolysis. In the crystal structure, GDP–AlF_4 is a transition-state analogue, and the conformation of Giα more closely resembles that found in the Giα-GDP–AlF_4 structure than in the Giα-GTPγS structure (126). Binding of RGS rotates the two lobes of the Gα-subunit 3.5° closer together, relative to the free Giα1-GDP–AlF_4 structure, forming a tighter interaction with the nucleotide bound in the cleft. RGS binding also seems to reduce the flexibility of the three Switch regions, as inferred by the 2.5-fold reduction in normalized B-factors for these regions. Although RGS4 does not insert any residues into the active site, the side chain of N128 is close to the active-site residue Q204 of Giα, and the nucleophilic water molecule, and could contribute to their catalytic properties.

An important inhibitor of heterotrimerization is the protein phosducin, which is best characterized in the context of the rod-cell visual transduction system, and which forms a complex with Gβγ-subunits. In rod cells, light photons stimulate the GPCR rhodopsin, which activates the associated heterotrimeric G protein, transducin. Gtα-GTP dissociates from the complex and activates cGMP phosphodiesterase, decreasing the intracellular concentration of cGMP. This in turn causes cGMP-gated cation channels to close, leading to cellular hyperpolarization and signal transmission. Long-term light adaptation of rod cells, which increases the amount of stimulation needed for a visual signal to be propagated, occurs at the G-protein level by the protein phosducin.

In its unphosphorylated form, phosducin binds tightly to the Gβγ heterodimer, blocking its interaction with Gα and sequestering it in the cytoplasm, away from the cell membrane. The 2.4-Å and 2.8-Å resolution structures of transducin βγ–phosducin complexes lend insights into how this process occurs (49, 50). Phosducin has two domains: an α-helical, N-terminal domain that binds to the base of the Gβγ heterodimer; and a C-terminal, thioredoxin-like domain. The N-terminal domain sterically blocks the Gβγ site where the Switch II region of Gα binds, while the thioredoxin-like domain, which possesses an overall negative charge, interacts favourably with the positively charged patches on β-blades 1 and 7, interfering with the putative membrane interaction region (Fig. 11). There is also evidence that the thioredoxin-like domain binds to the lipid modification on the Gγ-subunit. The N-terminal region thus appears to directly interfere with the formation of a Gα–Gβγ complex, while the C-terminal domain appears to function by sequestering the Gβγ-subunit away from the cell membrane. Phosducin also induces a conformational change in β-propeller blades six and seven, opening a cavity in between them. The electron density filling this space was interpreted as the farnesyl moiety of the vicinal C-terminus of the γ-subunit, providing clues as to how Gβγ is dissociated from the membrane (50). The phosphorylation site on phosducin, serine 73, lies within the

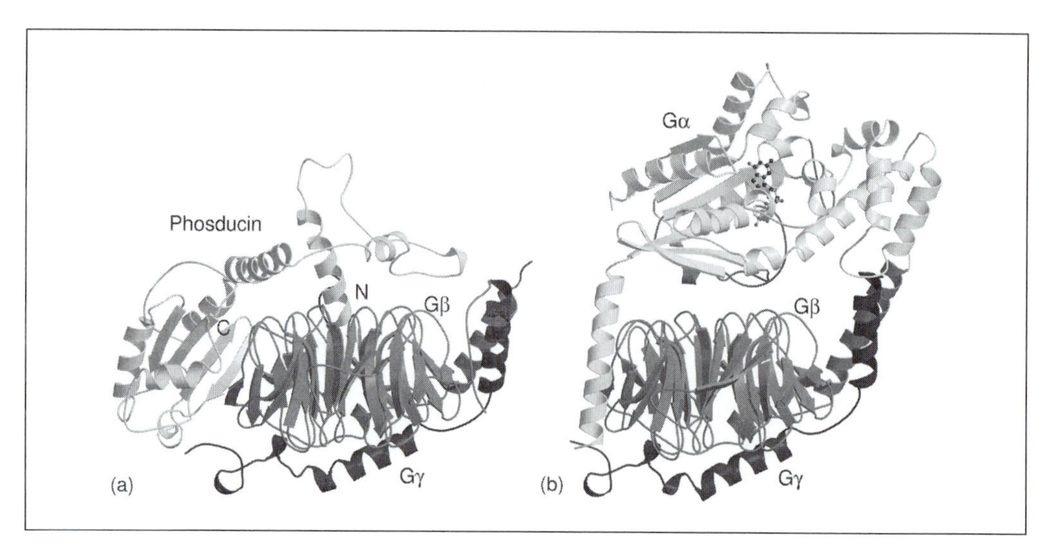

Fig. 11 Gβγ complexed with phosducin (PDB:2TRC). (a) A side view of the complex with phosducin in lighter grey and the Gβ- and Gγ-subunits shown in darker grey. (b) For comparison, the same view as in (a) of the heterotrimeric complex (1GOT). Shading as in Fig. 9.

N-terminal domain and points towards a large disordered portion of the structure. Phosphorylation of this residue is likely to cause conformational changes which decrease the affinity of phosducin for its target.

4. Cyclin-dependent kinases

Cyclin-dependent kinases (Cdks) are regulators of the cell cycle in eukaryotic cells. As their name implies, their function is dependent on proteins called cyclins, which they bind with high affinity. Cyclin binding to Cdk is required for the activation of these kinases, although full activity needs additional phosphorylation by Cdk-activating kinase (CAK) (127). Cdks are downregulated by the controlled degradation of cyclins, by interactions with other proteins, or by phosphorylation. The Cdk inhibitors (CdkIs) can be divided into two classes: the Cip/Kip family and the INK4 family. Other Cdk-interacting proteins are known, such as Cks1 and -2, but their function is still not clear.

Cdks are homologous to other protein kinases, having a bilobal structure consisting of an N-terminal β-stranded and a C-terminal α-helical domain. The catalytic site (and binding site for ATP) resides in a cleft between these domains. The geometry of the catalytic site, and consequently the activity of the enzyme, is affected by the relative orientation of the N- and C-terminal domains, a feature commonly employed in the regulation of protein kinases.

The structures of free and cyclin-bound Cdk2 are known, as are several complexes with inhibitors, and these will be discussed below (see ref. 128 for a more detailed discussion).

4.1 Cdk–cyclin interactions

The first step in the activation of most Cdks is binding of their activating cyclins. Crystal structures of free and cyclin A-bound Cdk2 illustrate the mechanism by which cyclin interactions activate Cdks (51, 129). The cyclins have two structurally related, α-helical domains which pack loosely against each other. The sequence motif that identifies cyclins, i.e. the cyclin box, corresponds to the N-terminal domain.

The cyclin A–Cdk2 interface is large, burying over 3500 Å^2. In the heart of the interface lies the so-called PSTAIRE helix (named after its sequence) of Cdk and α-helices 3, 4, and 5 of the cyclin (Fig. 12). The PSTAIRE helix is clamped by cyclin A forming extensive hydrophobic contacts with the central region, whereas the ends are involved in multiple hydrogen bonds. Both α-3 and α-4 are conserved between cyclins and similar interactions are likely to occur in other cyclin–Cdk complexes. Additional interactions in the complex involve both the N- and C-terminal domains of Cdk2. These are not, however, very extensive and are more likely to play a role in stabilizing the complex rather than determining the specificity of the interaction.

The structure of the cyclin does not change upon complexation but it seems to function as a template for the structural rearrangements in Cdk. The most important cyclin A-induced structural changes in Cdk2 involve the PSTAIRE helix and the so-called T loop (carrying the phosphorylated threonine 160). The PSTAIRE helix rotates and moves towards the catalytic site bringing residues involved in ATP binding together. At the same time it forces the N-terminal β-stranded domain out of its way. The T loop, which in free Cdk2 was blocking the substrate-binding site and restricting movements of the PSTAIRE helix, has now moved away and become more flexible. For full activity Cdk2 still needs to be phosphorylated on T160 by

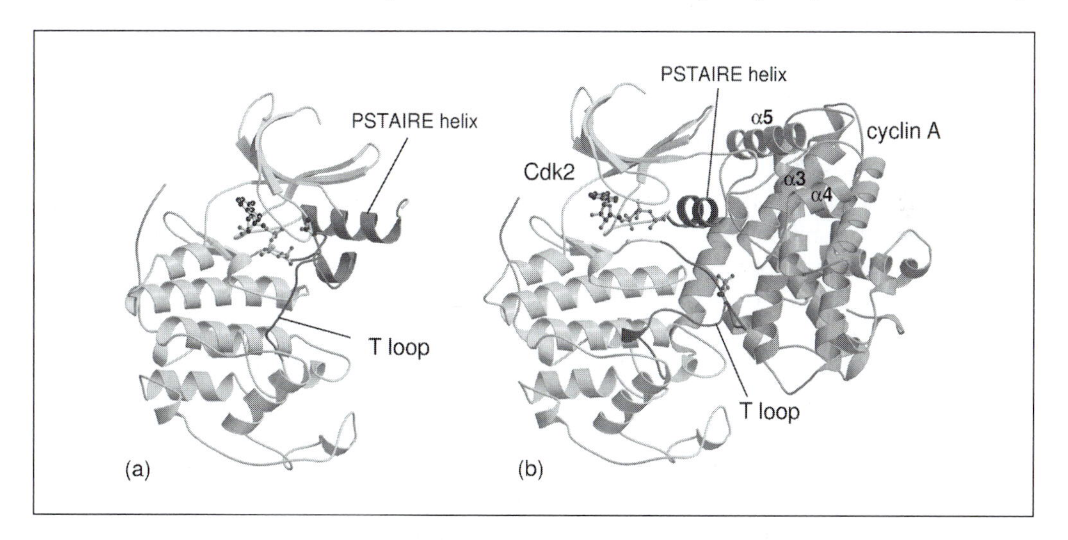

Fig. 12 Cdk2 (a) without (PDB:1HCK) and (b) with cyclin A (1JST). For these and other Cdk figures the C-terminal, α-helical domains were superimposed to illustrate the structural changes in the N-terminal domain. PSTAIRE helix and the T loop in Cdk2 are shown in dark grey. ATP molecules and, in (b), the phosphorylated threonine in the T loop are shown as ball-and-stick models.

CAK. In addition to activating the complex, phosphorylation can also increase the stability of Cdk–cyclin complex. Cyclin A binding to Cdk2 alters the conformation of the T loop and makes T160 accessible for phosphorylation. In the crystal structure of the phosphorylated Cdk2–cyclin A complex, the phosphate group acts as an organizing centre by coordinating three arginines from different parts of Cdk2: the PSTAIRE helix, the T loop, and the catalytic loop (54). These interactions stabilize this part of the kinase and induce conformational changes further away in the molecule. Phosphorylated T160 moves over 6 Å and pulls along the rest of the T loop, opening the catalytic site for a substrate. In addition, several residues in the T loop then interact with cyclin, possibly explaining the stabilizing effect that phosphorylation has on the complex.

4.2 Cdk interactions with inhibitors

There are several mechanisms for inhibiting Cdks, including the dephosphorylation of T160, phosphorylation of residues in the N-terminus, and the binding of inhibitory proteins of the Kip/Cip and INK4 families. Members of the Kip/Cip family have broad specificity and bind to Cdk–cyclin complexes, whereas the INK4 inhibitors are restricted to Cdk4/6 and bind kinases also in the absence of cyclins. The Kip/Cip family contains proteins of different sizes; these share a common 65-residue region in their N-terminal regions which are capable of interacting with Cdk–cyclins. A crystal structure of a 69-residue peptide of p27^{Kip1} (p27) complexed with phosphorylated (i.e. fully active) Cdk2–cyclin A shows that the inhibitor adopts an elongated structure which interacts both with the Cdk2 and the cyclin A (see Fig. 13a) (54). The interaction area is very large (5700 Å^2) due to the extended structure of the inhibitor. The peptide used in the structure determination is unstructured in solution and adopts its conformation upon binding to Cdk2–cyclin A.

p27 only interacts with the first half-domain (the cyclin box) of cyclin A. The N-terminus of p27 contains a fully conserved LFG sequence which binds as a rigid coil to a groove formed by conserved residues in helices-α1, -α3, and -α4 of cyclin A. The N-terminus of the coil forms multiple hydrogen bonds with cyclin A, followed by hydrophobic interactions of the LFG motif. These interactions do not impose structural changes on cyclin A. The coil is followed by an amphipathic α-helix which crosses from the surface of cyclin A to the surface of Cdk2.

The Cdk2–p27 interaction involves the N-terminal, β-stranded lobe and the ATP-binding site of the kinase. p27 clamps around the N-terminal lobe and induces structural changes. The amphipathic β-hairpin of p27 sandwiches with the β-sheet of Cdk2 and forms extensive van der Waals interactions. Next, the β-strand 1 of Cdk2 is replaced by p27 which forms a new antiparallel β-strand. The C-terminal end of the p27 peptide forms a 3_{10} helix and inserts into the catalytic cleft of the kinase, blocking the ATP-binding site. A tyrosine side chain mimics the purine base of adenosine and forms similar hydrogen bonds with Cdk2. Displacement of the first β-strand of Cdk2 induces large conformational changes in the N-terminal lobe that also affect the ATP-binding site. These conformational changes and blockade of the ATP-binding site are

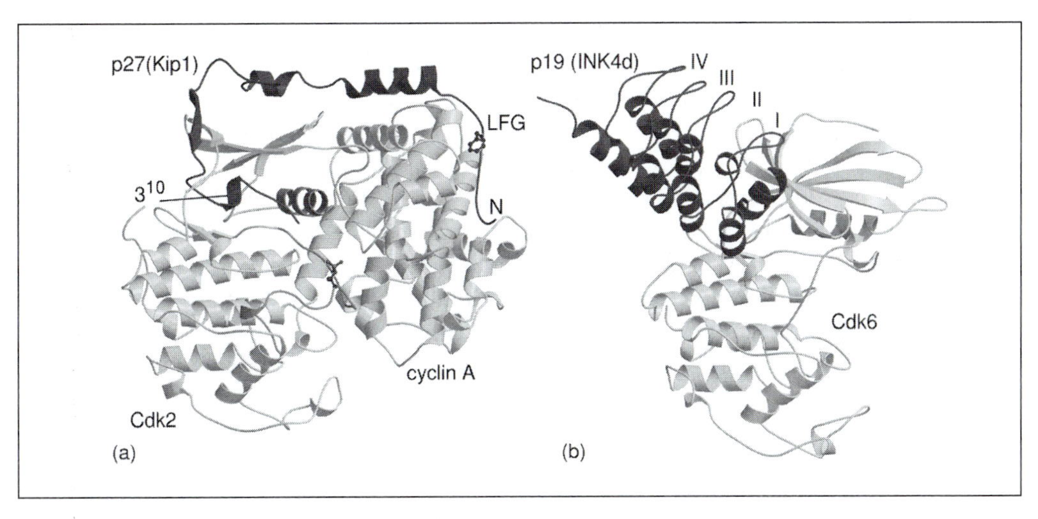

Fig. 13 Cdk2 interactions with inhibitors (a) p27 (PDB:1JSU) and (b) p19 (1BLX). In both cases the inhibitors are shaded in dark grey and Cdks and cyclin A in lighter grey. The LFG motif and the 3_{10} helix of p27 and the ankyrin repeats of p19 are labelled.

the mechanism whereby Cdk2 inhibition is achieved.

The INK4 family of Cdk inhibitors includes four members which are usually referred to by their molecular weight (p15, p16, p18, and p19). They consist of the so-called ankyrin repeats, i.e. small helix–loop–helix motifs connected together by short β-hairpins. p16 contains four and p19 five of these repeats. INK4s interact exclusively with Cdk4 and -6, either complexed with cyclin Ds or free. Inhibition of Cdk4/6 by the INK4 family blocks the cell cycle, and mutations both in p16 and Cdk4 have been associated with cancer.

Structures have been determined for both p16 and p19 bound to Cdk6 (55, 56). They show that INK4 proteins interact with both the N- and C-terminal lobes of Cdk6, on the opposite side to the binding site for cyclin D, and induce conformational changes affecting both the catalytic site and the PLSTIRE helix in the predicted cyclin D-binding site (see Fig. 13b). Three ankyrin repeats of p16 and p19 bind with helices-α1, -α3, and -α5 to the β-sheet of Cdk6 and with the connecting β-hairpins to the C-terminal domain of the kinase. The buried surface area is 1700 and 2200 $Å^2$ in p19 and p16, respectively. The interface is mostly polar, with a number of salt bridges which are believed to determine INK4s' specificity for Cdk4 and -6. Cancer-associated mutations both in p16 and Cdk4 map to interface residues, including several charged residues.

INK4 binding induces a change in the relative orientation of the N- and C-terminal domains (as compared to Cdk2, since no structure of free Cdk6 is available). This affects the orientation of the PLSTIRE helix that moves away from the catalytic site. In the p19 complex it rotates around the helical axis by 30°, but in a direction opposite to that induced by cyclin A in Cdk2. The T loop adopts different conformations in p16 and p19 complexes, possibly as a consequence of the crystallization conditions.

The conformational changes induced by INK4 molecules affect the cyclin-binding site and possibly prevent the binding of cyclin D at its normal site.

5. Conclusions

We have described some major classes of protein–protein interactions observed in intracellular signalling: small signalling modules recognizing linear epitopes, G proteins binding to various effectors and regulators in a nucleotide-dependent manner, and interactions of cyclin-dependent kinases with activators and inhibitors. Such protein–protein interactions are often transient and require careful control—mutations affecting the regulation of the system can have severe consequences, such as the induction of cancer. We now ask whether there are any special features that distinguish the signalling complexes from other protein complexes.

Linear epitope interactions often involve the binding of a polypeptide in a shallow groove on a globular protein; this allows accessibility of the polypeptide to the protein, reasonable recognition of residues in the continuous sequence, and optimization of interactions between the two proteins. The interacting residues are few in number so the relative contribution of each residue is likely to be high—in the ligands of SH2 domains the phosphotyrosine and up to three following residues are required for high-affinity interactions. Electrostatic interactions clearly determine the high-affinity and specificity of interactions with phosphopeptides, but they also play a major role in other cases. The SH3 domains form extensive van der Waals contacts with the proline-rich peptides, but the orientation of the ligand on the surface is determined by salt bridges between the domain and the peptide. The PDZ domains have evolved to recognize the C-terminal tails of their binding partners and the negative charge of the terminal carboxylate is essential for these interactions.

Interactions with linear peptide epitopes can be controlled in very different ways. The SH2 and PTB domains and 14–3–3 proteins recognize phosphorylated peptides, and their interactions with target proteins can be modulated by phosphorylation and dephosphorylation. Calmodulin binds to its targets in response to elevated intracellular calcium levels. But the regulation of domain–peptide interactions is not always so clear. Some may not be directly regulated, but instead form stable complexes. Alternatively, the regulation may be indirect, affecting the interactions of neighbouring domains and modulating the cooperative interactions in multiple domain complexes. The G proteins are molecular switches that change conformation when they are activated by exchanging the bound nucleotide from GDP to GTP. This is typically accomplished with the aid of nucleotide-exchange factors (GEFs) and requires a transient complex between the two components. The GTP induced changes on the surface of the molecules create binding sites for downstream effector molecules, that propagate the signals further. Other classes of molecules interact with G proteins, again to enhance their catalytic activity and thus facilitating GTP hydrolysis and effectively turning off the signal. The interactions are almost exclusively restricted to the conformationally sensitive areas of the molecules, emphasizing the importance of regulation.

The cyclin-dependent kinases are capable of interacting with several classes of molecules, most of which regulate Cdk activity. Whereas G proteins undergo conformational changes on nucleotide binding thereby creating interaction sites, conformational changes in Cdks are caused by the interacting proteins. Cyclin binding activates Cdks by positioning active-site residues correctly, whereas Kip and INK4 inhibitors do the opposite and disrupt the active site.

Many of the structures described in this review contain only the minimal interacting domains and additional contacts might be formed between full-length proteins. Their role is impossible to predict and further structural studies are needed to answer this question.

5.1 Interface-accessible surface area

In Fig. 14 we have plotted the interface-accessible surface area against the number of intramolecular hydrogen bonds per unit surface area for all the crystallographic structures of signalling complexes. In the left part of the plot cluster most of the adaptor complexes involving linear epitopes: the SH2, SH3, PTB, and PDZ domains. The interface surface area is typically around 1000 $Å^2$, well below the average values (between 1500 and 2000 $Å^2$) for heteromeric protein–protein complexes (130; see Chapter 1). The only clear exceptions are the two CaM–peptide complexes which bury rather large (>2000 $Å^2$) surfaces and have relatively small numbers of hydrogen bonds. SH3 domain complexes contain similarly small number of hydrogen bonds, but bury smaller surface areas. The phosphopeptide complexes cluster together due to the high number of hydrogen bonds between the domain and the phosphate group in the peptide. The surface areas are not, however, correlated with affinities, as complexes with similar sized interfacial areas can have very different affinities.

In the far right of the plot there are complexes that bury very large surface areas (>4000 $Å^2$). These include Gβγ dimers, phosducin complexed with Gβγ, and the Cdk2–cyclin A complexed with p27. These involve extensive interactions with elongated peptides that wrap around the other components and often form coiled-coils.

The complexes with intermediate buried surface areas include the G protein (Ras and Gα) complexes with effectors and regulators, and the cyclin–Cdk interactions. These are often transient interactions requiring careful regulation. A common feature of the signalling complexes in this category is the structural rearrangements that occur either before or during the complexation.

In all, the buried surface area between the interacting components can be used for the broad classification of signalling complexes. It alone can not, however, be used to distinguish high- and low-affinity complexes. Fine details of the contacts between the molecules determine the specificity of interactions. Mutagenesis studies have revealed that only a few side chains in the interface can contribute most of the binding energy, and that additional interactions surrounding these key residues play less important role (35, 117, 133). In oncogenic proteins, single point mutations can abolish interactions *in vivo* and subsequently cause cancer (56, 134).

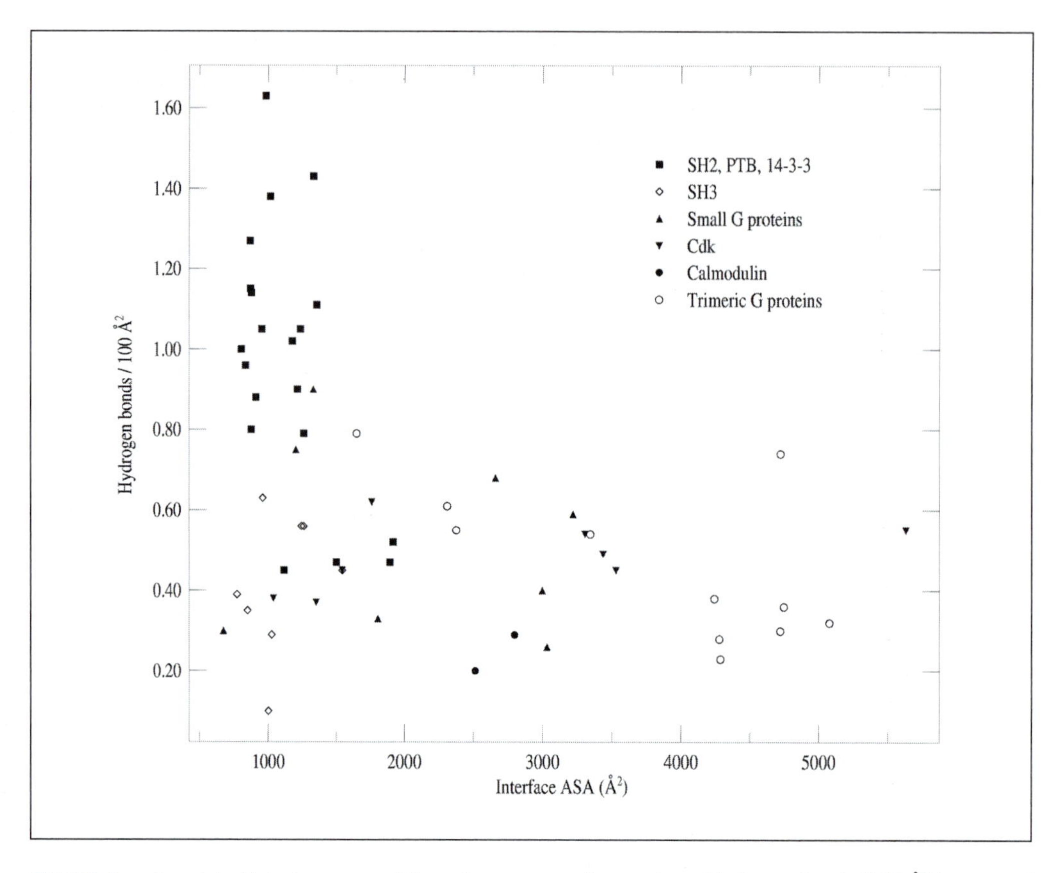

Fig. 14 A scatter plot of interface-accessible surface area vs. the number of hydrogen bonds/100 Å^2 for most of the signalling complexes. The interface ASA was calculated using the Grasp program (131) and the hydrogen bonds using the HBPLUS program (132).

We have discussed cooperativity of the interactions mainly in the context of the linear epitopes and multidomain proteins. The low affinity of the domain–peptide interactions can be compensated for by the formation of multiple interactions with the target protein, giving a subsequent increase in specificity. Cooperative interactions also play a role in larger complexes, cyclin-regulated Cdks are a fine example of this.

5.2 Effects on membrane localization

Very often the function of protein–protein interactions is to link several components of the pathway together for the duration of the signal. The correct subcellular localization of participating molecules is very important and relocalization is an essential part of the activation process. Extracellular signals are transmitted by transmembrane receptors, and the plasma membrane plays an important role as a meeting point for signalling proteins. Special adaptor proteins, such as Grb2 and Shc, have even evolved to serve this purpose. Direct membrane binding of proteins without transmembrane sequences is usually achieved by covalent lipid modification of the proteins, or with the aid of lipid-binding protein modules (135, 136). Although most of the structural work has been carried out with modified proteins lacking the lipid tails, many of them are modified *in vivo*: Ras, Src, and Gα are myristoylated, Gγ is prenylated. Lipidated proteins bound to the membrane surface are restricted in mobility to two dimensions on the plane of the membrane and are therefore more likely to interact with each other. The connecting membrane can be considered to be an additional interaction surface which stabilizes the complex. Further, the complex formation can stabilize the membrane interaction of the complex compared to the individual components. Interactions with a lipidated, membrane-bound protein can also recruit soluble proteins to the membrane. Activation of Raf kinase is achieved by membrane localization when Ras binds to the Ras-binding domain of Raf. Conversely, occlusion of a lipid, as in the phosducin–Gβγ complex, may decrease the affinity of a component for other key components and so serve to downregulate the transduction of signals.

Acknowledgements

We would like to thank all the structural biologists who made their coordinates available before they were released in the PDB. Marko Hyvönen is supported by a long-term fellowship from the European Molecular Biology Organization. Jake Begun is grateful for the support of the Winston Churchill Foundation. Tom Blundell is supported by the Wellcome Trust.

References

1. Berridge, M. (1993). Inositol trisphosphate and calcium signalling. *Nature*, **361**, 315–24.
2. Hunter, T. (1997). Oncoprotein networks. *Cell*, **88**, 333–46.
3. Bork, P., Schultz, J., and Ponting, C. P. (1997). Cytoplasmic signalling domains: the next generation. *Trends Biochem. Sci.*, **22**, 296–8.
4. Eck, M. J., Shoelson, S. E., and Harrison, S. C. (1993). Recognition of a high-affinity phosphotyrosyl peptide by the Src homology-2 domain of $p56^{lck}$. *Nature*, **362**, 87–91.
5. Tong, L., Warren, T. C., King, J., Betageri, R., Rose, J., and Jakes, S. (1996). Crystal structures of the humans $p56^{lck}$ SH2 domain in complex with two short phosphotyrosyl peptides at 1.0 Å and 1.8 Å resolution. *J. Mol. Biol.*, **256**, 601–10.
6. Eck, M. J., Atwell, S. K., Shoelson, S. E., and Harrison, S. C. (1994). Structure of the regulatory domains of the Src-family tyrosine kinase Lck. *Nature*, **368**, 764–9.
7. Waksman, G., Shoelson, S. E., Pant, N., Cowburn, D., and Kuriyan, J. (1993). Binding of a high affinity phosphotyrosyl peptide to the Src SH2 domain: crystal structures of the complexed and peptide-free forms. *Cell*, **72**, 779–90.
8. Waksman, G., Kominos, D., Robertson, S. C., Pant, N., Baltimore, D., Birge, R. B., Cowburn, D., Hanafusa, H., Mayer, B. J., Overduin, M., Resh, M. D., Rios, C. B., Silverman, L.,

and Kuriyan, J. (1992). Crystal structure of the phosphotyrosine recognition domain SH2 of the v-*src* complexed with tyrosine-phosphorylated peptides. *Nature*, **358**, 646–53.
9. Gilmer, T., Rodriguez, M., Jordan, S., Crosby, R., Alligood, K., Green, M., Kimery, M., Wagner, C., Kinder, D., and Charifson, P. (1994). Peptide inhibitors of src SH3–SH2–phosphoprotein interactions. *J. Biol. Chem.*, **269**, 31711–19.
10. Nolte, R. T., Eck, M. J., Sclessinger, J., Shoelson, S. E., and Harrison, S. C. (1996). Crystal structure of the PI 3-kinase p85 amino-terminal SH2 domain and its phosphopeptide complexes. *Nature Struct. Biol.*, **3**, 364–74.
11. Piccione, E., Case, R. D., Domchek, S. M., Hu, P., Chaudhuri, M., Baker, J. M., Sclessinger, J., and Shoelson, S. E. (1993). Phosphatidylinositol 3-kinase p85 SH2 domain specificity defined by direct phosphopeptide / SH2 domain binding. *Biochemistry*, **32**, 3197–202.
12. Rahuel, J., Gay, B., Erdmann, D., Strauss, A., Garcia-Echeverría, G., Furet, P., Caravatti, G., Fretz, H., Schoepfer, J., and Grütter, M. G. (1996). Structural basis for specificity of GRB2–SH2 revealed by a novel ligand binding mode. *Nature Struct. Biol.*, **3**, 586–9.
13. Lee, C.-H., Kominos, D., Jacques, S., Margolis, B., Schlessinger, J., Shoelson, S. E., and Kuriyan, J. (1994). Crystal structures of peptide complexes of the amino-terminal SH2 domain of the Syp tyrosine phosphatase. *Structure*, **2**, 423–38.
14. Hatada, M. H., Lu, X., Laird, E. R., Green, J., Morgenstern, J. P., Lou, M., Marr, C. S., Phillips, T. B., Ram, M. K., Theriault, K., Zoller, M. J., and Karas, J. L. (1995). Molecular basis for interaction of the protein tyrosine kinase ZAP-70 with the T-cell receptor. *Nature*, **377**, 32–8.
15. Eck, M. J., Pluskey, S., Trüb, T., Harrison, S. C., and Shoelson, S. E. (1996). Spatial constraints on the recognition of phosphoproteins by the tandem SH2 domains of the phosphatase SH-PTP2. *Nature*, **379**, 277–80.
16. Musacchio, A., Saraste, M., and Wilmanns, M. (1994). High-resolution crystal structures of tyrosine kinase SH3 domains complexed with proline-rich peptides. *Nature Struct. Biol.*, **1**, 546–51.
17. Viguera, A. R., Arrondo, J. L. R., Musacchio, A., Saraste, M., and Serrano, L. (1994). Characterization of the interaction of natural proline-rich peptides with five different SH3 domains. *Biochemistry*, **33**, 10925–33.
18. Wu, X., Knudsen, B., Feller, S. M., Zheng, J., Sali, A., Cowburn, D., Hanafusa, H., and Kuriyan, J. (1995). Structural basis for the specific interaction of lysine containing proline-rich peptides with the N-terminal SH3 domain of c-Crk. *Structure*, **3**, 215–26.
19. Pisabarro, M. T., Serrano, L., and Wilmanns, M. (1998). Crystal structure of the abl-SH3 domain complexed with a designed high-affinity peptide ligand: implications for SH3–ligand interactions. *J. Mol. Biol.*, **281**, 513–21.
20. Pisabarro, M. T. and Serrano, L. (1996). Rational design of specific high-affinity peptide ligands for the Abl–SH3 domain. *Biochemistry*, **35**, 10634–40.
21. Lim, W. A., Richards, F. M., and Fox, R. O. (1994). Structural determinants of peptide-binding orientation and of sequence specificity in SH3 domains. *Nature*, **372**, 375–9.
22. Lee, C.-H., Saksela, K., Mirza, U. A., Chait, B. T., and Kuriyan, J. (1996). Crystal structure of the conserved core of HIV Nef complexed with a Src family SH3 domain. *Cell*, **85**, 931–42.
23. Lee, C.-H., Leung, B., Lemmon, M. A., Zheng, J., Cowburn, D., Kuriyan, J., and Saksela, K. (1995). A single amino acid in the SH3 domain of Hck determines its high affinity and specificity in binding to HIV-1 Nef protein. *EMBO J.*, **14**, 5006–15.
24. Arold, S., Franken, P., Strub, M.-P., Hoh, F., Benichou, S., Benarous, R., and Dumas, C. (1997). The crystal structure of HIV-1 Nef protein bound to the Fyn kinase SH3

domain suggests a role for this complex in altered T cell receptor signalling. *Structure*, **5**, 1361–72.

25. Eck, M. J., Dhe-Paganon, S., Trub, T., Nolte, R. T., and Shoelson, S. E. (1996). Structure of the IRS-1 PTB domain bound to the juxtamembrane region of the insulin receptor. *Cell*, **85**, 695–705.
26. Zhou, M. M., Huang, B., Olejniczak, E. T., Meadows, R. P., Shuker, S. B., Miyazaki, M., Trub, T., Shoelson, S. E., and Fesik, S. W. (1996). Structural basis for IL-4 receptor phosphopeptide recognition by the IRS-1 PTB domain. *Nature Struct. Biol.*, **3**, 388–93.
27. Zhou, M. M., Ravichandran, K. S., Olejniczak, E. F., Petros, A. M., Meadows, R. P., Sattler, M., Harlan, J. E., Wade, W. S., Burakoff, S. J., and Fesik, S. W. (1995). Structure and ligand recognition of the phosphotyrosine binding domain of Shc. *Nature*, **378**, 584–92.
28. Zhang, Z., Lee, C.-H., Mandiyan, V., Borg, J.-P., Margolis, B., Schlessinger, J., and Kuriyan, J. (1997). Sequence-specific recognition of the internalization motif of the Alzheimer's amyloid precursor protein by the X11 PTB domain. *EMBO J.*, **16**, 6141–50.
29. Doyle, D. A., Lee, A., Lewis, J., Kim, E., Sheng, M., and MacKinnon, R. (1996). Crystal structures of a complexed and peptide-free membrane protein-binding domain: molecular basis of peptide recognition by PDZ. *Cell*, **85**, 1067–76.
30. Schultz, J., Hoffmüller, U., Krause, G., Ashurst, J., Macias, M. J., Schmieder, P., Schneider-Mergener, J., and Oschkinat, H. (1998). Specific interactions between the syntrophin PDZ domain and voltage-gated sodium channels. *Nature Struct. Biol.*, **5**, 19–24.
31. Meador, W. E., Means, A. R., and Quincho, F. A. (1992). Target enzyme recognition by calmodulin: 2.4 Å structure of a calmodulin-peptide complex. *Science*, **257**, 1251–5.
32. Meador, W. E., Means, A. R., and Quincho, F. A. (1993). Modulation of calmodulin plasticity in molecular recognition on the basis of X-ray structures. *Science*, **262**, 1718–21.
33. Yaffe, M. B., Rittinger, K., Volinia, S., Caron, P. R., Aitken, A., Leffers, H., Gamblin, S. J., Smerdon, S. J., and Cantley, L. C. (1998). The structural basis for 14–3–3: phosphopeptide binding specificity. *Cell*, **91**, 961–71.
34. Nassar, N., Horn, G., Herrmann, C., Scherer, A., McCormick, F., and Wittinghofer, A. (1995). The 2.2 Å crystal structure of the Ras binding domain of the serine/threonine kinase c-Raf1 in complex with Rap1A and a GTP analogue. *Nature*, **375**, 554–60.
35. Nassar, N., Horn, G., Herrmann, C., Block, C., Janknecht, R., and Wittinghofer, A. (1996). Ras/Rap effector specificity determined by charge reversal. *Nature Struct. Biol.*, **3**, 723–9.
36. Scheffzek, K., Ahmadian, M. R., Kabsch, W., Wiesmüller, L., Lautwein, A., Schmitz, F., and Wittinghofer, A. (1997). The Ras–RasGAP complex: structural basis for GTPase activation and its loss in oncogenic Ras mutants. *Science*, **277**, 333–8.
37. Huang, L., Hofer, F., Martin, G. S., and Kim, S.-H. (1998). Structural basis for the interaction of Ras with RalGDS. *Nature Struct. Biol.*, **5**, 422–6.
38. Vetter, I. R., Nowak, C., Nishimoto, T., and Wittinghofer, A. (1999). Structure of a Ran-binding domain complexed with Ran bound to a GTP analogue: implications for nuclear transport. *Nature*, **398**, 39–46.
39. Ostermeier, C. and Brünger, A. T. (1999). Crystal structure of the small G protein Rab3A complexed with the effector domain of Rabphilin-3A. *Cell*, **96**, 363–74.
40. Boriak-Sjodin, P. A., Margarit, S. M., Bar-Sagi, D., and Kuriyan, J. (1998). The structural basis of the activation of Ras by Sos. *Nature*, **394**, 337–43.
41. Goldberg, J. (1998). Structural basis for activation of ARF GTPase: mechanisms of guanine nucleotide exchange and GTP-myristoyl Switching. *Cell*, **95**, 237–48.
42. Rittinger, K., Walker, P. A., Eccleston, J. F., Nurmahomed, K., Owen, D., Laue, E.,

Gamblin, S. J., and Smerdon, S. J. (1997). Crystal structure of a small G protein in complex with the GTPase-activating protein rhoGAP. *Nature*, **388**, 693–7.

43. Rittinger, K., Walker, P. A., Eccleston, J. F., Smerdon, S. J., and Gamblin, S. J. (1997). Structure at 1.65 Å of RhoA and its GTPase-activating protein in complex with a transition-state analogue. *Nature*, **389**, 758–61.
44. Tesmer, J. J. G., Berman, D. M., Gilman, A. G., and Sprang, S. R. (1997). Structure of RSG4 bound to AlF_4^- activated Gα: stabilization of the transition state for GTP hydrolysis. *Cell*, **89**, 251–261.
45. Tesmer, J. J. G., Sunahara, R. K., Gilman, A. G., and Sprang, S. R. (1997). Crystal structure of the catalytic domains of adenylyl cyclase in a complex with Gα-GTPγS. *Science*, **278**, 1907–16.
46. Lambright, D. G., Noel, J. P., Hamm, H. E., and Sigler, P. B. (1994). Structural determinants for activation of the Gα-subunit of a heterotrimeric G protein. *Nature*, **369**, 621–8.
47. Wall, M. A., Coleman, D. E., Lee, E., Iñiguez-Lluhi, J. A., Posner, B. A., Gilman, A. G., and Sprang, S. R. (1995). The structure of the G protein heterotrimer Gαβγ *Cell*, **83**, 1047–58.
48. Sondek, J., Bohm, A., Lambright, D., Hamm, H., and Sigler, P. (1996). Crystal structure of a G-protein βγ dimer at 2.1 Å resolution. *Nature*, **379**, 369–74.
49. Gaudet, R., Bohm, A., and Sigler, P. B. (1996). Crystal structure at 2.4 Å resolution of the complex of transducin βγ and its regulator, phosducin. *Cell*, **87**, 577–88.
50. Loew, A., Ho, Y.-K., Blundell, T., and Bax, B. (1998). Phosducin induces a structural change in transducin βγ. *Structure*, **6**, 1007–19.
51. Jeffrey, P. D., Russo, A. A., Polyak, K., Gibbs, E., Hurwitz, J., Massagué, J., and Pavletich, N. P. (1995). Mechanism of Cdk activation revealed by the structure of a cyclinA–CDK2 complex. *Nature*, **376**, 313–20.
52. Russo, A. A., Jeffrey, P. D., Patten, A. K., Massagué, J., and Pavletich, N. P. (1996). Crystal structure of the $p27^{Kip1}$ cyclin-dependent kinase inhibitor bound to the cyclinA–Cdk2 complex. *Nature*, **382**, 325–31.
53. Bourne, Y., Watson, M. H., Hickey, M. J., Holmes, W., Rocque, W., Reed, S. I., and Tainer, J. A. (1996). Crystal structure and mutational analysis of the human CDK2 kinase complex with cell cycle-regulatory protein CksHs1. *Cell*, **84**, 863–74.
54. Russo, A. A., Jeffrey, P. D., and Pavletich, N. P. (1996). Structural basis of cyclin-dependent kinase activation by phosphorylation. *Nature Struct. Biol.*, **3**, 696–700.
55. Russo, A. A., Tong, L., Lee, J.-O., Jeffrey, P. D., and Pavletich, N. P. (1998). Structural basis for inhibition of the cyclin-dependent kinase Cdk6 by the tumor suppressor $p16^{INK4a}$. *Nature*, **395**, 237–43.
56. Brotherton, D. H., Dhanaraj, V., Wick, S., Brizuela, L., Domaille, P. J., Volyanik, E., Parisini, E., Smith, B. O., Archer, S. J., Serrano, M., Brenner, S. L., Blundell, T. L., and Laue, E. D. (1998). Crystal structure of the complex of the cyclin D-dependent kinase Cdk6 bound to the cell-cycle inhibitor $p19^{INK4d}$. *Nature*, **395**, 244–50.
57. Kussie, P. H., Gorina, S., Marechal, V., Elenbaas, B., Moreau, J., Levine, A. J., and Pavletich, N. P. (1996). Structure of the MDM2 oncoprotein bound to the P53 tumor suppressor transactivation domain. *Science*, **274**, 948–53.
58. Gorina, S. and Pavletich, N. (1996). Structure of the P53 tumor suppressor bound to the ankyrin and SH3 domains of 53BP2. *Science*, **274**, 1001–5.
59. Lee, J. O., Russo, A. A., and Pavletich, N. P. (1998). Structure of the retinoblastoma tumour-suppressor pocket domain bound to a peptide from HPV E7. *Nature*, **391**, 859–65.

60. Mayer, B. J., Hamaguchi, M., and Hanafusa, H. (1988). A novel oncogene with structural similarity to phospholipase C. *Nature*, **332**, 272–5.
61. Sadowski, I., Stone, J. C., and Pawson, T. (1986). A noncatalytical domain conserved among cytoplasmic protein-tyrosine kinases modifies the kinase function and transforming activity of Fujinami sarcoma virus P130$^{gag\text{-}fps}$. *Mol. Cell. Biol.*, **6**, 4396–408.
62. Superti-Furga, G. and Courtneidge, S. (1995). Structure–function relationships in Src family and related protein tyrosine kinases. *BioEssays*, **17**, 321–30.
63. Russell, R. B., Breed, J., and Barton, G. J. (1992). Conservation analysis and structure prediction of the SH2 family of phosphotyrosine binding domains. *FEBS Lett.*, **304**, 15–20.
64. Songyang, Z., Shoelson, S. E., Chauduri, M., Gish, G., Pawson, T., Haser, W. G., King, F., Roberts, T., Ratnofsky, S., Lechleider, R. J., Neel, B. G., Birge, R. B., Fajardo, J. E., Chou, M. M., Hanafusa, H., Schaffhausen, B., and Cantley, L. (1993). SH2 domains recognize specific phosphopeptide sequences. *Cell*, **72**, 767–78.
65. Songyang, Z., Shoelson, S. E., McGlade, J., Olivier, P., Pawson, T., Bustelo, X. R., Barbacid, M., Sabe, H., Hanafusa, H., Yi, T., Ren, R., Baltimore, D., Ratnofsky, S., Feldman, R. A., and Cantley, L. (1994). Specific motifs recognized by the SH2 domains of Csk, 3BP2, fps/fes, GRB-2, HCP, SHC, Syk, and Vav. *Mol. Cell. Biol.*, **14**, 2777–85.
66. Gay, B., Furet, P., García-Echeverría, C., Rahuel, J., Chêne, P., Fretz, H., Schoepfer, J., and Caravetti, G. (1997). Dual specificity of Src homology 2 domains for phosphotyrosine peptide ligands. *Biochemistry*, **36**, 5712–18.
67. Wallace, A. C., Laskowski, R. A., and Thornton, J. M. (1995). LIGPLOT: a program to generate schematic diagrams of protein–ligand interactions. *Protein Eng.*, **8**, 135–42.
68. Musacchio, A., Gibson, T., Lehto, V.-P., and Saraste, M. (1992). SH3—an abundant protein module in search for a function. *FEBS Lett.*, **307**, 55–61.
69. Musacchio, A., Wilmanns, M., and Saraste, M. (1994). Structure and function of the SH3 domain. *Progr. Biophys. Mol. Biol.*, **61**, 283–97.
70. Musacchio, A., Noble, M., Pauptit, R., Wierenga, R., and Saraste, M. (1992). Crystal structure of a Src-homology 3 (SH3). domain. *Nature*, **359**, 851–4.
71. Noble, M. E., Musacchio, A., Saraste, M., Courtneidge, S. A., and Wierenga, R. K. (1993). Crystal structure of the SH3 domain of human Fyn; comparison of the three-dimensional structures of SH3 domains in tyrosine kinases and spectrin. *EMBO J.*, **12**, 2617–24.
72. Yu, H., Rosen, M. K., Shin, T. B., Seidel-Dugan, C., Brugge, J. S., and Schreiber, S. L. (1992). Solution structure of the SH3 domain of Src and identification of its ligand-binding site. *Science*, **258**, 1665–8.
73. Cicchetti, C., Mayer, B. J., Thiel, G., and Baltimore, D. (1992). Identification of a protein that binds SH3 region of Abl and is similar to Bcr and GAP-Rho. *Science*, **257**, 803–6.
74. Ren, R., Mayer, B. J., Cicchetti, P., and Baltimore, D. (1993). Identification of a ten-amino acid proline rich SH3 binding site. *Science*, **259**, 1157–61.
75. Feng, S., Chen, J. K., Yu, H., Simon, J. A., and Schreiber, S. L. (1994). Two binding orientations for peptides to the Src SH3 domain: development of a general model for SH3–ligand interactions. *Science*, **266**, 1241–7.
76. Blaikie, P., Immanuel, D., Wu, J., Li, N., Yajnik, V., and Margolis, B. (1994). A region in Shc distinct from the SH2 domain can bind tyrosine-phosphorylated growth factor receptors. *J. Biol. Chem.*, **269**, 32031–4.
77. van der Geer, P., Wiley, S., Lai, V. K.-M., Olivier, J. P., Gish, G. D., Stephens, R., Kaplan, D., Shoelson, S., and Pawson, T. (1995). A conserved amino-terminal Shc domain binds to phosphotyrosine motifs in activated receptors and phosphopeptides. *Curr. Biol.*, **5**, 404–12.

78. Gustafson, T. A., He, W., Craparo, A., Schaub, C. D., and O'Neill, T. J. (1995). Phosphotyrosine-dependent interaction of SHC and insulin receptor substrate 1 with the NPEY motif of the insulin receptor via novel non-SH2 domain. *Mol. Cell. Biol.*, **15**, 2500–8.
79. Kavanaugh, W. M. and Williams, L. T. (1994). An alternative to SH2 domains for binding tyrosine-phosphorylated proteins. *Science*, **266**, 1862–5.
80. Kavanaugh, W. M., Turck, C. W., and Williams, L. T. (1995). PTB domain binding to signaling proteins through a sequence motif containing phosphotyrosine. *Science*, **268**, 1177–9.
81. Trüb, T., Choi, W. E., Wolf, G., Ottinger, E., Chen, Y., Weiss, M., and Shoelson, S. E. (1995). Specificity of the PTB domain of Shc for β-turn-forming pentapeptide motifs amino-terminal to phosphotyrosine. *J. Biol. Chem.*, **270**, 18205–8.
82. Zhou, M. M., Harlan, J. E., Wade, W. S., Crosby, S., Ravichandran, K. S., Burakoff, S. J., and Fesik, S. W. (1995). Binding affinities of tyrosine-phosphorylated peptides to the COOH-terminal SH2 and NH2-terminal phosphotyrosine binding domains of Shc. *J. Biol. Chem.*, **270**, 31119–23.
83. Charest, A., Wagner, J., Jacob, S., McGlade, C. J., and Tremblay, M. L. (1996). Phosphotyrosine-independent binding of SHC to the NPLH sequence of murine protein–tyrosine phosphatase–PEST. Evidence for extended phosphotyrosine binding/phosphotyrosine interaction domain recognition specificity. *J. Biol. Chem.*, **271**, 8424–49.
84. Borg, J. P., Ooi, J., Levy, E., and Margolis, B. (1996). The phosphotyrosine interaction domains of X11 and FE65 bind to distinct sites on the YENPTY motif of amyloid precursor protein. *Mol. Cell. Biol.*, **16**, 6229–41.
85. Ponting, C. P. and Phillips, C. (1995). DHR domains in syntrophins, neuronal NO synthases and other intracellular proteins. *Trends Biochem. Sci.*, **20**, 102–103.
86. Kennedy, M. B. (1995). Origin of PDZ (DHR, GLGF) domains. *Trends Biochem. Sci.*, **20**, 350.
87. Cabral, J. H. M., Petosa, C., Sutcliffe, M. J., Raza, S., Byron, O., Poy, F., Marfatia, S. M., Chishti, A. H., and Liddington, R. C. (1996). Crystal structure of a PDZ domain. *Nature*, **382**, 649–52.
88. Kim, E., Niethammer, M., Rothschild, A., Jan, Y. N., and Sheng, M. (1995). Clustering of Shaker-type K^+ channels by interaction with family of membrane associated guanylate kinases. *Nature*, **378**, 85–8.
89. Kornau, H.-C., Schenker, L. T., Kennedy, M. B., and Seeburg, P. H. (1995). Domain interaction between NMDA receptor subunits and the postsynaptic density protein PSD-95. *Science*, **269**, 1737–40.
90. Brenman, J. E., Chao, D. S., Ges, S. H., McGee, A. W., Craven, S. E., Santillano, D. R., Wu, Z., Huang, F., Xia, H., Peters, M. F., Froehner, S. C., and Bredt, D. S. (1996). Interaction of nitric oxide synthase with the postsynaptic density protein PSD-95 and α1-syntrophin mediated by PDZ domain. *Cell*, **84**, 757–67.
91. O'Neil, K. T. and Grado, W. F. (1990). How calmodulin binds its targets: sequence independent recognition of amphipatic α-helices. *Trends Biochem. Sci.*, **15**, 59–64.
92. Hof, P., Pluskey, S., Dhe-Paganon, S., Eck, M. J., and Shoelson, S. E. (1998). Crystal structure of the tyrosine phosphatase SHP-2. *Cell*, **92**, 441–50.
93. Sicheri, F., Moarefi, I., and Kuriyan, J. (1997). Crystal structure of the Src family tyrosine kinase Hck. *Nature*, **385**, 602–9.
94. Xu, W., Harrison, S. C., and Eck, M. J. (1997). Three dimensional structure of the tyrosine kinase c-Src. *Nature*, **385**, 595–601.
95. Williams, J. C., Weijland, A., Gonfloni, S., Thompson, A., Courtneidge, S. A., Superti-

Furga, G., and Wierenga, R. K. (1997). The 2.35 Å structure of the inactivated form of chicken Src: a dynamic molecule with multiple regulatory interactions. *J. Mol. Biol.*, **274**, 757–75.

96. Superti-Furga, G., Fumagalli, S., Koegl, M., Courtnidge, S. A., and Draetta, G. (1993). Csh inhibition of c-Src activity requires both the SH2 and SH3 domains of c-Src. *EMBO J.*, **12**, 2625–34.
97. Aitken, A. (1996). 14–3–3 and its possible role in co-ordinating multiple signalling pathways. *Trends Cell. Biol.*, **6**, 341–7.
98. Liu, D., Bienkowska, J., Petosa, C., Collier, R., Fu, H., and Liddington, R. (1995). Crystal structure of the zeta isoform of the 14–3–3 protein. *Nature*, **376**, 191–4.
99. Xiao, B., Smerdon, S. J., Jones, D. H., Dodson, G. G., Soneji, Y., Aitken, A., and Gamblin, S. J. (1995). Structure of a 14–3–3 protein and implications for coordination of multiple signalling pathways. *Nature*, **376**, 188–91.
100. la Cour, T. F. M., Nyborg, J., Thirup, S., and Clark, B. F. C. (1985). Structural details of the binding of guanosine diphosphate to elongation factor Tu from E. coli as studied by X-ray crystallography. *EMBO J.*, **4**, 2385–8.
101. Jurnak, F. (1985). Structure of the GDP domain of EF-Tu and location of the amino acids homologous to ras oncogene proteins. *Science*, **230**, 32–6.
102. Sprang, S. R. (1997). G protein mechanism: insights from structural analysis. *Annu. Rev. Biochem.*, **66**, 639–78.
103. Pai, E. F., Krengel, U., Petsko, G. A., Goody, R. S., Kabsch, W., and Wittinghofer, A. (1990). Refined crystal structure of the triphosphate conformation of H-ras p21 at 1.35 Å resolution: implications for the mechanism of GTP hydrolysis. *EMBO J.*, **9**, 2351–9.
104. Milburn, M. V., Tong, L., de Vos A. M., Brünger, A., Yamaizumi, Z., Nishimura, S., and Kim, S. H. (1990). Molecular Switch for signal transduction: structural differences between active and inactive forms of proto-oncogenic ras proteins. *Science*, **247**, 939–45.
105. Kraulis, P. J., Domaille, P. J., Campbell-Burk, S. L., Van Aken, T., and Laue, E. D. (1994). Solution structure and dynamics of Ras p21. GDP determined by heteronuclear three- and four-dimensional NMR spectroscopy. *Biochemistry*, **33**, 3515–31.
106. McCormick, F. and Wittinghofer, A. (1996). Interactions between Ras proteins and their effectors. *Curr. Opin. Biotech.*, **7**, 449–56.
107. Boguski, M. and McCormick, F. (1993). Proteins regulating Ras and its relatives. *Nature*, **366**, 643–54.
108. Quilliam, L. A., Khosravi-Far, R., Huff, S. Y., and Der, C. J. (1995). Guanine nucleotide exchange factors: activators of the Ras superfamily of proteins. *BioEssays*, **17**, 395–404.
109. Schlessinger, J. (1994). How receptor tyrosine kinases activate Ras. *Trends Biochem. Sci.*, **18**, 273–5.
110. Wittinghofer, A. and Nassar, N. (1996). How Ras-related proteins talk to their effectors. *Trends Biochem. Sci.*, **21**, 488–91.
111. Marshall, M. S. (1994). Ras target proteins in eukaryotic cells. *Faseb J.*, **9**, 1311–18.
112. McCormick, F. (1994). Activators and effectors of Ras p21 proteins. *Curr. Opin. Genet. Dev.*, **4**, 71–6.
113. Marshal, M. S. (1993). The effector interactions of p21ras. *Trends Biochem. Sci.*, **18**, 250–4.
114. Polakis, P. and McCormick, F. (1993). Structural requirements for the interaction of p21ras with GAP and exchange factors and and its biological effector target. *J. Biol. Chem.*, **268**, 9157–60.
115. Herrmann, C., Martin, G., and Wittinghofer, A. (1995). Quantitative analysis of the

complex between p21ras and the Ras-binding domain of the human Raf-1 protein kinase. *J. Biol. Chem.*, **270**, 2901–5.

116. Calés, C., Hancock, J. F., Marshall, C. J., and Hai, A. (1988). The cytoplasmic protein GAP is implicated as the target for regulation by the ras gene product. *Nature*, **332**, 548–51.
117. Block, C., Janknecht, R., Herrmann, C., Nassar, N., and Wittinghofer, A. (1996). Quantitative structure-activity analysis correlating Ras/Raf interaction *in vitro* to Raf activation *in vivo*. *Nature Struct. Biol.*, **3**, 244–51.
118. Herrmann, C., Horn, G., Spaargaren, M., and Wittinghofer, A. (1996). Differential interaction of the Ras family GTP-binding proteins H-Ras and Rap1A and R-Ras with the putative effector molecules Raf-kinase and Ral-guanine nucleotide exchange factor. *J. Biol. Chem.*, **271**, 6794–800.
119. Huang, L., Hofer, F., Martin, G. S., and Kim, S.-H. (1997). Three-dimensional structure of the Ras-interacting domain of RalGDS. *Nature Struct. Biol.*, **4**, 609–15.
120. Trahey, M. and McCormick, F. (1987). Biochemical and biological properties of the human N-ras p21 protein. *Mol. Cell. Biol.*, **7**, 541–4.
121. Mittal, R., Ahmadian, M. R., Goody, R. S., and Wittinghofer, A. (1996). Formation of a transition-state analog of the Ras GTPase reaction by RasGDP and tetrafluoroaluminate and GTPase-activating proteins. *Science*, **273**, 115–17.
122. Eccleston, J. F., Moore, K. J., Morgan, L., Skinner, R. H., and Lowe, P. N. (1993). Kinetics of interaction between normal and proline 12 Ras and the GTPase-activating proteins and p120-GAP and neurofibromin. The significance of the intrinsic GTPase rate in determining the transforming ability of ras. *J. Biol. Chem.*, **268**, 27012–19.
123. Neer, E. J. (1995). Heterotrimeric G proteins: organizers of transmembrane signals. *Cell*, **80**, 249–57.
124. Lambright, D. G., Sondek, J., Bohm, A., Skiba, N. P., Hamm, H. E., and Sigler, P. B. (1996). The 2.0 Å crystal structure of a heterotrimeric G protein. *Nature*, **379**, 311–19.
125. Bohm, A., Gaudet, R., and Sigler, P. B. (1997). Structural aspects of heterotrimeric G-protein signaling. *Curr. Opin. Biotech.*, **8**, 480–7.
126. Coleman, D. E., Berghuis, A. M., Lee, E., Linder, M. E., Gilman, A. G., and Sprang, S. (1994). Structures of active conformations of Giα1 and the mechanism of GTP hydrolysis. *Science*, **265**, 1405–12.
127. Nigg, E. A. (1995). Cyclin-dependent protein kinases: key regulators of the eukaryotic cell cycle. *BioEssays*, **17**, 471–80.
128. Morgan, D. O. (1997). Cyclin-dependent kinases: engines, clocks and microprocessors. *Annu. Rev. Cell. Dev. Biol.*, **13**, 261–91.
129. de Bondt, H. L., Rosenblatt, J., Jancarik, J., Jones, H. D., Morgan, D. O., and Kim, S. H. (1993). Crystal structure of cyclin-dependent kinase 2. *Nature*, **363**, 595–602.
130. Jones, S. and Thornton, J. M. (1996). Principles of protein–protein interaction. *Proc. Natl. Acad. Sci. USA*, **93**, 13–20.
131. Nicholls, A., Sharp, K. A., and Honig, B. (1991). Protein folding and association: insights from the interfacial and thermodynamic properties of hydrocarbons. *Proteins: Struct., Funct., Genet.*, **11**, 281–96.
132. McDonald, I. and Thornton, J. (1994). Satisfying hydrogen bonding potential in proteins. *J. Mol. Biol.*, **238**, 777–93.
133. Clackson, T. and Wells, J. A. (1995). A hot spot of binding energy in a hormone-receptor interface. *Science*, **267**, 383–6.

134. Shi, Y., Hata, A., Lo, R. S., Massagué, J., and Pavletich, N. P. (1997). A structural basis for mutational inactivation of the tumor suppressor Smad4. *Nature*, **388**, 87–93.
135. Bhatnagar, R. and Gordon, J. I. (1997). Understanding covalent modifications of proteins by lipids: where cell biology and biophysics mingle. *Trends Cell Biol.*, **7**, 14–20.
136. Rebecchi, M. (1998). Pleckstrin homology domains: a common fold with diverse functions. *Annu. Rev. Biochem.*, **27**, 503–28.
137. Ettmayer, P., France, D., Gounarides, J., Jarosinski, M., Martin, M. S., Rondeau, J. M., Sabio, M., Topiol, S., Weidmann, B., Zurini, M. and Bair, K. W. (1999). Structural and conformational requirements for high affinity binding to the SH2 domain g Grb2. *J. Med. Chem.* **42**, 971–80.

8 Interaction of standard mechanism, canonical protein inhibitors with serine proteinases

MICHAEL LASKOWSKI, Jr, M. A. QASIM, and STEPHEN M. LU

1. Introduction

Proteinases have a cradle-to-grave relationship with proteins. They assist in the birth of the proteins by removing the initiating methionine (Met) residues. They further aid in delivering them to an appropriate destination by removing the signal peptides. At death, they convert both exogenous proteins (food digestion) and endogenous proteins (protein turnover) to amino acids, which are then utilized for new protein synthesis. However, the most striking is the interaction of proteins with proteinases in their adult life. Protein processing turns numerous activities on or off. These activities in turn are responsible for a large range of biological phenomena such as blood clotting, clot dissolution, protein hormone action, emergence of the silk moth from cocoons, penetration of outer layers of ova by sperm, differentiation, cell death, and apoptosis (1). Along with so many important functions there are also many dangers associated with proteolysis. It must be rigidly controlled in time and place in order to be effective. Clotting from head to toe from a cut or digesting one's own pancreas rather than that of one's victim would be just a few untoward examples of uncontrolled proteolysis.

Nature has evolved two major mechanisms for control beyond the regulation of enzyme synthesis and breakdown. Most proteinases are biosynthesized as inactive, or only slightly active, precursors. Removal of the amino terminal region of these precursors is required for turning on full proteolytic activity. Because proteins are synthesized from the N-terminus to the C-terminus, the partially synthesized products are never active. The mechanisms of activation are very different for different enzymes (2). In some cases, a covalently attached inhibitor is released. This seems to merge into the second mechanism, although it appears that these inhibitors do not follow the standard mechanism and are not canonical (see Section 2.2).

The second major mechanism employed for the control of proteolysis is the ubiquitous presence of proteinase inhibitors in all known organisms and in various tissues of these organisms, sometimes in truly massive amounts. For example, avian ovomucoids are present at a concentration of 3×10^{-4} M in egg whites. The inhibitors inhibit their cognate proteinases through a large number of mechanisms. Since there are a wide variety of inhibitors and different types of proteinases, classification is needed to initiate a useful discussion. Hartley (3) grouped proteinases (endopeptidases) into four mechanistic classes: serine, cysteine, metallo and aspartyl proteinases. While this division has certainly stood the test of time, the huge number of newly discovered proteinases still keeps the classifiers busy (4). For our inhibitor classification (5) we retain the Hartley (3) categories. We thus divide all inhibitors into: (a) inhibitors devoid of significant class specificity and (b) class-specific inhibitors.

The inhibitors in the first group are all homologues of human α_2-macroglobulin, and, in our opinion, are not inhibitors at all but rather proteinase traps. α_2-Macroglobulin, with a mass of 720 kDa, is a homotetramer of four identical chains. These are associated pairwise by disulfide bridges into homodimers, the two covalent homodimers are strongly but non-covalently associated. Each chain contains a flexible and accessible bait region (6). The bait region has on it sequences that are specific for most common endoproteinases, e.g. Lys–X and Arg–X for trypsin, Trp–X, Tyr–X, and Phe–X for chymotrypsin, Glu–X for glutamic-acid specific enzymes, etc. Hydrolysis of any peptide bond in the bait region induces a major conformational change which aids in trapping the proteinase. The trapped proteinases do not have their active sites blocked, but the access of large substrates to the active site is cut off. Thus, the complexes still turn over small but not large substrates. The peptide-bond hydrolysis in the bait region also activates the α_2-macroglobulin's thioester bonds between the side chains of Glu (originally Gln) and Cys residues. These react with free ϵ-amino groups on the Lys side chains of the trapped enzyme. Thus, the trapping becomes covalent, but the active site of the enzyme still remains open (7). Because only hydrolysis of peptide bonds in the bait region is a requirement, it seems that α_2-macroglobulin-like inhibitors show little class-specificity or individual enzyme specificity. They inhibit enzymes from all four classes and of many different specificities within each class. However, they do not inhibit all proteinases. Strikingly, when different α_2-macroglobulins are compared, hypervariability is evident. The hypervariability of reactive sites of class-specific inhibitors is extensively discussed below.

Most of the predominantly class-specific inhibitors block the active site of the enzyme. There are two corollaries of this statement. In enzyme inhibitor complexes, all the enzymatic activity towards all substrates is lost. Some rather interesting exceptions to this rule (inhibition of tryptase and of thrombin) are discussed separately (see Section 2.19). Again, with just a few exceptions, all these inhibitors are strictly competitive. However, many workers new to this field conclude that most strong inhibitors are non-competitive. They inhibit about half of the enzyme activity by adding an appropriate amount of inhibitor and add substrate, but they do not note enzyme release from the complex. It was pointed out by N. M. Green (8) that unless a very long time is allowed in this test, the test is flawed. The dissociation rate constants of complexes are very small (for a discussion see below). Thus, after an addition of excess substrate, a long time must be allowed for a significant dissociation of the complex—in a few extreme cases, several weeks.

Laskowski and Kato (5) threw down a challenge that the huge number of predominantly class-specific inhibitors are absolutely class-specific. It was surprising that while many workers attempted to prove this wrong, and some even provided mistaken proofs, which were later overturned, it took more than a decade to be shown to be conclusively wrong. Komiyama *et al.* (9) found that CrmA—a serpin, but not a standard mechanism canonical inhibitor—with P_1 Asp at its reactive site, inhibits the interleukin converting enzyme (ICE), which is a cysteine proteinase specific for P_1 Asp. More recently, cysteine/serine proteinase cross-class specificity was extended to several other serpins (10).

It should also be noted that in the soybean trypsin inhibitor (STI) (Kunitz) family, some inhibitors are reported to inhibit both a serine proteinase and an aspartyl proteinase. Possibly more important and surprisingly little explored is the finding that SSI (the *Streptomyces* subtilisin inhibitor) strongly inhibits a metalloproteinase from *Streptomyces nigrescens* at a site overlapping the reactive site for serine proteinases (11). (To view a comprehensive list of standard mechanism inhibitors involved in other functions, visit our website at http://www.chem.purdue.edu/LASKOWSKI/)

In spite of several important exceptions, class specificity dominates for protein inhibitors other than the macroglobulin related ones. It still serves as a superb basis for nomenclature. We retain the Hartley classification and divide the non-macroglobulin inhibitors into protein inhibitors of serine, cysteine, metallo- and aspartyl proteinases.

We know of several examples of each kind, but the overwhelming majority of well-characterized inhibitors inhibit serine proteinases and only a few aspartyl proteinases. The last statement seems very surprising. A very large number of peptide-mimetic inhibitors for aspartyl proteinases were prepared and studied intensively. This work was driven by various medical applications, predominantly by the desire to inhibit HIV (human immunodeficiency virus) proteinase.

The huge number of well-characterized serine proteinase inhibitors is probably a consequence of the dominant importance of serine proteinases and of their inhibitors in life processes. However, a serious case could be made for an investigational bias. The first few protein proteinase (indeed protein enzyme) inhibitors to be described were all inhibitors of trypsin—a serine proteinase. For example, one of us (M. Laskowski, Jr) was a junior author of a major review (12) called, 'Naturally occurring trypsin inhibitors'. The title properly reflected the field at that time. Once the belief that many trypsin inhibitors were there to be found,[1] screening produced many more. Similar waves of discovery followed the finding of chymotrypsin and elastase inhibitors, etc. It is clear that high expectations of success drove many searches for inhibitors, and in most cases these high expectations existed for serine proteinases.

The workers in the field of trypsin inhibitors were fascinated by how inhibitors interact with enzymes. However, early studies gave very confusing results. Some of the inhibitors lost their activity on acetylation with acetic anhydride, others did not (14). This result was later easily explained by Ozawa and Laskowski (15), who suggested that some trypsin inhibitors have Lys–X and thus lose activity, while others have Arg–X reactive sites and do not. Jirgensons *et al.* (16) perceived that the inhibitors they studied interact with trypsin in an identical or closely similar manner. However, Jirgensons' principal interest was in protein conformation, which he studied by optical rotatory dispersion. He was troubled by his finding that chicken ovomucoid, bovine pancreatic trypsin inhibitor (Kunitz), and lima bean trypsin inhibitor all have strikingly different structures. This result was later explained when it became known that the three inhibitors Jirgensons studied are members of three different families: Kazal, Kunitz BPTI, and Bowman–Birk (Table 1, Fig. 1).

Table 1 The families of protein inhibitors of serine proteinases

Family	**Nearest preceding Cys**	**Nearest following Cys**	**Pro at P3′**[g]	**3D structure**		**Standard mechanism**	**Number of domains**	**Ref.**[j]
				Free	**Complex**			
				Standard mechanism inhibitors[i]				
Antistasin[a]	P_2	P'_3	No	1 SKZ	1 HIA	Yes	1, 2	17
Arrowhead[b]								
site 1	P_2	P'_{45}	No	None	None	Probably	1	18
site 2	—	P'_{13}	No					
Ascaris[a]	P_3	P'_2	No	1 ATA	1 EAI	Yes	1	19
BBI[b]	P_3	P'_6[f]	Yes[h]	1 PI2	1 SMF	Yes	2, 4	20
Chelonianin[a]	P_2	P'_8	No	2 REL	1 FLE	Yes	1, 2	21
Ecotin[c]	P_{35}	P'_3	No	1 ECY	1 AZZ	Yes	1	22
Grasshopper[a]	P_3	P'_3	No	1 PMC	None	Yes	1, 9	23
Kazal[a]	P_3	P'_6	No	2 OVO	1 CHO	Yes	1, 2, 3, 4, 7, 9, 15	24-26
Kunitz (BPTI)[a]	P_2	P'_{15}	No	1 BPI	2 PTC	Yes	1, 2, 3, 5	27, 28
Kunitz (STI)[b]	P_{25}	P'_{23}	No	1 BA7	1 AVW	Yes	1	29
Marinostatin[c, d]	—	—	Yes	None	None	Yes	1	30
Potato I[b, e]								
Eglin c	—	—	No	1 EGL	1 ACB	Yes	1	31
CMTI-V	P_{42}	P'_4	No	1 MIT	None	Yes	1	32
Potato II[b]	P_3	P'_2	No	1 TIH	4 SGB	Yes	1, 2, 3, 4, 6	33, 34
Cereal[b]	P_6	P'_{10}	No	1 B1U	None	Yes	1	35
Rapeseed[b]	P_4	P'_6	Yes	None	None	Yes	1	36
Silkworm[a]	P_3	P'_6	No	None	None	Yes	1	37
SSI[c]	P_3	P'_{28}	No	3 SSI	2 SIC	Yes	1	38
Squash[b]	P_3	P'_5	No	2 CTI	1 PPE	Yes	1	39

Other protein inhibitors of serine proteinases

Hirudin family[a]

Serpin family[a]

Streptomyces proteinase group propeptide family[c]

Subtilin clan propeptide family[c]

TAP family[a]

[a]Animal, [b]plant, [c]microorganism. [d]It is postulated that ester bonds play the role of disulfides in these molecules; [e]Potato I has members with one or no disulfide bonds; [f]some Bowman–Birk inhibitors have Cys at P'_8; [g]exceptions: in Kunitz (BPTI), one Pro/99; in Kazal, three Pro/456; in Ascaris, one Pro/10; in Kunitz (STI): seven Pro/29 and in BBI, one Ser/52; [h]P'_3 Pro is *cis* in all known BBI structures. [i]See Fig.1 for the disulfide bonding pattern in each family and the sequence around the reactive site for a member in each family. [j]Minimal number of references are given; they, or references therein, provide most of the information given here.

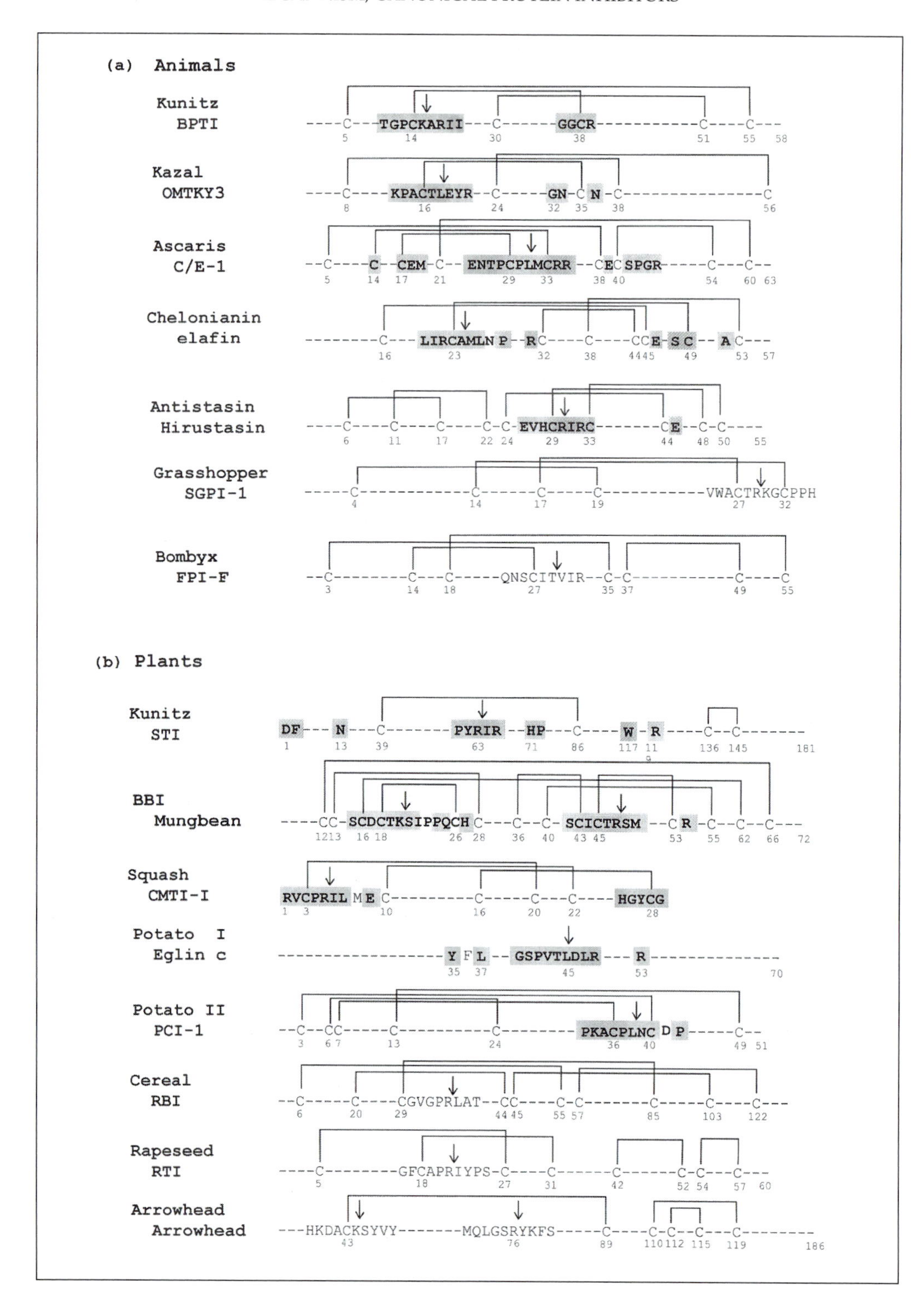
(a) Animals
Kunitz BPTI
Kazal OMTKY3
Ascaris C/E-1
Chelonianin elafin
Antistasin Hirustasin
Grasshopper SGPI-1
Bombyx FPI-F
(b) Plants
Kunitz STI
BBI Mungbean
Squash CMTI-I
Potato I Eglin c
Potato II PCI-1
Cereal RBI
Rapeseed RTI
Arrowhead Arrowhead

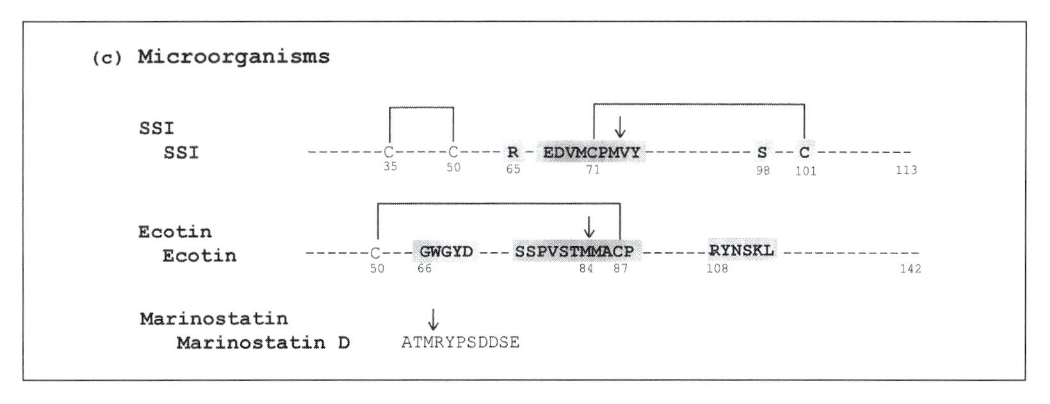

Fig. 1 Families of standard mechanism inhibitors. For each family a partial sequence of one of its member is shown. The arrow indicates the reactive-site peptide bond. The shaded residues are the consensus contact residues obtained from the X-ray structures of inhibitor–serine proteinase complexes in each family.

Because the protein inhibitors of serine proteinases are the most and best studied, we have limited this chapter only to those inhibitors. Even that group is very diverse and within it we select the standard mechanism, canonical inhibitors for detailed discussion.

2. Standard mechanism inhibitors

2.1 Historical

The inhibitors belonging to this group often exhibit extremely large values for the association equilibrium constant K_a: $K_a = [C]/[E][I]$, where C is the inactive complex, E is free enzyme, and I is free inhibitor. Classically, the largest value of K_a is that for the interaction of bovine β-trypsin with bovine pancreatic trypsin inhibitor (40, 41). The measurements were indirect and since the time they were made, they have often been quoted but seldom, if ever, repeated. It turns out that many other systems show very large equilibrium constants. For example, *Streptomyces griseus* proteinases A and B interact with eglin c with K_a values greater than 10^{13} M^{-1} (42). Constants larger than 10^{11} M^{-1} are quite common (see collections of natural K_a values in references 43–45). Such values are still daunting to many investigators.

Wishing to devise a facile technique for the measurement of large K_a values, Lebowitz and Laskowski (46) realized that for most associations between enzyme and inhibitor, the equilibrium constants sharply decline with pH at pH values lower than neutral (Fig. 2).

According to the Le Chatelier principle, this implies that in the reaction:

$$E + I \underset{k_{off}}{\overset{k_{on}}{\rightleftharpoons}} C + \bar{q}\, H^{+}; \qquad [1]$$

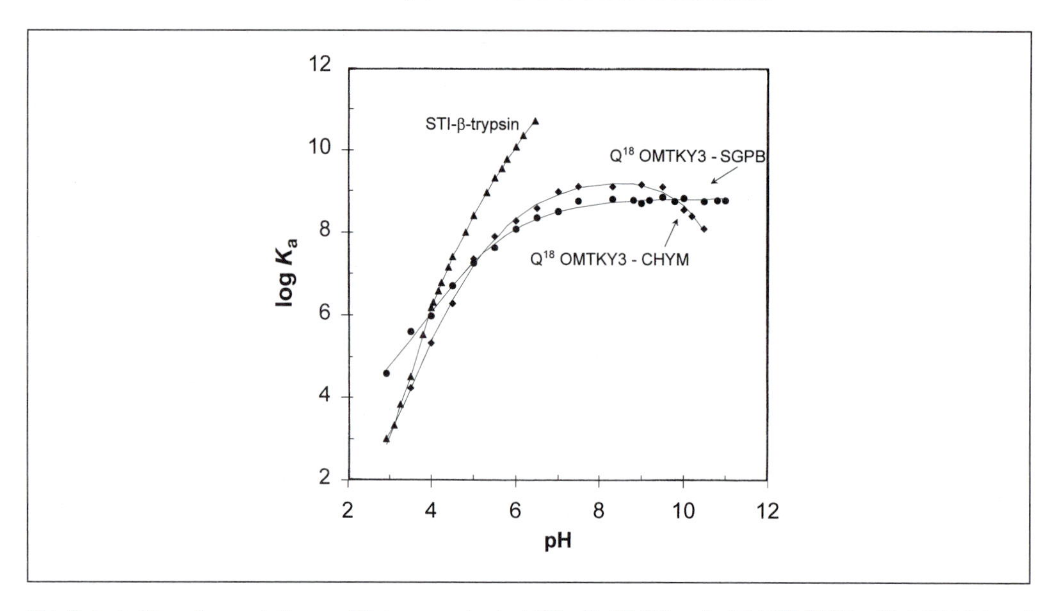

Fig. 2 Logarithm of association equilibrium constant of STI with BT (▲) and of Q18OMTKY3 with SGPB (●) and CHYM (◆) as a function of pH. STI–BT data and Q18OMTKY3–SGPB data are taken from refs 41 and 47, while Q18OMTKY3–CHYM are the unpublished results from this lab. (Qasim and Laskowski, unpublished.)

$\bar{q}$, the average number of protons released per mole of complex formed, is large and positive. The quantitative statement of the Le Chatelier principle in this system (46) is:

$$\frac{\mathrm{dlog}\, K_{\mathrm{a}}}{\mathrm{dpH}} = -\bar{q}. \qquad [2]$$

Wyman (48) and Tanford (49) who are often credited for this equation, published it later.

$$\log K_{\mathrm{a\ pH_2}} = \log K_{\mathrm{a\ pH_1}} + \int_{\mathrm{pH_1}}^{\mathrm{pH_2}} \bar{q}\ \mathrm{dpH}. \qquad [3]$$

It is regarded as a special case of linkage relations. From the above equations, it follows that, eqn 3 can be used to measure very large $K_{\mathrm{apH_2}}$ by utilizing the much smaller and easier to measure $K_{\mathrm{apH_1}}$. All that is needed is the average value of protons released upon complex formation as a function of pH over the $\mathrm{pH_1}$ to $\mathrm{pH_2}$ range. When this technique was applied to the STI (Kunitz)/β-trypsin interaction (46) and to the BPTI (Kunitz)/β-trypsin interaction (41), the results were found to be highly satisfactory. It is still a superb technique for the measurement of high K_{a} values, but is indirect and requires large amounts of material.

When enzyme and inhibitor are mixed at a constant pH the pH drops. Titrating back to the original pH determines $\bar{q}$. However, below pH 4.0, a very rapid pH drop was reproducibly followed by a much slower pH rise, which decreased but did not eliminate the total drop (46). An analysis of the results showed that more than an equilibrium amount of complex was formed and this then relaxed toward equilibrium. The complex could dissociate (50, 51) to the enzyme and to the inhibitor in one of the two forms: virgin, denoted by I and modified, denoted by I* (Fig. 3).

$$E + I \underset{k_{off}}{\overset{k_{on}}{\rightleftharpoons}} C \underset{k^{*}_{on}}{\overset{k^{*}_{off}}{\rightleftharpoons}} E + I^{*}. \quad [4]$$

The same mechanism was shown to apply to the inhibition of trypsin by chicken ovomucoid. I* differs from I by having one specific peptide bond hydrolysed (15, 50). In the case of STI, the bond was Arg63–Ile, in the case of chicken ovomucoid, it was Arg89–Ala. In both of these examples the P_1 residues were Arg, consistent with trypsin specificity and consistent with the earlier finding that neither of them was inactivated by acetylation. Other trypsin inhibitors, where P_1 was Lys, were inactivated by acetylation. All was consistent with inhibitors being substrate analogues, in which the enzyme–inhibitor complex was especially stable.

The modified inhibitors, I*, could be readily prepared by hydrolysis at low pH. Such modified inhibitors differ from virgin inhibitors in a variety of ways.

- The MW of I* is 18 daltons higher than that of I, as one internal peptide bond in I is hydrolysed in I*.
- I generally consists of a single polypeptide chain, while I* consists of two fragments held together either (rarely) by non-covalent forces or (frequently) by disulfide bridges.
- I* and I have approximately the same charge in the neutral pH range. However, at low pH, I* is more positive by one charge. At high pH, it is one charge more negative. This behaviour is consistent with the presence of a split peptide bond in a modified inhibitor (Fig. 3). At either low or at high pH, the ratio of I to I* can be accurately determined by either electrophoretic (54) or by ion-exchange chromatographic (55) techniques.
- With the exception of P_1 Lys trypsin inhibitors, acylation does not inactivate I. On the other hand, I* is always inactivated by acylation (56, 57). This is because the reactive-site peptide bond can no longer be re-formed when the newly formed NH_2 terminal is blocked by acylation.
- At neutral pH, subtilisin Carlsberg reacts with virgin and modified inhibitors at approximately equal rates, $k_{on} \approx k^{*}_{on}$ (see eqn 4). On the other hand, bovine chymotrypsin A-α, CHYM, reacts 10^6 times faster with I than with I* (58). Bigler *et al.* (59) used this to determine the total concentration of I and I* by rapid reaction with subtilisin Carlsberg and the concentration of I by rapid reaction with CHYM. It should be stressed that when long times (days) are allowed, CHYM reacts approximately equally with I and I*.

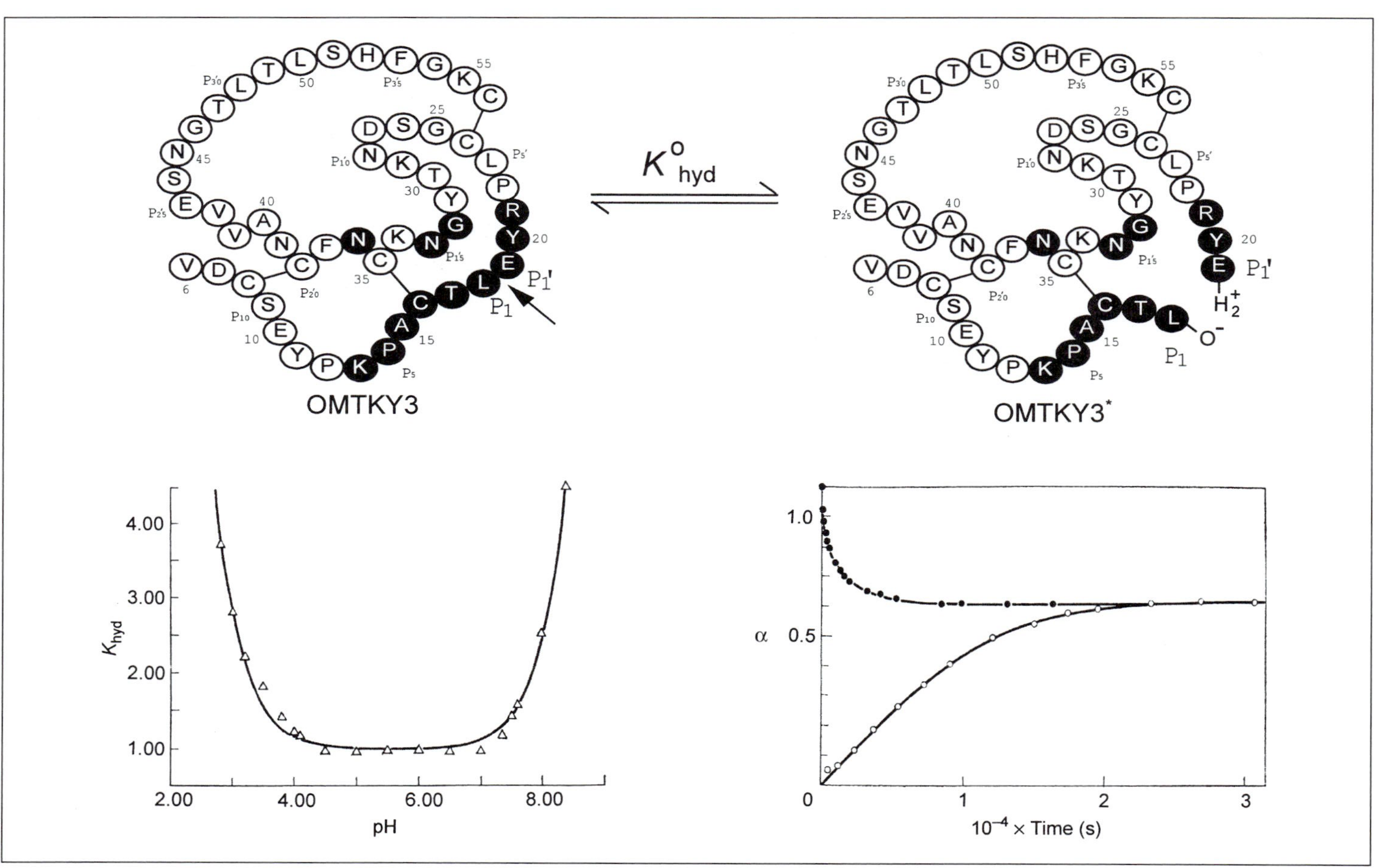

Fig. 3 The covalent structure of virgin (upper left) and of modified (upper right) turkey ovomucoid third domain. The three disulfide bridges are indicated by solid lines. The hydrolysis of the reactive-site peptide bond (indicated by an arrow on upper left) does not lead to the separation of fragments as they are held together by two disulfides. The residues are numbered from the NH_2 terminus of the first third domain whose structure was determined (52) and by Schechter and Berger notation (53) where residues following the reactive-site peptide bond are called P_1', P_2',...P_n' and those preceding it P_1, P_2,...P_n. The consensus set of 12 contact residues which form contacts in complexes with SGPB, CHYM, and HLE are highlighted in black. At lower left the pH-dependence of K_{hyd} (54, 55) is shown. The rise at its left limb is due to protonation of COO^- of P_1Leu. The rise at its right limb is due to deprotonation of NH_3^+ of P_1'Glu. Lower right shows how single points in lower left were obtained. This is the approach to equilibrium catalysed by asperogillopeptidase B at pH 7.6. The variable α is the fraction of virgin inhibitor, and $K_{hyd} = \alpha_{eq}/1-\alpha_{eq}$.

- The melting temperature, T_m, is always lower for I* than for I. This is consistent with a near unity value of K_{hyd}, the equilibrium constant for reactive-site, peptide bond hydrolysis (60–63)
- The three-dimensional structures of virgin ovomucoid third domain from Japanese quail (52, 64) and from silver pheasant (65) were determined by X-ray crystallography and can be compared to the structures of modified inhibitors from the same species (66). The three-dimensional structures of virgin and modified inhibitors are very similar and the newly formed COO^- and ^+H_3N-termini are close to one another. We believe that the results will be generally applicable to all standard mechanism, canonical inhibitors. On the other hand, reactive-site hydrolysis of serpins leads to a very large conformational change (67).

Most protein chemists believe that proteolysis of single peptide bonds in globular proteins always goes to completion or is irreversible (1). This mistaken view suggests that in the mechanism of eqn 4, $k^*_{on} = 0$. Such a mechanism would be strictly analogous to the reaction of trypsins with their designed burst substrate—NPGB, *p*-nitrophenylguanidinobenzoate (68). In the burst phase, trypsin is acylated to inactive guanidinobenzoate–trypsin. This deacylates very slowly to active trypsin and guanidinobenzoic acid. The latter is not an effective inhibitor of trypsin. Such is also the case in serpins. However, for standard mechanism inhibitors, the modified inhibitor, I*, reacts with the enzyme and inhibits it completely as $k^*_{on} > 0$ and $K_{hyd} \sim 1$.

The large differences between the properties of I* and I suggest to many, that stable complexes made from enzyme and I, which we have already called C, and made from enzyme and I*, provisionally called C* differ. They do not. They are in fact the same substance. They cannot be separated either by electrophoresis or by ion-exchange chromatography. Instead, a single sharp band is observed (62). C and C* made with most serine proteinases yield predominantly I, not I*, upon kinetically controlled dissociation of the complexes at low pH (58). The most impressive proof of the identity was the experiment of Miyaki and Kainosho (69). They labelled the carbon atom with ^{13}C and the nitrogen atom with ^{15}N in the reactive site of *Streptomyces* subtilisin inhibitor, SSI. The double-labelled SSI was enzymatically converted to modified inhibitor SSI*. The conversion could be readily seen by a 5.56 ppm downfield shift in the ^{13}C NMR spectrum of the reactive-site carbon. Complexes with subtilisin BPN′ were made both from virgin SSI and from modified SSI*. The entire NMR spectra were indistinguishable. The ^{13}C, ^{15}N splitting indicates an intact reactive-site peptide bond in both complexes. It is a major assertion of the standard mechanism that C and C* are the same substance. The other is the substrate-like nature of the inhibitor.

2.2 Canonical inhibitors

The term canonical inhibitors was introduced (70) for the subset of inhibitors which share a common three-dimensional structure of residues in the exposed loop surrounding the reactive site (71). In a gross formulation, the residues in the loop have a β-strand configuration, although the P_1 residue has the 3_{10} helix conformation. Upon complex formation (Fig. 4) this loop forms a two-stranded β-sheet with serine proteinases of the chymotrypsin family (the S1, S2, and S29 families of the SA clan (4)) and a triple-stranded antiparallel β-sheet (72) with the enzymes of clan SB, family S8, the subtilisin family. Other classifiers (73) call this broad group the subtilin clan in order to retain the subtilisin name for a family within the clan. Note in Fig. 4 that the hydrogen-bonding pattern between the enzyme and the inhibitor's residues P_3, P_1, and P_2' is common for the interaction with the SA and SB clan members. The SB clan members make the additional interactions involving the P_4 and P_2 residues of the inhibitor to the third β-sheet strand contributed by the enzymes. In all these cases, the dominant interaction is that of the O atom of P_1 with the oxyanion hole of the enzyme. In the chymotrypsin clan, SA, this consists of the main-chain –NH of Ser195 and Gly193. In the subtilin clan, SB, it consists of the main-chain –NH of

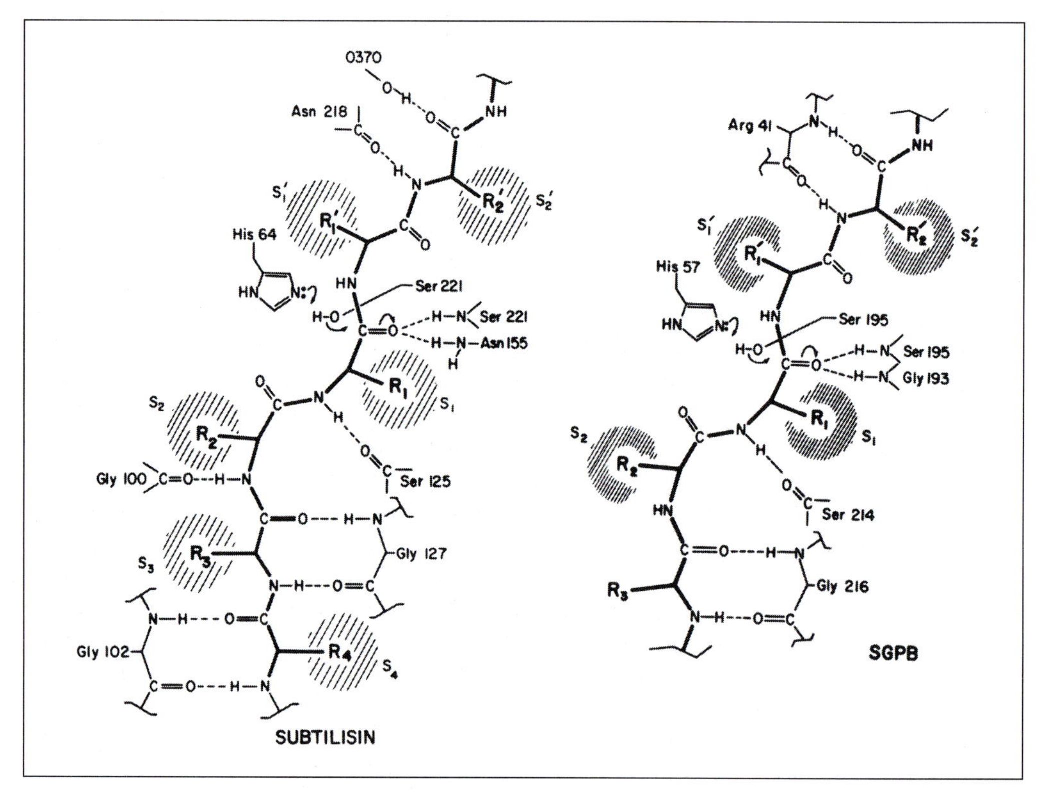

Fig. 4 Diagrammatic representation of the interaction of the reactive site of canonical inhibitors with the active sites of subtilisin Carlsberg (left) and SGPB (adapted from ref. 72). The differences in interaction exemplify the differences between subtilin (SB) and chymotrypsin (SA) clan enzymes. In both cases, an imperfect antiparallel β-structure is formed between the enzyme and the inhibitor. The hydrogen bonds between P_3, P_1, and P_2' residue of the inhibitor and the enzyme are common in both SA and SB clan enzymes. In SB clan enzymes additional hydrogen bonds are formed involving P_2 and P_4 residues of the inhibitor.

Ser221 (functional equivalent of Ser195 in chymotrypsins) and the side chain –NH of Asn155.

The need to make the common β-sheet conformations in the complexes with minimal conformational change dictates that the loop in the free inhibitor has an appropriate conformation. The canonical hypothesis can be broken down into several subhypotheses.

- Within one inhibitor family, the Ramachandran ϕ and ψ angles of all the canonical residues are identical for all family members.
- For any two canonical inhibitors, the Ramachandran ϕ and ψ angles of all the canonical residues are identical, in spite of the fact that the remainder of the conformation may be entirely different (Plate 12) (42).
- The ϕ and ψ angles do not change upon complex formation.
- (Corollary) In complexes of one inhibitor (e.g. turkey ovomucoid third domain) with all its cognate enzymes, the canonical ϕ and ψ angles are the same.

The canonical behaviour is illustrated in Table 2 for three especially well-studied families.

Two questions are immediately raised. The first is which are the canonical residues. It was common to talk about P_4 to P'_3. However, a more detailed analysis (70) suggests that this is too broad and P_3 to P'_3 is better. The finding that the P'_3 Pro is *cis* in Bowman–Birk family inhibitors (75) suggests that this residue should also be omitted. Besides the Bowman–Birk family there are only two other families which have a Pro at P'_3. They are rapeseed and marinostatin. Very few inhibitors are known to belong to these two families and their three-dimensional structures are not available. It appears that P'_3 Pro is good for Bowman–Birk inhibitors and presumably for rapeseed and marinostatin families. Table 1 shows that P'_3 is (almost) never Pro in 15 of the 18 inhibitor families. The P'_3 residue is always *trans* in these inhibitors. In the turkey ovomucoid third domain, replacement of P'_3 Arg by any residue other than Pro has only a relatively small effect on K_a except for CHYM. However, replacement of P'_3 by Pro drops the K_a by at least four orders of magnitude for all the six enzymes studied (45). The extension of canonical residues to P'_3 seems suspect if the Bowman–Birk family is to be included. Even if we select the region from P_3 to P'_2, some exceptions would still exist. For example, the BPTI family is somewhat exceptional (70). The P_3 ϕ and ψ angles in this inhibitor differ from the consensus. It appears that all inhibitors that obey the standard mechanism are canonical.

An important corollary of the canonical hypothesis is that a single mutation of any of the canonical residues will not shift the position of P_1. This was shown by many chemical experiments (58), but most directly by X-ray crystallography—10 P_1 BPTI variants all bind to the same S_1 pocket of trypsin (76); 17 P_1 OMTKY3 variants bind to the same S_1 pocket of SGPB (77, 78, and Huang, Bateman, James, Lu, Anderson, and Laskowski, unpublished). In both cases, the best-binding and the worst-binding P_1 residues are included in the sets that were studied (79).

Table 2 The main-chain angles (ϕ, ψ) of residues surrounding the reactive-site peptide bond

Inhibitor	Enzyme	P4 ϕ	P4 ψ	P3 ϕ	P3 ψ	P2 ϕ	P2 ψ	P1 ϕ	P1 ψ	P1′ ϕ	P1′ ψ	P2′ ϕ	P2′ ψ	P3′ ϕ	P3′ ψ	PDB entry
Kazal																
OMSVP3	free	–149	163	–131	155	–87	174	–96	9	–58	139	–99	93	–130	69	2ovo
OMTKY3	chymotrypsin	–129	136	–131	150	–68	160	–107	32	–74	159	–113	107	–142	76	1cho
OMTKY3	SGPB	–158	157	–126	147	–69	162	–119	45	–84	155	–100	114	–148	92	3sgb
OMTKY3	HLE	–127	148	–119	143	–73	156	–101	31	–84	155	–102	107	–138	82	1ppf
BPTI-Kunitz																
BPTI	free	81	–178	–93	–3	–87	162	–115	19	–64	162	–121	90	–106	118	1bpi
BPTI	β–trypsin	79	175	–77	–29	–70	156	–117	39	–88	164	–113	79	–106	122	2ptc
BPTI	kallikrein	77	179	–64	–32	–91	168	–112	46	–88	167	–115	84	–110	108	2kai
BPTI	chymotrypsin	99	172	–84	–16	–69	154	–103	21	–81	–169	–150	79	–104	134	1cbw
Potato-I																
eglin c	free	–71	153	–132	163	–74	150	–82	12	–99	159	–111	128	–129	123	Ref. (74)
eglin c	chymotrypsin	–81	156	–144	166	–79	153	–97	40	–90	166	–131	121	–129	129	1acb
eglin c	subtilisin Carlsberg	–71	140	–139	168	–62	143	–115	45	–97	169	–117	110	–121	112	1cse

2.3 Specificity of inhibitors

It is often said that standard mechanism, canonical inhibitors of serine proteinases are highly specific. This is not true, unless what is meant is the remarkably strong discrimination for inhibition of serine proteinases against the inhibition of other enzymes.

2.3.1 Many inhibitors inhibit the same serine proteinase

Among the 18 standard mechanism, canonical families, 16 contain at least one reported strong inhibitor of trypsin. The two families in which no trypsin inhibitors are known, Bombyx and marinostatin, consist of just one member each. It is quite likely that some of the as yet undiscovered members of these two families are also trypsin inhibitors. CHYM is also inhibited by at least one member of the majority of the 18 families. It is difficult to make similar statements for other enzymes as relatively few workers check for the inhibition of PPE, HLE, SGPA, and SGPB by newly discovered inhibitors.

2.3.2 Many serine proteinases are inhibited by the same inhibitor

The notion of the broad specificity of many inhibitors is quite old. In the classic book by Vogel, Trautschold, and Werle (80), BPTI-Kunitz is referred to as a polyvalent inhibitor from bovine organs—since it was then known to inhibit both bovine and human trypsin, plasmin, chymotrypsin, and various kallikreins. Since that time, BPTI's inhibitory spectrum was extended to several other enzymes. The relatively newly described squash inhibitors inhibit not only trypsin but coagulation factors IXa, Xa, and XIIa, human plasmin, human cathepsin G, and plasma kallikrein (81). In our laboratory, we have measured K_a values greater than 10^8 M^{-1} for the inhibition of 14 different enzymes by the leech inhibitor, eglin c (potato I family). These are bovine chymotrypsins A and B and mosquito *Aedes aegypti* chymotrypsin, porcine pancreatic, human leucocyte, and *Schistosoma mansoni* cercarial elastases, subtilisins Carlsberg, BPN′, Savinase and proteinase K, human cathepsin G, SGPA and SGPB, and α-lytic proteinase from *Lysobacter enzymogens*. We are aware of several other enzymes that are described in the literature which are inhibited by eglin c but we do not know what the K_a values are. It may well be worth noting that the physiological function of eglin c is not known. Overshadowing these broad specificities is the panspecific ecotin, the only thus far described member of the ecotin family (82)

2.4 Cognate enzymes and target enzymes

The research on protein inhibitors of serine proteinases originated by assaying them with enzymes readily available in the laboratory: bovine trypsin and chymotrypsin in the USA and porcine trypsin and chymotrypsin in Europe. The increase in the availability of serine proteinases led to an extension of this repertory. However, as the repertory grew, it became very clear that the enzyme–inhibitor pairs that were studied were often totally non-physiological. Many laboratories made great efforts to find the true physiological targets of various inhibitors. This proved very difficult. The major difficulty arose when it was realized that inhibitors control not only endogenous proteinases of its own organism but also exogenous ones with which the organism interacts. Several general approaches for the identification of target enzymes have been employed. One is the finding of genetic diseases arising from low levels of a specific inhibitor. Another is studying inhibitors in organisms with unusual ecological adaptations involving the control of enzyme activity. Intestinal parasites are responsible for many of the members of the *Ascaris* inhibitor family. Another example is provided by haematophagous animals such as the tick (TAP inhibitor family—not standard mechanism), the black fly (rhodniin—Kazal family), the leech (hirudin—not standard mechanism). In all such cases, it is simple to think of a specific set of target enzymes. In contrast, the plant inhibitors, whose principal role is postulated to be one of defence against bacterial, mould, insect, and vertebrate parasites are a problem. The parasites that now thrive on the plants probably evolved their enzymes to escape inhibition. The role of inhibitors has been to discourage the parasites that abandoned foraging on the plant in question.

2.5 Hypervariability

Avian ovomucoids consist of three tandem homologous domains (83). All the domains belong to the Kazal family of inhibitors and each has a reactive-site peptide bond, indicated in Fig. 5 by an arrow. In turkey ovomucoid (Fig. 5), the first domain with P_1–P_1' Glu–Asp inhibits GluSGP (glutamic acid-specific *Streptomyces griseus* proteinase), the second domain with Lys–Ala inhibits trypsin-like enzymes, and the third domain with Leu–Glu inhibits various chymotrypsins, elastases, and subtilisins. Rhodes *et al.* (84) gave impetus to the comparative study of ovomucoids by discovering that when tested for inhibition of bovine trypsin and bovine chymotrypsin:

- Chicken ovomucoid inhibits only one molecule of trypsin (on the second domain) and no CHYM. It was therefore called single-headed.
- Turkey ovomucoid inhibits one molecule of trypsin (on the second domain) and one of CHYM (on the third domain). It is therefore called double-headed.
- Duck ovomucoid inhibits two molecules of trypsin (one weakly on the first domain, the second strongly on the second domain) and a single molecule of CHYM (on the third domain). It is therefore called triple-headed.

The finding of single-, double-, and triple-headed inhibitors in such studies is beautifully consistent with later sequencing results, showing that all avian ovomucoids have three tandem Kazal domains (83, 85). Most of them are in fact triple-headed, if the range of enzymes whose inhibition is tested is increased (e.g. turkey gains another head when GluSGP is added, chicken when Lys-specific enzyme is added).

As a specific example of general nomenclature, the first domains of ovomucoids

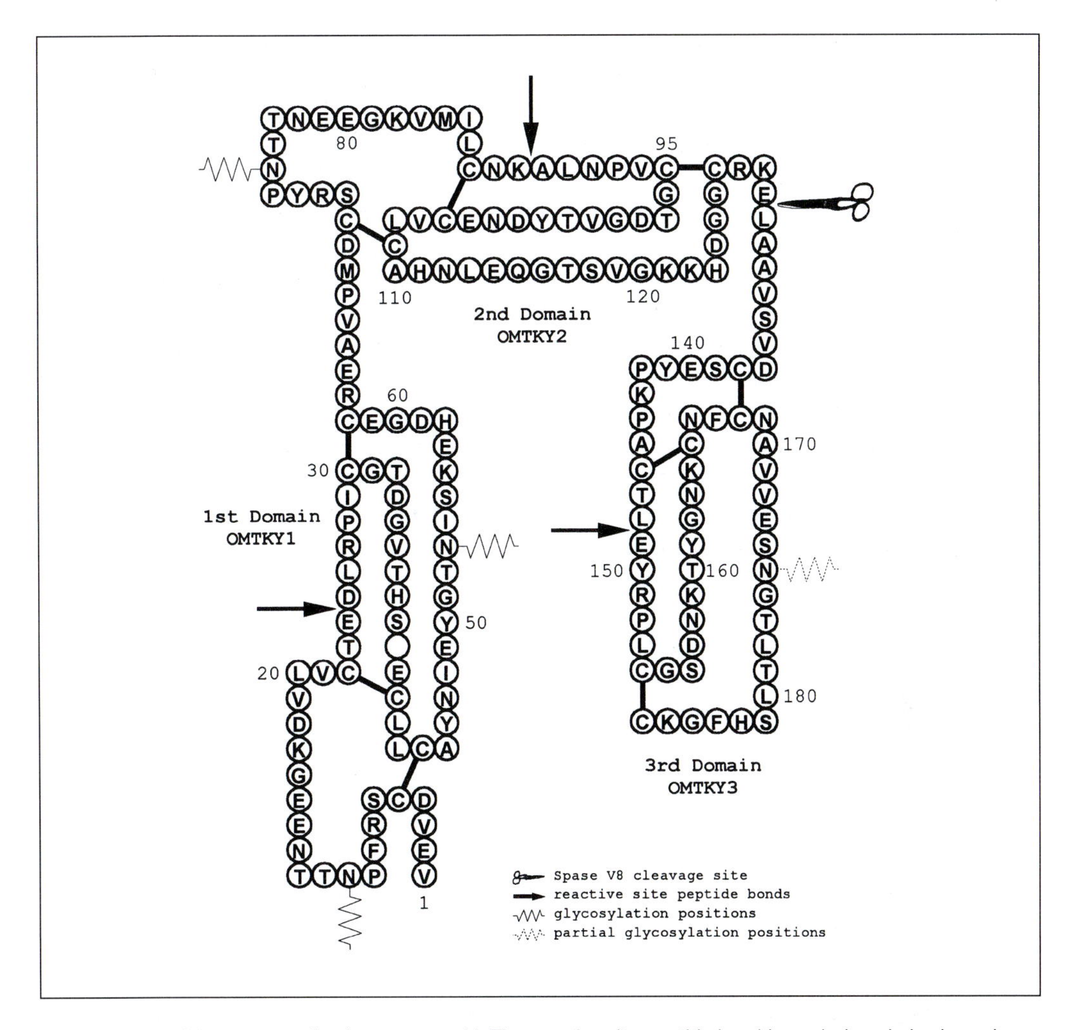

Fig. 5 Amino-acid sequence of turkey ovomucoid. The reactive-site peptide bond in each domain is shown by an arrow. The third domain of ovomucoid is prepared by enzymatic treatment of ovomucoid. The most commonly used enzyme is *Staphylococcus aureus* protease V8 which cleaves the Glu–Leu peptide bond between the second and third domain. Thus, in general, natural ovomucoid third domains are 56 residues long. In the expression of ovomucoid third domains, the first five residues are deleted and hence they are 51 residues long (see Fig. 3).

are said to be paralogous (86) to second and third domains, so are second domains to the third domains. However, the first domains from different species are said to be orthologous, so are the second domains and the third domains when compared only among themselves. Most biochemists believe that in the evolution of orthologous proteins the biological function of the molecules is strictly conserved (see ref. 87 for a more extensive discussion). Furthermore, they believe that orthologous proteins evolved predominantly by drift, with the structurally and functionally important residues strongly conserved. Exposed surface residues are free to accept mutations as these do not affect either the structure or function of the molecules. These ideas are based on a large body of sequence data and a much smaller number of quantitative functional comparisons. The comparisons involve housekeeping proteins such as triosephosphate isomerase or cytochrome *c* functioning in different organisms on identical or nearly identical substrates and in closely similar environments. This is, however, not true for serine proteinases, especially if the functional parameter that is examined is the equilibrium association constant with inhibitors rather than catalysis of peptide-bond hydrolysis in substrates. It is strikingly not correct for some standard mechanism, canonical protein inhibitors of serine proteinases (87). One of the best demonstrated cases of functional differences between orthologous proteins is the behaviour of avian ovomucoid third domains. After sequencing several ovomucoids (83, 85, Fig. 5), the Purdue group realized that the study of enzyme inhibition by ovomucoids, with their three reactive sites, was needlessly complex. The three domains could be separated in favourable cases by limited CNBr cleavage and/or by limited proteolysis. While entire ovomucoids are much too long, the separated domains, especially the third, which is the shortest, could be sequenced in a single pass in a protein sequencer. Furthermore, carbohydrate-free or even carbohydrate-containing, ovomucoid third domains could be isolated and crystallized. Their structures both in free form (52, 64, 65) and in complexes with serine proteinases have been determined (77, 88–91; Huang, Bateman, Anderson, Lu, Laskowski, and James, unpublished data).

We have isolated ovomucoid third domains from 153 species of birds and obtained their sequences. These are summarized in Table 3. In all, 92 different sequences were obtained. For six species, clear evidence was obtained for the presence of two allelic forms (92). In some related species, predominantly ornithologically closely related ones, the sequences are identical. It is seen in Table 3 that positions 15 (P_4), 17 (P_2), 18 (P_1), 20 (P_2'), 21 (P_3'), 32 (P_{14}'), and 36 (P_{18}') are the most variable in the whole molecule. As is seen from the right side of Table 3, these seven positions are all among the 12 consensus contact residues of avian ovomucoid third domains. It is an unexpected result, since as is shown in the remainder of this chapter, these contact residues determine most of the strength of enzyme–inhibitor interaction. Therefore, they are functional. They are also hypervariable (87, 93).

X-ray crystallography shows us that the hypervariable residues are mostly exposed surface residues. Such residues are expected to accept more than their share of mutations, provided that their substitution does not affect the function (i.e. ΔG_a°) of the protein. For example, switching between Arg and Lys at the P_1 position of trypsin inhibitors would be expected to be and is isofunctional. Indeed, most ovomucoid second domains inhibit bovine trypsin. The 28 we sequenced have at P_1, 20 Lys, 7 Arg, and 1 Asp. Turning to Table 3, we note the large variation at P_1 as evidenced by the presence of 11 different residues there. However, note that of the 153 species, 65 have Leu and 60 Met at this position. For most enzymes we have studied (CHYM, PPE, CARL, SGPA, SGPB, and HLE) Leu and Met are nearly isofunctional at P_1. Thus, the huge variability at P_1 of ovomucoid third domains does not greatly contribute to the variation of ΔG_a° among species. This seems at first a

Table 3 Composite amino acid sequence of ovomucoid third domains from 153 species (from Ref. 24)

												No. of contacts with		
												HLE	**CHYM**	**SGPB**
6	V_{151}	I_2												
7	D_{150}	N_3												
8	C_{153}													
9	S_{153}													
10	$E_{67.5}$	$D_{58.5}$	G_{27}											
11	Y_{145}	H_8												
12	P_{153}													
13	K_{142}	Q_6	R_2	M_1	T_2								2	5
14	P_{150}	H_2	S_1									4	11	2
15	A_{107}	V_{30}	D_8	E_2	G_3	T_2	S_1					5	7	8
16	C_{153}											9	10	9
17	T_{111}	S_{28}	L_5	P_4	M_3	A_1	R_1					12	11	16
18	$L_{65.5}$	M_{60}	$A_{6.5}$	Q_5	$V_{4.5}$	T_3	P_3	K_2	$S_{1.5}$	G_1	I_1	39	30	29
19	E_{139}	D_{13}	L_1									10	15	9
20	Y_{124}	D_8	Q_5	H_4	E_4	R_2	S_2	L_2	F_1	N_1		11	12	17
21	R_{63}	M_{72}	K_7	F_5	V_4	L_1	T_1					2	8	5
22	P_{153}													
23	L_{130}	V_9	F_8	I_6										
24	C_{153}													
25	G_{153}													
26	S_{153}													
27	D_{151}	N_2												
28	N_{119}	S_{34}												
29	K_{132}	Q_{13}	I_4	T_3	E_1							1		
30	T_{150}	S_2	I_1									2		
31	Y_{153}													
32	$G_{79.5}$	$S_{46.5}$	D_{13}	$A_{5.5}$	N_5	V_1	H_1	R_1	$P_{0.5}$			1		4
33	$N_{151.5}$	D_1	$S_{0.5}$									3	1	
34	K_{151}	R_1	E_1											
35	C_{153}													
36	N_{118}	D_{18}	S_8	A_4	G_2	Y_1	T_1	I_1				6	3	4
37	F_{153}													
38	C_{153}													
39	N_{151}	S_2											1	
40	A_{153}													
41	V_{149}	A_3	F_1											
42	V_{141}	A_{10}	M_1	L_1										
43	E_{71}	D_{74}	K_5	Q_2	H_1									
44	S_{148}	K_4	R_1											
45	N_{152}	S_1												
46	G_{148}	V_5												
47	T_{148}	del_5												
48	L_{144}	I_4	del_5											
49	T_{148}	N_2	S_2	I_1										
50	L_{150}	V_2	F_1											
51	S_{133}	N_{11}	R_5	G_4										
52	H_{146}	R_5	Y_1	N_1										
53	F_{148}	L_3	I_1	P_1										
54	G_{150}	E_3												
55	K_{126}	E_{20}	Q_2	R_2	V_1	T_1	N_1						2	
56	C_{153}													

paradoxical conclusion and makes one wonder whether large ΔG°_a variability exists. It exists indeed. ΔG°_a values have been measured for ovomucoid third domains for most, but not all, of the 153 species studied in Table 3 against the six enzymes listed above (Park, Tashiro, Wynn, Lu, and Laskowski, Jr, unpublished). Compared to the strict neutralist expectation, the variation is huge. However, as pointed out above, it is not predominantly due to P_1 but to other positions. As an example, at P'_3 we have 72 Met and 63 Arg (Table 3). For CHYM, Arg21 binds 80 times better than Met21 (45), thus accounting for a significant part of K_a variation for this enzyme.

The hypervariability of sequences surrounding the reactive sites of protein inhibitors of proteinases and even of active sites of proteinases attracted some attention (86, 87, 93–97). However, most of such work is concerned solely with the variability of amino-acid sequences or (better) of coding nucleotide sequences. All such work, we feel, has a very serious drawback. It faces a daunting task of distinguishing between isofunctional substitutions (such as Arg and Lys at P_1 for trypsin inhibitors or Leu and Met at P_1 of ovomucoid third domains) on the one hand and functional variability on the other. Data such as that obtained on natural ovomucoid third domains (43, 44) and on OMTKY3 variants (45) are absolutely needed, but even with such data we cannot be sure that relevant enzymes are in the set.

Analysis of potato I inhibitors (97) and of squash inhibitors (39) indicate the lack of hypervariability in these families. However, the relatively small sets of published ΔG°_a data for these families show considerable variation in ΔG°_a, albeit smaller than for ovomucoid third domains. One cause of such variation in squash inhibitors is particularly intriguing. As the reactive site is very close to the N-terminus, the amino terminal processing has a significant effect on ΔG°_a values (98, 99). Thus, significant ΔG°_a variation among molecules can arise even without genetic sequence variability. In conclusion, we urge more and better ΔG°_a measurements.

2.6 What affects the K_a values?

Because the inhibitors are generally more stable and smaller, the production of their variants proved easier than the production of enzyme variants. For example, total chemical synthesis of inhibitors is now common. In contrast, it is rare for proteolytic enzymes. It is convenient to divide the inhibitors' interactions with the enzyme into main-chain and side-chain components. The reason for such a division is that site-specific modifications, as well as most chemical modifications and ionizations, change only the side chains and leave the main chain intact. Changing to or from a prolyl residue is a notable exception, as both the side chain and the main chain are affected.

2.6.1 Main-chain/main-chain interactions

R. M. Jackson (100) pointed out that these are far more important for the association of proteinases with their inhibitors (Fig. 4) than for antigen–antibody association. We believe that these main-chain/main-chain interactions are particularly important in preserving the register in which binding occurs.

The replacement of an amino-acid residue by the corresponding hydroxy acid residue replaces the original peptide bond by an ester bond. Such a replacement no longer allows for the formation of a hydrogen bond between the appropriate N atom and the enzyme. This was done by semisynthesis for the P_2' residue of BPTI (101). The resultant complex with trypsin was 25 times weaker than the original. Semisynthesis, while very elegant, has many limitations. Lu *et al.* (102) replaced the P_1 Leu18 in OMTKY3 with leucic acid by total chemical synthesis. This approach is far more general. The introduction of the P_2 to P_1 ester bond weakened complex formation by factors ranging from 4 to 29 depending upon the partner enzyme.

In preliminary experiments (Lu, Kent, Qasim, Laskowski, Jr, Bateman, and James, unpublished) the reactive-site peptide bonds of OMTKY3 and of eglin c were replaced by reduced peptide bonds.

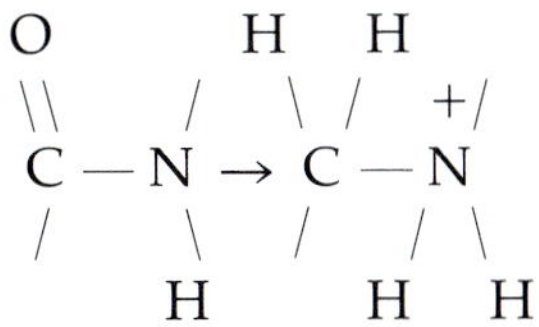

```
O        H    H
\\      /  \  /   +/
C — N  →  C — N          [5]
/      \  /    / \
       H     H   H
```

Note that not only is the O atom of P_1 taken out of the oxyanion hole in the complex but the N atom of P_1' and carbonyl carbon of P_1 change from trigonal to tetrahedral geometry. The NH of the reduced peptide seems to be in the protonated form both in the free inhibitor and in the complex. The free energy of association of the two reduced peptide-bond inhibitors with SGPA and SGPB decreases by 8–11 $kcal\,mol^{-1}$ compared to the wild-type inhibitors. The three-dimensional structure of the reduced peptide bond OMTKY3 has been determined (103). With the exception of atoms constituting P_1–P_1' residues, the rest of the structure superimposed on the wild-type structure quite well. Regrettably, useful crystals of the complexes have not yet been obtained. In general, the detailed studies of main-chain/main-chain interactions are still in its infancy. It seems, however, to be a promising approach.

2.6.2 Side-chain interactions

In contrast to the main-chain/main-chain studies, the side-chain studies are considerably advanced. The principal quantity of interest in such studies is the quantity $-\Delta\Delta G_a^\circ$ (X_1 P_n X_2), which is the difference in the standard free-energy changes on association of two inhibitor variants with a serine proteinase. The variants differ by changing one residue at position P_n from X_1 to X_2. The additivity assertion that underlies the following analysis states that the value of $-\Delta\Delta G_a^\circ$ (X_1 P_n X_2) is independent of the sequence in which it is located.

The strongest formulation of the additivity statement is interscaffolding additivity. This statement asserts that as long as P_n is in the canonical combining loop, the term is independent of the family of inhibitors (i.e. replacing Gly at P_1 by some other residue, such as P_1 Leu, will improve the binding by the same free-energy increment). At first, this statement seems absurd. Inhibitors that belong to different inhibitor families have a totally different global three-dimensional structure. However, the main-chain conformation of the combining loop is nearly identical. Once this is realized, the interscaffolding additivity does not seem to be as paradoxical as it first appeared.

Most tests of interscaffolding additivity were made for the P_1 residue. The reason was undoubtedly the exaggerated importance assigned to this residue by many investigators, including ourselves (see below). However, P_1 is the most exposed residue in the inhibitors (Table 2 in ref. 104). Tamura and Sturtevant (105) call it hyperexposed. Therefore, upon association, the P_1 side chain is transferred from solvent to the S_1 pocket of the enzyme. Table 4 summarizes the available results. If interscaffolding additivity were perfect, the entries in each row would be identical. They are not. How close to identical must they be to be judged additive? The Purdue group (43–45) has long been operating with an error band of ± 400 cal mol^{-1} based on their error estimate of the determination of four equilibrium constants involved in the simplest additivity cycle. This was more formally stated by Qasim *et al.* (42, 106). Krowarsch *et al.* (107), who determined most of the X15 BPTI data in Table 4, disagree. The estimation of experimental errors in protein chemistry is notoriously

Table 4 The side–chain contribution of P_1 residues to binding to chymotrypsin

	$-\Delta\Delta G^\circ_a$ (Gly P_1 X), kcal mol^{-1}, pH 8.3, 21 ± 2°C		
X	**X^{18}OMTKY3**[a]	**X^{45} eglin c**[b]	**X^{15} BPTI**[c]
Gly	0	0	0
Ala	1.29	1.03	1.91
Ser	1.07	1.08	0.72
Val	1.78	NA	1.96
Thr	1.57	NA	2.01
Pro	−2.67	−0.82	NA
Leu	5.97	6.03	5.65
Ile	1.57	1.79	1.18
Met	5.60	NA	5.49
Asn	2.68	NA	2.77
Asp	−1.10	−0.74	−1.09
Lys°	4.66	NA	5.38
Lys$^+$	0.35	NA	3.97
Gln	3.07	NA	3.78
Glu	−0.57	−0.52	0.25
His°	3.03	NA	4.00
Phe	7.45	7.36	6.03
Arg$^+$	2.04	NA	4.69
Tyr	8.16	NA	6.68
Trp	7.63	NA	6.50

[a]Refs 79, 106
[b]Refs 42, 106
[c]Ref. 107

hard and they suggest ±800 cal mol^{-1} as an error band. They may be right. In any case, we are of the opinion that even if the experimental error were to be lowered to say ±100 cal mol^{-1} band, a much larger uncertainty band would remain. This can possibly be better appreciated after analysis of the next section.

2.7 Isoenergetic is isostructural

Armed with the ±800 cal mol^{-1} band, we conclude that Ala, Ser, Val, Thr, Leu, Ile, Met, Asn, Asp, Lys°, Gln, and Glu are interscaffolding additive in the three inhibitors interacting with CHYM (Table 4). There is only one case where three-dimensional structures of complexes of CHYM with two of the inhibitors in Table 4, and with P_1 being the same residue from the above set, have been determined. The residue is P_1 Leu and the complexes are with OMTKY3 (90), 1CHO, and eglin c (31), 1ACB. The CHYM S_1 pockets and the P_1 residues in these two structures are compared in Plate 13. A numerical comparison is made in Qasim *et al.* (42). All conclusions agree. The P_1–S_1 regions of the complexes of CHYM with these two strikingly different inhibitors are isostructural (Plate 13). The data of Table 4 tell us that they are isoenergetic. Since P_1 Leu is also isoenergetic in the Leu15 BPTI–CHYM complex, we predict that the S_1 pocket structure will be exactly the same as the two given on the left side of Plate 13.

2.8 Not isostructural implies not isoenergetic

Another case where structures of two inhibitors with the same P1 have been obtained in complex with CHYM are OMTKY3 and BPTI with P_1 residue being Lys^+ (Table 4). There is a 3.62 kcal mol^{-1} difference between the two $-\Delta\Delta G^\circ_a$ (Gly P_1 Lys^+) value. This is clearly greatly outside the ±800 cal mol^{-1} error band. We suspect that the structures are quite different, and we are not disappointed. On the top right (Plate 13), the Lys^+ side chain of (Lys^+)18 OMTKY3 (see pH-dependence for a discussion) stretches out into the hydrophobic and ionophobic[2] S_1 pocket of CHYM (Ding, Lu, Qasim, Laskowski, Jr, Anderson, and James, unpublished) 1HJA. On the bottom right (Plate 13), the P_1 Lys^+ in the BPTI complex bends sharply upward into the up position (108, 109). It now makes hydrogen bonds with its own Pro13 (P_3 residue) and with Ser217 of the enzyme. It is also clear that the local dielectric constant near its charge is much higher than that near the charge of (Lys^+)18 in Lys18 OMTKY3 (Plate 13, right).

It is worth noting that while the P_1 Lys^+ free-energy changes differ greatly, the P_1 Lys° ones do not. On this basis, we predict that the P_1 Lys° side chains of both inhibitors adopt the down position. This is a testable experiment. Raising the pH of the crystals to pH 10.5 should leave the Lys18 OMTKY3 side chain in essentially the same position but stretch out the Lys15 side chain of Lys15 BPTI into the down position.

Note that Arg^+ values differ by 2.65 cal mol^{-1} in the same direction as Lys^+ values.

There are no exactly appropriate X-ray structures. However, Scheidig *et al.* (108) obtained the structure of a complex of CHYM with the inhibitor domain of Alzheimer's β-amyloid protein precursor, APPI. This domain is a member of the BPTI (Kunitz) inhibitor family and it has Arg15 at P_1. The Arg15 is in the up position in the complex. It seems highly likely that Arg15 BPTI would also assume the up position in complex with CHYM, while Arg18 OMTKY3 would be in the same down position as is shown for Lys18.

A more challenging problem is that of the three aromatic amino acids. We expect Phe18 OMTKY3 and Phe45 eglin c to assume the same conformations in complexes with CHYM. However, all three P_1 aromatic side chains bind 1–2 kcal mol^{-1} more weakly to CHYM when in BPTI as opposed to OMTKY3 and eglin c (Table 4). We expect the conformations of these residues to differ. Finally, the $-\Delta\Delta G°$ (Gly P_1 Pro) terms differ strongly between OMTKY3 and eglin c. Here, we expect that P_1 Pro is likely to lead to different distortions in the main chain of OMTKY3 and of eglin c on binding to CHYM (42).

The comparison of relative binding strengths of P_1 residues to S_1 pockets of enzymes is not limited to inhibitors. Determination of the relative k_{cat}/K_M values allows it to be extended to the transition-state complexes of substrates (Table 5). Many comparisons between inhibitors and substrates are made in the literature (59, 79). The two systems differ. Enzyme–inhibitor complexes are particularly suitable for X-ray crystallography studies. Because of their short half-lives, transition-state complexes are not. Therefore, it may prove useful to predict their structures from the known ones of enzyme–inhibitor complexes (106).

Table 5 Comparison of $-\Delta\Delta G°$ (Ala P_1 X) for two substrates to $-\Delta\Delta G°$ (Ala P_1 X) for binding of X^{18} OMTKY3 to chymotrypsin (kcal mol^{-1}), pH 8.3, 21 ± 2°C

	$-\Delta\Delta G°$ (Ala P_1 X)	$-\Delta\Delta G°$ (Ala P_1 X)	
X	X^{18} OMTKY3[a]	Ac X O Me[b]	SucAAPXpna[c]
Gly	−1.29	−1.52	NA
Ala	0	0	0
Abu	1.68	1.43	NA
Ape	3.43	3.09	NA
Val	0.49	0.06	−0.23
Ahx	4.18	4.32	NA
Leu	4.68	NA	4.59
Ile	0.28	0.19	−0.42
Asp	−2.39	NA	−1.88
Lys°	3.37	NA	2.93
Lys^+	−0.94	NA	−1.29
Phe	6.16	6.36	6.33

[a]Refs 59, 79, 106
[b]Ref. 110
[c]Qasim and Laskowski, Jr, unpublished data

2.9 The interscaffolding additivity at P_1 for other enzymes

Interscaffolding additivity for P_1 residues interacting with CHYM in three inhibitor families has been discussed in Section 2.6.2. Similar data for other enzymes are limited. Unfortunately, the most data (exactly the same amount as shown for CHYM in Table 4) is available for HLE. It is also well known that the S_1 pocket of HLE is very easy to distort. For example, HLE is the only well-studied serine proteinase that prefers Ile to Leu at P_1. In view of the plasticity of the S_1 pocket (111), it should not be surprising that HLE is non-additive in most comparisons even though the rank order of residue preferences is almost the same in X18 OMTKY3, X45 eglin c, and X15 BPTI.

For the other four enzymes we have studied, SGPA, SGPB, PPE, and CARL, there is considerably less data, but what is available indicates substantial additivity (42). Additivity was also observed for α-lytic proteinase between some X45 eglin c and X18 OMTKY3 variants (Qasim and Laskowski, unpublished), as well as between some X18 OMTKY3 and X73 SSI variants interacting with subtilisin BPN′ (112; Qasim and Laskowski, unpublished)

2.10 Additivity at positions other than P_1

The historical focus on P_1 greatly reduced the number of available variants. In our laboratory (45), ΔG°_a values for six enzymes for the whole set of coded P'_3 X21 variants of OMTKY3 were obtained, but there is relatively little to compare this with. It was found that for all the enzymes, P'_3 Pro was very bad. This is consistent with the virtual absence of P'_3 Pro in 15 out of the 18 families (Table 1). In contrast, P'_3 Pro is present in the great majority of Bowman–Birk family members and in all (very few) known members of rapeseed and marinostatin families. Lu (45) confirmed our earlier (113) finding that P'_3 Arg is the optimal residue for interaction with CHYM among the X21 OMTKY3 variants. Many inhibitors in other families, e.g. eglin c in potato I family and chymotrypsin/elastase inhibitor from *Ascaris* (114), have Arg at P'_3 and are powerful inhibitors of CHYM, while other members of the same families without P'_3 Arg are much weaker for this enzyme. This is far from conclusive, but is strongly suggestive of interscaffolding additivity. Based on information supplied by our laboratory, the specificity toward CHYM was greatly improved by the introduction of P'_3 Arg analogue into a small synthetic inhibitor (115).

Recently, P_2 residues were varied (20) in cyclic nonapeptides based on the combining region of Bowman–Birk inhibitors and their ΔG°_as were measured. There is a surprisingly good correlation, and even some additivity, between these results and a full set of coded P_2 variants of X17 OMTKY3 (116, 117).

In spite of the sporadic successes one should not be excessively hopeful. Most contact residues in protein inhibitors are exposed but not hyperexposed as is P_1. Some of these residues are involved in intrainhibitor interactions (104) and are less likely to be interscaffolding additive than the hyperexposed P_1 residue. Such interactions do not necessarily eliminate the possibility of intrascaffolding additivity as they are generally conserved within a scaffold. It is probably too early to form any firm conclusions, as too little work has been done.

2.11 Steric effects modulate the interaction of P_1 side chains with hydrophobic, ionophobic S_1 enzyme pockets

Binding of a homologous (in a chemical sense) series of straight-chain, aliphatic side chains: Gly, Ala, Abu, Ape, Ahx, and Ahp to the six enzymes with hydrophobic S_1 pockets is shown in Fig. 6. The last four of the series are obviously non-coded. They were introduced into X18 OMTKY3 by enzymatic semisynthesis (59). Regrettably, the series ends with Ahp due to technical difficulties. However, higher homologues

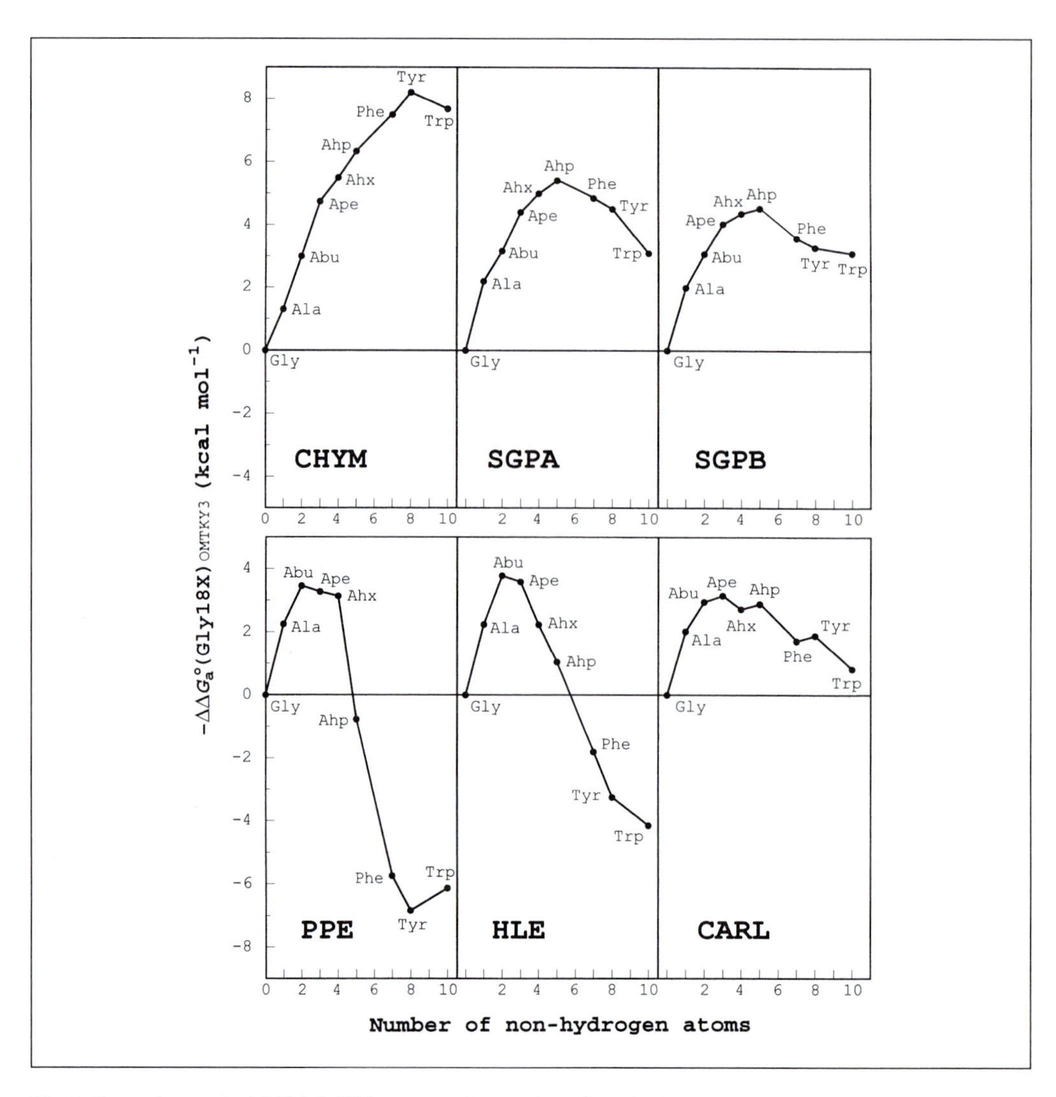

Fig. 6 Dependence of $-\Delta\Delta G_a^\circ$ (Gly18X)$_{OMTKY3}$ on the number of non-hydrogen atoms in the side chain. For clarity only, members of homologous series Gly, Ala, Abu, Ape, Ahx, Ahp, and the three aromatic residues are shown (79). Readers are referred to Table 6 for a listing of binding free energies for all coded and many non-coded P_1 variants of OMTKY3 with six serine proteinases.

were clearly needed. The $-\Delta G_a^\circ$ values for all coded and non-coded P_1 variants of OMTKY3 interacting with six enzymes are shown in Table 6.

CHYM has the most accommodating S_1 pocket. As hydrophobicity increases with increasing side-chain size, it is not surprising that so does the binding to CHYM except for Trp18. The S_1 pockets of SGPA and SGPB are known to be smaller and a maximum at Ahp is apparent for both. It appears that the aromatics are slightly too large for these pockets. This is not the case for substrates. Aromatic P_1s are better substrates for SGPA and SGPB than the straight-chain aliphatics (118; Qasim and Laskowski, unpublished). It is universally accepted that the PPE and HLE S_1 pockets are much smaller than that of CHYM. Therefore, maxima for these enzymes between Abu and Ape do not seem surprising. With increasing side-chain size, the binding becomes greatly weaker but for PPE it seems to level off. This obviously leads to the double-valued character of the graphs. It explains why for PPE, the binding of Ala and of Leu is equally strong (79). We call such residue pairs isofunctional and will return to them later. For CARL, it has been known for a long time from studies on substrates (119) that it does not exhibit very strong specificity at P_1. CARL's S_1 cavity

Table 6 $-\Delta G_a^\circ$ for recombinant P_1 variants of OMTKY3 interacting with six serine proteinases[a] at pH 8.3, 21 ± 2°C

P1	CHYM	PPE	CARL	SGPA	SGPB	HLE
Gly	9.18	12.05	11.85	10.30	9.52	9.88
Ala	10.47	14.29	13.86	12.48	11.51	12.11
Abu	12.21	15.55	14.85	13.50	12.64	13.70
Ser	10.26	12.02	12.62	11.04	10.36	10.12
Cys	12.62	13.99	15.50	15.50	14.47	13.25
Ape	13.96	15.36	15.05	14.74	13.61	13.50
Val	10.96	13.38	12.04	12.54	11.46	13.65
Hse	12.69	13.33	14.51	13.25	12.46	11.36
Thr	10.75	14.08	14.01	12.51	11.30	12.26
Pro	6.49	7.76	6.59	6.27	6.13	7.10
Ahx	14.72	15.23	14.62	15.34	13.93	12.14
Leu	15.17	14.29	14.17	15.44	14.46	13.17
Ile	10.76	13.15	10.96	11.00	10.04	13.89
Met	14.80	13.61	14.68	15.36	14.03	11.92
Asn	11.87	10.55	12.69	11.32	11.11	8.01
Asp	8.07	6.54	9.69	9.02	8.87	5.66
Ahp	15.55	11.31	14.78	15.76	14.10	10.95
Lys	10.76	6.27	10.87	11.57	11.32	7.51
Gln	12.26	10.27	13.61	12.22	11.89	9.90
Glu	8.61	6.63	12.02	9.28	8.56	6.21
His	12.22	6.15	13.61	12.67	12.77	6.78
Phe	16.66	6.31	13.56	15.14	13.10	8.07
Arg	11.22	4.95	11.27	10.65	11.14	6.08
Tyr	17.37	5.21	13.73	14.80	12.81	6.62
Trp	16.84	5.92	12.67	13.39	12.62	5.73

[a]Data taken from Refs 59, 79

is not a pocket but a cleft, and therefore residues of different sizes can be more readily accommodated.

The ascending slopes of these graphs are quite similar for all six enzymes. In the top three (Fig. 6), the slopes are long enough to merit quantitative analysis. Huang *et al.* (77) plotted the $-\Delta\Delta G_a^\circ$ (Gly18X) values vs. the buried apolar surface area (Fig. 7) and obtained slopes between 34 and 40 cal Å^{-2}. It should be noted that some H_2O molecules remain in the pockets even after the largest hydrophobic side chains are bound (77; Bateman and James, personal communication).

The descending slopes, especially those for PPE and HLE were not analysed in detail. The question here seems to be, 'how does a side chain that is much too big for the pocket, bind to it none the less?'. The possible answers are bending of the side chain such as seen for Lys^+ of BPTI (108, 109) or a conformational change in the enzyme opening up the restricted binding pocket. More structural data are needed to decide.

Thus far, we have considered only hydrophobic residue size. However, the shape also matters. The binding of Ahx and Leu (Table 6; see Fig. 4 in ref. 79) is similar. Now compare straight-chain Ape and Ahx to their β-branched isomers Val and Ile. The straight chains are clearly favoured. When the coded Leu and Ile are compared, the γ-branched Leu is 2000 times better than the β-branched Ile. The deleterious effect of β-branching is a consequence of the narrow entrance to the S_1 pockets of serine proteinases (78).

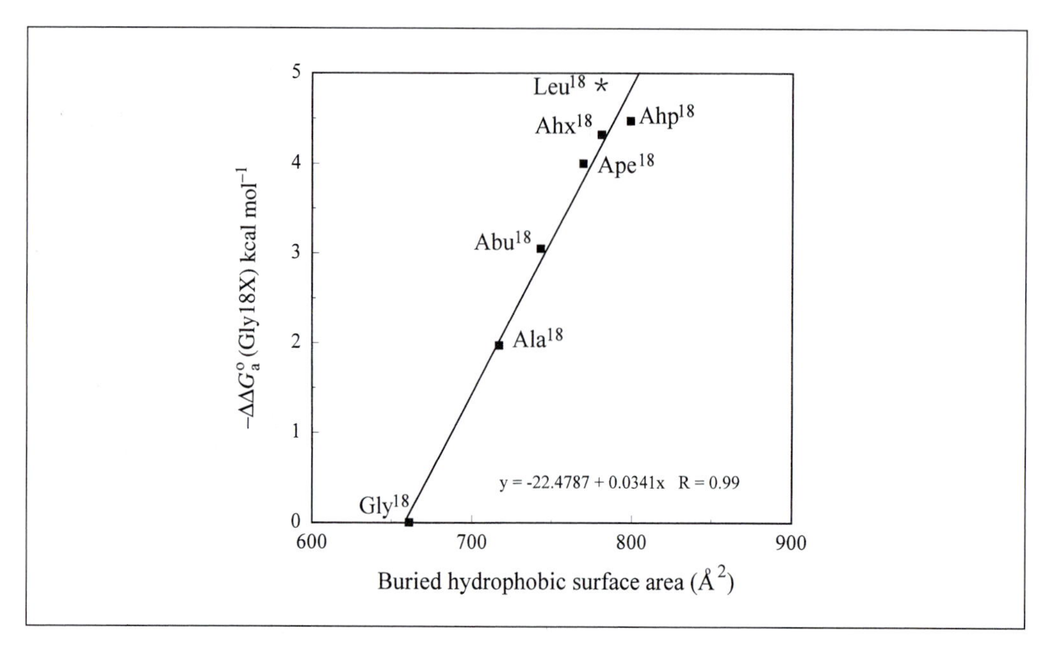

Fig. 7 Correlation of the change in binding free energy $-\Delta\Delta G_a^\circ$ (Gly18X) with buried hydrophobic surface area in the interaction of homologous series of P_1 variants of OMTKY3 with SGPB. The Leu18 variant (shown by *) was not included in the line fitting. (Adapted from ref. 77.)

2.12 Side-chain polarity is not very deleterious

Many comparisons of polar (but uncharged) side chains have been made (79). Even more will be available for comparison when the analysis of all ionizable side chains is completed and the binding constants for their uncharged counterparts can be compared. The polar side chains we looked at are Ser, Thr, Hse, CySH, Asn, and Gln. A global conclusion is at hand. The replacement of a terminal methyl group by a polar group is deleterious but turns out to be only slightly deleterious, except for the Abu/Ser comparison and Abu/CySH comparison (Table 6). P_1 Ser binds sizeably more poorly than P_1 Abu. In contrast, P_1 CySH binds much better than P_1 Abu for CARL, SGPA, and SGPB. Indeed, P_1 CySH is the optimally coded P_1 residue for these enzymes. This observation is somewhat curious. The three enzymes are all microbial. Did their strong preference for P_1 CySH evolve in a reducing environment? On the other hand, it is striking that P_1 CySH and P_1 Trp have not been found as P_1 residues of any natural inhibitor. Here, the surprise may not be great, as P1CySH inhibitors can form disulfide-bridged homodimers which are devoid of inhibitory activity. This may be too bothersome for nature's P_1 repertory.

2.13 In ionophobic S_1 pockets, charged residues are highly deleterious

Introduction of an ionizable group into a pocket sometimes changes its pK from its free state (pK_f) to the one in complex (pK_c). The most complete study of pK_f values in any inhibitor was carried out for OMTKY3 by Forsyth *et al.* (120). J. Song and J. L. Markley at the University of Wisconsin–Madison are building up a very large library of pK_f values in numerous single replacement variants of OMTKY3 provided by our laboratory at Purdue. From the limited data we have, it appears the pK_f values of ionizable residues at P_1 are the unperturbed values for the kind of residue present there. If this is correct, pK_f values would be entirely independent of the inhibitor. These values may also serve as ideal model values for ionizable groups in proteins. They may have certain advantages over the model values reported earlier in the literature (121, 122).

What happens when the complex forms and P_1 side chain inserts into the S_1 pocket of the enzyme? If the pocket is hydrophobic and ionophobic, as is the case for elastases, chymotrypsins, and subtilisins, the charged form will be strongly disfavoured as opposed to the neutral form. This will lead to a very big change in the $-\Delta G^\circ_a$ between the charged and uncharged residue which is equivalent to a very large pK shift (47).

To treat the problem quantitatively, consider first the behaviour of $-\Delta\Delta G^\circ_a$ as a function of pH. If the residues are both non-ionizable and the size and shape of the S_1 pocket is not pH-dependent, we expect $-\Delta\Delta G^\circ_a$ (Ala P_1 Phe), for example, to be pH-independent. This was extensively tested for many enzymes and many non-ionizable residue pairs and found to be pH-independent (see Fig. 8B). This was also found to hold for P_4–S_4 interaction in CARL. However, the pH-independence of the

$-\Delta\Delta G^\circ_a$ term involving non-ionizable residues is not general. Ranjbar (117) surmises that in the complexes of P_2 variants, X17 OMTKY3 with CHYM, the large non-ionizable P_2 residues touch the catalytic His57 residue of the enzyme, while the small ones (e.g. Gly and Ala) do not. He further surmises that the P_2 contact is responsible for most of the downward pK shift of His57 on complex formation. Specifically, he finds that $-\Delta\Delta G^\circ_a$ (Gly P_2 Ile) has a large pH-dependence.

Returning now to the $-\Delta\Delta G^\circ_a$ (X_1 P_1 X_2), consider the cases where X_1 is non-ionizable and X_2 is ionizable. Both the X_1 and X_2 systems experience the global pH-dependence for enzyme–inhibitor complex formation (P_1 Q in Fig. 8A). However, the uncharged (P_1 E in Fig. 8A) and cationic (P_1 H in Fig. 8A) residues experience both the global pH-dependence and the specific pH-dependence due to the introduction of an ionizable residue at P_1. Note that the P_1 Leu residue serves as a control. Its $-\Delta\Delta G^\circ_a$ (Gln P_1 Leu) is pH-independent (Fig. 8B). It can be seen that the specific pH-dependence is qualitatively different for uncharged and charged ionizable residues. This is very simple to understand. The uncharged ionizable residues; Asp, Glu, CySH, and Tyr, are neutral in their protonated form. It is that form that binds well to the hydrophobic, ionophobic pockets. On the other hand, the ionized or deprotonated form does not, as it is negative. Its binding involves burial of a charge. The opposite behaviour is expected and found for cationic ionizable residues, Arg, Lys, and His. Here the protonated form is positively charged. It is the form that binds badly. Deprotonation converts it to a neutral side chain which binds better.[3]

For the cationic residues we can write:

$$-\Delta\Delta G^\circ_a (X_1\, P_1\, X_2) = -\Delta\Delta G^\circ_a (X_1\, P_1\, X_2^+) + RT \ln \frac{1 + 10^{(\text{pH} - \text{p}K_c)}}{1 + 10^{(\text{pH} - \text{p}K_f)}}. \quad [6]$$

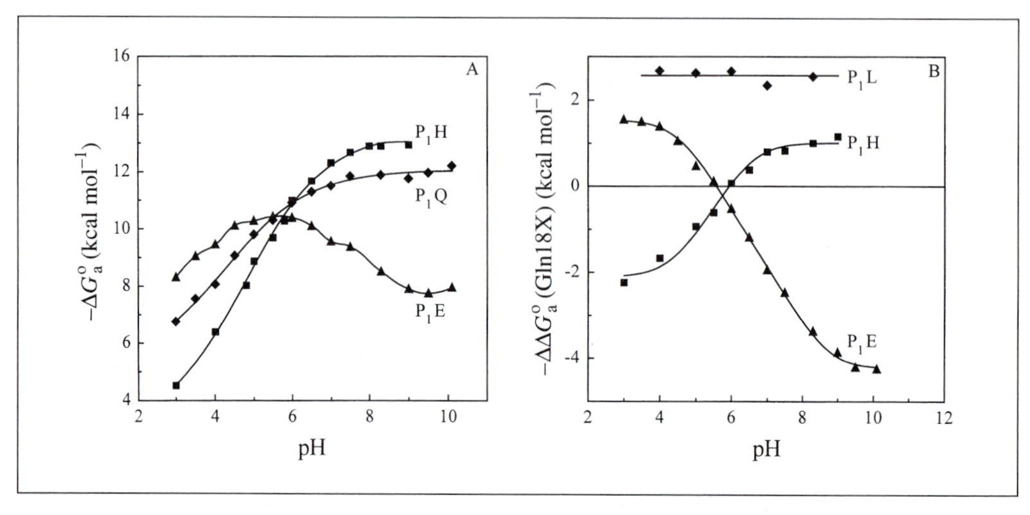

Fig. 8 pH-dependence of the interaction of P_1 Q18 OMTKY3, P_1 H18 OMTKY3, and P_1 E18 OMTKY3 with SGPB (47) is shown in the left panel (a). The right panel (b) shows the corresponding $-\Delta\Delta G^\circ_a$ $(\text{Gln18X})_{\text{OMTKY3}}$ plot vs. pH plot. The data for the two ionizable groups were fitted into eqns 6 and 7 to calculate pK_c.

Note that at very low pH, the second term of this equation drops out as one would anticipate. On the other hand, at very high pH, the second term reaches its asymptote and the sum of both terms is $-\Delta\Delta G^\circ_a$ (X_1 P_1 X_2°). These terms can be obtained from the hypothetical equilibrium constants. The appropriate constant for the low pH value deals with a protonated P_1 side chain in free inhibitor forming a protonated side chain in complex. The global part of the constant can be obtained from the reference. As we measure the ΔG°_a for the non-ionizable residue at pH 8.3, we set that global part at pH 8.3. The second term in eqn 6 is the free-energy penalty that the system must pay for changing the pK. Please note that it is zero if $pK_c = pK_f$. Otherwise, it is cheaper to pay the penalty than to bury a charge until pH = pK_c.

Figure 8 illustrates this type of analysis for the interaction of P_1 His (a cationic residue) variant of OMTKY3 with SGPB. The inherent pH-dependence of $-\Delta G^\circ_a$ of OMTKY3 interacting with SGPB is seen in the P_1 Gln (non-ionizable residue curve). Note that the replacement of P_1 Gln by P_1His leads to steepening of this curve since P_1 His° binds much better than P_1 His^+. To measure this quantitatively, the two $-\Delta G^\circ_a$ curves are subtracted and the difference curve is fitted to eqn 6 yielding $-\Delta\Delta G^\circ_a$ (Gln P_1 His°) of –0.88 kcal mol^{-1} and $\Delta\Delta G^\circ_a$ (Gln P_1 His^+) of +2.25 kcal mol^{-1}. The pK_f of His18 is 6.63 and its pK_c is 4.31.

For the uncharged, ionizable residues (Asp, Glu, CySH, Tyr), we write:

$$-\Delta\Delta G^\circ_a (X_1\, P_1\, X_2) = -\Delta\Delta G^\circ_a (X_1\, P_1\, X_2^\circ) + RT \ln \frac{1 + 10^{(pH - pK_c)}}{1 + 10^{(pH - pK_f)}}. \qquad [7]$$

Note that the only change for eqn 7 is that the term $\Delta\Delta G^\circ_a$ (X_1 P_1 X_2^+) is now replaced by $\Delta\Delta G^\circ_a$ (X_1 P_1 X_2°) as both of these terms refer to protonated forms of the residues. The analysis for Glu is illustrated in Fig. 8B. The Glu curve on the left is complicated. However, after the subtraction of the reference curve, it becomes just as simple as the difference curve for the cationic His. There is an important difference, of course. For P_1 Glu, the P_1 Glu° form binds much better than the P_1 Glu^- form and the pK_c is much higher than the pK_f.

SGPB is an unusually stable enzyme. For all the other serine proteinases we study, obtaining $-\Delta G^\circ_a$ data below pH 4 and above pH 10 proved very difficult. However, measurement of pK_f need not involve the enzyme. It can be done on the inhibitor alone, for example by NMR spectroscopy.

In an extensive collaborative project involving several laboratories we have determined pK_f (by NMR) and pK_c values for a few ionizable P_1 residues of OMTKY3, eglin c, and BPTI interacting with different enzymes. The experiments involved measuring the pH-dependence of $-\Delta\Delta G^\circ_a$ (X_1 P_1 X_2) (where X_1 is a non-ionizable and X_2 is an ionizable (Asp, Glu, and Lys) residue) and fitting the data in eqn 6 or 7. In a few cases, variants at P_4, where CARL and PPE have hydrophobic pockets were also investigated. All six of our standard enzymes were studied. The final results are not yet complete but some semiquantitative conclusions can be stated.

- The pK shifts are very large. In most cases, they are 4–5 pK units for the carboxylic acids and 2–4 units for Lys.

- A particularly dramatic way to state the above result is that in the hydrophobic, ionophobic pockets, the pK_c of Asp residues is commonly higher than of Lys residues. This is strikingly against the normal intuition of protein chemists, who, based on pK_f values, expect the opposite.
- The binding affinities of Asp° and Glu° are invariably stronger than those of Asn and Gln. In fact, Asp° binds more strongly than Leu, to which it is isosteric, to the CARL and SGPB S_1 pockets.

It is our strong opinion that binding data for inhibitor variants with Asp, Glu, His, and Lys are almost impossible to interpret unless binding studies are carried out over a broad pH range. As an example, the analysis of this type (eqn 6) was carried out for P_1 Lys residues in BPTI and in Lys18 OMTKY3 in complexes with CHYM. The Lys° and Lys^+ values in Table 4 are a product of this analysis. It is clear that in the BPTI–CHYM complex (see Plate 13, bottom right), a lesser price is paid for burying the positive P_1 Lys^+ side chain as it bends up into a region of higher dielectric constant (106).

2.14 The cationophilic, hydrophobic S_1 pocket in trypsin

Krowarsch *et al.* (107) measured the ΔG°_a values for 18 coded variants (at P_1) for X15 BPTI interacting with bovine trypsin, BT, and with Atlantic salmon trypsin, AST (Fig. 9). In the same study, they also obtained data on bovine chymotrypsin (BCHYM) and human neutrophil elastase (HNE). These were already discussed and some were tabulated in Table 4. As expected, Arg and Lys bind very strongly due to their interaction with Asp^{189E}. We clearly expect that the pK_c for Lys15 of BPTI in these complexes is much higher than the pK_f, which is 10.68 (106). It is clear that the various trypsins' S_1 pockets are cationophilic.

It is the binding of the remaining residues that is of great interest. They range from least favourable Gly to most favourable Tyr (Fig. 9). In fact, the binding of all the residues other than Arg and Lys is strongly correlated between AST and BT (107) on the one hand and CHYM on the other. Therefore, the various trypsins' pockets are cationophilic and hydrophobic.

While this result might be implied by the structural similarity of the S_1 pockets in BT and in CHYM, the opposite prediction would also be possible. Placing large hydrophobic residues such as Phe in the pocket might lower the dielectric constant and make the environment less favourable for Asp^{189E}. The experiments are clear. It happens that the hydrophobic binding dominates.

An interesting unsolved problem in binding to the trypsins remains. At pH 8.3, P_1 Asp variants bind more poorly than the P_1 Asn variants. What is the pK_c of P_1 Asp? It is very high in binding to chymotrypsin. Here, we might expect either a bigger shift or large upward shifts in the pK values of both Asp189 of the enzyme and of P_1 Asp of the inhibitor.

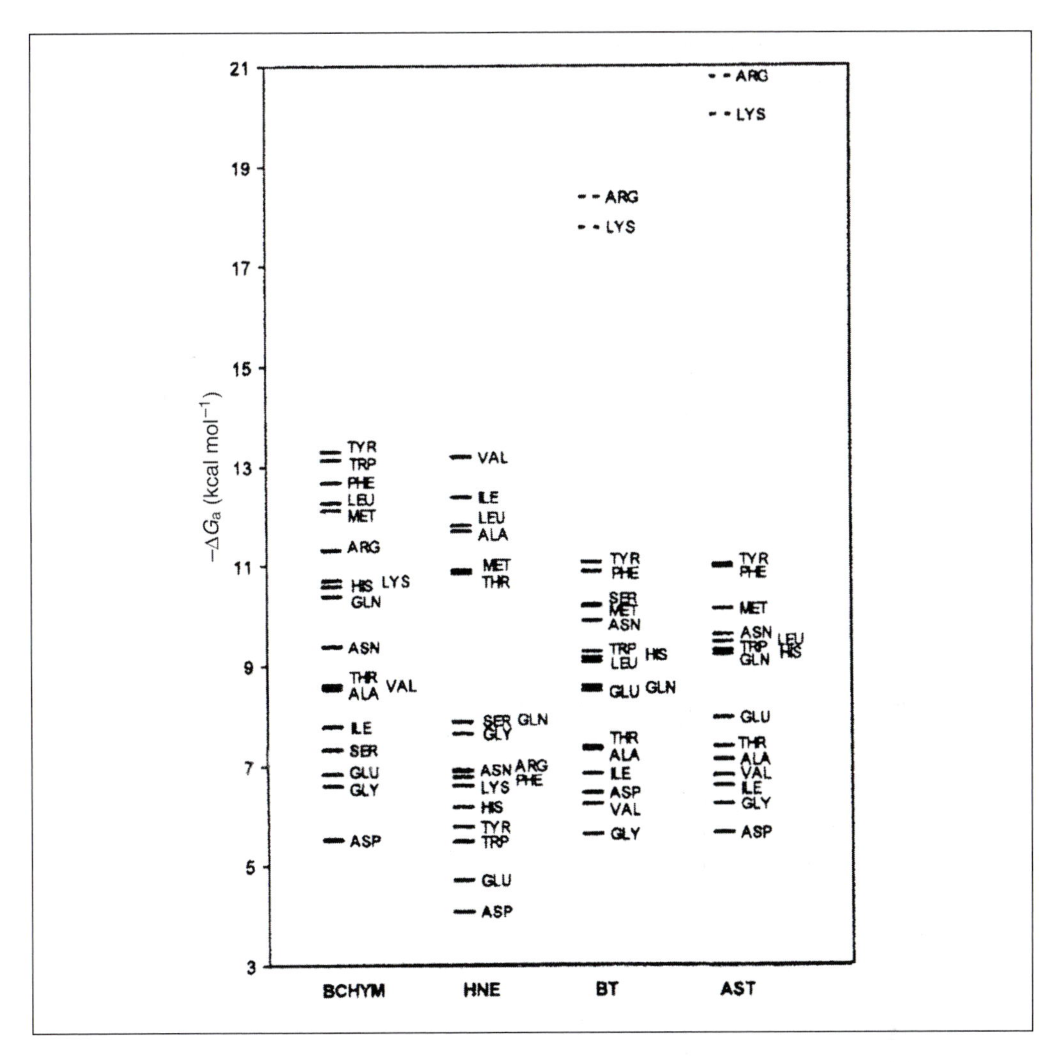

Fig. 9 Interaction of 18 P_1 mutants of BPTI with BCHYM, HNE, BT, and AST at pH 8.3, 22 °C. The values for Arg and Lys mutants were calculated from measurements performed at pH 5.0. (Adapted from ref. 107.)

2.15 The anionophilic, hydrophilic S_1 pocket of Glu SGP

Changing P_1 Leu18 in OMTKY3 to Glu18 produced a strong and efficient inhibitor of glutamic acid specific *Streptomyces griseus* proteinase (GluSGP) (124). The enzyme was discovered by Yoshida *et al.* (125) and further characterized by Birktoft and Breddam (126). The availability of the full set of coded P_1 variants of OMTKY3 (79) allowed us (unpublished results) to measure the inhibition of GluSGP at pH 8.3 by all of these. As expected, P_1 Glu and Asp were best, followed by P_1 Asn and P_1 Gln. However, surprisingly, P_1 Gly and Ala were next, with Phe, Tyr, and Trp the worst. Clearly, the P_1 residues other than Glu and Asp are anticorrelated to CHYM. The behaviour here is opposite to that of trypsin and the S_1 pocket of GluSGP seems to be anionophilic and hydrophilic. It seems clear that the set of all coded P_1 variants can be used to characterize and classify many other S_1 pockets.

2.16 Intrascaffolding additivity

The interscaffolding additivity discussed in the previous sections is an unusual idea in protein chemistry, since proteins with dramatically different sequences and global three-dimensional structures are being compared. The intrascaffolding additivity is quite familiar. When it holds, it dramatically simplifies the task of a protein engineer. It asserts that the free-energy change caused by one mutation made at one position in a wild-type protein can simply be added to the free-energy change caused by another mutation at another position, and that this sum properly predicts the free-energy change of the double mutant. We have seen in the previous sections that for standard mechanism, canonical protein inhibitors of serine proteinases the interscaffolding additivity holds relatively frequently. Therefore, one should expect, and indeed we find, that intrascaffolding additivity is commonly found (Table 7). It is regrettable that there are a few exceptions.

Table 7 Difference in measured and predicted free energy of binding $|\Delta G_i^\circ|$ for ovomucoid third domains[a]

$\|\Delta G_i^\circ\|$ range (cal mol^{-1})	No. of additivity comparisons	%	Cumulative %
<400	238	54.3	54.3
401–800	86	19.6	74.0
801–1200	56	12.8	86.8
1201–1600	23	5.3	92.0
>1600	35	8.0	100.0

[a]Various ovomucoid third domains (see text and Table 3) for which measured and predicted values are compared and differed from OMTKY3 by a maximum of 14 substitutions (maximum of 7 in the contact region). For details see Ref. 79.

2.17 Changes in residues not in contact do not matter

In Fig. 3, the consensus contact residues of OMTKY3 are shown in black. This is a very strong consensus because the contact residues in complexes with three different enzymes are nearly the same. Analyses of ΔG_a° values of numerous natural variants of ovomucoid third domains consistently showed (43–45) that changes of residues in contact positions generally caused large changes in ΔG_a° for most interchanges. In sharp contrast, changes in residues that are not in contact produced no or very small changes in ΔG_a°. In addition, elimination of the Cys30–Cys51 disulfide bridge or cleavage of the Met52–Arg bond in BPTI had no effect on ΔG_a° when it was measured at 5 °C (127), much below the T_m of the BPTI derivative. Similar results were obtained in bovine PSTI (127). This was further extended to the cleavage of the Met42–Asp bond (93) in laughing kookabura ovomucoid third domain (Qasim and Laskowski, unpublished data). The main result of these studies is that the nature of the residues in the scaffolding, and even the integrity of the polypeptide chain in the scaffolding, do not matter to binding as long as the inhibitor molecule is overwhelmingly folded under the conditions where measurements of ΔG°_a are made. The slogan 'residues not in contact do not matter' describes the situation.

2.18 Sequence to reactivity algorithm

In 1980, Laskowksi (128) proposed that his research group would obtain a database that, given additivity, would allow for the calculation of ΔG°_a values for the protein inhibitor–serine proteinase interaction on the basis of the amino-acid sequence of the inhibitor alone. Such calculations are now frequently made in other laboratories, but they are made on the basis of the available three-dimensional structures. The approach used here is entirely different.

Our test system is a set of 73 avian ovomucoid third domains that have different sequences and whose ΔG°_a values were determined for the six enzymes we study. These values have a very large range, thus providing a challenging test (43, 44). We proceed to develop a system which will allow us to determine the ΔG°_a for any Kazal inhibitor sequence. The first step is to choose a wild type. Turkey ovomucoid third domain was chosen because it is easily available. We have then obtained sets of all 19 coded variants at each of the 12 contact positions marked in black in Fig. 3. We temporarily skipped P_3 Cys and P_{15}'Asn as these residues are very strongly conserved in Kazal inhibitors and play predominantly a structural role. We anticipate returning to these later. This leaves us with 10 (positions) $\times$ 19 (variants per position) = 190 variants. We determined ΔG°_a (X) values for all these variants interacting with our six enzymes, 1140 ΔG°_a (X) values in all. These values were trivially converted to $\Delta\Delta G^\circ_a$ (X_{wt} i X) values. In this, i is an index designating the contact position and X_{wt} is the amino-acid residue at that position in the wild type (OMTKY3). The $\Delta\Delta G^\circ_a$ (X_{wt} i X) value is obtained by subtracting ΔG°_a (OMTKY3) from ΔG°_a (Xi). Examples of similar subtractions are given in Tables 4 and 5, except there the values for P_1 Gly and P_1 Ala are being subtracted.

We are now ready to calculate the ΔG°_a values for a Kazal family inhibitor of any sequence:

$$\Delta G^\circ_{a,\text{ input sequence}} = \Delta G^\circ_{a_{wt}} + \sum_{i=l}^{i=n} \Delta\Delta G^\circ_a (X_{wt}\, i\, X). \qquad [8]$$

This equation makes two very strong assertions. The first of these is that changes at positions that are not in contact with the enzyme have no effect. The second is that intrascaffolding additivity is strictly observed. However, even if both of these are fulfilled, the experimental errors in $\Delta\Delta G^\circ$ (X_{wt} i X) terms remain. It is clear that these increase as a greater number of residues differs from wild type. Given all these possible objections, it is truly remarkable (Table 7) that more than 70% of the cases agree experimentally to within ±800 cal mol^{-1} (107). It is clear, on the other hand, that about 15% of the cases are truly non-additive. We are investigating the causes of non-additivity in these cases.

Some of the cases are rather straightforward. In a majority of avian ovomucoid third domains, including OMTKY3, the P_2 residue is Thr and the P'_1 residue is Glu. These two residues are hydrogen-bonded to one another in the free inhibitor. The hydrogen bond tightens greatly upon complex formation. Such an interaction implies non-additivity, which is indeed observed. The current expedient is to avoid predicting ΔG°_a values for sequences where both P_2 is changed from Thr and P'_1 is changed from Glu. Note that the cases of a single change are in our database and thus present no problem. It turns out that relatively few P_2, P'_1 pairs must be determined in order to deal with all the sequences we have encountered. The interaction of chicken ovomucoid third domain with subtilisin is dramatically non-additive. Chicken third domain differs from turkey third domain at three positions (A P_4 D, L P_1 A, and Y P'_2 D). It is the two aspartyl residues that are non-additive. Each of them alone is highly deleterious for CARL. The combination is less deleterious. The non-additivity increases greatly with rising pH, suggesting that it is Asp^- not Asp° that is troublesome.

With this list of reservations out of the way, the sequence to reactivity algorithm can be put to use. It is particularly easy to design (129) the strongest possible inhibitor for each of the six enzymes we study. You simply use the best possible residue at each of the contact positions. Table 8 lists these designs. It is clear that these $-\Delta G^\circ_a$ values are larger than our ability to measure, and therefore preparing the best possible inhibitors is not an appealing task. However, consider a CHYM inhibitor with P_1 Gly18 and the remainder of the contact residues as 'best possible'. The expected $-\Delta G^\circ_a$ is slightly greater than 14 kcal mol^{-1}. Such an inhibitor is measurable and of obvious pedagogic value. It points out that residues other than P_1 have a very important role in inhibition. Furthermore, inhibitors of this general type with a single, highly deleterious residue at a single position and the remainder optimal should be highly prized for spectroscopic studies of the role of the deleterious residue. The dissociation of the enzyme–inhibitor complex at extreme pH values can be avoided by such designs.

The problem of designing the most specific inhibitors is less well defined and more challenging (129).[4] We have to identify the set of enzymes within which we wish the inhibitor to be specific. If the set has more than two members, the mathematical problem becomes complex. Finally, one might wish for the specific inhibitor to be of some specified strength. With these caveats in mind, a somewhat arbitrary set of the most specific inhibitors is given in Table 8. Most natural avian ovomucoid third domains inhibit SGPA more strongly than SGPB. Note that it is possible to achieve a significant reversal of this trend.

Having determined the ΔG°_a for all coded variants at all contiguous P_6–P'_3 (except P_3 Cys) contact positions, we are able to talk about the importance of all these positions to overall binding. Two measures of this occur to us. One is the entire range

Table 8 ΔG_a° for strongest predicted and most specific ovomucoid third domains (kcal mol^{-1}, pH 8.3, 21°C)

Inhibitor design *	CHYM	PPE	CARL	SGPA	SGPB	HLE
strongest CHYM	23.3	6.3	13.3	16.2	13.7	10.1
strongest PPE	8.7	18.4	13.9	16.3	14.7	13.7
strongest CARL	6.0	10.1	22.7	12.7	12.0	11.7
strongest SGPA	11.6	15.0	14.8	21.4	18.8	13.6
strongest SGPB	11.4	13.4	17.8	19.6	20.8	11.8
strongest HLE	11.7	13.3	8.5	13.1	11.3	22.0
OMTKY3	**15.2**	**14.3**	**14.2**	**15.4**	**14.5**	**13.2**
CHYM-specific	18.2	0.3	7.4	6.3	5.6	1.4
PPE-specific	0.5	11.4	3.7	4.0	2.5	6.2
CARL-specific	−0.2	2.6	18.7	5.9	5.1	3.0
SGPA-specific	4.2	−1.0	1.7	12.7	7.2	4.5
SGPB-specific	−1.0	−1.9	−0.1	7.0	11.2	0.3
HLE-specific	5.9	5.6	2.5	6.8	4.0	20.1

*** Strongest OM3 sequence for:**

```
        P6 –P5 –P4 –P3 –P2 –P1 –P1′–P2′–P3′      P14′P15′       P18′
CHYM    Gln–Cys–Trp–Cys–Thr–Tyr–Glu–Tyr–Arg–...–Ala–Asn–...–Ser
PPE     Thr–Glu–Tyr–Cys–Thr–Ala–Glu–Tyr–Met–...–Pro–Asn–...–Val
CARL    Trp–Asp–Phe–Cys–Tyr–Cys–Glu–Trp–Asp–...–Gly–Asn–...–Ala
SGPA    Ser–Tyr–Tyr–Cys–Thr–Cys–Glu–Tyr–Ser–...–Gly–Asn–...–Val
SGPB    Ser–Asp–Tyr–Cys–Thr–Cys–Ile–Tyr–Lys–...–Gly–Asn–...–Val
HLE     Arg–Leu–Trp–Cys–Thr–Ile–Glu–Tyr–Phe–...–Gly–Asn–...–Asp

OMTKY3  Lys–Pro–Ala–Cys–Thr–Leu–Glu–Tyr–Phe–...–Gly–Asn–...–Asn
```

Most specific OM3 sequence for:

```
        P6 –P5 –P4 –P3 –P2 –P1 –P1′–P2′–P3′      P14′P15′       P18′
CHYM    Asp–Ser–Gly–Cys–Thr–Trp–Glu–Ile–Arg–...–Ala–Asn–...–Ser
PPE     Lys–Pro–Gln–Cys–Gly–Ala–Glu–His–Trp–...–Pro–Asn–...–Asn
CARL    Trp–Asp–Leu–Cys–Tyr–Glu–Glu–Trp–Asp–...–Gly–Asn–...–Asn
SGPA    Pro–Trp–Lys–Cys–Pro–Phe–Glu–Phe–Ser–...–Gly–Asn–...–Val
SGPB    Ser–Pro–Pro–Cys–Thr–Lys–Lys–Ser–Leu–...–Gly–Asn–...–Asn
HLE     Arg–Leu–Arg–Cys–Val–Ile–Glu–Tyr–Leu–...–Gly–Asn–...–Asp
```

Thr17 and Glu19 are a coupled pair. Either one of these residues was changed, not both.

from worst to best. The other is the range from Gly to the best, which is taken as a measure of a constructive contribution to binding. The results are summarized in the bar graphs of Fig. 10 for the six enzymes we study. To many workers, the results are surprising. To emphasize the point they are summarized in Table 9. While P_1 is frequently the largest contributor, it is far from dominant. In the case of PPE, where the productive interactions at P_1 are small, the side chains of P_4, P_2, P_1', and P_2' all make a greater relative contribution than the side chains of P_1.

A characteristic of eqn 8 is that while the collection of the data that allows us to employ it is very hard work (measurement of ΔG_a° for over 1000 combinations), the calculation is exceedingly simple, and billions of ΔG_a° values can be calculated on a

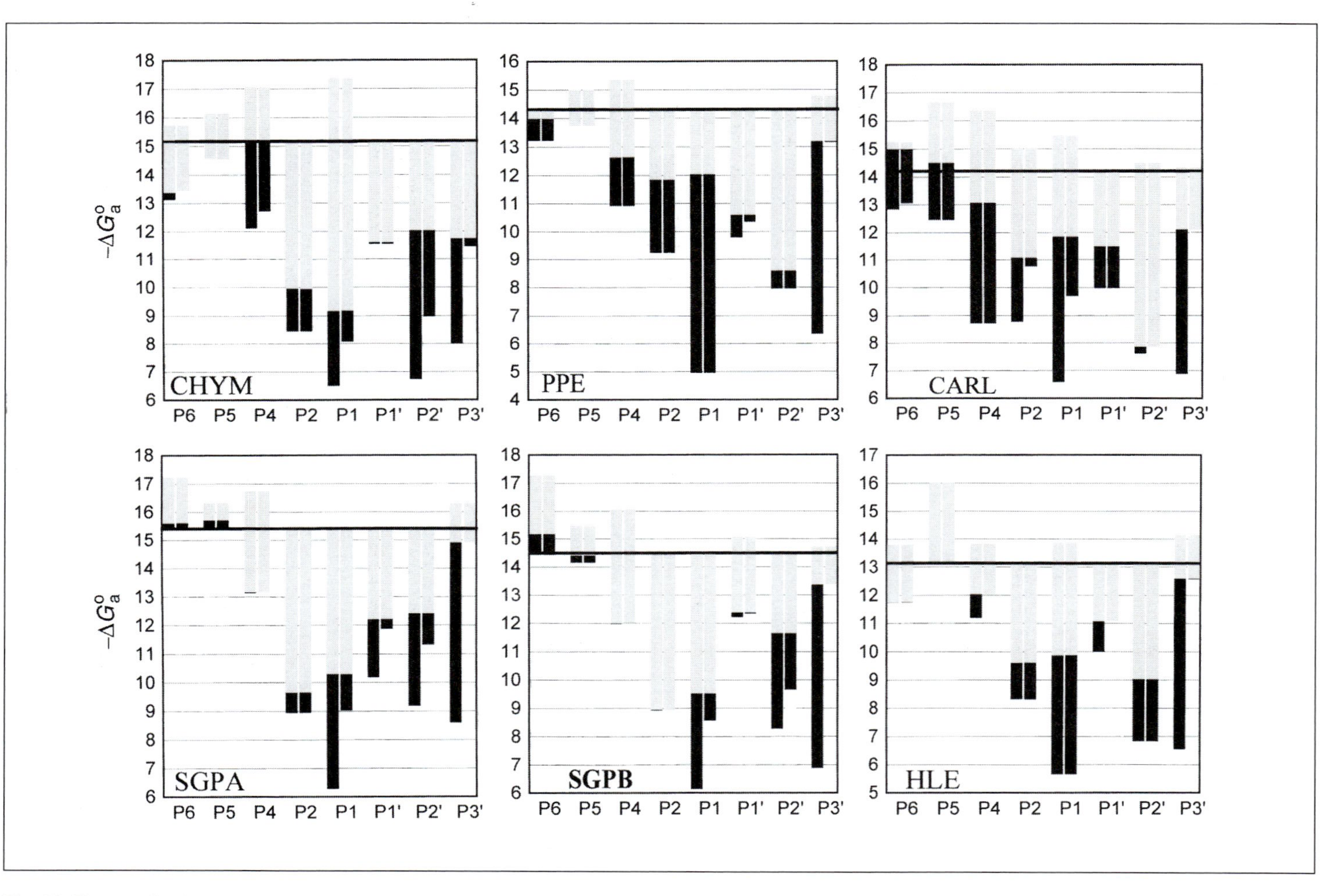

Fig. 10 The contributions of side chains to the free energy of association at all contiguous contact residues (see Fig. 3) except P_3 Cys, which was not varied. The length of each bar corresponds to $-\Delta\Delta G^{\circ}_{a}$ (X_{worst} P_i X_{best}) in kcal mol^{-1}. The grey (upper) part of each bar shows constructive contributions to binding, $-\Delta\Delta G^{\circ}_{a}$ (Gly P_1 X_{best}) as Gly has no side chain. The lower black part shows deleterious contributions. On the left of each pair all 20 coded amino-acid residues are included; on the right, Pro is omitted as it affects the main-chain interactions as well. The solid line indicates the contribution of the residue present in OMTKY3. If the solid line is at the tip of the bar, e.g. at P_2, P_1', P_2', P_3' in CHYM, the residue present is optimal. If the solid line crosses the bar, OMTKY3 could be improved by a mutation at that position. CHYM, SGPA, SGPB, PPE, HLE, and CARL, designate the six enzymes for which data were available.

Table 9 Constructive side-chain contributions (%)

	P_6	P_5	P_4	P_3	P_2	P_1	P_1'	P_2'	P_3'	$-\Delta\Delta G_a^{\circ}$ [a] (total)
CHYM	8.0	5.4	6.7	N/A	17.7	27.8	12.1	10.6	11.7	29.5
PPE	2.0	6.3	13.5	N/A	12.1	11.2	18.5	28.5	7.9	20.0
CARL	1.1	8.6	13.3	N/A	15.9	14.6	10.7	26.8	9.0	24.9
SGPA	6.7	2.6	14.7	N/A	23.6	21.2	13.2	12.3	5.7	24.5
SGPB	8.5	4.4	16.6	N/A	22.4	20.1	11.0	11.4	5.6	24.3
HLE	9.3	13.5	8.1	N/A	16.0	18.1	9.5	18.7	6.9	22.1

[a] in kcal mol^{-1}

laptop computer in a day. This opens up a series of new problems that previously could not be attacked. As this is very new, better problems may soon arise, but one is given here as an example. Ovomucoids as well as many standard mechanism, canonical protein inhibitors of serine proteinases, especially in plants, are clearly storage proteins. A number of critics occasionally speculate that their inhibitory role may be accidental. If this were so, the hypervariability of the reactive-site sequence would be easy to explain. For a number of reasons, we felt that the critics were wrong. Armed with the sequence to reactivity algorithm, we provide a proof. We calculate the ΔG_a° for all possible avian ovomucoid third domains with variation restricted to P_6 to P_3' and the caveats stated earlier applied. We plot the frequency of ΔG_a° as in Fig. 11. Unsurprisingly, it turns out Gaussian. We superimpose on it, as a solid line, the 132 known ΔG_a° of different ovomucoid third domains (24). These data show much stronger interaction than in a 'random' inhibitor. Figure 11 shows results for chymotrypsin. Similar results were also obtained for each of the other five enzymes we study.

The usefulness of eqn 8 depends upon how widely it can be applied. We hope that with appropriate modifications it can be applied to all or most of the families listed in Table 1. However, we are less sanguine that the interscaffolding additivity will come to our aid. Table 4 shows that for OMTKY3, eglin c, and BPTI non-additivities are seen even at P_1, which is probably the most interscaffolding-additive residue. It should be noted that eqn 8 can be employed only for the enzymes for which tables are available. For example, our group has no such tables for trypsin, so until this is remedied by another group (76, 107), we will be powerless in dealing with inhibitors which were used to establish our field.

2.19 Resistance to inhibition—special cases

A few serine proteinases fail to be efficiently inhibited by standard mechanism inhibitors. Some of these exceptions are discussed. A gene coding for human trypsin with a G193 R mutation (chymotrypsinogen numbering) has been identified (130). The trypsin product of this gene is not inhibited by human PSTI. A different scenario

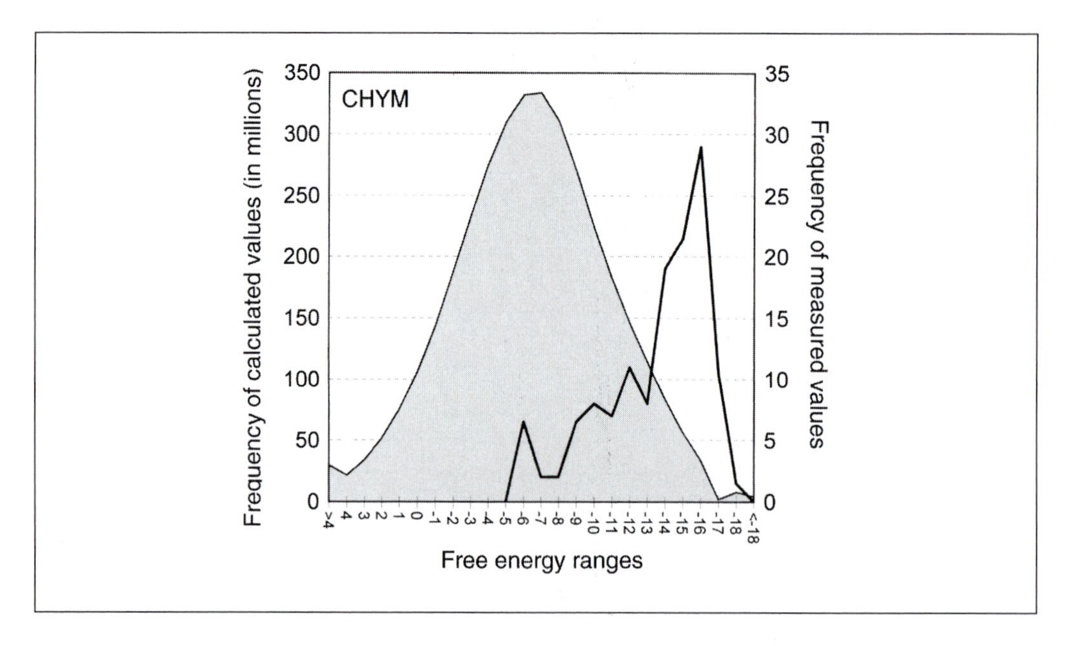

Fig. 11 Comparison of the distribution of calculated association free-energy changes of OMTKY3, ΔG°_a, for all random sequences of contact residues from P_6 to P_3' (grey area) and the distribution of ΔG°_a values for natural variants that were measured (solid line). In computing the random sequences, P_3 Cys was not varied. P_2 and P_1' were excluded since they are a non-additive subset. Only two of the three non-contiguous contact positions (P_{14}' and P_{18}') were taken into consideration. However, at the moment we have data only for 7 and 8 variants, respectively, at these positions (3.6×10^9 ΔG°_as were calculated). The probability of all coded amino acids was taken as 0.05. The distribution of measured values is weighted for the number of species sharing the same sequence (132 total). The great difference in the position of maxima of these functions appears to eliminate chance as the principal cause of inhibition of CHYM by avian ovomucoid third domain.

is presented by certain proteinases, which, instead of being inhibited, efficiently catalyse the cleavage of the reactive-site peptide bond in some inhibitors. The best-known examples are starfish trypsin and human cationic trypsin which hydrolyse the reactive-site peptide bond in STI and BPTI (131, 132). An interesting alternative approach to escape inhibition is adopted by thrombin and tryptase. Both of them show trypsin-like specificity in their selection of P_1 site. However, classical trypsin inhibitors fail to inhibit them. The reasons are similar but details are different. In the case of thrombin, the reactive site is located in a canyon surrounded by two insertion loops, the 60-loop and the 148-loop, which restrict access to most protein substrates and many natural inhibitors (133). Once these loops have been removed through mutations, the mutated thrombin can be strongly inhibited both by BPTI and TFPI (134–136). A similar result is obtained through a single mutation involving E192 Q mutation in thrombin. The only other mammalian serine proteinase which is known to have a Glu at 192 is activated protein C. Inhibition profiles of activated protein C and its Q192 mutant with BPTI are similar to the one described for thrombin and its Q192 mutant (137).

A second, well-documented example of steric inaccessibility is provided by human tryptase, which has a unique tetrameric structural arrangement (138). The active sites of the four monomeric units are aligned towards the centre of an oval and are shielded from large substrates and inhibitors. The only known protein inhibitor of tryptase is the non-classical Kazal inhibitor, leech-derived tryptase inhibitor (LDTI) which causes a 50% inhibition of tryptase. LDTI compared to classical inhibitors has an elongated shape and can access the reactive site of the enzyme. However, because of steric hindrance only two of the four monomeric units of tryptase can bind to LDTI at a time. Steric incompatibility is probably also responsible for the poor inhibition of subtilisins by BPTI and STI (71, 139) and of α-lytic proteinase by ovomucoid third domains (Qasim, Bachovchin, and Laskowski, unpublished).

2.20 Hypothetical scenarios for enzyme–inhibitor coevolution

HIV patients are now treated with inhibitors of the viral aspartyl proteinase. These inhibitors are mainly peptide mimetics that are substrate analogues of the HIV proteinase substrates. The treatment is at first highly successful, but resistance to the drug often develops after 10 or more weeks after the start of the treatment. This resistance results from mutations in HIV proteinase that weaken the K_a for enzyme–inhibitor interaction (140). Many possible mutations could accomplish this simple goal, but as inhibitors are substrate analogues most such mutations are strongly correlated to the loss of enzymatic activity. If this loss is very large, the virus can no longer replicate. Thus, mutations are selected such that they cause only a small drop in k_{cat}/K_M for the proteinase acting on its natural substrates but a large drop in K_a, to escape the action of the drug. The term 'vitality' has been coined to deal with such effects quantitatively (140).

We suggest that, except for the time scale (for a week, substitute about one million years), the coevolution of serine proteinases and of their inhibitors is very similar. We believe that comparisons of orthologous serine proteinases from different species (139, 141) or of closely related isoenzyme pairs such as bovine chymotrypsin A and B (Qasim, Wilusz and Laskowski, unpublished) will reveal a similar behaviour—that is the k_{cat}/K_M values within the related pairs vary much less than the K_a values for interaction with a common inhibitor.

The analogy between HIV proteinase resistance and the enzyme–inhibitor pairs can be pushed further. Two strategies used by pharmacologists to overcome the drug resistance also find their analogues in nature. Massive amounts of drugs are given to overcome the development of resistance, and cocktails of inhibitors are employed in order to thwart escape from inhibition. As already pointed out, numerous inhibitors often inhibit the same enzyme and often several are present in the same source (e.g. ovomacroglobulin, the seven domain ovoinhibitor and the three domain ovomucoid in egg whites as inhibitors of serine proteinases). This example also serves to indicate that inhibitors are often present in massive amounts.

An interesting case of enzyme–inhibitor coevolution is the case of the *Ascaris* intestinal parasites. Two of these parasites are especially intensively studied and both are believed to be host-specific. *Ascaris suum* lives in pigs and *Ascaris lumbricoides* in humans. It is clear that the parasites need proteinase inhibitors, and inhibitors of trypsin, chymotrypsin/elastase, carboxypeptidase A, and pepsin were isolated from them (142). The first two are members of the *Ascaris* inhibitor family dealt with in Table 1. Hawley and Peanasky (143) report the equilibrium constants for the four possible combinations of human and porcine trypsin and of *A. lumbricoides* (human host) and *A. suum* (porcine host) trypsin inhibitors (Table 10).

Porcine trypsin is effectively inhibited by both inhibitors, so is human trypsin by *A. lumbricoides* (human host) inhibitor. However, human trypsin is not effectively inhibited by *A. suum* (porcine host) inhibitor. The data are insufficient to infer an evolutionary sequence of events, but a plausible scenario is that human trypsin evolved to escape inhibition by the generalized *Ascaris* precursor. Subsequently, a subset of the *Ascaris* population became *Ascaris lumbricoides* by improving its trypsin inhibitor to become an efficient inhibitor of human trypsin.

Both forms of *Ascaris* thrive in an isosmotic salt medium. However, when a plant proteinase such as ficin or papain is added to the medium, the worms are killed by proteolysis. This is presumably because the worms have no inhibitors against the plant enzymes. *Ascaris suum* (porcine host) worms thrive when incubated with porcine trypsin but die very slowly (compared to ficin or papain) when incubated with human trypsin. These experiments appear to prove that *Ascaris suum* is host-specific because its trypsin inhibitor does not withstand human trypsin for long periods. It should be noted that the above experiments do not explain at all why *Ascaris lumbricoides* (human host) does not thrive in pigs. This could be due to the inhibitors of the other proteolytic enzymes or to entirely unrelated causes.

The situation described above seems simple. It becomes more complicated when we realize that all vertebrates store pancreatic secretory trypsin inhibitor (Kazal) in their zymogen granules. One of the roles of this inhibitor is to prevent the premature activation of pancreatic zymogens in the pancreas or in the pancreatic duct. Human trypsin that escaped inhibition by *Ascaris suum* trypsin inhibitor did so by becoming generally more difficult to be inhibited by standard mechanism trypsin inhibitors. As a result, human pancreatic secretory trypsin inhibitor evolved to become a better inhibitor of human trypsin than other mammalian trypsin inhibitors (139; Wilson and Laskowski unpublished).

The second complication of the above description is the facile use of the terms

Table 10 Association equilibrium constants for the interaction of human and porcine trypsins with *A. lumbricoides* and *A. suum*, pH 7.6

Parasite trypsin inhibitor	Ka (M^{-1})	
	Human trypsin	Porcine trypsin
A. lumbricoides	9.0×10^7	2.5×10^8
A. suum	2.0×10^5	6.0×10^7

porcine trypsin and human trypsin. In neither case is it a single species. For example, five different human trypsinogens have been sequenced in humans (144). It presumably suffices if only one of these trypsins escapes inhibition by *Ascaris suum* trypsin inhibitor. That will endow humans with resistance. However, our simple picture would make us expect that all five human trypsins would be efficiently inhibited by the human pancreatic trypsin inhibitor. This is apparently not the case for at least one of these (130).

The availability of several isoenzymes with similar catalytic efficiency and specificity but with different susceptibility to inhibition explains how animals can react to an inhibitor challenge on a short time-scale. Larvae of *Spodoptera exigua* have 78% of their proteolytic activity inhibited by the potato II family chymotrypsin/trypsin inhibitor. This is no great hindrance to them as tobacco leaves on which they prey do not contain this inhibitor. The inhibitor was incorporated into transgenic plants and successfully expressed there. However, it was little hindrance to the larvae. When reared on the transgenic leaves they induced a new activity and only 18% of their proteolytic activity was susceptible to inhibition (145). However, in some other plants transgenic inhibitors provided excellent protection against parasitic larvae.

Acknowledgments

Many of the ideas mentioned in this chapter were developed and supported by our many colleagues at Purdue. Their work can be loosely divided into two areas, although some of them participated in both. They are listed only once. Those who mostly contributed to the standard mechanism—STI era are: S. Crema, U. Devonis-Biddlingmeyer, R. Duran, D. Estell, P. Fankhauser, W. R. Finkenstadt, M. A. Hamid, H. F. Hixon, Jr, G. A. Homandberg, D. F. Kowalski, T. R. Leary, J. Lebowitz, J. Luthy, R. McKee, J. McKee, J. Mattis, A. Morawiecki, C. Niekamp, K. Ozawa, M. Praissman, R. W. Sealock, D. Wang, M. Welch Baillargeon, and K. A. Wilson. Those who were here in the ovomucoid era or are still here are: I. Apostol, W. Ardelt, T. Bigler, W. C. Bogard, J. Chen, C.-W. Chi, J. Cook, M. Empie, K. Forwith, A. Giletto, I. Kato, C. Kelly, W. J. Kohr, T. Komiyama, T.-Y. Lin, Y.-M. Lin, W. Lu, C. J. March, O. Ogunjobi-Green, J. Otlewski, S. J. Park, R. J. Peanasky, L. Price, S. Qasim, M. Ranjbar, O.E. Schoenberger, J. Schrode, M. Tashiro, N. Warne, H. Whatley, A. Wieczorek, M. Wieczorek, H. Wojciechowska, R. Wynn, and W. Zhang. We were further aided by many collaborations. X-ray structures were provided by groups led by W. Bode, R. Huber, and M. N. G. James. Molecular biology was done by S. Anderson's group. NMR data, especially p*K* values were provided by J. L. Markley's group. We received many variants of inhibitors from the groups of M. Kimura, C. Saunders, and J. Otlewski and enzymes from W. W. Bachovchin, D. Estell, M. Laskowski, Sr, J. McKerrow, L. Smillie, J. Travis, J. Wells, M. A. Wells, T. Wilusz, and N. Yoshida. J. Bieth, G. Salveson, and K. A. Wilson made useful comments on parts of this chapter. Most of the research on inhibitors at Purdue was supported by NIH grant GM10831.

References

1. Neurath, H. (1989). Proteolytic processing and physiological regulation. *Trends Biochem. Sci.*, **14**, 268–71.
2. Khan, A. R. and James, M. N. G. (1998). Molecular mechanisms for the conversion of zymogens to active proteolytic enzymes. *Protein Sci.*, **7**, 815–36.
3. Hartley, B. S. (1960). Proteolytic enzymes. *Annu. Rev. Biochem.*, **29**, 45–72.
4. Rawlings, N. D. and Barrett, A. J. (1994). Families of serine peptidases. In *Methods in enzymology*, Vol. 244 (ed. A. J. Barrett), pp. 19–61. Academic Press, New York.
5. Laskowski, M., Jr and Kato, I. (1980). Protein inhibitors of proteinases. *Annu. Rev. Biochem.*, **49**, 593–626.
6. Barrett, A. J. and Starkey, P. M. (1973). The interaction of α_2-macroglobulin with proteinases. Characteristics and specificity of the reaction, and a hypothesis concerning its molecular mechanism. *Biochem. J.*, **133**, 709–24.
7. Sottrup-Jensen, L. (1989). α-Macroglobulins: structure, shape, and mechanism of protein complex formation. *J. Biol. Chem.*, **264**, 11539–42.
8. Green, N. M. (1953). Competition among trypsin. *J. Biol. Chem.*, **205**, 535–51.
9. Komiyama, T., Ray, C. A., Piokup, D. J., Howard, A. D., Thornberry, N. A., Peterson, E. P., and Salvesen, G. S. (1994). Inhibition of interleukine-1 beta converting enzyme by the cowpox virus serpin CrmA. An example of cross-class inhibition. *J. Biol. Chem.*, **269**, 19331–7.
10. Schick, C., Pemberton, P. A., Shi, G. P., Kamachi, Y., Cataltepe, S., Bartuski, A. J., Gornstein, E. R., Bromme, D., Chapman, H. A., and Silverman, G. A. (1998). Cross-class inhibition of the cysteine proteinases cathepsin K, L, and S by the serpin squamous cell carcinoma antigen 1: a kinetic analysis. *Biochemistry*, **37**, 5258–66.
11. Kumazaki, T., Kajiwara, K., Kojima, S., Miura, K., and Ishii, S. (1993). Interaction of *Streptomyces* subtilisin inhibitor (SSI) with *Streptomyces griseus* metallo-endopeptidase II (SGMP II). *J. Biochem.*, **114**, 570–5.
12. Laskowski, M., Sr and Laskowski, M., Jr (1954). Naturally occurring trypsin inhibitors. *Adv. Protein Chem.*, **9**, 203–42.
13. Travis, J. and Salvesen, G. S. (1983). Human plasma proteinase inhibitors. *Annu. Rev. Biochem.* **52**, 655–709.
14. Fraenckel-Conrat, H., Bean, R. C., Ducay, E. D., and Olcott, H. S. (1952). Isolation and characterization of a trypsin inhibitor from lima beans. *Arch. Biochem. Biophysics*, **37**, 393–407.
15. Ozawa, K. and Laskowski, M., Jr. (1966). The reactive site of trypsin inhibitors. *J. Biol. Chem.*, **241**, 3955–61.
16. Jirgensons, V. B., Ikenaka, T., and Gorguraki, V. (1960). A study of trypsin inhibitors. *Makromol. Chem.*, **39**, 149–65.
17. Moser, M., Auerswald, E., Mentele, R., Eckerskorn, C., Fritz, H., and Fink, E. (1998). Bdellastasin, a serine protease inhibitor of the antistasin family from the medical leech (*Hirudo medicinalis*). Primary structure, expression in yeast, and characterization of native and recombinant inhibitor. *Eur. J. Biochem.*, **253**, 212–20.
18. Xie, Z.-W., Luo, M.-J., Xu, W.-F., and Chi, C.-W. (1997). Two reactive site locations and structure–function study of the arrowhead proteinase inhibitors, A and B, using mutagenesis. *Biochemistry*, **36**, 5846–52.
19. Huang, K., Strynadka, N. C., Bernard, V. D., Peanasky, R. J., and James, M. N. G. (1994).

The molecular structure of the complex of Ascaris chymotrypsin/elastase inhibitor with porcine elastase. *Structure*, **2**, 679–89.
20. McBride, J. D., Brauer, A. B., Nievo, M., and Leatherbarrow, R. J. (1998). The role of threonine in the P2 position of Bowman–Birk proteinase inhibitors: studies on P2 variation in cyclic peptides encompassing the reactive site loop. *J. Mol. Biol.*, **282**, 447–58.
21. Tsunemi, M., Matsura, Y., Sakakibara, S., and Katsube, Y. (1996). Crystal structure of an elastase-specific inhibitor elafin complexed with porcine pancreatic elastase determined at 1.9 A resolution. *Biochemistry*, **35**, 11570–6.
22. Yang, S. Q., Wang, C. I., Gillmor, S. A., Fletterick, R. J., and Craik, C. S. (1998). Ecotin: a serine protease inhibitor with two distinct and interacting binding sites. *J. Mol. Biol.*, **279**, 945–57.
23. Liang, Z., Sottrup-Jensen, L., Aspán, A., Hall, M., and Söderhall, K. (1997). Pacifastin, a novel 155-kDa heterodimeric proteinase inhibitor containing a unique transferrin chain. *Proc. Natl. Acad. Sci.*, **94**, 6682–7.
24. Apostol, I., Giletto, A., Komiyama, T., Zhang, W., and Laskowski, M., Jr. (1993). Amino acid sequences of ovomucoid third domains from 27 additional species of birds. *J. Protein Chem.*, **12**, 419–33.
25. Magert, H. J., Standker, L., Kreutzmann, P., Zucht, H. D., Reinecke, M., Sommerhoff, C. P., Fritz, H., and Forssmann, W. G. (1999). LEKTI, a novel 15 domain type of human serine proteinase inhibitor. *J. Biol. Chem.*, **274**, 21499–502.
26. Rupp, F., Payan, D. G., Magill-Solc, C., Cowan, D. M., and Scheller, R. H. (1991). Structure and expression of a rat agrin. *Neuron*, **6**, 811–23.
27. Burgering, M. J., Orbons, L. P., van der Doelen, A., Mulders, J., Theunissen, H. J., Grootenhuis, P. D., Bode, W., Huber, R., and Stubbs, M. T. (1997). The second Kunitz domain of human tissue factor pathway inhibitor: cloning, structure determination and interaction with factor Xa. *J. Mol. Biol.*, **269**, 395–407.
28. Du, H., Gu, G., William, C. M., and Chalfie, M. (1996). Extracellular proteins needed for C. elegans mechanization. *Neuron*, **16**, 183–94.
29. Song, H. K. and Suh, S. W. (1998). Kunitz-type soybean trypsin inhibitor revisited: refined structure of its complex with porcine trypsin reveals an insight into the interaction between a homologous inhibitor from Erythrina caffra and tissue-type plasminogen activator. *J. Mol. Biol.*, **275**, 347–63.
30. Takano, R., Imada, C., Kamei, K., and Hara, S. (1991). The reactive site of marinostatin, a proteinase inhibitor from marine *Alteromonas* sp. B-10–31. *J. Biochem.*, **110**, 856–8.
31. Frigerio, F., Coda, A., Pugliese, L., Lionetti, C., Menegatti, E., Amiconi, G., Schnebli, H. P., Ascenzi, P., and Bolognesi, M. (1992). Crystal and molecular structure of the bovine alpha-chymotrypsin–eglin c complex at 2.0 A resolution. *J. Mol. Biol.*, **225**, 107–23.
32. Liu, J., Prakash, O., Huang, Y., Wen, L., Wen, J. J., Huang, J. K., and Krishnamoorthi, R. (1996). Internal mobility of reactive-site-hydrolysed recombinant Cucurbita maxima trypsin inhibitor-V characterized by NMR spectroscopy: evidence for differential stabilization of newly formed C- and T- termini. *Biochemistry*, **35**, 12503–10.
33. Greenblatt, H. M., Ryan, C. A., and James, M. N. G. (1989). Structure of the complex of *Streptomyces griseus* proteinase B and polypeptide chymotrypsin inhibitor 1 from Russet burbank potato tubers at 2.1 A resolution. *J. Mol. Biol.*, **205**, 201–28.
34. Nielsen, K. J., Heath, R. L., Anderson, M. A., and Craik, D. J. (1995). Structures of a series of 6 kDa trypsin inhibitors isolated from the stigma of *Nicotiana alata*. *Biochemistry*, **34**, 14304–11.

35. Strobl, S., Muhlhahn, P., Bernstein, R., Wiltscheck, R., Maskos, K., Wunderlich, M., Huber, R., Glockshuber, R., and Holak, T. A. (1995). Determination of the three-dimensional structure of the bifunctional α-amylase/trypsin inhibitor from ragi seeds by NMR spectroscopy. *Biochemistry*, **34**, 8281–93.
36. Ceciliani, F., Bortolotti, F., Menegatti, E., Ronchi, S., Ascenzi, P., and Palmieri, S. (1994). Purification, inhibitory properties, amino acid sequence and identification of the reactive site of a new serine proteinase inhibitor from oil-rape (*Brassica napus*) seed. *FEBS Lett.*, **342**, 221–4.
37. Pham, T. N., Hayashi, K., Takano, R., Nakazawa, H., Mori, H., Ichida, M., Itoh, M., Eguchi, M., Matsubara, F., and Hara, S. (1996). Expression of Bombyx family fungal protease inhibitor F from *Bombyx mori* by baculovirus vector. *J. Biochem.*, **119**, 1080–5.
38. Terabe, M., Kojima, S., Taguchi, S., Momose, H., and Miura, K.-I. (1995). A subtilisin inhibitor produced by *Streptomyces bikiniensis* possesses a glutamine residue at reactive site P1. *J. Biochem.*, **117**, 609–13.
39. Otlewski, J. and Krowarsch, D. (1996). Squash inhibitor family of serine proteinases. *Acta Biochim. Pol.*, **43**, 431–44.
40. Vincent, J. P. and Lazdunski, M. (1972). Trypsin–pancreatic trypsin inhibitor association. Dynamics of the interaction and role of disulfide bridges. *Biochemistry*, **11**, 2967–77.
41. Finkenstadt, W. R., Hamid, M. A., Mattis, J. A., Schrode, J., Sealock, R. W., Wang, D., and Laskowski, M., Jr. (1974). Kinetics and thermodynamics of the interaction of proteinases with protein inhibitors. In *Proteinase inhibitors. Proceedings of Bayer Symposium V* (ed. H. Fritz, H. Tschesche, and L. J. Greene), pp. 389–411. Springer-Verlag, Heidelberg.
42. Qasim, M. A., Ganz, P. J., Saunders, C. W., Bateman, K. S., James, M.N. G., and Laskowski, M., Jr. (1997). Interscaffolding additivity. Association of P_1 variants of eglin c and turkey ovomucoid third domain with serine proteinases. *Biochemistry*, **36**, 1598–607.
43. Park, S. J. (1985). Effect of amino acid replacements in ovomucoid third domains upon their association with serine proteinases. PhD thesis, Purdue University, West Lafayette, IN.
44. Wynn, R. (1990). Design of a specific human leukocyte elastase inhibitor based on ovomucoid third domains. PhD thesis, Purdue University, West Lafayette, IN.
45. Lu, W. (1994). Energetics of the interactions of ovomucoid third domain variants with different serine proteinases. PhD thesis, Purdue University, West Lafayette, IN.
46. Lebowitz, J. and Laskowski, M., Jr. (1962). Potentiometric measurement of protein–protein association constants. Soybean trypsin inhibitor–trypsin association. *Biochemistry*, **1**, 1044–55.
47. Qasim, M. A., Ranjbar, M. R., Wynn, R., Anderson, S., and Laskowski, M., Jr. (1995). Ionizable P1 residues in serine proteinase inhibitors undergo large p*K* shifts on complex formation. *J. Biol. Chem.*, **270**, 27419–22.
48. Wyman, J., Jr. (1964). Linked functions and reciprocal effects in hemoglobin: a second look. *Adv. Protein Chem.*, **19**, 223.
49. Tanford, C. (1968). Protein denaturation. In *Advances in Protein Chemistry*, (eds C. B. Anfinsen, Jr., M. L. Anson, J. T. Edsall and F. M. Richards), pp. 122–282. Academic Press, New York.
50. Finkenstadt, W. R. and Laskowski, M., Jr. (1965). Peptide bond cleavage on trypsin–trypsin inhibitor complex formation. *J. Biol. Chem.*, **240**, PC963.
51. Finkenstadt, W. R. and Laskowski, M., Jr. (1967). Resynthesis by trypsin of the cleaved peptide bond in modified soybean trypsin inhibitor. *J. Biol. Chem.*, **242**, 771–3.
52. Weber, E., Papamokos, E., Bode, W., Huber, R., Kato, I., and Laskowski, M., Jr. (1981).

Crystallization, crystal structure analysis and molecular model of the third domain of Japanese quail ovomucoid, a Kazal type inhibitor. *J. Mol. Biol.*, **149**, 109–23.
53. Schechter, I. and Berger, A. (1967). On the size of the active site in proteases I. Papain. *Biochem. Biophys. Res. Commun.*, **27**, 157–62.
54. Mattis, J. A. and Laskowski, M., Jr. (1973). pH dependence of the equilibrium constant for the hydrolysis of the Arg^{63}–Ile reactive site peptide bond in soybean trypsin inhibitor (Kunitz). *Biochemistry*, **12**, 2239–49.
55. Ardelt, W. and Laskowski, M., Jr. (1991). Effect of single amino acid replacements on the thermodynamics of the reactive site peptide bond hydrolysis in ovomucoid third domain. *J. Mol. Biol.*, **220**, 1041–53.
56. Kowalski, D. and Laskowski, M., Jr. (1972). Inactivation of enzymatically modified trypsin inhibitors upon chemical modification of the α-amino group in the reactive site. *Biochemistry*, **11**, 3451–9.
57. Kowalski, D. and Laskowski, M., Jr. (1974). Replacements, insertions, and modifications of amino acid residues in the reactive site of soybean trypsin inhibitor (Kunitz). In *Proteinase inhibitors. Proceedings of Bayer Symposium V* (ed. H. Fritz, H. Tschesche, and L. J. Greene), pp. 311–24. Springer-Verlag, Heidelberg.
58. Ardelt, W. and Laskowski, M., Jr. (1985). Turkey ovomucoid third domain inhibits eight different serine proteinases of varied specificity on the same ...Leu^{18}–Glu^{19}... reactive site. *Biochemistry*, **24**, 5313–20.
59. Bigler, T. L., Lu, W., Park, S. J., Tashiro, M., Wieczorek, M., Wynn, R., and Laskowski, M., Jr. (1993). Binding of amino acid side chains to preformed cavities: interaction of serine proteinases with turkey ovomucoid third domains with coded and noncoded P_1 residues. *Protein Sci.*, **2**, 786–99.
60. Laskowski, M., Jr. (1967). Search for the reactive site of trypsin inhibitors. *7th International Congress Biochemistry Colloquia*, 3, Tokyo, pp. 417–18.
61. Laskowski, M., Jr, Duran, R. W., Finkenstadt, W. R., Herbert, S., Hixon, H. F., Jr, Kowalski, D., Luthy, J. A., Mattis, J. A., McKee, R. E., and Niekamp, C. W. (1970). Kinetics and thermodynamics of interaction between soybean trypsin inhibitor and bovine β trypsin. *Proceeding of the International Research Conference on Proteinase Inhibitors*, pp. 117–134.
62. Sealock, R. W. and Laskowski, M., Jr. (1973). Thermodynamics and kinetics of the reactive site peptide bond hydrolysis in bovine pancreatic secretory trypsin inhibitor (Kazal). *Biochemistry*, **12**, 3139–46.
63. Krokoszynska, I. and Otlewski, J. (1996). Thermodynamic stability effects of single peptide bond hydrolysis in protein inhibitors of serine proteinases. *J. Mol. Biol.*, **256**, 793–802.
64. Papamokos, E., Weber, E., Bode, W., Huber, R., Empie, M. W., Kato, I., and Laskowski, M., Jr. (1982). Crystallographic refinement of Japanese quail ovomucoid, a Kazal type inhibitor, and model building studies of complexes with serine proteases. *J. Mol. Biol.*, **158**, 515–37.
65. Bode, W., Epp, O., Huber, R., Laskowski, M., Jr, and Ardelt, W. (1985). The crystal and molecular structure of the third domain of silver pheasant ovomucoid (OMSVP3). *Eur. J. Biochem.*, **147**, 387–95.
66. Musil, D., Bode, W., Huber, R., Laskowski, M., Jr, Lin, T.-Y., and Ardelt, W. (1991). Refined X-ray crystal structures of the reactive site modified ovomucoid inhibitor third domains from silver pheasant (OMSVP3*) and from Japanese quail (OMJPQ3*). *J. Mol. Biol.*, **220**, 739–55.

67. Carrell, R. W., Pemberton, P. A., and Boswell, D. R. (1987). The serpins: evolution and adaptation in a family of protease inhibitors. *Cold Spring Harbor Symp. Quant. Biol.*, **52**, 527–35.
68. Chase, Jr, T. and Shaw, E. (1967). *p*-Nitrophenyl-*p*-guanidinobenzoate HCl: a new active site titrant for trypsin. *Biochem. Biophys. Res. Commun.*, **29**, 508–14.
69. Miyake, Y. and Kainosho, M. (1994). Nuclear magnetic resonance studies on *Streptomyces* subtilisin inhibitor and its complexes with proteinases. In *Molecular aspects of enzyme catalysis* (ed. T. Fukui and K. Soda), p. 37. VCH, New York.
70. Bode, W. and Huber, R. (1992). Natural protein proteinase inhibitors and their interaction with proteinases. *Eur. J. Biochem.*, **204**, 433–51.
71. Mitsui, Y., Satow, Y., Watanabe, Y., Hirono, S., and Iitaka, Y. (1979). Crystal structures of *Streptomyces* subtilisin inhibitor and its complex with subtilisin BPN'. *Nature*, **277**, 447–52.
72. McPhalen, C. A. and James, M. N. G. (1988). Structural comparison of two serine proteinase–protein inhibitor complexes: eglin c-subtilisin Carlsberg and CI-2-subtilisin Novo. *Biochemistry*, **27**, 6582–98.
73. Seizen, R. J. and Leunissen, J. A. M. (1997). Subtilases: the superfamily of subtilisin-like serine proteases. *Protein Sci.*, **6**, 501–23.
74. Hipler, K., Priestle, J. P., Rahuel, J., and Grütter, M. G. (1992). X-ray crystal structure of the serine proteinase inhibitor eglin c at 1.95 Å resolution. *FEBS Lett.*, **309**, 139–45.
75. Tsunogae, Y., Tanaka, I., Yamane, T., Kikkawa, J.-I., Ashida, T., Ishikawa, C., Watanabe, K., Nakamura, S., and Takahashi, K. (1986). Structure of the trypsin binding domain of Bowman–Birk type protease inhibitor and its interaction with trypsin. *J. Biochem.*, **100**, 1637–46.
76. Helland, R., Otlewski, J., Sundheim, O., Dadlez, M., and Smalas, A. O. (1999). The crystal structures of the complexes between bovine β-trypsin and ten P_1 variants of BPTI. *J. Mol. Biol.*, **287**, 923–42.
77. Huang, K., Lu, W., Anderson, S., Laskowski, M., Jr, and James, M. N. G. (1995). Water molecules participate in proteinase–inhibitor interactions: crystal structures of Leu^{18}, Ala^{18}, and Gly^{18} variants of turkey ovomucoid inhibitor third domain complexed with *Streptomyces griseus* proteinase B. *Protein Sci.*, **4**, 1985–97.
78. Bateman, K. S., Anderson, S., Lu, W., Qasim, M. A., Laskowski, M., Jr, and James, M. N. G. (2000). Deleterious effects of β-branched residues in the S_1 specificity pocket of *Streptomyces griseus* proteinase β: crystal structures of the turkey ovomucoid third domain variants Ile^{18I}, Val^{18I}, Thr^{18I} and Ser^{18I} in complex with *Streptomyces griseus* proteinase B. *Protein Sci.*, **9**, 83–94.
79. Lu, W., Apostol, I., Qasim, M. A., Warne, N., Wynn, R., Zhang, W.-L., Anderson, S., Chiang, Y.-W., Ogin, E., Rothberg, I., Ryan, K., and Laskowski, M., Jr. (1997). Binding of amino acid side-chains to S_1 cavities of serine proteinases. *J. Mol. Biol.*, **266**, 441–61.
80. Vogel, R., Trautschold, I., and Werle, E. (1968). *Natural proteinase inhibitors*. Academic Press, New York.
81. Wilson, K. A. (1997). The protease inhibitors of seeds. In *Cellular and molecular biology of plant seed development* (ed. B. A. Larkins and I. K. Vasil), pp. 331–74. Kluwer Academic, The Netherlands.
82. McGrath, M. E., Gillmor, S. A., and Fletterick, R. J. (1995). Ecotin: lessons on survival in a protease-filled world. *Protein Sci.*, **4**, 141–8.
83. Kato, I., Schrode, J., Kohr, W. J., and Laskowski, M., Jr. (1987). Chicken ovomucoid: determination of its amino acid sequence, determination of the trypsin reactive site, and preparation of all three of its domains. *Biochemistry*, **26**, 193–201.

84. Rhodes, M. B., Bennet, N., and Feeney, R. E. (1960). The trypsin and chymotrypsin inhibitors from avian egg whites. *J. Biol. Chem.*, **235**, 1686–93.
85. Kato, I., Schrode, J., Wilson, K. A., and Laskowski, M., Jr. (1976). Evolution of proteinase inhibitors. *Protides Biol. Fluids*, **23**, 235–43.
86. Laskowski, M., Jr and Fitch, W. M. (1989). Evolution of avian ovomucoids and of birds. In *The hierarchy of life* (ed. B. Fernholm, K. Bremer, and H. Jornvall), Chapter 27, pp. 371–87. Elsevier Science, Karlskoga, Sweden.
87. Laskowski, M., Jr, Kato, I., Kohr, W. J., Park, S. J., Tashiro, M., and Whatley, H. E. (1988). Positive Darwinian selection in evolution of protein inhibitors of serine proteinases. *Cold Spring Harbor Symp. Quant. Biol.*, **52**, pp. 545–53.
88. Fujinaga, M., Read, R. J., Sielecki, A., Ardelt, W., Laskowski, M., Jr, and James, M. N. G. (1982). Refined crystal structure of the molecular complex of *Streptomyces griseus* protease B, a serine proteinase, with the third domain of ovomucoid inhibitor from turkey. *Proc. Natl Acad. Sci.*, **79**, 4868–72.
89. Read, R. J., Fujinaga, M., Sielecki, A. R., and James, M. N. G. (1983). Structure of the complex of *Streptomyces griseus* protease B and the third domain of the turkey ovomucoid inhibitor at 1.8 Å resolution. *Biochemistry*, **22**, 4420–33.
90. Fujinaga, M., Sielecki, A. R., Read, R. J., Ardelt, W., Laskowski, M., Jr, and James, M. N. G. (1987). Crystal and molecular structures of the complex of α-chymotrypsin with its inhibitor turkey ovomucoid third domain at 1.8 Å resolutions. *J. Mol. Biol.*, **195**, 397–418.
91. Bode, W., Wei, A. Z., Huber, R., Meyer, E., Travis, J., and Neumann, S. (1986). X-ray crystal structure of the complex of human leukocyte elastase (PMN elastase) and the third domain of the turkey ovomucoid inhibitor. *EMBO J.*, **5**, 2453–8.
92. Bogard, W. C. Jr., Kato, I., and Laskowski, M., Jr. (1980). A Ser^{162}/Gly^{162} polymorphism in Japanese quail ovomucoid. *J. Biol. Chem.*, **255**, 6569–74.
93. Laskowski, M., Jr, Kato, I., Ardelt, W., Cook, J., Denton, A., Empie, M. W., Kohr, W. J., Park, S. J., Parks, K., Schatzley, B. L., Schoenberger, O. L., Tashiro, M., Vichot, G., Whatley, H. E., Wieczorek, A., and Wieczorek, M. (1987). Ovomucoid third domains from 100 avian species. Isolation, sequence and hypervariability of enzyme–inhibitor contact residues. *Biochemistry*, **26**, 202–21.
94. Hill, R. E. and Hastie, N. D. (1987). Accelerated evolution in the reactive center regions of serine protease inhibitors. *Nature*, **326**, 96–9.
95. Creighton, T. E. and Darby, N. J. (1989). Functional evolutionary divergence of proteolytic enzymes and their inhibitors. *Trends Biochem. Sci.*, **14**, 319–24.
96. Ohta, T. (1994). On hypervariability at the reactive center of proteolytic enzymes and their inhibitors. *J. Mol. Evol.*, **39**, 614–19.
97. Beuning, L. L., Spriggs, T. W., and Christeller, J. T. (1994). Evolution of the proteinase inhibitor I family and apparent lack of hypervariability in the proteinase contact loop. *J. Mol. Evol.*, **39**, 644–54.
98. Wieczorek, M., Otlewski, J., Cook, J., Parks, K., Leluk, J., Wilimowska-Pelc, A., Polanowski, A., Wilusz, T., and Laskowski, M., Jr. (1985). The squash family of serine proteinase inhibitors. Amino acid sequences and association equilibrium constants of inhibitors from squash, summer squash, zucchini, and cucumber seeds. *Biochem. Biophys. Res. Comm.*, **126**, 646–52.
99. Wynn, R. and Laskowski, M., Jr. (1990). Inhibition of human β-factor XIIA by squash family serine proteinase inhibitors. *Biochem. Biophys. Res. Comm.*, **166**, 1406–10.

100. Jackson, R. M. (1999). Comparison of protein–protein interactions in serine protease–inhibitor and antibody–antigen complexes: implications for the protein docking problem. *Protein Sci.*, **8**, 603–13.
101. Groeger, C. Wenzel, H. R., and Tschesche, H. (1994). BPTI backbone variants and implications for inhibitory activity. *Int. J. Pept. Protein Res.*, **44**, 166–72.
102. Lu, W., Qasim, M. A., Laskowski, M., Jr, and Kent, S. B. H. (1997). Probing intermolecular main chain hydrogen bonding in serine proteinase–protein inhibitor complexes: chemical synthesis of backbone engineered turkey ovomucoid third domain. *Biochemistry*, **36**, 673–9.
103. Bateman, K. S., Huang, K., Anderson, S., Lu, W., Qasim, M. A., Laskowski, M., Jr, and James, M. N. G. (2000). Contribution of peptide bonds to inhibitor binding: crystal structures of the turkey ovomucoid third domain backbone variants OMTKY3-Pro^{18I} and OMTKY3-COO-Leu^{18I} in complex with *Streptomyces griseus* proteinase β (SGPB) and the structure of the free inhibitor, OMYKY3-CH_2-Asp^{19I}. (Submitted)
104. Apostoluk, W. and Otlewski, J. (1998). Variability of canonical loop conformation in serine proteinase inhibitors and other proteins. *Proteins*, **32**, 459–74.
105. Tamura, A. and Sturtevant, J. M. (1995). A thermodynamic study of *Streptomyces* subtilisin inhibitor. 3. Replacement of hyper-exposed residue, Met 73. *J. Mol. Biol.*, **249**, 646–53.
106. Qasim, M. A., Lu, S. M., Ding, J., Bateman, K. S., James, M. N. G., Anderson, S., Song, J., Markley, J. L., Ganz, P. J., Saunders, C. W., and Laskowski, M., Jr. (1999). Thermodynamic criterion for the conformation of P_1 residues of substrates and of inhibitors in complexes with serine proteinases. *Biochemistry*, **38**, 7142–50.
107. Krowarsch, D., Dadlez, M., Buczek, O., Krokoszynska, I., Samalas, A. O., and Otlewski, J. (1999). Interscaffolding additivity: binding of P1 variants of bovine pancreatic trypsin inhibitor to four serine proteinases. *J. Mol. Biol.*, **289**, 175–86.
108. Scheidig, A. J., Hynes T. R., Pelletier, L. A., Wells, J. A., and Kossiakoff, A. A. (1997). Crystal structures of bovine chymotrypsin and trypsin complexed to the inhibitor domain of Alzheimer's amyloid beta-protein (APPI) and basic pancreatic trypsin inhibitor (BPTI): engineering of inhibitors with altered specificities. *Protein Sci.*, **6**, 1806–24.
109. Capasso, C., Rizzi, M., Menegatti, E., Ascenzi, P., and Bolognesi, M. (1997). Crystal structure of the bovine alpha-chymotrypsin: Kunitz inhibitor complex. An example of multiple protein:protein recognition sites. *J. Mol. Recog.*, **10**, 26–35.
110. Dorovska, V. N., Varfolomeyev, S. D., Kazanskaya, N. F., Klyosov, A. A., and Martinek, K. (1972). The influence of the geometric properties of the active center on the specificity of α-chymotrypsin catalysis. *FEBS Lett.*, **23**, 122–4.
111. Bode, W., Meyer, E., Jr, and Powers, J. C. (1989). Human leukocyte and porcine pancreatic elastase. X-ray crystal structures, mechanism, substrate specificity, and mechanism-based inhibitors. *Biochemistry*, **28**, 1951–63.
112. Kojima, S., Nishiyama, Y., Kumagai, I., and Miura, K-I. (1991). Inhibition of subtilisin BPN′ by reaction site P1 mutants of *Streptomyces* subtilisin inhibitor. *J. Biochem.*, **109**, 377–82.
113. Empie, M. W. and Laskowski, M. Jr. (1982). Thermodynamic and kinetics of single residue replacements in avian ovomucoid third domains: effect on inhibitor interactions with serine proteinases. *Biochemistry*, **21**, 2274–84.
114. Nguyen, T. T., Qasim, M. A., Morris, S., Lu, C.-C., Hill, D., Laskowski, M., Jr, and Sakanari, J. A. (1999). Expression and characterization of elastase inhibitors from the ascarid nematodes *Anisakis simplex* and *Ascaris suum*. *Mol. Biochem. Parasit.*, **102**, 79–89.

115. Imperiali, B. and Abeles, R. H. (1987). Extended binding inhibitors of chymotrypsin that interact with leaving group subsites S'_1–S'_3. *Biochemistry*, **26**, 4474–7.
116. Ranjbar, M., Wynn, R., Anderson, S. and Laskowski, M., Jr. (1998). Effects of the P_2 residue of the turkey ovomucoid third domain inhibitor on the catalytic histidine of serine proteinases. *Protein Sci.* **7** (Suppl. 1), 152.
117. Ranjbar, M. (1999). Energetic effects of the P2 residue in standard mechanism inhibitors on the strength of association with different serine proteinases. PhD thesis, Purdue University, West Lafayette, IN 47907.
118. Bauer, C.-A. (1978). Active centers of *Streptomyces griseus* protease 1, *Streptomyces griseus* protease3 and α-chymotrypsin: enzyme–substrate interactions. *Biochemistry*, **17**, 375–80.
119. Svendsen, I. B. (1976). Chemical modifications of the subtilisins with special reference to the binding of large substrates. A review. *Carls. Res. Commun.*, **41**, 237–91.
120. Forsyth, W. R., Gilson, M. K., Antosiewicz, J., Jaren, O. R., and Robertson, A. D. (1998). Theoretical and experimental analysis of ionization equilibria in ovomucoid third domain. *Biochemistry*, **37**, 8643–52.
121. Nozaki, Y, and Tanford, C. (1967). Examination of titration behavior. In *Methods in enzymology*, Vol. 11 (ed. C. H. W. Hirs), pp. 715–48. Academic Press, New York.
122. Bundi, A. and Wüthrich, K. (1979). ^{1}H-NMR parameters of the common amino acid residues measured in aqueous solutions of the linear tetrapeptides H–Gly–Gly–X–L–Ala–OH. *Biopolymers*, **18**, 285–97.
123. Homandberg, G. A., Mattis, J. A., and Laskowski, M., Jr. (1978). Synthesis of peptide bonds by proteinases. Addition of organic cosolvents shifts peptide bond equilibria toward synthesis. *Biochemistry*, **17**, 5220–7.
124. Komiyama, T., Bigler, T. L., Yoshida, N., Noda, K., and Laskowski, M., Jr. (1991). Replacement of P1 Leu^{18} by Glu^{18} in the reactive site of turkey ovomucoid third domain converts it into a strong inhibitor of Glu-specific *Streptomyces griseus* proteinase (GluSGP). *J. Biol. Chem.*, **266**, 10727–30.
125. Yoshida, N., Tsuruyama, S., Nagata, K., Hirayama, K., Noda, K., and Makisumi, S. (1988). Purification and characterization of an acidic amino acid specific endopeptidase of *Streptomyces griseus* obtained from a commercial preparation (Pronase). *J. Biochem.*, **104**, 451–6.
126. Birktoft, J. J. and Breddam, K. (1994). Glutamyl endopeptidases. In *Methods in enzymology*, Vol. 244 (ed. A. J. Barrett), pp. 114–26. Academic Press, New York.
127. Kelly, C. A. (1989). Structural and functional consequences of breaking non-reactive site bonds in protein inhibitors of serine proteinases. PhD thesis. Purdue University, West Lafayette, IN 47907.
128. Laskowski, M., Jr. (1980). An algorithmic approach to sequence leads to reactivity of proteins. Specificity of protein inhibitors of serine proteinases. *Biochem. Pharmacol.*, **29**, 2089–94.
129. Laskowski, M., Jr, Park, S. J., Tashiro, M., and Wynn, R. (1989). Design of highly specific inhibitors of serine proteinases. In *Protein recognition of immobilized ligands* (ed. T. W. Hutchens), pp. 149–68. Alan R. Liss, New York.
130. Nyaruhucha, C. N. M., Kito, M., and Fukuoka, S-I. (1997). Identification and expression of the cDNA-encoding human mesotrypsin(ogen), an isoform of trypsin with inhibitor resistance. *J. Biol. Chem.*, **272**, 10573–10578.
131. Estell, D. A. and Laskowski, M., Jr. (1980). *Dermasterias imbricata* trypsin 1: An enzyme which rapidly hydrolyses the reactive-site peptide bonds of porcine trypsin inhibitors *Biochemistry*, **19**, 124–131.

132. Estell, D. A., Wilson, K. A., and Laskowski, M., Jr. (1980). Thermodynamic and kinetics of the hydrolysis of the reactive-site peptide bond in pancreatic trypsin inhibitor (Kunitz). by *Dermasterias imbricata* trypsin 1 *Biochemistry*, **19**, 131–137.
133. van de Locht, A., Bode, W., Huber, R., Le Bonniec, B. F., Stone, S. R., Esmon, C. T., and Stubbs, M. T. (1997). The thrombin E192Q–BPTI complex reveals gross structural rearrangements: implications for the interaction with antitrhombin and thrombomodulin. *EMBO J.*, **16**, 2977–84.
134. Le Bonniec, B. F., Guinto, E. R., and Esmon, C. T. (1992). Interaction of thrombin des-ETW with antithrombin III, the Kunitz inhibitors, thrombomodulin and protein C. *J. Biol. Chem.*, **267**, 19341–8.
135. Le Bonniec, B. F., Guinto, E. R., MacGillivray, R. T., Stone, S. R., and Esmon, C. T. (1993). The role of thrombin's Tyr–Pro–Pro–Trp motif in the interaction with fibrinogen, thrombomodulin,protein C, antithrombin III, and the Kunitz inhibitors *J. Biol. Chem.*, **268**, 19055–61.
136. Guinto, E. R., Ye, J., Le Bonniec, B. F., and Esmon, C. T. (1994). Glu^{192}®Gln substitution in thrombin yields an enzyme that is effectively inhibited by bovine pancreatic trypsin inhibitor and tissue factor pathway inhibitor. *J. Biol. Chem.*, **269**, 18395–400.
137. Rezaie, A. R. and Esmon, C. T. (1993). Conversion of glutamic acid 192 to glutamine in activated protein C changes the substrate specificity and increases reactivity toward macromolecular inhibitors. *J. Biol. Chem.*, **268**, 19943–8.
138. Pereira, P. J. B., Bergner, A., Macedo-Ribeiro, S., Huber, R., Matschiner, G., Fritz, H., Sommerhoff, C. P., and Bode, W. (1998). Human β-tryptase is a ring-like tetramer with active sites facing a central pore. *Nature*, **392**, 306–11.
139. Laskowski, M., Jr, Kato, I., Leary, T. R., Schrode, J., and Sealock, R. W. (1974). Evolution of specificity of protein proteinase inhibitors. In *Proteinase inhibitors. Bayer Symposium V* (ed. H. Fritz, H. Tschesche, L. J. Green, and E. Truscheit), pp. 597–611. Springer-Verlag, New York.
140. Klabe, R. M., Bacheler, L. T., Ala, P. J., Erickson-Vitanen, S., and Meek, J. L. (1998). Resistance to HIV protease inhibitors: a comparison of enzyme inhibition and antiviral potency. *Biochemistry*, **37**, 8735–42.
141. Asao, T., Takahashi, K., and Tashiro, M. (1998). Interaction of second and third domains of Japanese quail ovomucoid with ten mammalian trypsins. *Biochim. Biophys. Acta*, **1387**, 415–21.
142. Peanasky, R. J., Martzen, M. R., Homandberg, G. A., Cash, J. M., Babin, D. R., and Litweiler, B. (1987). Proteinase inhibitors from intestinal parasitic helminths: structure and indications of some possible functions. In *Molecular paradigms for eradicating helminthic parasites* (ed. A. J. MacInnis), p. 349. Alan R. Liss, New York.
143. Hawley, J. H. and Peanasky, R. J. (1992). *Ascaris suum:* are trypsin involved in species specificity of *Ascarid* Nemotodes? *Exp. Parasitol.*, **75**, 112–18.
144. Roach, J. C., Wang, K., Gan, L., and Hood, L. (1997). The molecular evolution of the vertebrate trypsinogens. *J. Mol. Evol.*, **45**, 640–52.
145. Jongsma, M. A., Bakker, P. L., Peters, J., Bosch, D., and Stiekema, W. J. (1995). Adaptation of *Spodoptera exigua* larvae to plant proteinase inhibitors by induction of gut proteinase activity insensitive to inhibition. *Proc. Natl Acad. Sci.*, **92**, 8041–5.

Notes

1. The earlier notion that most or all proteinase inhibitors are trypsin inhibitors gave rise to many blunders in nomenclature. The most abundant serpin in human blood plasma was named α_1-antitrypsin. It was renamed α_1-proteinase inhibitor when it was realized that its major target enzyme was human neutrophil elastase (13). The earlier incorrect name persists, confusing the literature and continuing the long string of studies on trypsin–α_1-proteinase inhibitor interaction. Inter-α-trypsin inhibitor is a similar misnomer.
2. We are coining the word ionophobic for pockets in which the neutral form of a buried ionizable side chain is strongly preferred to the charged one, i.e. Asp° is better than Asp^-, Lys° is better than Lys^+.
3. This argument flies in the face of the well-known observation (see ref. 123 for a list of references) that uncharged, ionizable organic compounds such as carboxylic acids experience a large increase in p*K* upon the addition of large amounts of low dielectric-constant, organic co-solvants to their aqueous solutions. In contrast, cationic organic compounds such as alkylamines experience no p*K* shift on such additions. The explanation is that the uncharged compounds ionize with charge separation. Neutral reactants yield two charged products: an anion and a hydronium ion. A low dielectric constant opposes this reaction. In the case of cationic compounds, ionization occurs with charge transfer. The positively charged acid, such as alkyl ammonium ion, transfers its charge to a hydronium ion. To a close approximation, the reactants and the products experience the same adverse effects of low dielectric constant.

   In the ionization in the hydrophobic, ionophobic pocket, the hydronium ion escapes into free solution and only the side-chain ion experiences the low dielectric constant.
4. About 10 years ago, the notion that we could design truly very specific inhibitors that were active against only one serine proteinase occurred to us (129). We were briefly enchanted by the medical prospects of the idea. However, physicians of our acquaintance set us straight. Humans have a very large number of serine proteinases. Physicians can often diagnose excess proteolysis, but almost never can identify the single enzyme that is responsible. The problem is analogous to the use of broad-spectrum antibiotics in preference to narrow-spectrum ones.

9 Nuclease inhibitors

COLIN KLEANTHOUS and ANSGAR J. POMMER

1. Introduction

Nucleases are involved in a wide variety of biological functions in prokaryotes and eukaryotes including DNA repair, RNA maturation, and recombination as well as having less well-understood roles in processes such as angiogenesis in mammals and self-incompatibility in plants (1, 2). The hydrolysis of a phosphodiester bond has the potential to be toxic to the host cell unless controlled in some way. In the case of restriction enzymes, the recognition sequences of which may occur many times in a bacterial genome, toxicity is avoided by accompanying methylases which 'detoxify' the recognition sequence. In other instances—for example where the nuclease is relatively non-specific in its choice of cleavage sequence—such a strategy is not possible, in which case cell death can be avoided by the action of inhibitor proteins. Hence, nuclease inhibitors serve an important cellular role, exemplified by the inhibitor ICAD which inhibits the activity of caspase-activated deoxyribonuclease, an enzyme involved in the degradation of DNA during apoptosis (3).

Nuclease–inhibitor complexes have proven to be a rich source of information about protein–protein interactions. Some of the best-understood systems are those that form very stable complexes with equilibrium dissociation constants (K_d) $< 10^{-14}$ M. This review focuses on three high-affinity, nuclease inhibitor systems: ribonuclease inhibitor binding RNase A and angiogenin (Figs 1a and 1b, respectively); barstar binding the RNase barnase (Fig. 1c); and the immunity protein Im9 binding the colicin E9 DNase (Fig. 1d). For each system, three-dimensional structures are available for both the free proteins and their complexes, and detailed kinetic and thermodynamic investigations have been conducted aimed at dissecting complex formation. These data will be summarized to illustrate current thinking on how and why these proteins form specific and stable complexes, identify similarities (and differences) between them, and highlight general principles that have come from their study.

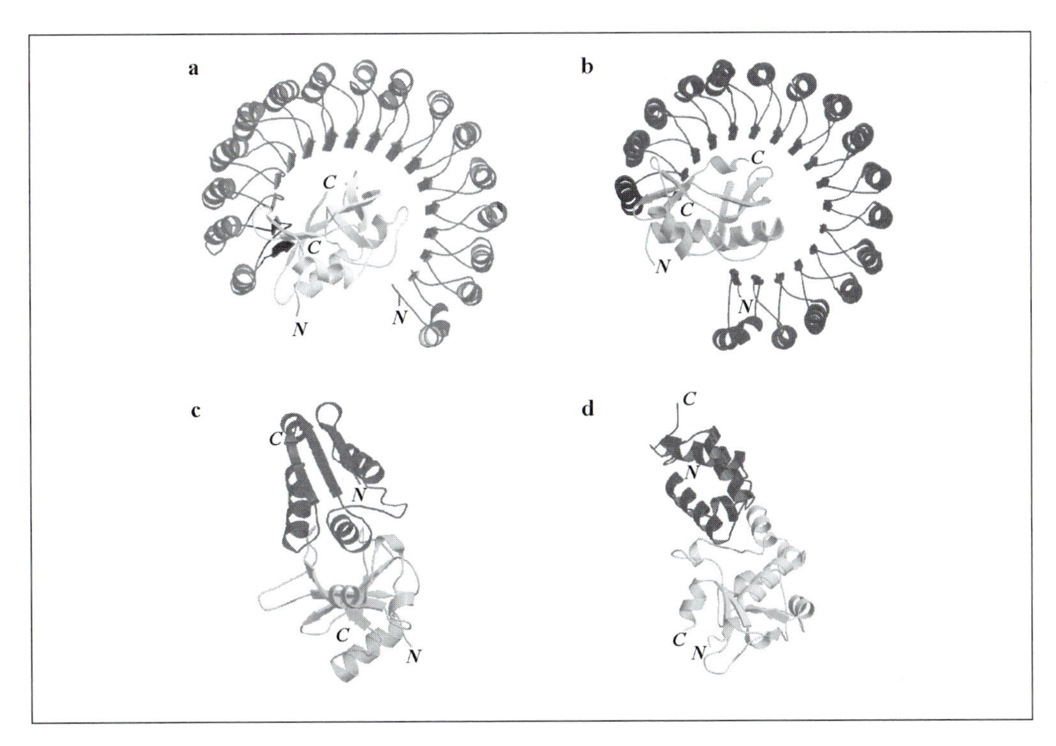

Fig. 1 Crystal structures of nucleases (shown in light shade) complexed to inhibitor proteins (dark shade). (a) 2.5-Å structure of the porcine ribonuclease inhibitor–RNase A complex (ref. 18; pdb accession number 1dfy). (b) 2.0-Å structure of the human ribonuclease inhibitor–angiogenin complex (ref. 19; pdb accession number 1a4y). (c) 2.0-Å structure of the barnase–barstar complex from *Bacillus amyloliquefaciens* (ref. 45; pdb accession number 1brs). (d) 2.05-Å structure of the colicin E9 DNase domain–Im9 complex from *Escherichia coli* (ref. 74; pdb accession number 1bxi). Figures were prepared using the program MOLSCRIPT (89).

2. Ribonuclease inhibitor

2.1 Ribonuclease-inhibitor biology

Ribonuclease inhibitor (RI) is a potent inhibitor of pancreatic-type RNases, such as RNase A, and is found in the tissues of mammals, amphibians, and insects. First discovered in 1952 by Pirotte and Desreux (4), the interaction of RI with different RNases has become a paradigm in studies of molecular recognition. Ironically, however, even though RI is thought to be an essential protein in mammals (it is found in essentially every tissue tested for its presence) its physiological function is still debated, with three proposed biological roles (5):

(1) regulation of cytoplasmic RNA levels through its ability to inhibit RNases (and this is supported, for example, by changes in RI levels which follow changes in RNA metabolism that occur during development);

(2) protection against secreted RNases inappropriately targeted to the cytoplasm;

(3) regulation or termination of the physiological roles of RNases such as angiogenin, a potent inducer of angiogenesis.

This latter activity, in particular, is of importance in medicine since increased neovascularization is associated with the growth of solid tumours, and so inhibitors of this activity would be potential antitumour agents (5).

2.2 Kinetic and thermodynamic studies of RI binding to RNases

RI forms very tight one-to-one complexes with RNases (6). Kinetic investigations have been reported for both human ribonuclease inhibitor (hRI) and porcine RI (pRI) binding to RNase A, and hRI binding to angiogenin (7–10). Capitalizing on a 50% enhancement in tryptophan emission fluorescence (which results from the association of angiogenin with hRI) and using stopped-flow experiments, Lee *et al.* (7) found that there was a hyperbolic protein concentration-dependence of the pseudo first-order rate constant for association, consistent with a two-step association mechanism of the form:

$$E + I \underset{k_{-1}}{\overset{k_1}{\rightleftharpoons}} EI \underset{k_{-2}}{\overset{k_2}{\rightleftharpoons}} EI^*; \qquad [1]$$

where E is the enzyme and I is the inhibitor. In this two-step mechanism, also observed in other nuclease–inhibitor associations (Sections 3.4 and 4.3), the first step is the rapid formation of an encounter complex EI ($k_1 \sim 10^8\ M^{-1}\ s^{-1}$), which then slowly isomerizes to form a tight enzyme–inhibitor complex (EI*) with a rate constant (k_2) $\sim 100\ s^{-1}$. The equilibrium constant of the encounter complex (K_1) was estimated as approximately micromolar. The slow dissociation rate constant (k_{-2}) was determined by subunit-exchange experiments, in which a chase or scavenger is added to preformed complexes and the appearance of released enzyme quantitated either by enzymatic activity or by the radioactivity of labelled protein following chromatography (8, 11). This rate is particularly slow in the angiogenin–hRI complex, equivalent to a half-life of about 60 days ($k_{-2} = 1.3 \times 10^{-7}\ s^{-1}$) (see ref. 8). The inhibition constant, K_i, for hRI binding to angiogenin and RNase A for the two-step mechanism was obtained from eqn 2:

$$K_i = \frac{K_1 k_{-2}}{k_{-2} + k_2}\ . \qquad [2]$$

Since $k_2 >> k_{-2}$ for these complexes, the inhibition constant (K_i, equivalent to K_d in this context) reduces to the ratio of dissociation (k_{-2}) and association (k_2/K_1) rate constants. The K_d for the RI–RNase A complex is 10^{-14}M, while that for the RI–angiogenin complex is 10^{-16} M. Of particular note is the association rate constant for RI–RNase complexes which approaches the diffusion controlled limit (estimated as $7 \times 10^9\ M^{-1}\ s^{-1}$; ref. 12). This is common in nuclease–inhibitor associations, observed in both the barnase–barstar complex and in colicin DNase–immunity protein complexes, and reflects the importance of electrostatics in these associations.

2.3 RI structure

Most molecular studies have focused on ribonuclease inhibitor from human placenta and porcine liver. The amino-acid sequences of both are known (13, 14), showing that they share 77% sequence identity. The proteins (both of which have been expressed either in bacteria or yeast) are ~50-kDa monomers; the human form is 460 amino acids in length, while that from pig is 456 amino acids. Two distinguishing features characterize the RI primary amino-acid sequence. First, it is made up almost exclusively of leucine-rich repeats (LRRs) and, second, it contains a large number of cysteine residues (30 in pRI) that are all in the sulfhydryl form, which explains the sensitivity of the protein toward oxidation and thiol-modifying reagents (15).

LRRs are consensus sequences of 20–29 amino acids (the most common being 24) where leucines or other aliphatic residues occur at positions 2, 5, 7, 12, 16, 21, and 24, with asparagine, cysteine, or threonine at position 10. Thus, porcine RI, the first LRR-containing protein for which a structure was determined (16), is composed of 15 LRRs alternating in repeat sequence length (the so-called A-type repeat being 28 amino acids and the B-type 29 amino acids in length; Fig. 2). The majority of the protein (90%) is built from these repeats, which correspond to right-handed β–α structural units arranged so that all the β-strands and helices are parallel to a common axis. This results in a non-globular, horseshoe-shaped molecule with a curved, parallel β-sheet lining the inner surface and helices lying on the outer surface (Fig. 1a). The inner diameter of the molecule is about 21 Å, while the outer diameter is 67 Å. The consensus residues of the LRRs play structural roles, with the side chains of leucines and other hydrophobic amino acids making up the core of the protein. Many of the cysteine residues are also buried in the structure forming hydrogen bonds with the protein backbone. LRRs are a very adaptable structural motif present in prokaryotic and eukaryotic proteins, serving functions as diverse as signal transduction, cell adhesion, development, and DNA repair as well as inhibiting RNases (17). In

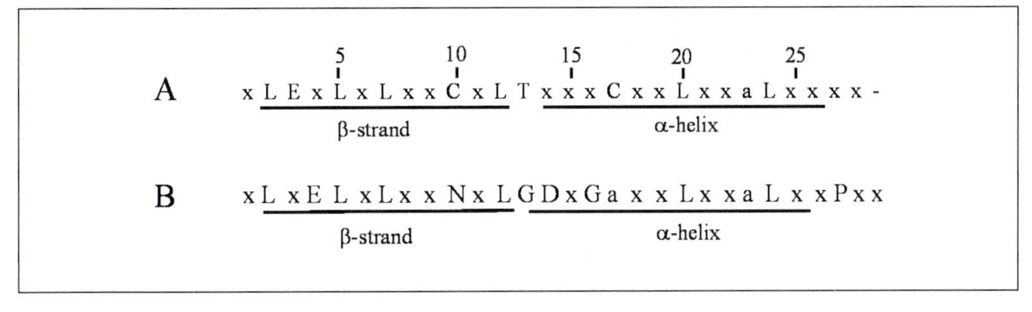

Fig. 2 Consensus sequences and secondary structure of the alternating A- and B-type LRR repeat sequence from pRI. The single letter amino-acid code is used: 'x' indicates any amino acid and 'a' an aliphatic amino acid. (Adapted from Kobe and Deisenhofer (17).)

nearly all these biological guises, LRRs are predicted to be involved in protein–protein recognition, presumably because of the large solvent-exposed areas that are available for binding interactions.

2.4 RI complexes and the mechanism of inhibition of RNase activity

The structures of two RI–RNase complexes are known at present which display interesting similarities and differences. The structure for a complex was first reported by Kobe and Deisenhofer (18) for RNase A bound to pRI (Fig. 1a). RNase A is a 124-amino acid, kidney bean-shaped molecule, composed primarily of an antiparallel β-sheet and stabilized by four disulfide bonds, with the active site located between the two lobes. In the complex with pRI, RNase binds to the concave region of pRI and straddles the C-terminus of the inhibitor with its active site, one lobe of the enzyme within the cavity of the horseshoe the other lying over its face. While RNase A does not undergo any significant conformational changes, the inhibitor undergoes a 'plastic reorganization' on binding that takes the form of small shifts along the entire length of the chain, resulting in a 2-Å enlargement of the concave cavity (18). This contrasts with the crystal structure of hRI bound to angiogenin where the inhibitor dimensions are similar to those of unbound pRI (19). The orientation of the enzyme with respect to the inhibitor, however, is the same in both structures (Fig. 1b). Roughly equivalent solvent-accessible surface areas are buried in the two complexes (>2550 Å^2), which is significantly larger than typical protein–protein interaction surfaces and indeed significantly larger than both barstar–barnase and Im9–E9 DNase complexes. Also, RNase–RI complexes contain more charged amino acids than are normally observed at protein–protein interfaces, consistent with the electrostatically driven association. In both RI complexes, there are between 110 and 120 contacts between the bound enzymes and the inhibitor, of which 11 and 18 are hydrogen bonds in the RNase–pRI and angiogenin–hRI complexes, respectively. The interfaces involve 24 amino acids coming from each of the enzymes, and 28 residues from pRI and 26 from hRI. Interestingly, the shape complementarity of the two interfaces, as determined by the method of Lawrence and Colman (20), are not identical with the pRI–RNase A complex being significantly poorer than that of hRI–angiogenin ($Sc = 0.58$ and 0.70, respectively, with 1.0 being a perfect match).

In both RI complexes the inhibitor covers the active cleft of the enzyme and makes several contacts with active-site residues, although some of these differ in molecular detail. Hence, inactivation of the enzyme is through competitive inhibition with the substrate and this is supported by many earlier solution experiments. Lee and Vallee (21) showed that cytidine 2′-phosphate decreased the rate of association of hRI for RNase A, while chemical modification experiments with RNase A and site-directed mutagenesis of angiogenin both showed that the same critical active-site lysine residue (Lys40 in angiogenin, Lys41 in RNase A) was important for inhibitor binding, both results implying direct competition (21, 22). The ε-amino group of

Lys41 in RNase A is part of the catalytic P1 site thought to be involved in stabilizing the trigonal bipyramidal transition state formed during the hydrolysis of RNA, since structural studies have shown that it makes direct contact with one of the oxygens of uridine vanadate, a transition-state mimic (23, 24). This residue is completely masked by ribonuclease inhibitor; in the structure of angiogenin in complex with hRI, the ε-amino group of Lys40 forms two salt bridges with the Asp435 carboxylate of the inhibitor (equivalent to Asp431 in porcine RI), while its alkyl chain forms numerous van der Waals contacts with the ring of Tyr434 (equivalent to Tyr430 in pRI) (Fig. 3). As well as coordinating active-site amino acids through a variety of interactions, RI also inhibits the enzyme by partially mimicking substrate RNA, a strategy also adopted by the RNase inhibitor barstar when binding to barnase. This is nicely illustrated in the pRI–RNase A structure, where the aromatic ring of Tyr433 superimposes closely with the ring of the T^4 thymine base from the crystal structure of RNase A in complex with a thymidilic acid tetramer (Fig. 4) (16, 25).

2.5 Protein–protein interaction specificity of RI

Many different RNases can be bound and inhibited by RI even though they share limited sequence identity, raising the question of how the inhibitor recognizes such diverse proteins yet with similarly high binding affinities? RNase A and angiogenin are only 27% identical in sequence, but both bind to hRI with very high affinities that differ by only two orders of magnitude. Kobe and Deisenhofer (18) argued that the ability of RI to bind diverse RNases so effectively could be a consequence of the small conformational adjustments that occur in the inhibitor to accommodate the bound enzyme, as well as the unusually large surface that is buried so compensating for the poor complementarity of the resulting complex. However, as noted above, neither the conformational adjustments nor the poor complementarity are seen in the

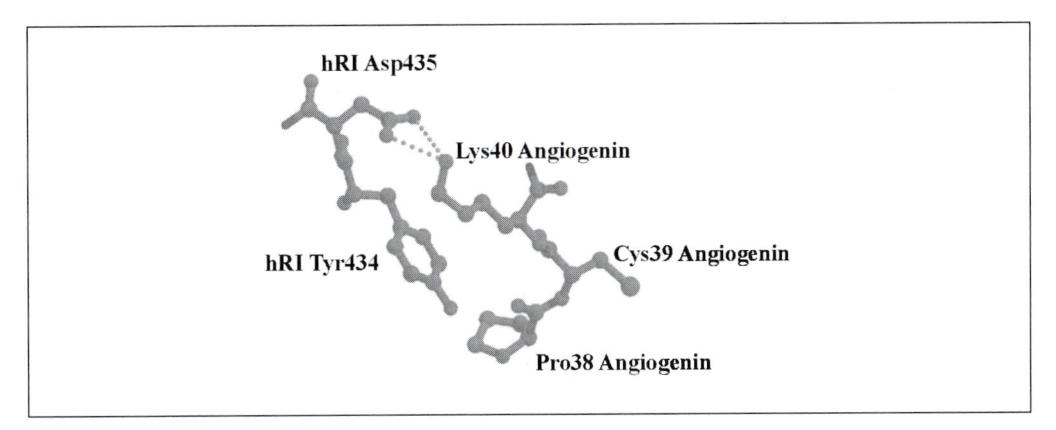

Fig. 3 Human RI ligates the active-site lysine residue of angiogenin. The ε-amino group of the active-site lysine, Lys40, of angiogenin forms a double salt bridge with the carboxylate of hRI Asp435, while its alkyl chain forms van der Waals interactions with hRI Tyr434. (Adapted from Papageorgiou *et al.* (19).)

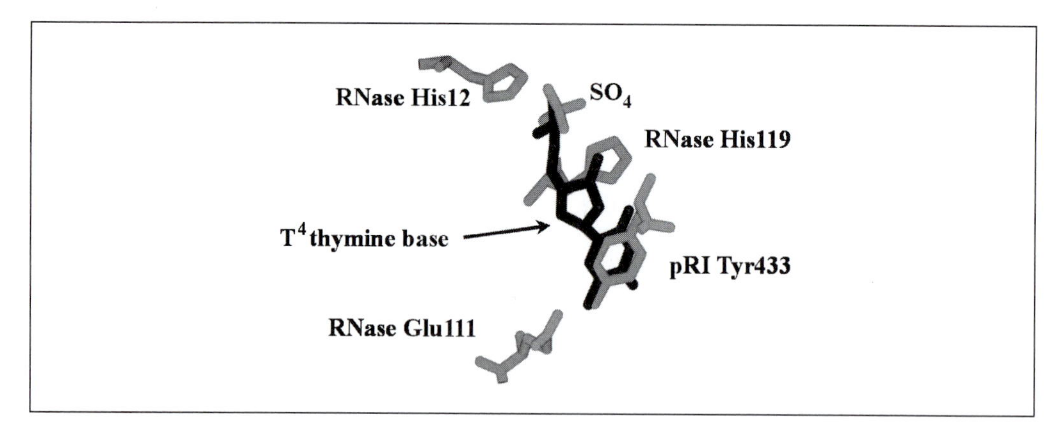

Fig. 4 Porcine RI mimics part of the nucleotide substrate of RNase A. The figure shows selected active-site residues from an overlay of the pRI–RNase A structure (shown in light shading), with the position of the T4 thymine base from the thymidilic acid tetramer $d(pT)_4$ bound to RNase A (shown in dark shading) (25). Note how Tyr433 of pRI binds in the same pocket as the aromatic ring of the base, with the tyrosine hydroxyl group pointing in the same direction as the methyl group of the base. (Adapted from Kobe and Deisenhofer (18).)

complex of angiogenin with hRI (19). Papageorgiou *et al.* (19) made a detailed comparison of the two RI complexes, concluding that recognition of the bound RNases by the inhibitor is achieved largely by different amino acids and a distinct complement of interactions. This parallels previous conclusions from Acharya *et al.* (26) who suggested, based on the crystal structure of angiogenin compared to that of RNase A, that RI recognized the two enzymes in different ways.

Looking more closely at the residues involved at the interfaces of RNase enzyme complexes with RI and using this as a basis for protein engineering experiments has revealed some important insights into the molecular basis of recognition in this system. Of the 24 amino acids from each of the RNases involved in the protein–protein interaction, only 9 are identical or conservatively substituted; while 17 residues from RI of the 28 and 26 of RNase A and angiogenin, respectively, are similarly conserved (19). In addition, conserved pairs of amino acids from each binding partner in many instances do not engage in similar interactions when the two complexes are compared. For example, Trp438 in hRI hydrogen-bonds to the Arg5 of angiogenin, whereas the equivalent residue in pRI is only involved in van der Waals interactions with Glu111 of RNase A. Interactions between RI residues and either angiogenin and RNase A have been explored by mutagenesis and show broad agreement with the differing interactions observed in the crystal structures. For example, both Tyr430 and Tyr433 of pRI, make extensive interactions with RNase A, the former with four residues in the enzyme (Leu35, Arg39, Lys41, and Pro42), the latter hydrogen-bonding Asn71 and Glu111. Mutation of their counterparts in hRI to alanine (Tyr434 and Tyr437; Fig. 5) produced significant effects on RNase A binding energy as determined from $\Delta\Delta G$, calculated according to eqn 3:

$$\Delta\Delta G = -\mathrm{RT}\ln(K_{\mathrm{d}}^{\mathrm{wild\ type}} / K_{\mathrm{d}}^{\mathrm{mutant}}). \qquad [3]$$

In the case of the hRI Tyr434Ala mutant binding RNase A, $\Delta\Delta G$ was 5.9 kcal mol^{-1}, while that for Tyr437Ala was 2.6 kcal mol^{-1} (11). The same hRI mutations had less of an impact when complexed to angiogenin ($\Delta\Delta G$ for Tyr434Ala was 3.3 kcal mol^{-1}, while Tyr437Ala was 0.8 kcal mol^{-1}), with which the original residues make fewer interactions (11). Nevertheless, some of the conserved residues clearly form similar types of interactions, such as hRI Asp435 which coordinates the active-site lysine of angiogenin (Fig. 3). Mutation of this residue to alanine affects the binding of both RNase A and angiogenin to similar extents ($\Delta\Delta G \sim 3.5$ kcal/ mol^{-1}), implying the same interactions may be lost. It should be noted, however, that the interaction towards the catalytic lysine residue is not observed in the crystal structure of pRI bound to RNase A due, it is argued by Papageorgiou *et al.* (19), to the crystallization conditions used by Kobe and Deisenhofer (18). In summary, most protein engineering experiments on RI–RNase binding have focused on the interactions of the C-terminal residues of RI (434–437 in hRI, Fig. 5) which are thought to constitute a binding energy hot-spot for the active sites of their target enzymes (11). Recent studies have shown general support for this notion, but have indicated that caution be used when assigning such hot-spots by single-site mutations since these may hide cooperative effects which can only be identified by multiple residue substitutions (27). The issue of cooperativity in protein–protein recognition is further explored in the other nuclease–inhibitor complexes discussed in this chapter.

Similar surface areas are buried in the hRI–angiogenin and pRI–RNase A complexes, but, as summarized above, the distribution of contact sites is not identical. Figure 6 shows the contact sites of each RI from the crystal structures. This depiction

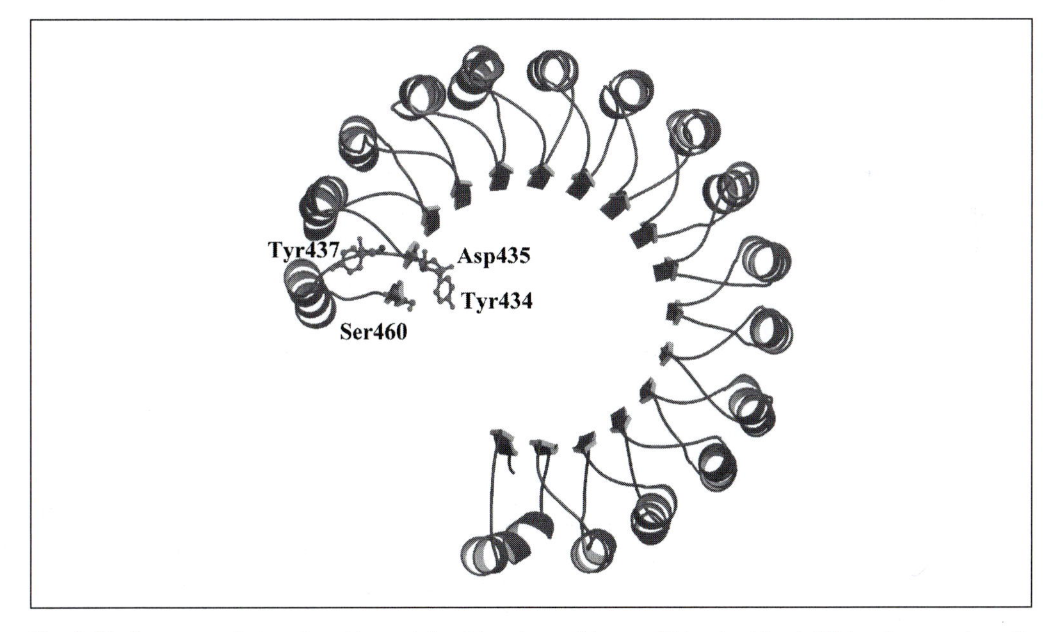

Fig. 5 Binding energy hot-spot residues at the C-terminus of human RI involved in stabilizing the complex with RNase A (11). See text for details.

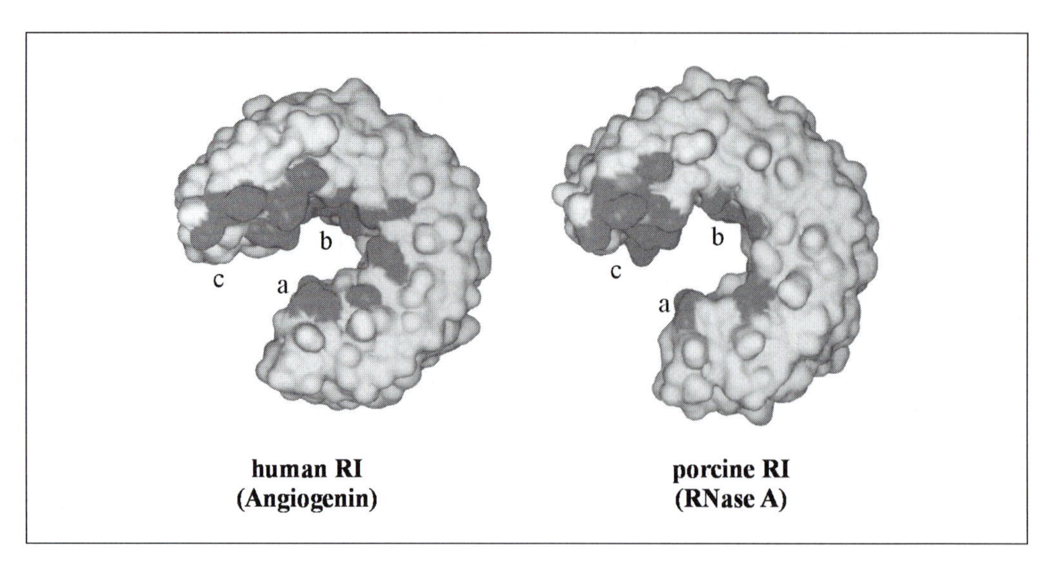

Fig. 6 Buried surfaces in RI–RNase complexes. Molecular surface representations of human and porcine RIs showing the three areas (a–c) of the inhibitor that become buried on binding angiogenin and RNase A, respectively. The orientation of the inhibitors is the same as that in Fig. 1. Some of the residues in region c for hRI are identified in Fig. 5.

makes it clear that three distinct regions of the inhibitor are used in forming interactions with the enzymes (denoted 'a–c', with region 'c' encompassing the hot-spot region identified by Chen and Shapiro (11, 27)). Although similar, these regions are not identical in the two complexes—a situation reminiscent of complexes formed between human growth hormone binding the prolactin and growth hormone receptors where the epitopes on the hormone for binding each receptor overlap but are not identical (28). A systematic analysis of all the contact sites, particularly region 'b' which involves a number of conserved tryptophan residues in both human and porcine RI, have yet to be reported and so it will be of interest to see whether any of these conserved amino acids are important in stabilizing both complexes.

3. Barnase and its inhibitor barstar

3.1 Barnase–barstar biology

Barnase (12.4 kDa) is an extracellular enzyme secreted by *Bacillus amyloliquefaciens*, and is a member of the family of small RNases found in bacteria and fungi (29). Barnase is secreted into the periplasmic space and then released into the environment where it may act as a digestive enzyme degrading extracellular RNA for nutritional use by the microorganism (30). Barstar (10.2 kDa) is an intracellular inhibitor protein of barnase activity (31). Although barnase and barstar occupy separate cellular compartments and so would not be expected to meet, it is thought that occasional mistargeting and intracellular folding of barnase requires the presence of an inhibitor protein to protect the organism. In keeping with this supposition, bacterial cells are not viable when secreting active barnase in the absence of the barstar gene, although inactive mutants of barnase can be secreted from such a genetic background (32, 33). The latter observation also demonstrates that barstar is not required for secretion *per se* but for protection against the intracellular RNase activity of barnase. The lethal effect of expressing active barnase in bacteria can be overcome by coexpressing active barstar, resulting in high levels of free and active barnase in the extracellular medium (33).

Barnase and barstar have proven to be excellent targets for protein structure–function studies. They are both small proteins that can be overexpressed in bacteria, neither contain disulfide bonds (although wild-type barstar possesses two cysteine residues, see below), and both can be reversibly denatured. This and the many structural studies of the two proteins have led to them both being used as models in protein folding studies (12, 34). The barnase–barstar system has also provided many insights into protein–protein recognition, and it is these studies that are described in the following sections.

3.2 Structure and mechanism of barnase

Barnase catalyses the hydrolysis of RNA through essentially the same in-line mechanism as RNase A, involving a rapid *trans*-esterification reaction converting a 3′,5′-phosphate into a 2′,3′-cyclic phosphate intermediate which is then slowly hydrolysed to yield the 3′-phosphate product. In common with other members of the microbial RNase family, barnase shows a preference for hydrolysing bonds on the 3′ side of guanine bases (35, 36) and is able to hydrolyse dinucleotide substrates, although it is significantly more active against longer RNA molecules. This rate enhancement is due to subsite interactions between the enzyme and regions of the substrate flanking the hydrolysed bond; for example, occupancy of the phosphate site at P2 (where P1 is the site of hydrolysis) improves the catalysis of a dinucleotide substrate by 1000-fold (37).

Several structures for barnase have been reported, the original being a 2-Å resolution crystal structure determined by Mauguen *et al.* (38). The enzyme consists of a five-stranded, antiparallel β-sheet and three α-helices connected by turns and loops with the active-site groove—which conforms to that seen in other guanyl-specific microbial RNases (39)—located on the β-sheet between two loops and a helix (Fig. 1c). In addition to an NMR solution structure (40), there are also two crystal structures of the enzyme bound to deoxynucleotide analogues neither of which cause substantial changes in the structure of the enzyme. Baudet and Janin (41) solved a 1.9-Å crystal structure of barnase bound to the dinucleotide d(GpC) in which the ligand bound in a seemingly non-productive manner, while Buckle and Fersht (42) reported the 1.76-Å structure of the enzyme bound to the tetranucleotide CGAC (Fig. 7) where the guanine binds in the characteristic recognition site observed in other microbial RNases (39). The scissile phosphate in this complex is surrounded by active-site residues Lys27, Glu73, and His102 (Fig. 8), shown previously to be essential for catalysis (36). Glu73 and His102 are general base–acid catalysts in the hydrolysis of RNA, where the glutamic acid substitutes for one of the two histidines in the RNase A mechanism, and Lys27 is implicated in transition-state stabilization (equivalent to Lys41 in RNase A). The scissile phosphate in the barnase–tetranucleotide complex is also coordinated by two arginine residues, Arg83 and Arg87 (Fig. 8).

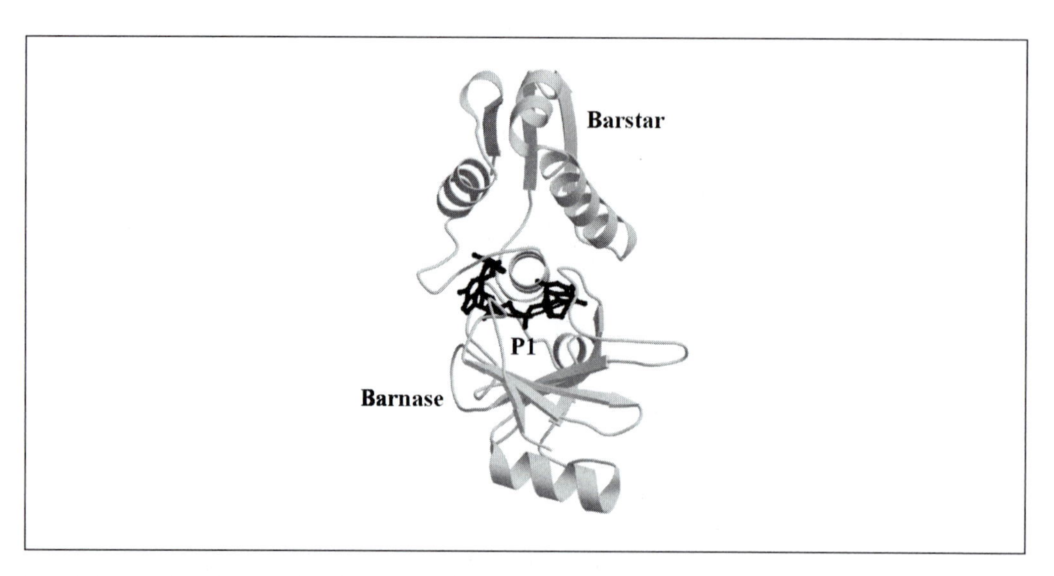

Fig. 7 Barstar binds in the active-site groove of barnase. Structure superposition of the 2.0-Å structure of the barnase–barstar complex (45) and the complex of barnase with a tetranucleotide ligand, solved at 1.76 Å, of which only the nucleotide is shown (ref. 42; pdb accession number 1brn). The figure illustrates how barstar binds in the same position as the RNA substrate. P1 indicates the position of the cleaved phosphodiester bond.

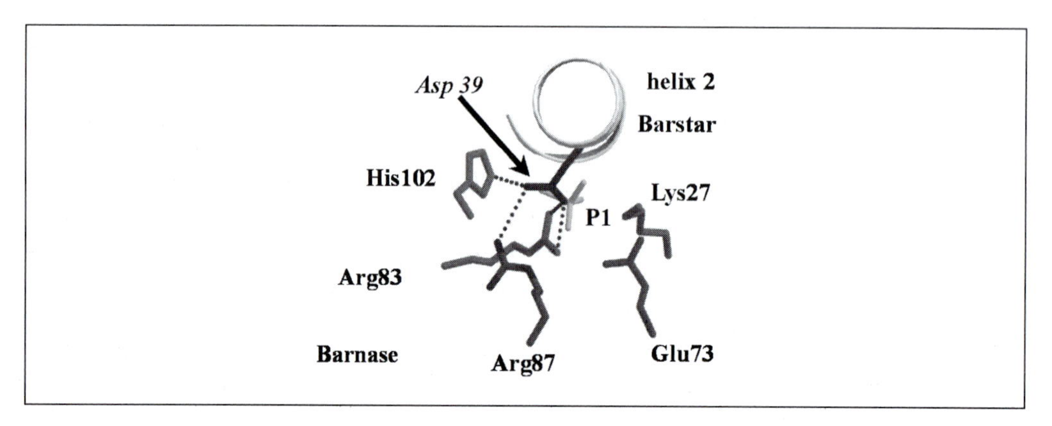

Fig. 8 Barstar mimics the scissile phosphate of the substrate. Close-up of the superposition shown in Fig. 7. Asp39 from helix 2 of barstar mimics the interactions made by the P1 phosphate of the substrate (the only part of the tetranucleotide ligand from Fig. 7 shown in this figure). Asp39 forms salt bridges to two active-site arginine residues and a hydrogen bond to His102, the general acid in the mechanism of RNA hydrolysis. The figure also shows the positions of other active-site residues of barnase that are not contacted by Asp39.

3.3 Structural studies of barstar and its complex with barnase

The structure of unbound barstar has been solved by NMR spectroscopy (43) and its complex with barnase by X-ray crystallography at 2.6-Å (44) and 2.0-Å resolution (45). The form of barstar used in these studies was a double mutant in which its two cysteines (at positions 40 and 82) were converted to alanine, resulting in negligible effects on function but avoiding the problems of oxidation previously encountered by these laboratories when working with the protein. Barstar is composed of three parallel α-helices stacked against a three-stranded, parallel β-sheet, with a two-turn helix connecting the central strand of the β-sheet with the third helix of the bundle (Fig. 1c).

Barstar binds to the active-site cleft of barnase, but it also forms interactions with residues outside the cleft (Fig. 1c). The main structural elements from barstar that are inserted into the active site are α-helix 2 and the loop connecting it to helix 1, occupying essentially the same site as would substrate RNA (Fig. 7). On forming the barnase–barstar complex, 1590 $Å^2$ of solvent-accessible surface area is buried. The structure of the enzyme changes very little, while that of barstar is more expanded compared to that of the unbound inhibitor, equivalent to an increase in molecular volume of 502 $Å^2$. A least-squares superposition of all Cα atoms of bound and free barstar shows a root-mean-squares (rms) deviation of 0.9 Å (45).

The surfaces of the inhibitor and the enzyme are highly complementary in both shape and charge. The shape complementarity (Sc) of the complex, determined by the method of Lawrence and Colman (20), was 0.70, i.e. similar to that found in the RI–angiogenin complex (45). When this calculation included 12 interfacial water molecules the *Sc* increased to 0.82, emphasizing the important role of solvent in stabilizing the complex by filling in the gaps where no protein side chains are located. Electrostatic interactions play an important role in stabilizing the barnase–barstar protein complex, with the clustered basic residues of the nuclease being met by acidic residues from the inhibitor (Plate 14). Of the 14 hydrogen bonds at the interface, 6 involve charged donor and acceptor groups, while 4 involve one charged partner (45). Importantly, the catalytically important residues of barnase, Lys27, Arg83, Arg87, and His102 all interact with barstar through salt bridges and hydrogen bonds, although some of these are mediated by water (Figs 8 and 9).

As well as occupying the substrate binding site barstar also mimics one key feature of the substrate, the scissile phosphate at P1 (44). This is accomplished by Asp39 from α-helix 2 of barstar which forms electrostatic interactions that are very similar to those formed by the P1 phosphate (Fig. 8). In addition, the P2 phosphate subsite is partially occupied by the main-chain carbonyl of Gly43 and a buried water molecule. Barstar does not present an amino-acid side chain to mimic interactions made by the guanine base towards the specificity pocket of barnase (residues 56–60). This results in poor protein–protein complementarity in this region of the interface, with the resulting gaps filled with buried water molecules that form a network of hydrogen bond interactions connecting residues of barstar, such as Asp35, with the residues of barnase involved in base recognition (Fig. 9).

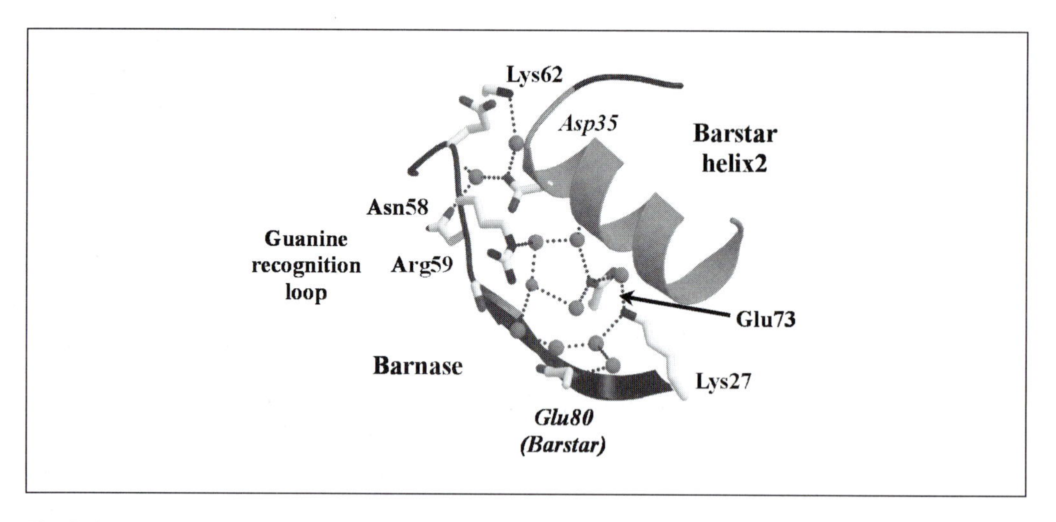

Fig. 9 Ordered solvent molecules improve the complementarity of the barnase–barstar complex. Buried water molecules (shown as spheres) fill in the gaps between barnase and barstar in the region of the guanine recognition loop. (Adapted from Buckle *et al.* (45).)

3.4 Kinetic and thermodynamic studies of the barnase–barstar interaction

Barstar forms a tight complex with barnase, the equilibrium dissociation constant for the complex having been measured as 10^{-14} M by several laboratories (46–48). Schreiber and Fersht (47) determined the affinity of the complex from the ratio of the individual association (k_{on}) and dissociation (k_{off}) rate constants (where $K_d = k_{off}/k_{on}$). The association rate constant, measured at pH 8 and at 25 °C and using stopped-flow tryptophan emission fluorescence (barstar binding to barnase results in a quench in tryptophan fluorescence), was determined as 6×10^8 M^{-1} s^{-1}, while the dissociation rate constant (measured by radioactive subunit-exchange kinetics in which tritiated barstar produced by *in vivo* radiolabelling was chased from the complex by unlabelled protein) was 8×10^{-6} s^{-1}. They also found that binding was at least a two-step process, with the first rate-determining association step approaching the diffusion controlled rate followed by subsequent slower steps that were not readily observed in stopped-flow experiments.

As for other nuclease–inhibitor complexes, the association of barstar and barnase is very sensitive to the ionic strength of the medium, consistent with the strongly charged surfaces that come together in the complex (Plate 14). Through mutagenesis of charged interface residues and through the effects of salt and glycerol on the association kinetics of the barnase–barstar complex, Schreiber and Fersht (49) demonstrated that association of the two oppositely charged proteins proceeds via an early transition state where long-range, charge–charge interactions predominate and serve to align the proteins prior to collision. The relative orientations of the associating proteins are envisaged to be constrained in the transition state, with a layer of solvent separating them and with their side chains retaining considerable flexibility (50). Precise docking of the proteins then takes place to form the high-affinity complex. Another important conclusion from the work of Schreiber and Fersht (49) was that the basal level of a protein–protein association rate constant in the absence of long-range electrostatic effects is of the order of 10^5 M^{-1} s^{-1}.

The thermodynamics of the barnase–barstar interaction have been dissected by microcalorimetry which demonstrated that the association is both enthalpy and entropy driven, although the enthalpy term dominates (51). This agrees with the overall structure of the interface which is highly charged with many polar interactions, salt bridges, and several ordered water molecules, the latter in agreement with the small entropic contribution to the association. Taking the calorimetric analysis of the barnase–barstar complex further, Frisch *et al.* (52) investigated several interface mutants and highlighted the difficulty in using enthalpies in structure– activity relationships because of the enthalpy–entropy compensation effects which seem to be characteristic of weak intermolecular interactions. It was argued that since these tend to cancel each other out, measurements of free energy is the preferred quantity for the simple analysis of macromolecular interactions.

3.5 Protein engineering of the barnase–barstar complex

The interface residues of the barnase–barstar complex have been analysed extensively by protein engineering, coupled with kinetic and thermodynamic analysis of the resulting mutants. This has provided one of the most detailed pictures of the energetics of a protein–protein interaction, equivalent to that of antibody–antigen and protease–protease inhibitor complexes (see Chapters 5 and 8). Even before the crystal structure for the complex was available, protein engineering data indicated that the inhibitor bound in the catalytic centre. Schreiber and Fersht (47) and Hartley (48) found that the mutation of several barnase active-site residues to alanine, particularly Lys27, Arg59, Arg87, and His102, not only diminished catalytic activity significantly, in agreement with their proposed involvement in catalysis, but also had a dramatic effect on the binding of barstar ($\Delta\Delta G > 5$ kcal mol^{-1}, ref. 47) indicating that barstar made direct contact with enzyme active-site residues.

An interesting extension of this site-directed approach was subsequently reported by Jucovic and Hartley (53) in which genetic selection was used to isolate mutants in barstar that compensated for barnase mutations that destabilized the complex. Their investigation focused on His102 which forms many interactions with barstar including three hydrogen bonds and a stacking interaction, the latter with Tyr29. Its pK_a is highly perturbed on binding barstar, falling from 6.3 in the free enzyme to < 5 in the complex, and this accounts for the pH-dependence of complex formation (47). The structural explanation for this pH-dependence is that His102 forms a main-chain hydrogen bond with Gly31 of barstar that can only occur when the imidazole side chain is unprotonated (45). The study of Jucovic and Hartley (53) revolved around barnase His102Lys. This mutant reduces barstar binding affinity by almost nine orders of magnitude and is also severely compromised as an enzyme, but it possesses enough residual RNase activity to select for compensating suppressor mutations in barstar. In the barnase–barstar complex, six barstar residues that surround His102 were randomly mutagenized and suppressors of the His102Lys mutation identified. Mutations at positions 29 (Tyr29Asp) and 30 (Tyr30Trp) were selected by this strategy and shown to improve binding to barnase His102Lys by several orders of magnitude (53); these improvements were later postulated to be largely the result of improved water solvation of the lysine positive charge (54).

Using the crystal structure of the complex as a guide, Schreiber and Fersht (55) conducted an exhaustive mutagenic analysis of barnase–barstar interactions. Having earlier identified five barnase residues whose mutation (primarily to alanine) significantly affected binding energy (using eqn 3 to determine $\Delta\Delta G$s for mutants relative to wild-type binding), they also identified seven residues in barstar that had a significant effect on binding, many of which were residues that formed interactions with active-site residues. Importantly, they used mutant and wild-type proteins in various combinations in a double-mutant cycle analysis (The use of double-mutant cycles is also discussed in Chapters 1 and 5). A traditional mutagenic approach such as alanine scanning does not, in general, measure the intrinsic binding energy of a given side chain X when mutated to residue A ($\Delta\Delta G_{X\rightarrow A}$), but rather provides a quantitative measure of the change in binding energy as a result of the change in side-chain identity. In the double-mutant cycle method, pioneered by the Fersht laboratory in Cambridge (12, 56), pairs of residues (X and Y) are mutated both singly and in combination to give the coupling energy (ΔG_{int}) which is a measure of the cooperativity between them. ΔG_{int} is defined by:

$$\Delta G_{int} = \Delta\Delta G_{X\rightarrow A,\, Y\rightarrow B} - \Delta\Delta G_{X\rightarrow A} - \Delta\Delta G_{Y\rightarrow B}; \qquad [4]$$

where $\Delta\Delta G_{Y\rightarrow B}$ is the change in binding energy on mutation of Y to B, and $\Delta\Delta G_{X\rightarrow A,\, Y\rightarrow B}$ the change on the simultaneous mutation of X to A and Y to B. If the effects of the two mutations are independent of each other (non-cooperative) the change in free energy for the double mutant is the sum of the two single mutations, but if the mutated residues are coupled then the change in free energy for the double mutant differs from the sum of the two single mutations. The overall conclusions from the double-mutant cycle analysis of Schreiber and Fersht (55) were:

1. Residues separated by less than 7 Å interact cooperatively, while at greater distances they do not show cooperativity.
2. The highest coupling energies (1.6–7 kcal mol^{-1}) were between charged residues, and many of these were interactions towards the Asp39 of barstar by active-site residues of barnase (Fig. 8).
3. Several important interactions detected by the analysis (with coupling energies of >3 kcal mol^{-1}) had not been identified by crystallography.

This latter point was further explored by a double-mutant cycle study centring around Glu73, the general base in the active site of barnase which does not contact barstar (Figs 8 and 9). Coupling energy between the Glu73 of barnase and Asp39 of barstar was detected but presumed to be indirect, since Glu73 organizes neighbouring positively charged residues of barnase such as Lys27, Arg83, and Arg87 (Plate 14) all of which interact with Asp39 (57). The crystal structures of several of the barnase–barstar, double-mutant complexes from these studies have been solved and shown to generate only local perturbations in the structures of the two proteins (58). Where truncation of amino acids generated cavities these tend to be filled with solvent water, which can mimic the hydrogen bonding patterns of the deleted side chains.

4. Immunity protein inhibitors of colicin endonucleases

4.1 The biology of E-group nuclease colicins and their immunity proteins

Colicins are plasmid-borne toxins produced by the Enterobacteriacae during times of nutrient or environmental stress that have lethal action against other related strains, while immunity proteins are their inhibitors that prevent suicide of the producing organism. The specificity of colicin–immunity protein complexes can be studied both *in vivo* and *in vitro*, and so these complexes have proven to be a powerful system with which to investigate specificity in protein–protein recognition (59). Recent work has also demonstrated that immunity proteins are a simple and amenable system for studying protein folding (60).

Colicins invade host cells by first recognizing outer membrane receptors, often involved in the import of essential nutrients, and then delivering their cytotoxic activity into the cell by recruiting several outer membrane and periplasmic proteins by mechanisms similar to those adopted by bacteriophage. In the case of the 60-kDa, E-group colicins, the BtuB receptor, normally used for the import of vitamin B_{12}, is bound specifically and the toxin then enters the cell via the Tol protein system (61, 62). Colicins have a characteristic arrangement of three domains responsible for each step of the cytotoxic mechanism: a central domain binds the receptor, an N-terminal domain is involved in binding Tol proteins and translocating across the outer membrane, and the C-terminus carries the cytotoxic activity, most often a nuclease in the E-colicin family (61). Colicin E3 (ColE3) is an RNase that site-specifically cleaves 16S ribosomal RNA thereby inactivating protein synthesis (63), ColE5 is an anticodon tRNA nuclease (64), and ColE2, ColE7, ColE8, and ColE9 are all non-specific endonucleases that kill bacterial cells by randomly cleaving chromosomal DNA (61).

DNase–colicin immunity proteins, the best studied of the colicin inhibitors, are small (~9.5 kDa) acidic proteins which associate with and neutralize the activity of the C-terminal cytotoxic domain, forming a 70-kDa heterodimeric complex that is released into the environment (59). The colicin complex must dissociate as the colicin penetrates a susceptible cell, but how this occurs is not known. The four DNase-specific immunity (Im) proteins (Im2, Im7, Im8, and Im9) share around 50% sequence identity, have all been overexpressed in bacteria (65, 66), and the structures for two of these determined: Im9 by NMR spectroscopy (67) and Im7 by X-ray crystallography

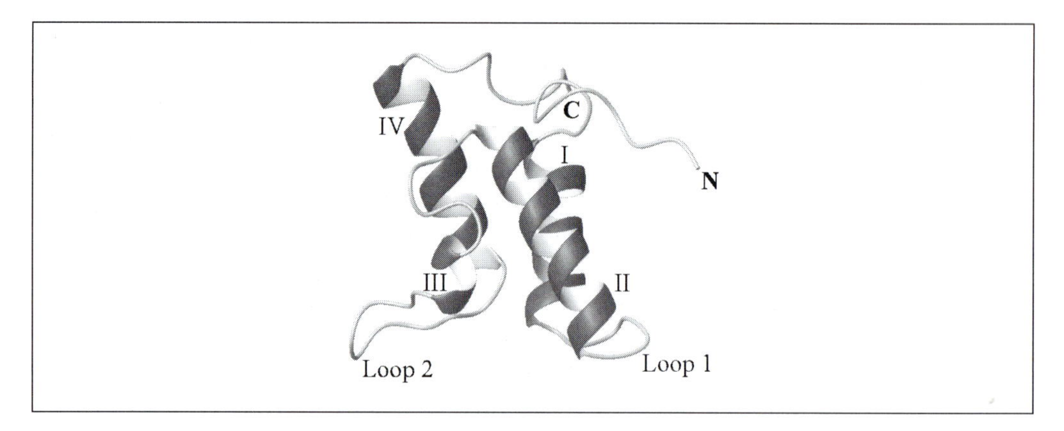

Fig. 10 Solution structure of the 86 amino-acid DNase colicin immunity protein Im9 (ref. 67; pdb accession number 1imq).

(68). The two immunity proteins adopt similar distorted, antiparallel, four-helical bundle structures where the rms deviation from a structure superposition is 1.7 Å for αC atoms in elements of regular secondary structure (69). While the overall topology of the immunity protein structure is the same in both Im9 and Im7 this is accomplished through a significant degree of sequence variation in two of the helices (helices I and II, Fig. 10). The other two helices are highly conserved, albeit that one (helix III) is only 1.5 turns in length due to the presence of a conserved proline residue at its C-terminus. These architectural nuances play an important part in immunity protein recognition of colicin DNases (see Section 4.4).

4.2 Enzymology of the colicin E9 DNase

ColE9 is 582 amino acids in length with the C-terminal 134 amino acids (15 kDa), defined by proteolysis experiments (70), forming the functional endonuclease (E9 DNase) domain; ColE9 shares about 65% sequence identity with the other colicin DNases in this family. Overexpression of the E9 DNase domain in bacteria, in the absence of the rest of the toxin, has been achieved by tandemly expressing its inhibitor Im9 (70). Moreover, using histidine-tagged Im9, Garinot-Schneider *et al.* (71) showed that the truncated 25-kDa E9 DNase–Im9 complex can be purified in a single step and the DNase subsequently isolated by denaturation on a nickel affinity column, followed by renaturation by dialysis. The refolded E9 DNase domain shows the same properties as the enzyme in the intact colicin, namely magnesium-dependent endonuclease activity (72, 73) and stoichiometric binding of its cognate immunity protein Im9 (70).

The mechanism of DNA hydrolysis by colicin endonuclease is unknown; however, the location of the active site has been inferred from a number of observations. First, the crystal structures of the E7 and E9 DNases (at 2.3 Å and 2.05 Å, respectively) bound to their cognate immunity proteins shows them to be α/β proteins of novel fold, with a very positively charged concave cleft in the molecule that is wide enough to accommodate double-stranded DNA (Fig. 11) (74, 75). Second, random mutagenesis of colicin E9, in which mutants were selected from colicin-producing bacteria by their inability to kill susceptible strains, identified three residues, Arg96, Glu100, and His127 (numbering according to the isolated E9 DNase domain), at which mutations destroyed both colicin toxicity and catalytic activity of the DNase (Fig. 12; ref. 71). These residues form a triad within the active-site cleft with Glu100 at its centre, flanked by Arg96 (through a salt bridge) and His127 (by a hydrogen bond). Third, most of the residues that line the concave cleft, including Arg96, Glu100, and His127 are conserved in all DNase colicins, as might be expected for active-site residues. Finally, much of the sequence encompassing these active-site residues shows identity to the so-called HNH motif found in homing endonucleases (76). Homing endonucleases are encoded by self-splicing introns (or inteins) that promote the homing of mobile genetic elements containing the genes for the nucleases into

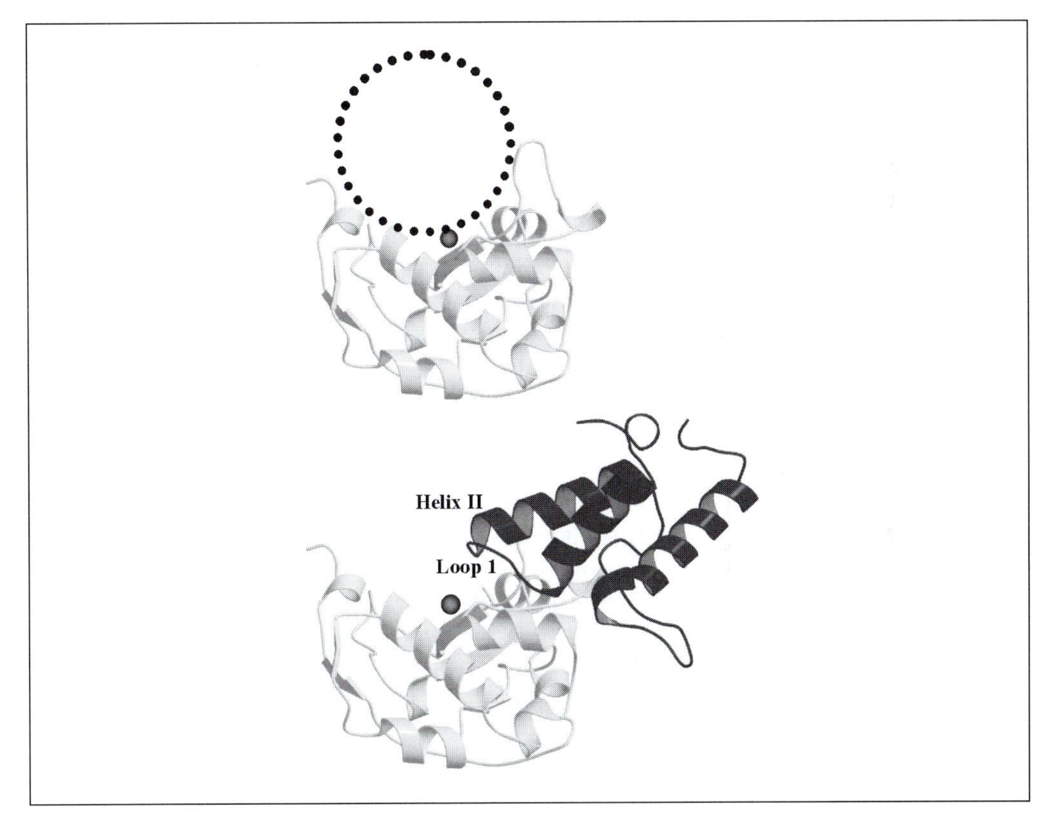

Fig. 11 Im9 inhibits the E9 DNase by blocking DNA binding. The top figure shows the E9 DNase from which the bound Im9 has been removed, to emphasize the location of the bound metal ion in the concave cleft believed to be the active-site cleft. Bound substrate is depicted by a circle representing a view down the helical axis of dsDNA. The lower figure shows the same orientation of E9 DNase but with Im9 replaced into the complex and the DNA removed. Helix II and loop 1 of Im9 form an 'acidic elbow' that would impede access of dsDNA into the active site at an approximate distance of one helical turn from DNA bound within the site (74).

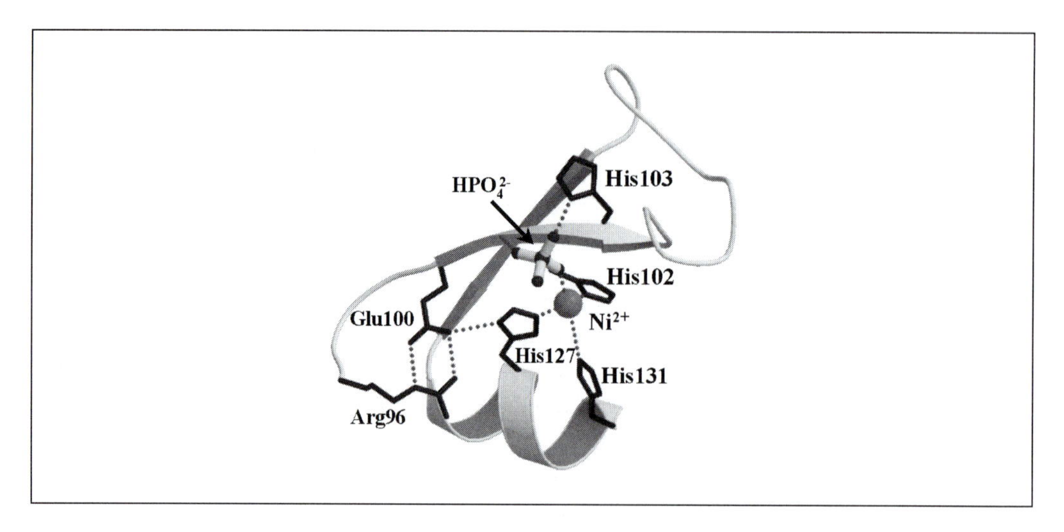

Fig. 12 The HNH motif is the core of the E9 DNase active site. The motif spans the C-terminal 33 amino acids of the E9 DNase domain and encompasses the bound nickel ion, tetrahedrally co-ordinated by three protein ligands, and a phosphate molecule. The figure highlights the interactions of several active-site residues, many of which were identified previously by random mutagenesis (71).

intronless (or inteinless) alleles in prokaryotes and eukaryotes. The HNH motif has been identified as a characteristic active-site motif in one of the four homing endonuclease families (77).

The crystal structures of both the E7 and E9 DNases also revealed that they are metalloproteins. In the E7 structure the metal was zinc (75), while in the E9 structure it was nickel (74). The metal ion is located within the HNH motif of the DNase and is tetrahedrally coordinated in both structures but with only three protein ligands (Fig. 12): His102, His127, and His131. The remaining ligation site is taken by either a phosphate molecule (in the E9 DNase) or a water molecule (in the E7 DNase). The likely reason for the different metal centres in the two structures is simply due to the different purification protocols used to prepare the complexes. Isothermal titration calorimetry has shown that zinc binds to the E9 DNase with a K_d of $\leq$ nM, while that for nickel is ~μM, suggesting zinc is the physiological metal ion (73). The location and geometry of the metal site in the active site of colicin DNases is suggestive of a role in catalysis. However, apoenzyme retains magnesium-dependent endonuclease activity (the magnesium ions do not bind to the colicin HNH motif), and this activity is not significantly affected by transition metals (73). The role of the transition metal is most likely structural since its removal affects the stability of the DNase domain, reducing its thermal denaturation temperature from 58 °C to 36 °C (73).

4.3 Mechanism of DNase inhibition by immunity proteins

Immunity proteins abolish the endonuclease activity of a colicin DNase. Binding of Im9 to either the whole toxin or the isolated E9 DNase domain can be monitored by intrinsic tryptophan-fluorescence spectroscopy or gel-filtration chromatography and shown to produce stoichiometric complexes (65, 70). A number of kinetic and thermodynamic studies indicate that this is a highly evolved antisuicide system:

1. At 25 °C and at pH 7.0, Im9 folds into its native state with a rate constant of 2200 s^{-1} (equivalent to a half-life of 0.3 ms), placing it among the fastest folding proteins observed to date (74).
2. Under the same conditions, folded Im9 binds the E9 DNase with an association rate constant (k_{on}) of 4×10^9 $M^{-1} s^{-1}$ (from stopped-flow, tryptophan-emission fluorescence experiments) essentially equivalent to the rate of diffusion in solution (66). This is an electrostatically controlled association that shows a strong salt-dependence, similar to the association kinetics of RI with RNase A and barnase with its inhibitor barstar (see Sections 2.2 and 3.4).
3. Using radioactive subunit exchange kinetics the dissociation rate constant (k_{off}) was determined as $3.7 \times 10^{-7} s^{-1}$, and so the resulting K_d (k_{off}/k_{on}) estimated as 9.3×10^{-17} M for the colicin E9–Im9 complex, equating to a change in free energy of -22 kcal mol^{-1} (66), similar to that of RI binding angiogenin.

The kinetics and thermodynamics of Im9 binding colicin E9 or the isolated E9 DNase domain are identical, showing that the immunity protein does not interact with toxin residues outside of its binding site on the endonuclease.

As with other nuclease–inhibitor systems discussed in this chapter, association of Im9 with the E9 DNase involves more than one step (66). Evidence for this came from the stopped-flow, tryptophan fluorescence experiments used to monitor the association. The initial fluorescence enhancement, due to the rapid collision of the two proteins, is followed by a slow, first-order quench (~ 4–5 s^{-1}), due to a conformational change in the E9 DNase. These data were originally interpreted according to the kinetic scheme shown in eqn 1 where the enzyme and inhibitor form an encounter complex which rearranges to form the final high-affinity complex (66). Subsequent NMR experiments caused a reassessment of this scheme, since it was found that the E9 DNase exists in solution as an equilibrium of two distinct conformers that interconvert on a time-scale similar to that seen in the stopped-flow experiments. This suggested that Im9 bound and stabilized one form of the enzyme, presumably that seen in the crystal structure (78), since such a kinetic scheme would also yield biphasic stopped-flow traces (66). The slow conformational dynamics in the DNase are centred around two regions that are not part of the Im9 binding site (nor active site) and which, more recent NMR data indicate, persist (albeit differently) even in the complex with Im9 (91). Hence the role of DNase conformational dynamics in immunity protein binding remains unresolved.

Gel-retardation assays and heparin chromatography show that Im9 inhibits the DNase of colicin E9 by blocking DNA binding (74). Based on the mode of action of most protein inhibitors of enzymes, and particularly the nuclease inhibitors described thus far, both observations suggest direct active-site binding by the inhibitor. This being the case, active-site mutations in which enzymatic activity is abolished should also affect inhibitor binding, reflecting interactions to both substrate and inhibitor (for example, see Sections 2.4 and 3.5). It became apparent even before the crystal structure of a DNase–immunity protein complex was solved that this was not the case, since active-site mutants of the E9 DNase did not affect immunity protein binding (71). The reason for this anomaly is explained by the E9 DNase–Im9 complex crystal structure (Fig. 13) which shows that the immunity protein does not bind in the active-site cleft of the enzyme but at an adjacent site (74). The immunity-protein binding site, formed by a contiguous stretch of 30 amino acids (residues 70–100), folds into a short helix, an extended strand, and two loops and contains only five conserved amino-acid residues none of which contact the immunity protein.

Two possible mechanisms suggest themselves as to how the immunity protein inactivates the colicin DNase. It might stabilize an inactive conformation of the enzyme, but since the active site remains exposed this does not explain why DNA binding is completely abolished. Alternatively (or in addition), the location of the immunity protein may sterically hinder access of DNA into the active site, and this is supported by modelling studies (74). Modelling B-form DNA into the active site shows that the location of Im9 in the complex places a cluster of acidic residues (an 'electrostatic elbow'), formed by the N-terminal end of helix II along with loop 1, in a position that would impede access of DNA to the enzyme (Fig. 11). Thus, although the Im9 binding site and active site do not overlap, inhibition occurs through steric

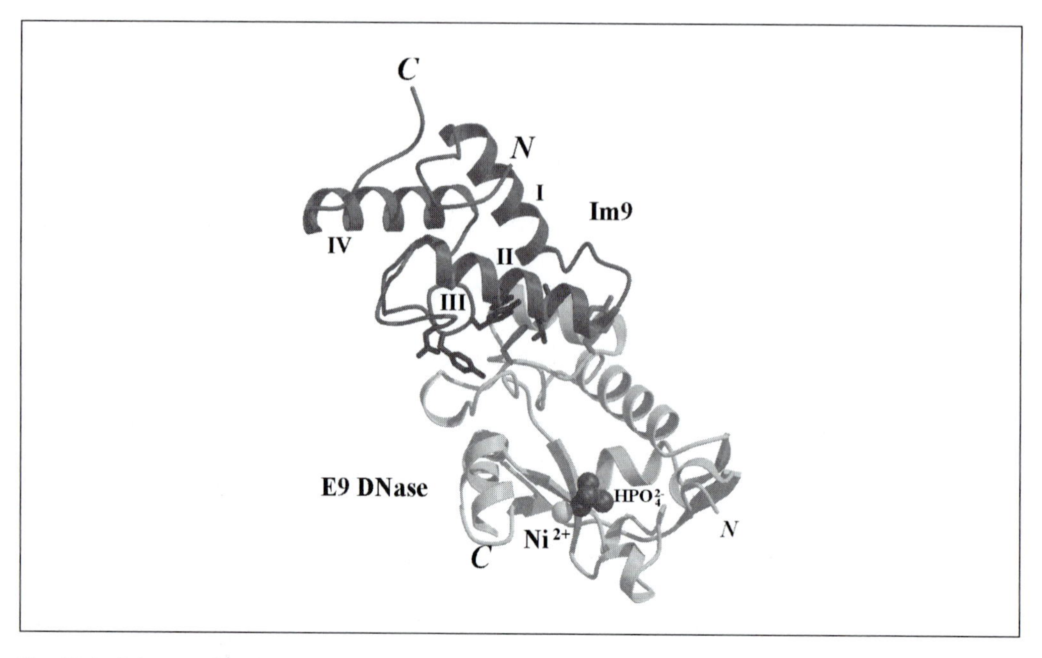

Fig. 13 Im9 does not bind in the active site of the E9 DNase. The figure shows the relative positions of the active site (denoted by the HNH-bound metal ion and phosphate molecule) and the position of bound Im9. A number of amino acids at the protein–protein interface are shown which are identified in Fig. 14.

and electrostatic repulsion because the substrate (bacterial DNA) is a large, negatively charged polymer that extends beyond the boundaries of the active site.

4.4 The colicin DNase–immunity protein interface

The two crystal structures of the E7 and E9 DNases in complex with their immunity proteins are very similar. In both studies the immunity protein (the structure of which expands slightly on binding to the enzyme) presents a negatively charged surface to counteract the very positively charged surface of the endonuclease, as expected for an electrostatically driven association. Similar solvent-accessible surface areas are buried in the two complexes, 1473 $Å^2$ in the E7 DNase–Im7 complex, 1575 $Å^2$ in the E9 DNase–Im9 complex, and both exhibit a high degree of steric and electrostatic complementarity. Present at the DNase–immunity interface are several buried water molecules, many of which seem to be conserved in the two complexes, and these serve to increase the complementarity yet further and aid in forming inter-protein hydrogen-bonding networks. From a recent structural survey of 36 protein–protein complexes, Lo Conte *et al.* (79) found one hydrogen bond per 170 $Å^2$ of buried surface. The frequency of hydrogen bonding in DNase–immunity protein complexes is much higher, with one per 92 $Å^2$ in the E7 complex and one per 131 $Å^2$ in the E9 complex, which is similar to the frequency of hydrogen bonding seen in protein–DNA interactions (80). Ko *et al.* (75) have proposed that the large number of hydrogen bonds, many involving charged partners, and salt bridges explains the high affinities of colicin DNase–immunity protein complexes, yet the affinities of most protein–DNA complexes are not as high as those of colicin–immunity protein complexes. Protein engineering experiments demonstrate that while both types of interactions certainly contribute to binding energy in these complexes, hydrophobic interactions are also a major stabilizing force (see Section 4.6).

Since the E7 and E9 structures are similar, the remaining description focuses on the E9 DNase–Im9 complex for which both kinetic and thermodynamic information are available. In the protein–protein interaction, 24 residues from Im9 and 21 from the E9 DNase are involved. Prior to the crystal structure, NMR perturbation analysis (where ^{15}N-labelled Im9 was bound to unlabelled E9 DNase and the perturbed amides assigned) and protein engineering experiments had already shown that the main binding interactions on Im9 are focused around the conserved residues of helix III and the variable residues of helix II (81). The crystal structure of the complex confirmed these results and also showed interactions of the DNase with residues at the C-terminus of helix I and the adjoining loop in Im9. One of the most striking features of the interface is the distinct type of interactions made by each of the helices of the immunity protein toward the DNase. Helix II forms mostly side-chain to side-chain interactions with specificity residues of the enzyme identified previously by site-directed mutagenesis (82), including salt bridges, hydrogen bonds, and hydrophobic interactions, while the conserved residues of helix III interact predominantly through hydrogen-bonding interactions with the backbone of the enzyme and van der Waals interactions (Fig. 14). The strict distinction between backbone and side-

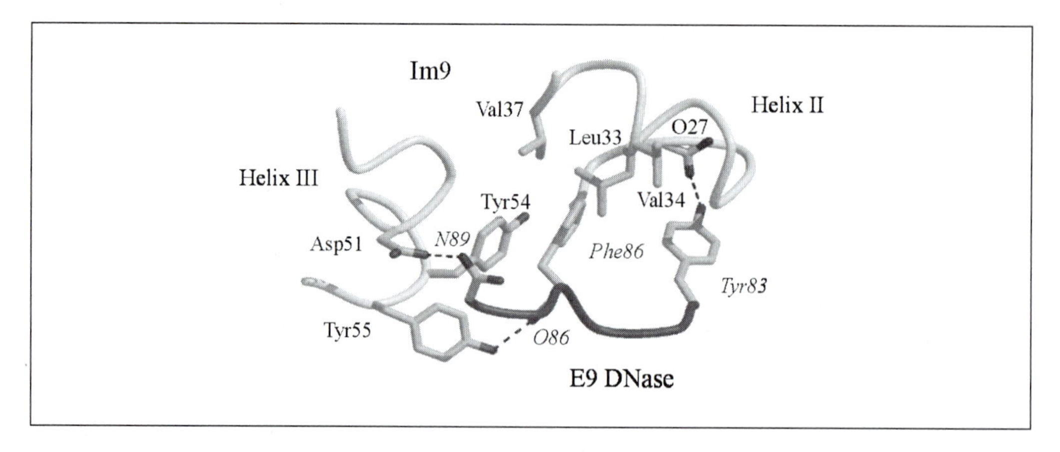

Fig. 14 The Im9–E9 DNase interface. Residues that define specificity for each protein co-localize at the interface; immunity specificity is controlled primarily by residue 33 in helix II along with residues 34 and 38 (the latter not shown) and these contact the side chains of specificity residues in the DNase (Phe86 and Tyr83). By contrast, the conserved residues of helix III from Im9 interact with the DNase predominantly through backbone interactions, and some of these are shown. See text for details of how the chemistry of the interface is consistent with the proposed dual-recognition mechanism. (Adapted from Kleanthous *et al.* (74).)

chain interactions exhibited by the two helices of the immunity protein lies at the heart of their ability to interact with colicin endonuclease domains with both high affinity and specificity (see Sections 4.5 and 4.6).

4.5 *In vivo* and *in vitro* specificity of DNase immunity proteins

The classical view of immunity protein specificity is one in which colicin-producing bacteria only show resistance against their specific toxin. However, non-cognate immunity proteins can provide *in vivo* protection towards colicins when over-expressed in bacteria. In the case of colicin E9, three out of the four DNase-specific immunity proteins provide bacterial cells with protection, the order of this cross-reactivity being Im9 (the cognate immunity protein) >> Im2 > Im8, with Im7 offering no protection against the toxin (83). It was subsequently found that *all* the immunity proteins are able to inhibit the endonuclease activity of the ColE9 *in vitro* with binding affinities that mirrored the biological data; the K_ds for Im2, Im8, and Im7 were, 10^{-8} M, 10^{-6} M, and 10^{-4} M, respectively. A closer examination of the kinetic parameters for the non-cognate immunity proteins showed that the association rate constants for Im2 and Im8 were almost identical to Im9, but their dissociation rate constants were more than 6 orders of magnitude faster and this explained the very much weaker affinities of these complexes (83).

The wide spectrum of affinities observed in colicin DNase–immunity protein complexes offers great potential to understand specificity in protein–protein recognition, and thus far these studies have centred on the E9 DNase binding either Im9, a cognate immunity protein, or Im2, a non-cognate immunity protein. In the presence of 200 mM salt (and at pH 7.0 and at 25 °C) the affinities of these proteins for the E9 DNase differ by six orders of magnitude (18 fM for Im9 vs. 15 nM for Im2). Homologue-scanning mutagenesis, first used in the study of growth-hormone receptor specificity (84), demonstrated that helix II, one of the variable helices in the immunity protein family, determined the specificity of the protein–protein interaction (85). Site-directed mutagenesis narrowed down the specificity determinants to three amino acids within this 15-residue helix, at positions 33 (Asp in Im2/Leu in Im9), 34 (Asn in Im2/Val in Im9), and 38 (Arg in Im2/Thr in Im9) (86). An Im2 triple mutant (Asp33Leu, Asn34Val, Arg38Thr) bound the E9 DNase with the same affinity as a helix-II chimera and completely switched the biological specificity of the protein to that of Im9 (86). Interestingly, the specificity determinants are not equivalent in their influence, with position 33 being the dominant site. This residue lies very close to the conserved residues of helix III of Im9 and interacts with specificity residues from the DNase at the interface of the complex (Fig. 14).

4.6 Alanine-scanning mutagenesis of immunity proteins

The relative contributions of helices II and III of Im9 toward E9 DNase binding energy have been analysed by alanine-scanning mutagenesis (87). Although the scan was completed before the crystal structure became available, the NMR data had successfully identified the residues involved in the interaction (81). Hence, nearly all the residues that are buried in the complex had been substituted for alanine and their relative effects on DNase binding energy quantitated. Of the 24 residues that are buried, only 10 affect binding by more than 1 kcal mol^{-1} when substituted for alanine, and of these only five significantly affect binding energy (2.5–6 kcal mol^{-1}). The five residues cluster together on the surface of Im9 forming a binding energy hot-spot (88), and includes two variable amino acids from helix II (two of the key specificity determinants, residues 33 and 34) and three conserved residues from helix III (residues 51, 54, and 55) (Fig. 14). The conserved residues of Im9 dominate the interaction, accounting for the majority of the relative binding energy (approximately two-thirds) with the remainder coming from the residues of helix II.

Li *et al.* (86) went on to conduct a limited alanine scan of 12 E9 DNase binding residues in helices II and III of the non-cognate immunity protein Im2 in order to compare them to the equivalent residues of Im9 (Plate 15). Of the 12 residues, 8 are conserved between the two proteins and 4 are variable in sequence. On substitution for alanine, the $\Delta\Delta G$ for conserved residues was qualitatively similar in the two sets of complexes (residues labelled red and in the right-hand section of the graph), in particular the hot-spot residues Asp51, Tyr54, and Tyr55, implying that they form similar DNase interactions in each of the complexes. Conversely, the specificity sites only contribute to DNase binding energy in the cognate complex (residues labelled blue in the left-hand side of Plate 15 and rising up the y-axis), in particular residues 33 and 34, whereas in the non-cognate complex they either do not contribute to binding energy or they destabilize the complex (residues 38 and 42; Plate 15). Future double-mutant cycle analysis should reveal more detailed information about the interface such as cooperative effects between conserved and variable amino acids, although such an analysis will necessarily be limited since many of the hydrogen bonds of helix III are directed towards backbone atoms of the DNase.

4.7 The dual-recognition model

The structural and protein engineering data on immunity proteins binding colicin DNases have been interpreted in terms of a dual-recognition model, in which a hot-spot of conserved immunity protein residues contribute the bulk of the protein–protein interaction binding energy while the variable amino acids modulate this binding by positive, negative, or neutral contributions to binding energy (59, 86, 87). A full contribution to binding energy by specificity sites signifies a cognate interaction and produces a tight binding complex ($K_d < 10^{-14}$ M) which protects the organism against the action of the colicin. If the specificity residues of an immunity protein do not dock in appropriate sites on the colicin DNase then a non-cognate complex is formed of much lower affinity resulting in the death of the bacterium.

The conserved and variable amino acids of the immunity protein that make up the DNase binding site are met by a similarly constructed binding site on the endonuclease. The specificity sites from helix II of the immunity protein are bound by specificity residues of the DNase as well as making packing interactions with conserved residues. Similarly, the conserved residues of helix III interact with conserved features of the DNase, although not with conserved amino-acid side chains but through the formation of hydrogen bonds to the protein backbone (Fig. 14). This form of backbone recognition is reminiscent of MHC molecules and their ability to bind many different peptide sequences by virtue of the sequence-independent interactions they form with backbone atoms (see Chapter 6).

The interdigitation of conserved and variable amino acids at the E9 DNase–Im9 interface is best illustrated by the interactions made by the two conserved tyrosine residues Tyr54 and Tyr55 from helix III of Im9 (Fig. 14). Both are essential for the stability of the complex, since mutating either to alanine destabilizes the complex by 4.8 and 4.6 kcal mol^{-1}, respectively (87). Tyr54 forms a stacking interaction with Phe86 of the E9 DNase (a variable specificity residue), while the hydroxyl of Tyr55 forms a hydrogen bond with the main-chain carbonyl oxygen of Phe86. Phe86 along with Tyr83, another known specificity residue from the DNase, in turn pack against the variable, colicin specificity-determining residues Leu33 and Val34 from Im9.

Many protein–protein interactions in biology involve the interplay of conserved and non-conserved structural elements which together define the specificity of the interaction. This is true, for example, of intracellular signalling complexes involving SH2 or SH3 domains which recognize specific peptide motifs through the conserved domains but in the context of neighbouring specificity residues (see Chapter 7). The rationale for including conserved amino acids or structures within a protein–protein interface (or indeed within a protein–DNA interface) is simple—the stable and specific association of macromolecular complexes usually involves large surfaces (>1600 Å^2) becoming buried in the complex, some of which may be conserved for functional or even structural reasons. The opposite of this type of recognition is antibody–antigen interactions (Chapter 5), wherein specificity is achieved almost entirely by variable loops whose sequences are selected for the optimal recognition of antigens. This form of affinity selection does not occur in most macromolecular complexes within cells, which instead rely on a limited number of pre-existing proteins forming complexes that may necessitate the burial of conserved sequences at protein–protein interfaces. Hence, dual recognition is a logical mechanism of selectivity in evolutionary terms, since the appearance of novel specificities need only require the mutation of a select few residues, in a surface made up of dozens of amino acids, which modulate the major binding interactions emanating from the conserved protein scaffold.

5. Conclusions

Barstar and Im9 are both similar in size to the enzymes they inhibit and form a single intermolecular surface, while RI is much larger, enveloping the enzyme and forming contacts in three places. Inhibition of nuclease activity is accomplished either by binding to the active site directly, with the inhibitor mimicking part of the nucleic acid substrate with amino-acid side chains (RI and barstar), or indirectly by binding to an adjacent site and sterically blocking substrate binding (Im9). Regardless of these differences the three systems are remarkably similar in terms of the kinetics and thermodynamics of complex formation, reflecting in large part the charge states of the proteins—in each case the enzyme is a basic protein bound by an acidic inhibitor. The charge states of the proteins also account for the strong salt-dependence of the association rate constants for the three complexes, all of which approach the diffusion controlled limit. Hence, electrostatic steering plays an important part in preorienting the binding sites during the association of nucleases and their inhibitors. Once bound, the complexes dissociate very slowly. While the overall modes of recognition are different in the three complexes (Fig. 1), as indeed are the solvent-accessible surface areas that are buried, the resulting K_ds are essentially the same (10^{-14} to 10^{-16} M). The complexes are stabilized by many charged hydrogen bonds and salt links, as well as hydrophobic interactions in some instances, with water molecules often serving to increase the steric and charge complementarity of the interface.

Protein engineering studies of two of the systems (RI complexes and E9 DNase-Im9) indicate the presence of binding energy hot-spots where amino acids that make substantial contributions to binding energy are clustered together, whereas the binding epitope is more extensive for the barnase–barstar complex. Further work on the barnase–barstar complex (and RI complexes) has shown that the interactions at the protein–protein interface are highly cooperative, and this is observed even for residues buried at the interface that are not involved in making direct contact with the partner protein. In RI–RNase complexes, the energetic contribution of inhibitor residues is context-dependent, with both negative and positive cooperativity evident. A similar situation is seen in the dual-recognition mechanism proposed for colicin DNase–immunity protein complexes, where conserved residues account for the bulk of the binding energy but high-affinity binding is controlled by specificity sites that make negative, positive, or neutral contributions to binding energy. Hence, cooperativity plays an important role in both the stability and specificity of nuclease–inhibitor protein complexes and most likely protein–protein interactions in general.

Acknowledgements

The Colicin Research Group at UEA, which includes the laboratories of Richard James, Geoff Moore, Andrew Hemmings, and Colin Kleanthous, is funded by BBSRC and the Wellcome Trust. Many thanks to Ryan Bingham, Toni Georgiou, Li Wei, Dan Walker, and Anthony Keeble for their critical reading of the manuscript and helpful suggestions.

References

1. Linn, S. M., Lloyd, R. S., and Roberts, R. J. (ed.) (1993). *Nucleases* (2nd edn). Cold Spring Harbor Laboratory Press, New York.
2. D'Alessio, G. and Riordan, J. F. (ed.) (1997). *Ribonucleases: structures and functions*. Academic Press, New York.
3. Enari, M., Sakahira, H., Yokoyama, H., Okawa, K., Iwamatsu, A., and Nagata, S. (1998). A caspase-activated DNase that degrades DNA during apoptosis, and its inhibitor ICAD. *Nature*, **391**, 43–50.
4. Pirotte, M. and Desreux, V. (1952). Distribution de la ribonucléase dans les extrait de granules cellulaire du foie. *Bull. Soc. Chim. Belg.*, **61**, 167–80.
5. Hofsteenge, J. (1997). Ribonuclease inhibitor. In *Ribonucleases: structures and functions* (ed. G. D'Alessio and J. F. Riordan), pp. 621–58. Academic Press, New York.
6. Blackburn, P., Wilson, G., and Moore, S. (1977). Ribonuclease inhibitor from human placenta. *J. Biol. Chem.*, **252**, 5904–10.
7. Lee, F. S., Auld, D. S., and Vallee, B. L. (1989). Tryptophan fluorescence as a probe of placental ribonuclease inhibitor binding to angiogenin. *Biochemistry*, **28**, 219–24.
8. Lee, F. S., Shapiro, R., and Vallee, B. L. (1989). Tight-binding inhibition of angiogenin and ribonuclease A by placental ribonuclease inhibitor. *Biochemistry*, **28**, 225–30.
9. Vicentini, A. M., Kieffer, B., Matthies, R., Meyhack, B., Hemmings, B. A., Stone, S. R., and Hofsteenge, J. (1990). Protein chemical and kinetic characterization of recombinant porcine ribonuclease inhibitor expressed in *Saccharomyces cerevisiae*. *Biochemistry*, **29**, 8827–34.
10. Shapiro, R. and Vallee, B. L. (1991). Interaction of human placental ribonuclease with placental ribonuclease inhibitor. *Biochemistry*, **30**, 2246–55.
11. Chen, C.-Z. and Shapiro, R. (1997). Site-specific mutagenesis reveals differences in the structural bases for tight binding of RNase inhibitor to angiogenin and RNase A. *Proc. Natl. Acad. Sci. USA*, **94**, 1761–6.
12. Fersht, A. R. (1999). *Structure and mechanism in protein science; a guide to enzyme catalysis and protein folding*. Freeman, New York.
13. Lee, F. S., Fox, E. A., Zhou, H.-M, Strydom, D. J., and Vallee, B. L. (1988). Primary structure of human placental ribonuclease inhibitor. *Biochemistry*, **27**, 8545–53.

14. Hofsteenge, J., Kieffer, B., Matthies, R., Hemmings, B. A., and Stone, S. R. (1988). Amino acid sequence of the ribonuclease inhibitor from porcine liver reveals the presence of leucine-rich repeats. *Biochemistry*, **27**, 8537–44.
15. Fominaya, J. M. and Hofsteenge, J. (1992). Inactivation of ribonuclease inhibitor by thiol-disulfide exchange. *J. Biol. Chem.*, **267**, 24655–60.
16. Kobe, B. and Deisenhofer, J. (1993). Crystal structure of porcine ribonuclease inhibitor, a protein with leucine-rich repeats. *Nature*, **366**, 751–6.
17. Kobe, B. and Deisenhofer, J. (1995). Proteins with leucine-rich repeats. *Curr. Opin. Struct. Biol.*, **5**, 409–16.
18. Kobe, B. and Deisenhofer, J. (1996). Mechanism of ribonuclease inhibition by ribonuclease inhibitor protein based on the crystal structure of its complex with ribonuclease A. *J. Mol. Biol.*, **264**, 1028–43.
19. Papageorgiou, A. C., Shapiro, R., and Acharya, K. R. (1997). Molecular recognition of human angiogenin by placental ribonuclease inhibitor—an X-ray crystallographic study at 2.0 Å resolution. *EMBO J.*, **16**, 5162–77.
20. Lawrence, M. C. and Colman, P. M. (1993). Shape complementarity at protein–protein interfaces. *J. Mol. Biol.*, **234**, 946–50.
21. Lee, F. S. and Vallee, B. L. (1989). Binding of placental ribonuclease inhibitor to the active site of angiogenin. *Biochemistry*, **28**, 3556–61.
22. Blackburn, P. and Gavilanes, J. G. (1980). The role of lysine-41 of ribonuclease A in the interaction with RNase inhibitor from human placenta. *J. Biol. Chem.*, **255**, 10959–65.
23. Wlodawer, A., Miller, M., and Sjölin, L. (1983). Active site of RNase: neutron diffraction study of a complex with uridine vanadate, a transition-state analog. *Proc. Natl. Acad. Sci. USA*, **80**, 3628–31.
24. Cuchillo, C. M., Vilanova, M., and Nogués, M. V. (1997). Pancreatic ribonucleases. In *Ribonucleases: structures and functions* (ed. G. D'Alessio and J. F. Riordan), pp. 217–304. Academic Press, New York.
25. Birdsall, D. L. and McPherson, A. (1992). Crystal structure disposition of thymidilic acid tetramer in complex with ribonuclease A. *J. Biol. Chem.*, **267**, 22230–6.
26. Acharya, K. R., Shapiro, R., Allen, S. C., Riordan, J. F., and Vallee, B. L. (1994). Crystal structure of human angiogenin reveals the structural basis for its functional divergence from ribonuclease. *Proc. Natl. Acad. Sci. USA*, **91**, 2915–19.
27. Chen, C.-Z. and Shapiro, R. (1999). Superadditive and subadditive effects of 'hotspot' mutations within the interfaces of placental ribonuclease inhibitor with angiogenin and ribonuclease A. *Biochemistry*, **38**, 9273–85.
28. Cunningham, B. C. and Wells, J. A. (1991). Rational design of receptor-specific variants of human growth hormone. *Proc. Natl. Acad. Sci. USA*, **88**, 3407–11.
29. Hill, C., Dodson, G., Heinemann, U., Saenger, W., Mitsui, Y., Nakamura, K., Borisov, S., Tischenko, G., Polyakov, K., and Pavlovsky, S. (1983). The structural and sequence homology of a family of microbial ribonucleases. *Trends Biochem. Sci.*, **8**, 364–9.
30. Hartley, R. W. (1989). Barnase and barstar: two small proteins to fold and fit together. *Trends Biochem. Sci.*, **14**, 450–4.
31. Hartley, R. W. and Smeaton, J. R. (1973). On the reaction between extracellular ribonuclease of *Bacillus amyloliquefaciens* (Barnase) and its intracellular inhibitor (Barstar). *J. Biol. Chem.*, **248**, 5624–6.
32. Paddon, C. J. and Hartley, R. W. (1987). Expression of *Bacillus amyloliquefaciens* extracellular ribonuclease (barnase) in *Escherichia coli* following an inactivating mutation. *Gene*, **53**, 11–19.

33. Hartley, R. W. (1988). Barnase and barstar. Expression of its cloned inhibitor permits expression of a cloned ribonuclease. *J. Mol. Biol.*, **202**, 913–15.
34. Hartley, R. W. (1997). Barnase and barstar. In *Ribonucleases: structures and functions* (ed. G. D'Alessio and J. F. Riordan), pp. 51–100. Academic Press, New York.
35. Rushizky, G. W., Greco, A. E., Hartley, R. W., and Sober, H. A. (1963). Studies on the characterisation of ribonucleases. *J. Biol. Chem.*, **239**, 2165–9.
36. Mossakowska, D. E., Nyberg, K., and Fersht, A. R. (1989). Kinetic characterisation of the recombinant ribonuclease from *Bacillus amyloliquefaciens* (Barnase) and investigation of key residues in catalysis by site-directed mutagenesis. *Biochemistry*, **28**, 3843–50.
37. Day, A. G., Parsonage, D., Ebel, S., Brown, T., and Fersht, A. R. (1992). Barnase has subsites that give rise to large rate enhancements. *Biochemistry*, **31**, 6390–5.
38. Mauguen, Y., Hartley, R. W., Dodson, E. J., Dodson, G. G., Bricogne, G., Chothia, C., and Jack, A. (1982). Molecular structure of a new family of ribonucleases. *Nature*, **297**, 162–4.
39. Sevcik, J., Sanishvili, R. G., Pavlovsky, A. G., and Polyakov, K. M. (1990). Comparison of active sites of some microbial ribonucleases; structural basis for guanylic specificity. *Trends Biochem. Sci.*, **15**, 158–62.
40. Bycroft, M., Ludvigsen, S., Fersht, A. R., and Poulsen, F. M. (1991). Determination of the three dimensional structure of barnase using nuclear magnetic resonance spectroscopy. *Biochemistry*, **30**, 8697–701.
41. Baudet, S. and Janin, J. (1991). Crystal structure of a barnase–d(GpC) complex at 1.9 Å resolution. *J. Mol. Biol.*, **219**, 123–32.
42. Buckle, A. M. and Fersht, A. R. (1994). Subsite binding in an RNase: structure of a barnase–tetranucleotide complex at 1.76 Å resolution. *Biochemistry*, **33**, 1644–53.
43. Lubienski, M. J., Bycroft, M., Freund, S. M. V., and Fersht, A. R. (1994). Three-dimensional solution structure and ^{13}C assignments of barstar using nuclear magnetic resonance spectroscopy. *Biochemistry*, **33**, 8866–77.
44. Guillet, V., Lapthorn, A., Hartley, R. W., and Mauguen, Y. (1993). Recognition between a bacterial ribonuclease, barnase, and its natural inhibitor, barstar. *Structure*, **1**, 165–77.
45. Buckle, A. M., Schreiber, G., and Fersht, A. R. (1994). Protein–protein recognition: crystal structural analysis of a barnase–barstar complex at 2.0 Å resolution. *Biochemistry*, **33**, 8878–89.
46. Mariani, C., Gossele, V., de Beuckeleer, M., Deblock, M., Goldberg, R. B., Degreef, W., and Leemans, J. (1992). A chimaeric ribonuclease-inhibitor gene restores fertility to male sterile plants. *Nature*, **357**, 384–7.
47. Schreiber, G. and Fersht, A. R. (1993). Interaction of barnase with its polypeptide inhibitor barstar studied by protein engineering. *Biochemistry*, **32**, 5145–50.
48. Hartley, R. W. (1993). Directed mutagenesis and barnase–barstar recognition. *Biochemistry*, **32**, 5978–84.
49. Schreiber, G. and Fersht, A. R. (1996). Rapid, electrostatically assisted association of proteins. *Nature Struct. Biol.*, **3**, 427–31.
50. Vijayakumar, M., Wong, K.-Y., Schreiber, G., Fersht, A. R., Szabo, A., and Zhou, H.-X. (1998). Electrostatic enhancement of diffusion-controlled protein–protein association: comparison of theory and experiment on barnase and barstar. *J. Mol. Biol.*, **278**, 1015–24.
51. Martinez, J. C., Filimonov, V. V., Mateo, P. L., Schreiber, G., and Fersht, A. R. (1995). A calorimetric study of the thermal stability of barstar and its interaction with barnase. *Biochemistry*, **34**, 5224–33.
52. Frisch, C., Schreiber, G., Johnson, C. M., and Fersht, A. R. (1997). Thermodynamics of the

interaction of barnase and barstar: changes in free energy versus changes in enthalpy on mutation. *J. Mol. Biol.*, **267**, 696–706.

53. Jucovic, M. and Hartley, R. W. (1996). Protein–protein interaction: a genetic selection for compensating mutations at the barnase–barstar interface. *Proc. Natl. Acad. Sci. USA*, **93**, 2343–7.
54. Martin, C., Hartley, R., and Mauguen, Y. (1999). X-ray structural analysis of compensating mutations at the barnase–barstar interface. *FEBS Lett.*, **452**, 128–32.
55. Schreiber, G. and Fersht, A. R. (1995). Energetics of protein–protein interactions: analysis of the barnase–barstar interface by single mutations and double mutant cycles. *Biochemistry*, **248**, 478–86.
56. Carter, P. J., Winter, G., Wilkinson, A. J., and Fersht, A. R. (1984). The use of double mutants to detect structural changes in the active site of the tyrosyl-tRNA synthetase (*Bacillus stearothermophilus*). *Cell*, **38**, 835–40.
57. Schreiber, G., Frisch, C., and Fersht, A. R. (1997). The role of Glu73 of barnase in catalysis and the binding of barstar. *J. Mol. Biol.*, **270**, 111–22.
58. Vaughan, C. K., Buckle, A. M., and Fersht, A. R. (1999). Structural responses to mutation at a protein–protein interface. *J. Mol. Biol.*, **286**, 1487–506.
59. Kleanthous, C., Hemmings, A. M., Moore, G. R., and James, R. (1998). Immunity proteins and their specificity for endonuclease colicins: telling right from wrong in protein–protein recognition. *Mol. Microbiol.*, **28**, 227–33.
60. Ferguson, N., Capaldi, A. P., James, R., Kleanthous, C., and Radford, S. E. (1999). Rapid folding with and without populated intermediates in the homologous four-helix proteins Im7 and Im9. *J. Mol. Biol.*, **286**, 1597–608.
61. James, R., Kleanthous, C., and Moore, J. R. (1996). The biology of E colicins: paradigms and paradoxes. *Microbiology*, **142,** 1569–80.
62. Lazdunski, C. J., Bouveret, E., Rigal, A., Journet, L., Lloubès, R., and Bénédetti, H. (1998). Colicin import into *Escherichia coli* cells. *J. Bacteriol.*, **180**, 4993–5002.
63. Bowman, C. M., Dahlberg, J. E., Ikemura, T., Konisky, J., and Nomura, M. (1971). Specific inactivation of 16S ribosomal RNA induced by colicin E3 *in vivo*. *Proc. Natl. Acad. Sci. USA*, **68**, 964–8.
64. Ogawa, T., Tomita, K., Ueda, T., Watanabe, K., Uozumi, T., and Masaki, H. (1999). A cytotoxic ribonuclease targeting specific tRNA anticodons. *Science*, **283**, 2097–100.
65. Wallis, R., Reilly, A., Rowe, A., Moore, G. R., James, R., and Kleanthous, C. (1992). *In vivo* and *in vitro* characterisation of overproduced colicin E9 immunity protein. *Eur. J. Biochem.*, **207**, 687–95.
66. Wallis, R., Moore, G. R., James, R., and Kleanthous, C. (1995). Protein–protein interactions in colicin E9 DNase–immunity protein complexes. Diffusion controlled association and femtomolar binding for the cognate complex. *Biochemistry*, **34**, 13743–50.
67. Osborne, M. J., Breeze, A., Lian, L.-Y., Reilly, A., James, R., Kleanthous, C., and Moore, G. R. (1996). Three-dimensional solution structure and ^{13}C NMR assignments of the colicin E9 immunity protein Im9. *Biochemistry*, **35**, 9505–12.
68. Chak, K.-F., Safo, M. K., Ku, W.-Y., Hsieh, S.-Y., and Yuan, H. S. (1996). The crystal-structure of the immunity protein of colicin E7 suggests a possible colicin-interacting surface. *Proc. Natl. Acad. Sci. USA*, **93**, 6437–42.
69. Dennis, C. A., Videler, H., Pauptit, R. A., Wallis, R., James, R., Moore, G. R., and Kleanthous, C. (1998). A structural comparison of the colicin immunity proteins Im7 and Im9 gives new insights into the molecular determinants of immunity protein specificity. *Biochem. J.*, **333**, 183–91.

70. Wallis, R., Reilly, A., Barnes, K., Abell, C., Campbell, D. G., Moore, G. R., James, R., and Kleanthous, C. (1994). Tandem overproduction and characterisation of the nuclease domain of colicin E9 and its cognate inhibitor protein Im9. *Eur. J. Biochem.*, **220**, 447–54.
71. Garinot-Schneider, C., Pommer, A. J., Moore, G. R., Kleanthous, C., and James, R. (1996). Identification of putative active-site residues in the DNase domain of colicin E9 by random mutagenesis. *J. Mol. Biol.*, **260**, 731–42.
72. Pommer, A. J., Wallis, R., Moore, G. R., James, R., and Kleanthous, C. (1998). Enzymological characterisation of the nuclease domain from the bacterial toxin colicin E9 from *Escherichia coli. Biochem. J.*, **334**, 387–92.
73. Pommer, A. J., Kühlmann, U. C., Cooper, A., Hemmings, A. J., Moore, G. R., James, R., and Kleanthous, C. (1999). Homing-in on the role of transition metals in the HNH motif of colicin endonucleases. *J. Biol. Chem.* **274**, 27153–60.
74. Kleanthous, C., Kühlmann, U. C., Pommer, A. J., Ferguson, N., Radford, S. E., Moore, G. R., James, R., and Hemmings, A. M. (1999). Structural and mechanistic basis of immunity towards endonuclease colicins. *Nature Struct. Biol.*, **6**, 243–52.
75. Ko, T.-P., Liao, C.-C., Ku, W.-Y., Chak, K.-F., and Yuan, H.-S. (1999). The crystal structure of the DNase domain of colicin E7 in complex with its inhibitor Im7 protein. *Structure*, **7**, 91–102.
76. Shub, D. A., Goodrich-Blair, H., and Eddy, S. R. (1994). Amino acid sequence motif of group I intron endonucleases is conserved in open reading frames of group II introns. *Trends Biochem. Sci.*, **19**, 402–4.
77. Belfort, M. and Roberts, R. J. (1997). Homing endonucleases: keeping the house in order. *Nucl. Acid Res.*, **25**, 3379–88.
78. Whittaker, S. B.-M., Boetzel, R., MacDonald, C., Lian, L.-Y., Pommer, A. J., Reilly, A., James, R., Kleanthous, C., and Moore, G. R. (1998). NMR detection of slow conformational dynamics in an endonuclease toxin. *J. Biomol. NMR*, **12**, 145–59.
79. Lo Conte, L., Chothia, C., and Janin, J. (1999). The atomic structure of protein–protein recognition sites. *J. Mol. Biol.*, **285**, 2177–98.
80. Nadassy, K., Wodak, S. J., and Janin, J. (1999). Structural features of protein–nucleic acid recognition sites. *Biochemistry*, **38**, 1999–2017.
81. Osborne, M. J., Wallis, R., Leung, K.-Y., Williams, G., Lian, L.-Y., Kleanthous, C., and Moore, G. R. (1997). Identification of critical residues in the colicin E9 DNase binding region of the Im9 protein. *Biochem. J.*, **323**, 823–31.
82. Curtis, M. D. and James, R. (1991). Investigation of the specificity of the interaction between colicin E9 and its immunity protein by site-directed mutagenesis. *Mol. Microbiol.*, **11**, 2727–33.
83. Wallis, R., Leung, K.-Y., Pommer, A. J., Videler, H., Moore, G. R., James, R., and Kleanthous, C. (1995). Protein–protein interactions in colicin E9 DNase–immunity protein complexes. Cognate and noncognate interactions that span the millimolar to femtomolar affinity range. *Biochemistry*, **34**, 13751–9.
84. Cunningham, B. C., Jhurani, P., Ng, P., and Wells, J. A. (1989). Receptor and antibody epitopes in human growth hormone identified by homolog-scanning mutagenesis. *Science*, **243**, 1330–6.
85. Li, W., Dennis, C. A., Moore, G. R., James, R., and Kleanthous, C. (1997). Protein–protein interaction specificity of Im9 for the endonuclease toxin colicin E9 defined by homologue scanning mutagenesis. *J. Biol. Chem.*, **272**, 22253–8.
86. Li, W., Hamill, S. J., Hemmings, A. M., Moore, G. R., James, R., and Kleanthous, C. (1998). Dual recognition and the role of specificity-determining residues in colicin E9 DNase–immunity protein interactions. *Biochemistry*, **37**, 11771–9.

87. Wallis, R., Leung, K.-Y., Osborne, M. J., James, R., Moore, G. R., and Kleanthous, C. (1998). Specificity in protein–protein recognition: conserved Im9 residues are the major determinants of stability in the colicin E9 DNase–Im9 complex. *Biochemistry*, **37**, 476–85.
88. Clackson, T. and Wells, J. A. (1995). A hot spot of binding energy in a hormone-receptor interface. *Science*, **267**, 383–6.
89. Kraulis, P. J. (1991). MOLSCRIPT—a program to produce both detailed and schematic plots of protein structures. *J. Appl. Crystallog.*, **24**, 946–50.
90. Christopher, J. A. (1992). SPOCK: The Structural Properties Observation and Calculation Kit, Centre for Macromolecular Design, Texas A and M University, Texas.
91. Whittaker, S. B., Czisch, M., Wechselberger, R., Kaptein, R., Hemmings, A. M., James, R., Kleanthous, C., and Moore, G. R. (2000). Slow conformational dynamics of an endonuclease persist in its complex with its natural inhibitor. *Protein Sci.* **9**, 713–20.

Index